CONTAMINATED SOILS, SEDIMENTS AND WATER:

Science in the Real World

Volume 9

This page intentionally left blank

CONTAMINATED SOILS, SEDIMENTS AND WATER:
Science in the Real World
Volume 9

Edited by

Edward J. Calabrese
Paul T. Kostecki
James Dragun

Springer

ISBN: 978-1-4419-3568-7
eBook ISBN: 978-0-387-23079-5
eBook ISBN: 0-387-23079-3

Contents

Part II: Chemical Oxidation

Part III: Heavy Metals

This page intentionally left blank

Foreword

Increasingly in the environmental industry we are reminded how much we are part of the real world. Global economic pressures, geopolitical uncertainty, business ethics, and other world dynamics exert enormous pressure on our academic, industrial, governmental or consulting work venues. Yet, often these pressures have challenged us to produce scientific and engineering solutions that are more focused, cost-effective and readily implementable. The environmental remediation industry in general--and the greater UMASS Soils Conference community in particular--has responded well.

Perhaps the growing alliance movement in business today is a signpost for another important opportunity. All around us we see companies forging alliances and partnerships to improve performance, streamline the time and reduce the costs of getting their important work or commerce done, and ensure their viability. Such collaborations enable entities to focus on what they do best and rely on others to do the same. Such collaborations build trust and encourage more strategic thinking. However, sometimes the synergistic benefits from such creative partnerships is not immediately apparent.

This is not the case with us. The UMASS Soils Conference presents us with a unique opportunity for scientific and engineering collaboration—and progress on remediation. When scientists, engineers, environmental consultants, industry and regulators focus on what they do best and bring a collaborative mindset, good things happen.

Contaminated Soils Volume 9 contains 38 technical papers, covering a wide range of environmental issues presented at the 19th Annual Contaminated Soils Conference, University of Massachusetts, Amherst, October 2003. Volume discussion includes: Part I Bioremediation; Part II Chemical Oxidation; Part III Heavy Metals; Part IV MTBE; Part V Phytoremediation; Part VI Radiation; Part VII Regulatory and Legal issues; Part VIII Remediation; Part IX Risk Based Cleanup; and Part X Site Assessment.

The environmental investigation and restoration papers presented here represent platform speakers and poster sessions. The authors who contributed to this volume are from government agencies, academic institutions, the consulting community and industrial companies. This important volume documents the state of environmental science and provides perspective on where we as an industry have come from. Increased scientific, engineering and consulting alliances and collaboration should drive our efforts going forward.

Christopher B. Mitchell
ENSR International
Westford, Massachusetts

Contributing Authors

George R. Alther, Biomin, Inc., P.O. Box 20028, Farndale, MI 48220

Rakesh Bajpai, University of Missouri, Dept of Chemical Engineering, Columbia, MO 65211

Jiri Barta, Institute of Chemical Technology, Dept of Fermentation Chemistry and Bioengineering, Technicka 5, CZ-166 28 Prague, Czech Republic

Nicholas Bauer, Saltire Industrial, Inc., 12030 Sunrise Valley Drive, Ste 300, Reston, VA 20190

Paul G. Beaulieu, Tighe & Bond Consulting Engineers, 53 Southampton Road, Westfield, MA 01085

James F. Begley, MT Environmental Restoration, 24 Bay View Avenue, Plymouth, MA 02360

Gary M. Birk, P.E., EOS Remediation, Inc., 3722 benson Drive, Raleigh, NC 27609

Reid H. Bowman, Ph.D., Applied Process Technology, Inc., 3333 Vincent Road, Suite 222, Pleasant Hill, CA 94523

Robert L. Burns, Jr., Esq., Thorp Reed & Armstrong, LLP, One Oxford Centre, 301 Grant Street 14th Floor, Pittsburgh, PA 15219

Raymond Cadorette, Shaw Environmental, 88C Elm Street, Hopkington, MA 01748

Alessia Cao, DIGITA, Dept. of Geoengineering and Environmental Technologies, University of Cagliari, Piazza d'Armi, 09123 Cagliari, Italy

Alessandra Carucci, Professor, DIGITA, Dept. of Geoengineering and Environmental Technologies, University of Cagliari, Piazza d'Armi, 09123 Cagliari, Italy

E. Cervantes-Gonzalez, National School of Biological Sciences, Carpio Y Plan de Ayala S/N Col. Casco de Santo Tomas, ZP 11340 Mexico City, Mexico

A.R. Clemente, Campinas State University, Food Science Dept, Food Engineering Faculty, Campinas- SP, Brazil

Edmund Crouch, Cambridge Environmental Inc., 58 Charles Street, Cambridge, MA 02141

R. Cruz-Camarillo, National School of Biological Sciences, Carpio Y Plan de Ayala S/N Col. Casco de Santo Tomas, ZP 11340 Mexico City, Mexico

Wm. Gordon Dean, PE, WRS Infrastructure & Environment, Inc., 625 E. Tennessee Street, Suite 100, Tallahassee, FL 32308

Ambrose Donovan, McPhail Associates, 30 Norfolk Street, Cambridge, MA 02139

M.F. Dos Reis- Gorny, Campinas State University, Food Science Department, Food Engineering Faculty, Campinas-SP, Brazil

Tom Dragoo, Parsons, 1700 Broadway, Ste. 900, Denver, CO 80290

L. R. Durrant, Campinas State University, Food Science Dept, Food Engineering Faculty, Campinas- SP, Brazil

Stephen Finn, Golder Associates, 1951 Old Cuthbert Rd., Ste 301, Cherry Hill, NJ 08034

Giuseppe Fois, Botanical Garden, Department of Botanical Sciences, University of Cagliari, Viale S. Ignazio da Laconi 11, 09123 Cagliari, Italy

T.T. Franco, Campinas State University, Food Science Dept, Food Engineering Faculty, Campinas -SP, Brazil

Steve Goins, Tennessee Div of Superfund, 4th Floor LNC Annex, 401 Church Street, Nashville, TN 37243-1538

R.M.A. Gondim-Tomaz, Campinas State University, Food Science Dept, Food Engineering Faculty, Campinas -SP, Brazil

James R. Greacen, The RETEC Group, Inc., 300 Baker Avenue, Suite 302, Concord, MA 01742

James J. Gusek, P.E., Golder Associates, Inc., 44 Union Blvd #300, Lakewood, CO 80228

John Hicks, Parsons, 1700 Broadway, Ste. 900, Denver, CO 80290

David Hill, Remediation Specialist, National Guard Bureau, Groundwater Program at Camp Edwards, P.B. 565, 567 West Outer Road, , Camp Edwards, MA 02542

Allen Kane, Golder Associates, 1951 Old Cuthbert Rd., Ste 301, Cherry Hill, NJ 08034

Rita Kaszala, University of Szeged, Dept of Climatology and Landscape Ecology, PO Box 653, 6701 Szeged, Hungary

Ilona Kevel-Barany, University of Szeged, Dept of Climatology and Landscape Ecology, PO Box 653, 6701 Szeged, Hungary

Mohammod Ruhul Amin Khan, Tokyo Institute of Technology, 2-12-1 Ookayama Meguro, Tokyo, 152-8552, Japan

Todd D. Kirton, Tighe & Bond Consulting Engineers, 53 Southampton Road, Westfield, MA 01085

Michael Komelasky, George Mason University, Chemistry Dept, Fairfax, VA 22030

Sadao Komemushi, Kinki University, Dept of Agricultural Chemistry, Faculty of Agriculture, 3327-204 Nakamachi, Nara, 631-8508, Japan

Itzchak E. Kornfeld, Oayses Environmental Systems, 16 Easyt Roumfort Road, Philadelphia, PA 19119-1618

M. Kuchanur, Texas A&M University - Kingsville, Dept of Environmental Engineering, MSC 213, Kingsville, TX 78363

Osamu Kusakabe, Tokyo Institute of Technology, 2-12-1 Ookayama, Meguro, Tokyo, 152-8552, Japan

Richard Lewis, CPG, Program Manager, ERM, 2 Commercial Drive, Sharon, MA 02067

M. Tony Lieberman, RSM, Solutions Industrial & Environmental Services, Inc., 3722 Benson Drive, Raleigh, NC 27609

Joy Lozano, Booz Allen & Hamilton, 700 North St. Mary's St., Ste. 700, San Antonia, TX 78205

Robert Luhrs, LSP, Raytheon Company, 141 Spring Street, Mail Stop 1-2-303, Lexington, MA 02173

David Major, GeoSyntec Consultants, 160 Research LaneSte.206, Guelph, Ontario, Canada, N1G 5B2

Shihabudheen M Maliyekkal, Indian Institute of Technology Madras, Dept of Civil Engineering, Chennai, 600 036, India

Michael A. Marando, GEI Consultants, Inc., 1021 Main Street, Winchester, MA 01890

Nick Mathis, O&G Environmental, Inc., Englewood, CO 80112

Sadayoshi Matsumoto, Kinki University, Dept of Agricultural Chemistry, Faculty of Agriculture, 3327-204 Nakamachi, , Nara, 631-8508, Japan

Douglas Mose, George Mason University, Chemistry Dept, Fairfax, VA 22030

Walter S. Mulica, Global Technologies, Inc., 4808 Westridge Drive, Fort Collins, CO 80526

Aldo Muntoni, Professor, DIGITA, Dept. of Geoengineering and Environmental Technologies, University of Cagliari, Piazza d'Armi, 09123 Cagliari, Italy

George Mushrush, George Mason University, Chemistry Dept, Fairfax, VA 22030

Denice Nelson, ARCADIS G&M, 420 North Fifth Street, Suite 1035, Minneapolis, MN 55401

Lawrence Nesbitt, Shaw Environmental, 88C Elm Street, Hopkington, MA 01748

James C. O'Shaughnessy, Ph.D., PE, Worcester Polytechnical Institute, 100 Institute Road, Worcester, MA 01609

Timothy Pac, CPG, Sr. Project Manager, ERM, 2 Commercial Drive, Sharon, MA 02067

Jan Paca, Institute of Chemical Technology, Dept of Fermentation Chemistry and Bioengineering, Technicka 5, CZ-166 28 Prague, Czech Republic

Frederick Payne, ARCADIS G&M, 25200 Telegraph Road, Southfield, MI 48034

Ligy Philip, Indian Institute of Technology Madras, Dept of Civil Engineering, Chennai, 600 036, India

Kandiah Ramanitharan, Tulane University, River Water Quality & Contaminated Sediments Research Group, Department of Civil & Environmental Engineering, Walter E. Blessey Hall, New Orleans, LA 70118

Elizabeth Rasmussen, P.E., Sr. Engineer, Raytheon Company, 528 Boston Post Road, Mail Stop 1880, Sudbury, MA 01776

Brian Ray, ENSR Corporation, 7041 Old Wake Forest Road, Suite 103, Raleigh, NC 27616

Eldon R. Rene, Indian Institute of Technology Madras, Dept of Chemical Engineering, Chennai, 600 036, India

David G. Ries, Esq., Thorp Reed & Armstrong, LLP, One Oxford Centre, 301 Grant Street 14th Floor, Pittsburgh, PA 15219

L.I. Rojas-Avelizapa, National School of Biological Sciences, Carpio Y Plan de Ayala S/N Col. Casco de Santo Tomas, ZP 11340 Mexico City, Mexico

N.G. Rojas-Avelizapa, Mexican Institute of Petroleum, Program of Petroleum Biotechnology, Eje Central Lazaro Cardenas 152, Mexico City, Mexico 07730

Shailesh Sahay, Cambridge Environmental Inc., 58 Charles Street, Cambridge, MA 02141

George Saiway, George Mason University, Chemistry Dept, Fairfax, VA 22030

Javier Santillan, AFCEE/ERT, 3207 Sidney Brooks, Brooks City Base, TX 78235

Akiyoshi Sawabe, Kinki University, Dept of Agricultural Chemistry, Faculty of Agriculture, 3327-204 Nakamachi, Nara, 631-8508, Japan

Fiorella Simoni, George Mason University, Chemistry Dept, Fairfax, VA 22030

Ronald Slager, Restoration Program Manager, Raytheon Company, 528 Boston Post Road, Mail Stop 1880, Sudbury, MA 01776

Laura J. Steinberg, Tulane University, Department of Civil & Environmental Engineering, New Orleans, LA 70118

Alton Day Stone, PE, Alton Engineer, 10 Rugg Road, Sterling, MA 01564

Suthan Suthersan, ARCADIS G&M, 3000 Cabot Blvd. W., Suite 3004, Langhorne, PA 19047

T. Swaminathan, Indian Institute of Technology Madras, Dept of Chemical Engineering, Chennai, 600 036, India

Ryuji Takeda, Kinki University, Dept of Agricultural Chemistry, Faculty of Agriculture, 3327-204 Nakamachi, Nara, 631-8508, Japan

Maria Antonia Tanchuling, Tokyo Institute of Technology, 2-12-1 Ookayama, Meguro, Tokyo, 152-8552, Japan

Chris TenBraak, Parsons, 1700 Broadway, Ste. 900, Denver, CO 80290

Dale Trippler, Minnesota Pollution Control Agency, 520 Lafayette Road N., St. Paul, MN 55155

John Tunks, Parsons, 1700 Broadway, Ste. 900, Denver, CO 80290

Venki Uddameri, Ph.D., Texas A&M University - Kingsville, Dept of Environmental Engineering, MSC 213, Kingsville, TX 78363

Eirlys Vanderhoff, McPhail Associates, 30 Norfolk Street, Cambridge, MA 02139

Raphael Vazquez, AFCEE/ERT, 3207 Sidney Brooks, Brooks City-Base, TX 78235

Scott Veenstra, AMEC Earth & Environmental, Inc., 215 Mistletoe Drive, Greensboro, NC 27403

John Vidumsky, E.I. Dupont de Nemours and Company, Barley Mill Plaza 27/2267, P.O. Box 80027, Wilmington, DE 19880-0027

David Walker, Shaw Environmental, 88C Elm Street, Hopkington, MA 01748

Edward J. Walsh, Lyme Properties LLC, 101 Main Street, 18th Floor, Cambridge, MA 02142

Kristen Wandland, ENSR Corporation, 7041 Old Wake Forest Road, Suite 103, Raleigh, NC 27616

S. Grant Watkins, ENSR Corporation, 7041 Old Wake Forest Road, Suite 103, Raleigh, NC 27616

Katherine Weeks, Project Manager, AMEC Earth & Environmental, Inc., 239 Littleton Road, Suite 1B, Westford, MA 01886

Noriyoshi Yoshimuri, Kinki University, Dept of Agricultural Chemistry, Faculty of Agriculture, 3327-204 Nakamachi, Nara, 63108505, Japan

Stephen Zemba, Cambridge Environmental Inc., 58 Charles Street, Cambridge, MA 02141

Acknowledgments

We wish to thank all agencies, organizations and companies that sponsored the conference. Without their generosity and assistance, the conference and this book would not have been possible.

BENEFACTORS

American Chemistry Council
Department of the Army
ENSR International
Environmental Business Solutions International, Inc.
Lyondell Chemical
Massachusetts Dept of Environmental Protection
Naval Facilities Engineering Command
Naval Facilities Engineering Service Center

SPONSORS

AMEC Earth and Environmental, Inc.
American Petroleum Institute
Regenesis
RETEC

SUPPORTERS

3M
Alpha Analytical Labs
Battelle
CRC/Lewis Press
EA Engineering, Science & Technology
Exponent
Geovation Technologies, Inc.
K-V Associates
LSP Association
New York Dept of Environmental Conservation
Shaw Environmental & Infrastructure
Texaco, Inc.
The Dragun Corporation
Tighe & Bond, Inc.

In addition, we express our deepest appreciation to the members of the Scientific Advisory Boards. The tremendous success of the conference has been the result of the dedication and hard work of our Board members.

SCIENTIFIC ADVISORY BOARD

Steve Goodwin	*University of Massachusetts Amherst*
Michael Gorski	*MA Department of Environmental Protection*
Peter R. Guest	*Parsons*
John Gustafson	*Equilon Enterprises LLC*
Eric Hince	*Geovation Technologies, Inc.*
Ihor Hlohowskyj	*Argonne National Labs*
Duane B. Huggett	*Pfizer, Inc.*
Barry L. Johnson	*Assistant Surgeon General (ret.)*
Evan Johnson, LSP	*Tighe & Bond*
William B. Kerfoot	*K-V Associates, Inc.*
Stephen S. Koenigsberg	*Regenesis Bioremediation Products*
Bill Kucharski	*Ecology & Environment, Inc.*
Cindy Langlois	*Journal of Children's Health*
Steven C. Lewis	*ExxonMobil Biomedical Sciences, Inc.*
Dave Ludwig	*Blasland, Bouck & Lee, Inc.*
Ronald J. Marnicio	*TetraTech FW, Inc.*
Rick McCullough	*MA Turnpike Authority*
Chris Mitchell	*ENSR International*
Nihar Mohanty	*MA Department of Environmental Protection*
Robert Morrison	*R. Morrison & Associates, Inc.*
Ellen Moyer	*Tighe & Bond, Inc*
Willard Murray	
Lee Newman	*University of South Carolina*
Gopal Pathak	*Birla Institute of Technology, India*
Tom A. Pedersen	*Camp Dresser & McKee, Inc.*
Frank Peduto	*Soils, Sediment and Water Magazine*
Ioana G. Petrisor	*University of Southern California*
David Reckhow	*University of Massachusetts Amherst*
William E. Robinson	*University of Massachusetts Boston*
Brian J. Rothschild	*University of Massachusetts Dartmouth*
Corinne E. Schultz	*Handex Environmental*
Alex Sherrin	*LSPA*
Nicholas P. Skoulis	*Arch Chemicals, Inc.*
Frank Sweet	*ENSR International*
Christopher Teaf	*Florida State University*
James C. Todaro	*Alpha Analytical Labs*
Allen D. Uhler	*Battelle Environmental Forensic Investigation Group*
Mark Vigneri	*Environmental Business Solutions International, Inc.*
A. Dallas Wait	*Gradient Corporation*
Richard Waterman	*EA Engineering, Science, and Technology, Inc.*

Jason C. White	*The Connecticut Agricultural Experiment Station*
Katie Winogroszki	*3M*
Peter Woodman	*Risk Management Incorporated*
Baoshan Xing	*University of Massachusetts Amherst*
Ed Zillioux	*FPL Environmental Services*

FEDERAL ADVISORY BOARD

John Abraham	*ATSDR/DHAC/E1CB*
John Cullinane	*US Army Engineer Waterways Exp. Sta.*
John Glaser	*US Environmental Protection Agency, ORD*
Stephen J. Grady	*US Dept of the Interior, USGS*
Leslie Karr	*Naval Facilities Engineering Service*
Lynn Kucharski	*AMEC Earth and Environmental*
Jeffrey Marqusee	*SERDP and ESTCP*
Paul Rakowski	*Naval Facilities Engineering Command*
Mike Reynolds	*USA - Cold Regions Research and Engineering Laboratory*
Brian D. Rogers	*Department of the Army*
Michael Sivak	*US EPA, Region 2*
Henry H. Tabak	*US EPA, ORD*

About the Editors

Edward J. Calabrese is a board certified toxicologist and professor of toxicology at the University of Massachusetts School of Public Health at Amherst. Dr. Calabrese has researched extensively in the area of host factors affecting susceptibility to pollutants and has authored more than three hundred papers in scholarly journals, as well as twenty-four books, including: Principles of Animal Extrapolation; Nutrition and Environmental Health, Vols. 1 and 2; Ecogenetic: Safe Drinking Water Act: Amendments. Regulations, and Standards; Soils Contaminated by Petroleum: Environmental and Public Health Effects; Petroleum Contaminated Soils, Vols. 1,2 and 3; Ozone Risk Communication and Management; Hydrocarbon Contaminated Soils, Vols. 1, 2, 3, 4 and 5; Hydrocarbon Contaminated Soils and Groundwater, Vols. 1, 2, 3, and 4; Multiple Chemical Interactions; Air Toxics and Risk Assessment; Alcohol Interactions with Drugs and Chemicals; Regulating Drinking Water Quality; Biological Effects of Low Level Exposures to Chemicals and Radiation; Contaminated Soils; Diesel Fuel Contamination; Risk Assessment and Environmental Fate Methodologies; Principles and Practices for Petroleum Contaminated Soils, Vols. 1, 2, 3, 4, and 5; Contaminated Soils, Vol. 1; and Performing Ecological Risk Assessments. He has been a member of the U.S. National Academy of Sciences and NATO Countries Safe Drinking Water Committees, and the Board of Scientific Counselors for the Agency for Toxic Substances and Disease Registry (ATSDR). Dr. Calabrese also serves as Director of the Northeast Regional Environmental Public Health Center at the University of Massachusetts and Chairman of the BELLE Advisory Committee.

Paul T. Kostecki, Associate Director, Northeast Regional Environmental Public Health Center, School of Public Health, University of Massachusetts at Amherst, received his Ph.D. from the School of Natural Resources at the University of Michigan in 1980. He has been involved with human and ecological risk assessment and risk management research for the last 12 years. Dr. Kostecki has co-authored and co-edited over fifty articles and sixteen books on environmental assessment and cleanup including: remedial Technologies for Leaking Underground Storage Tanks; Soils Contaminated by Petroleum Products; Petroleum Contaminated Soils, Vols. 1, 2, and 3; Hydrocarbon Contaminated Soils and Groundwater, Vols. 1, 2, 3 and 4; Hydrocarbon Contaminated Soils, Vols. 1, 2, 3, 4 and 5; Principles ad Practices for Petroleum Contaminated Soils; Principles and Practices for Diesel Contaminated Soils, Vols. 1, 2, 3, 4 and 5; SESOIL in Environmental Fate and Risk modeling; Contaminated Soils, Vol. 1; and Risk Assessment and Environmental Fate Methodologies. Dr. Kostecki also serves as Associate Editor for the Journal of Soil Contamination, Chairman of the Scientific Advisory Board for Soil and Groundwater Cleanup Magazine, as well as an editorial board member for the journal Human and Ecological Risk Assessment.

In addition, Dr. Kostecki serves as Executive Director for the Association for the Environmental Health of Soils (AEHS). He is a member of the Navy's National Hydrocarbon Test Site Advisory Board and a member of the Steering Committee for the Total Petroleum Hydrocarbon Criteria Working Group and the Association of American Railroads Environmental Engineering and Operations Subcommittee.

James Dragun, as a soil chemist (Ph.D. Penn State University in Agronomy), has a rich background in the fate of hazardous materials in soil systems and groundwater. He has assessed the migration and degradation of chemicals and waste of national concern in soil-groundwater systems such as dioxin, PBB, Radionuclides at Three Mile Island Nuclear Power Plant, PCB and petroleum spills, organics and inorganics at the Stringfellow Acid Pits, pesticides in San Joaquin Valley groundwater, and solvents in Silicon Valley groundwater. In addition, he has analyzed engineering controls designed to prevent the leakage of chemicals and wastes from landfills, surface impoundments, deepwell injection systems, underground storage tanks, land treatment systems, manufacturing and processing facilities, and hazardous waste sites. He has analyzed the chemical integrity and reactivity of materials used to treat and store hazardous and nonhazardous wastes. He has served as an expert reviewer of over 40 projects and programs involving the siting, design, construction, performance, and failure mechanisms of landfills, land treatment systems, surface impoundments, and waste piles. In addition, he has authored exposure assessments for over 100 chemicals and wastes.

Widely recognized for his expertise, Dr. Dragun was appointed the primary technical advisor on exposure assessment to the Interagency Testing Committee, a consortium of 14 federal agencies that selects chemicals for potential regulatory control. He directs the Association of Official Analytical Chemist's development of standard methods to measure the migration and degradation of chemicals and wastes, and has authored test methods that are used today by environmental laboratories in North America, Europe, and Asia. His counsel and scientific findings have been disseminated and utilized by 24 nations including Japan, Canada, the United Kingdom, Australia, West Germany, Switzerland, Italy, France, Spain, Scandinavia, and the Netherlands.

Dr. Dragun is a member of Phi Kappa Phi and Sigma Xi, both international honorary scientific societies, and was awarded the U.S. EPA Bronze Medal for distinguished service in 1980.

This page intentionally left blank

PART I: BIOREMEDIATION

CHAPTER 1

EFFECT OF BACTERIA AUGMENTATION ON AROMATIC AND ASPHALTENIC FRACTION REMOVAL IN SOLID CULTURE

E. Cervantes-Gonzalez[1], L.I. Rojas-Avelizapa[1], R. Cruz-Camarillo[1] and N.G. Rojas-Avelizapa[2]
National School of Biological Sciences, Microbiology Dept., Mexico City [1]; Mexico; Mexican Institute of Petroleum, Mexico City, Mexico[2]

Abstract: Bioaugmented-assisted remediation of a hydrocarbon-polluted soil was performed in sealed serum bottles using a native and well adapted mixed culture consisting on *S. liquefaciens* and *B. coagulans*. The soil for this study was acidic (pH 3.8) with a total petroleum hydrocarbon (TPH) concentration of 220,000 mg/kg soil and a native microbial population of 2.0×10^5 CFU/g soil. This study examines a biological treatment to clean up petroleum-polluted soil using a mixed culture composed by *S. liquefaciens* and *B. coagulans*. C/N/P ratio, moisture and pH were adjusted to appropriate values to assist the biodegradation process. Results demonstrated that heterotrophic activity in microcosms was higher (1.9 mg/g) in bioaugmented microcosms compared with non-bioaugmented (1.3 mg/g). During all the incubation period only three strains were cultivated *S. liquefaciens, B. coagulans* and a non-identified yellow bacteria. The TPHs removal increased 12% in bioaugmented microcosms at the end of 74 days of treatment. In the case of aromatic and asphaltenic fractions the removals increased 12 and 10% respectively. During the first 35 days of incubation, TPHs removal correlates with the exponential phase of microbial growth. Our results showed that bioaugmented-assisted remediation of hydrocarbon-polluted soils in highly polluted and weathered sites are feasible.

Key words: aromatic, asphaltenic, bioaugmentation, removal

1. INTRODUCTION

Biological treatment of hydrocarbon-polluted sites (bioremediation) use mainly two strategies: bioaugmentation and biostimulation (Vogel, 1996). Bioaugmentation consists of introducing microorganisms in order to increase or to accelerate pollutant degradation. Bioaugmentaion is considered when the native microorganism population is insufficient or does not posses the ability to degrade pollutants. The added population can be autochthonous or not. Thus, microorganism addition to polluted sites is effective in the degradation of chemical compounds; however; microorganisms should demonstrate survival and degradative capacity (Kerr, 1994). In some cases, bioaugmentation can increase pollutant removal or the biodegradation rate, although other studies have reported inoculation failure (Vogel, 1996), due to other factors influencing the biodegradation of hydrocarbons in soil. These factors include weathering, bioavailability, pollutant structure, organic matter content, nutrient concentration, electron acceptor presence, pH and salinity (Leahy and Colwell, 1990). Sanjeet *et al.,* (2001), reported an enhancement of TPHs removal (78%) when *Acinetobacter baumannii* was inoculated in a soil containing a TPHs concentration of 180,000 mg/kg. Moreover, while bioaugmentation can increase hydrocarbon removal, it can also accelerate the rate of biodegradation as reported by Jitnuyanont. *et al.,* (2001) when butane was removed by cometabolism of 1,1,1-trichloroethane at an extent of 0.04 mg/mg butane by a consortia in culture liquid. Previous work carried out in our laboratory (Cervantes, *et al.,* 2003) demonstrated the removal of TPHs, aromatic and asphaltenic fraction in solid culture when it was inoculated with a mixed culture composed by *S. liquefaciens* and *B. coagulans.* However the behavior or monitoring of several parameters (TPHs, aromatic and asphaltenic removal, C/N and C/P, heterotrophic activity and microbial count) was not considered during the incubation period, so the aim of this study was to evaluate the removal kinetic of TPH, aromatic and asphaltenic fractions in a polluted soil with 220,000 mg TPH/kg using a mixed culture composed by *S. liquefaciens* and *B. cougulans* at 3%. Other important parameters were followed up during this study.

2. MATERIALS AND METHODS

2.1 Soil and Microorganisms Source

An old hydrocarbon-polluted soil was used for this study with a concentration of 220,000 mg TPH/kg soil. Physical-chemical characteristics corresponded to pH 3.8, moisture content 6.03%, total organic carbon 183,000mg/kg soil and a C/N/P ratio of 11300000/170800/1 (Cervantes, *et al.,* 2003). Native microorganisms used for augmentation were *Serratia liquefaciens* and *Bacillus coagulans.*

2.2 Studies of hydrocarbon removal

Three experimental sets were prepared in 125 ml sealed serum bottles containing 15 g of soil. Experimental sets were prepared as follows: a) C/N/P ratio 100/11/1 and inocula size (bioaugmented) of 3%, b) C/N/P ratio 100/11/1, without inocula (non-bioaugmented) and c) C/N/P ratio 100/11/1 and sterilized with $HgCl_2$ (5%) (Control). C/N/P ratio was adjusted adding urea and monobasic potassium phosphate. The moisture and pH were adjusted at 36% and 7 respectively.

All the microcosms were prepared by duplicate and incubated in the dark at 28°C over a period of 74 days. Each 48 h, CO_2 production was measured, and each 7 days TPHs, aromatic and asphaltenic fraction removal, microbial count, pH, moisture, phosphorous and nitrogen concentration were quantified. During all the incubation period, moisture content was maintained constant with wet air. Data were statistically analyzed by the Kruskal-Wallis test to determine significant difference among treatments.

2.3 Inocula preparation

Inocula were prepared in 500 ml Erlenmeyer flasks, containing 250 ml of mineral medium containing per liter: 1 g KNO_3, 0.02 g $FeCl_3$; 0.2 g $MgSO_4$; 0.1 g $CaCl_2$; and 1 g K_2HPO_4. The pH was adjusted at 6.8. Petroleum was added as the sole carbon source at a concentration of 5000 ppm, and finally, the flasks were inoculated with 5 ml of a fresh microbial culture and then incubated at 28°C, 180 rpm during 24 h. Afterwards, the culture media was centrifuged and a cellular package was obtained which was added to the microcosms at 3% (w/w).

2.4 CO_2 Production

Each 48 h, 2 ml of the microcosm atmosphere were taken out using a 5 ml syringe and injected in a conductivity detector-gas chromatograph Gow Mac 550 with an Alltech CTRI column at the following conditions: oven and injector temperature 30°C, detector 125°C. Helium was used as carrier gas (65 ml/min). Data were integrated to get CO_2 accumulative production. CO_2 production was considered as a measurement of soil heterotrophic activity.

2.5 Microbial count

Each 7 days, soil samples were collected and placed in contact with a 0.85% NaCl solution and shaken to extract the bacteria present in the soil. Nutritive agar plates were inoculated and incubated during 24 h at 28°C. The total heterotrophic bacteria were counted according to the surface spread method.

2.6 Hydrocarbons extraction

The TPHs extraction was carried out by a shaking–centrifugation modified method (Arce, 2003) in 15 ml centrifuge tubes using 0.5 to 2 g soil and 3 g of $NaSO_4$ anhydrous. The mixture was homogenized perfectly and then added with 5 ml of dichloromethane, shaken and centrifuged 10 min at 6000 rpm; afterwards, supernatant was taken out and collected in 25 ml vials. This procedure was repeated two more times. Once all the solvent volume had been collected, dichloromethane was evaporated and the TPHs were determined by gravimetry.

2.7 Separation of aromatic and asphaltenic fractions

The TPHs residue was suspended in 15 ml of hexane (Burdick & Jackson) and sonicated 10 min to precipitate asphaltenes, which were obtained by filtration (Whatman No.1) and weighed. Asphalthenes were analyzed by infrared spectroscopy (IR) in a Nicolet Nexus 470, where the solid sample was blended with KBr anhidrous at a 1:131 ratio. Lecture was done in the range of 4000 a 400 cm^{-1}.

Hexane-soluble fraction, containing the aromatic fraction, was solvent evaporated and weighted. Aromatic fraction was analyzed by gravimetry and by FID-GC where 1 μl of sample was injected in an Agilent Technologies GC System with a HP-1 capillar column (30 m x 320 μm x 0.25 μm). GC operates at the following conditions: initial temperature 30°C/ min., 30°-100°C at 15°/min., 100°-200°C at 7°C/min., 200°-250°C at 6°C/min, injector temperature 250°C, detector 280°C. Helium was used as a carrier gas (1.5 ml/min). Hydrocarbon removal was determined by comparing areas of treated samples to those from initial samples and controls.

3. RESULTS AND DISCUSSION

3.1 Heterotrophic Activity (CO_2)

CO_2 evolution was used as a measure of soil heterotrophic activity. In Fig. 1 the profiles of CO_2 evolution of the three experimental sets are shown: as can be observed treatments that were added with biomass *(B. coagulans* and *S. liquefaciens)* had a higher CO_2 production (1.9 mg/g) after 74 days of incubation Those non-bioaugmented treatments presented a lower CO_2 production (1.4 mg/g) but statistically different from the sterilized control (0.43 mg/g). Results of the statistical analysis demonstrated that bioaugmentation caused an increase in the heterotrophic activity of the microcosms.

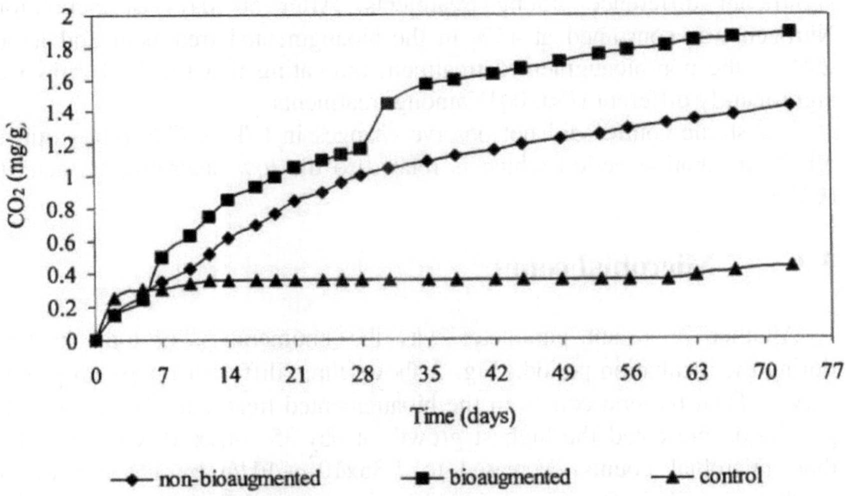

Figure 1. CO_2 evolution of microcosm bioaugmented, non-bioaugmented and $HgCl_2$ control at 28 °C during 74 days of incubation

Fig. 1, also shows that higher CO_2 evolution was obtained during the first 30 days of incubation, as much in the bioaugmented treatments (1.5 mg CO_2/g soil) as in the non-bioaugmented (1.0 mg CO_2/g soil). It is important to point out that CO_2 evolution in the bioaugmented treatments (1.9 mg/Kg) is lower than that obtained in a soil sample with 20,000 mg TPHs/kg (5.4 mg CO_2/g), suggesting the negative effect of high hydrocarbon concentration on heterotrophic activity (Cervantes, *et al.,* 2003). Perhaps this situation could cause a decrease in the degradative efficiency (Braddock, *et al.,* 1997).

The sterile controls produced only 0.34 mg CO_2/g soil indicating low microbial activity in these microcosms.

3.2 C/P and C/N ratio

Phosphorous and nitrogen sources are essential nutrients for microbial growth, so they should be present at an appropriate ratio with respect to the carbon sources to assist hydrocarbon removal (Braddock, *et al.,* 1997). The results showed that during the first 7 days, the phosphorous was consumed at a extent of 45% in the bioaugmented and the non-bioaugmented treatments. After this period the C/P ratio was constant, with the C and P sources consumed at the same ratio. Statistical analysis showed that there was not a significant difference among treatments. After 50 days of incubation, Nitrogen was consumed at 43% in the bioaugmented treatment and about 29% in the non-bioaugmented treatment indicating that the C/N ratio was significantly different ($P<0.001$) among treatments.

The sterile control did not observe changes in C/P or C/N ratios during all the incubation period which is related to the low heterotrophic activity (CO_2).

3.3 Microbial count

Another important parameter was the enumeration of total bacteria during the incubation period. Fig. 2 shows three different microbial growth curves. Total bacteria counts in the bioaugmented treatment showed that the population presented the highest growth at day 35 (4.7×10^8 CFU/g). After that, microbial counts decreased to 1.35×10^7 CFU/g by the end of the incubation period. Significant increases in the total bacteria for the non-bioaugmented treatment was not observed and the native population was remained constant (1.3×10^7 CFU/g). Since it was possible to identify *S. liquefaciens* and *B. coagulans,* Fig. 2 shows separate counts for each of the inoculated bacteria. *S. liquefaciens* presented a higher population throughout the incubation period. During the 21st and 42nd day, a yellow pigmented bacteria appeared (Fig. 2), coinciding with the asphaltene consumption phase suggesting that this bacteria could be involved in or use some of the metabolic products forming at this time.

Currently, there are few reports regarding the ability of *B. coagulans* to metabolize pollutants although Langlois *et al,* (1970) have reported the ability of this bacterium to degrade the pesticide DDT. Nevertheless, other reports show that the *Bacillus* genera demonstrate potential as a petroleum hydrocarbon degrader (Aitken, *et al.*, 1998; Greene, *et al.*, 2000; Kazunga and Aitken, 2000).

Serratia genera has also received little attention as a hydrocarbon degrader but important findings indicate the ability to this genera to oxidize hydrocarbons such as **α-pineno** (Wright, *et al.*, 1986), and some aromatic aldehydes (De la Fuente, *et al.*, 1991).

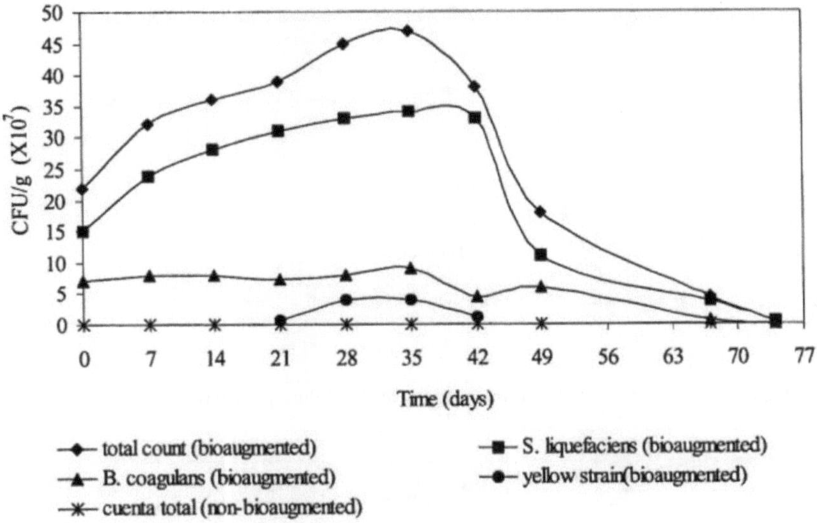

Figure 2. Microbial counts in soil microcosms at 28°C during 74 days of incubation

3.4 Removal of TPH, aromatic and asphaltenic fractions

The monitoring of hydrocarbon removal during the incubation period was the most important parameter in this study and the results are shown in Figures 3, 4 and 5. As can be observed from Figures 3 and 4, higher TPHs and aromatic removals were obtained during the first 35 days of treatment reaching removals of 28% and 33% respectively in the bioaugmented treatments. In the non-bioaugmented treatments, removals within 16% and 21% only were observed. Higher hydrocarbon removal (Figure 3) coincides with the period of higher heterotrophic activity (Figure 1), and higher microbial count (Figure 2), indicating that the hydrocarbons were consumed preferentially during this period and used as a carbon source. This finding is important since the aromatic fraction was composed of C_{14}-C_{21} compounds, branched and two-ringed or three-ringed polycyclic aromatic hydrocarbons, and heterocyclic compounds containing sulfur such as dibenzothiophene (data not shown): compounds that have been reported as recalcitrant and difficult to remove in a polluted soil.

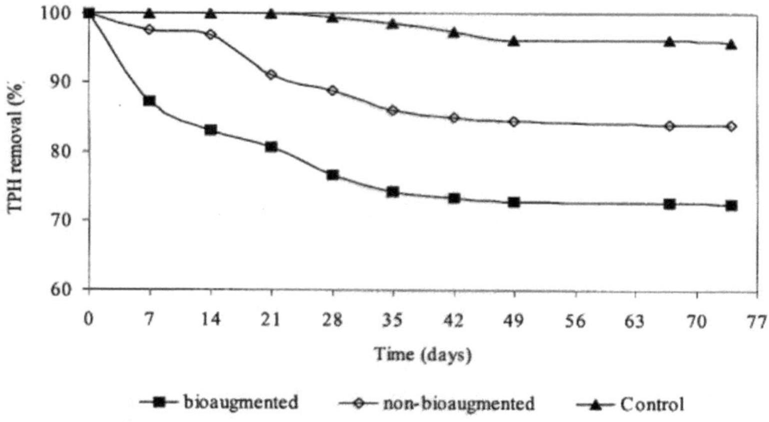

Figure 3. Removal of TPHs in microcosms at 28°C during 74 days of incubation

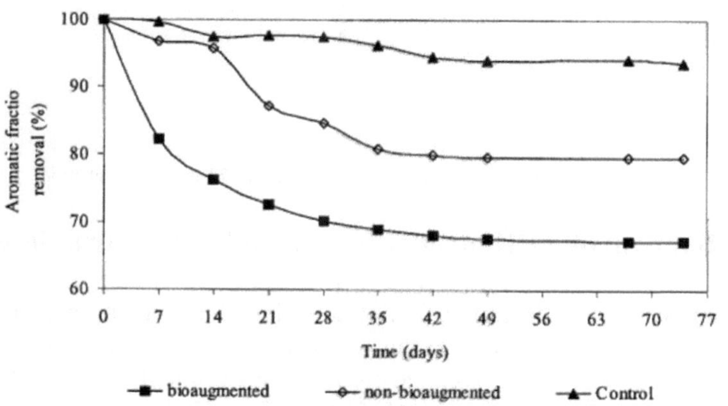

Figure 4. Removal of aromatic fraction in microcosms at 28°C during 74 days of incubation

Other important data obtained in this study were the correlation between CO_2 evolution and hydrocarbon removal (r=0.99) where higher CO_2 production indicates higher hydrocarbon removal.

Asphaltenic fraction removal (Figure 5) it began from the 21st day and continued until the 35th day with a removal of 17% in the bioaugmented treatment and 7% in the non-bioaugmented. Previous experiments suggest that microorganisms first consume compounds present in the aromatic fraction and them consume the asphaltenic one, but all of them are totally consumed. The same behavior was observed in the bioaugmented and non-bioaugmented treatments, except that the removal was higher by 10% in the bioaugmented treatment.

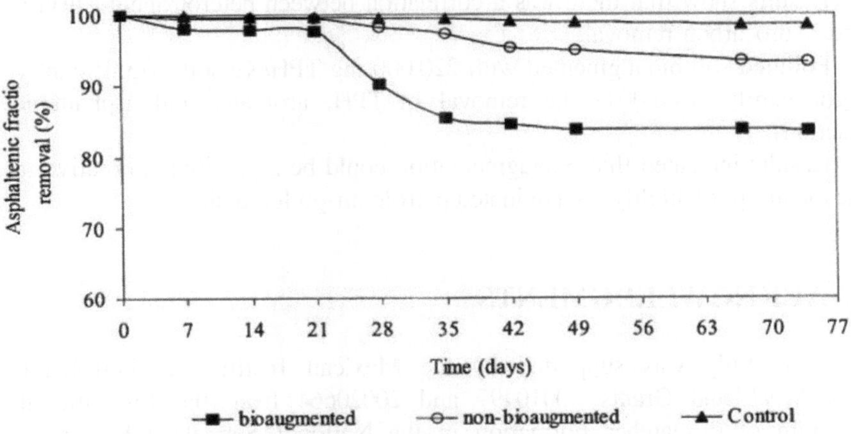

Figure 5. Removal of asphaltenic fraction in microcosms at 28°C during 74 days of incubation

IR analysis showed only small changes in the region of 1200-1700 cm^{-1} indicating minimal modification to the asphaltene structure by the effects of the bioaugmentation.

In comparison, the removal difference between the bioaugmented and non-bioaugmented treatments was 12% for TPHs removal (25,740 mg/kg soil), 12% for the aromatic fraction (17,652 mg/kg soil) and 10% for the asphaltenic fraction (6,899 mg/kg soil). The sterile control had an abiotic loss within 5% to 6%.

Statistical analysis indicated that there was a significant difference between treatments ($P<0.001$); thus the effect of bioaugmentation caused an increase in hydrocarbon removal. Some authors have also reported that bioaugmentation enhance hydrocarbon removal rates. Gentry, *et al.,* (2001) reported an increase of 50% in the biodegradation of 3-chlorobenzoate (1000 μg/g soil) when *Comamonas testosteroni* BR60 was inoculated in soil. Also, Schwart & Scow, (2001) reported an enhancement of the phenanthrene mineralization rate when the soil was inoculated repeatedly with *Arthrobacter sp.*

4. CONCLUSIONS

Results show that there was a correlation between heterotrophic activity and hydrocarbon removal.

Polluted soil bioaugmented with 220,000 mg TPH/Kg soil, significantly significantly (P<0.001) the removal of TPH, aromatic and asphaltenic fractions.

Results indicated that bioaugmentation could be a possible alternative to the clean-up of highly contaminated petroleum-polluted sites.

ACKNOWLEDGMENTS

This study was supported by the Mexican Institute of Petroleum, CONACyT and Grants 20010497 and 20020664 from the Institutional Program of Researcher Formation in the National School of Biological Sciences, National Polytechnic Institute, Mexico.

REFERENCES

Aitken, M. D., Stringfellow, W. T., Nagel, R, D. Kazunga, C. and Chen, S. H. 1998. Characteristics of phenanthrene-degrading bacteria isolated from soils contaminated with polycyclic aromatic hydrocarbons. Can. J. Microbiol. 44, 743 -752.

Arce, O. J. M. 2003. Biotransformaci6n de la fracci6n policiclica, nitrogenada y azufrada durante un tratamiento por biopilas. PH. D. Thesis (in process), Department of Biotechnology and Bioengineering. Research and Advances Studies of National Politechnic Institute, Mexico.

Braddock, J. F., Ruth, M. L., Catterall, P. H., Walworth, J. L. and McCarthy, K. A. 1997. Enhancement inhibition of microbial activity in hydrocarbon-contaminated artic soils: implications for nutrient-amended bioremediation. Environ. Sci. Technol. 31, 2078-2084.

Cervantes-Gonzalez E., Rojas-Avelizapa L.I., Cruz-Camarillo R., and Rojas-Avelizapa N.G. 2003. Microbial removal of recalcitrant hydrocarbons using a Box-Behnken experimental design. In-situ and on-site bioremediation: Proceedings of the seventh International Symposium. Batelle Bioremediation Symposium. In Press.

De la Fuente, G., Perestelo, F., Rodriguez, P. A., and Falcon, M. A. 1991. Oxidation of aromatic aldehydes by Serratia marcescens. Appl. Microbiol. Biotechnol. 57, 1275-1276.

Gentry, T.J., Newby, D.T., Josephson, K.L., and Pepper, I.L. 2001. Soil microbial population dynamics following bioaugmentation with a 3-chlorobenzoate-degrading bacterial culture. Biodegradation. 349, 349-357.

Greene, E. A., Kay, J. G., Stehmeier, L. G. and Voorgouw, G. 2000. Composition of soil microbial communities enriched on a mixture of aromatic hydrocarbons. Appl. Environ. Microbiol. 66, 5282-5289.

Jitnuyanont, P., Sayavedra, L. A., and Semprini, L. 2001. Bioaugmentation of butane-utilizing microorganisms to promote cometabolism of 1,1,1 -trichloroethane in groundwater microcosms. Biodegradation. 12, 11 -22.

Kazunga, C. and Aitken, M. D. 2000. Products from the incomplete metabolism of pyrene by polycyclic aromatic hydrocarbon-degrading bacteria. Appl. Environ. Microbiol. 66, 1917 - 1922.

Kerr, R. S. 1994. Handbook of bioremediation. Boca Raton, Fla., Lewis Publishers, pp. 85 - 96.

Langlois, B. E., Collins, J. A. and Sides, K. G. 1970. Some factors affecting degradation of organochlorine pesticides by bacteria. J. Dairy Sci. 53, 1671-1675.

Leahy, J.G., and Colwell, R. R. 1990. Microbial degradation of hydrocarbons in the environment. Microbiol. Rev. 54, 305 -315.

Sanjeet, M., Jeevan, J., Ramesh, C.K., and Banwari, L. 2001. Evaluation of inoculum addition to stimulate in situ bioremediation of oily-sludge-contaminated soil. Appl. Environ. Microbiol. 67, 1675-1681.

Schwart, E. and Scow, K. M. 2001. Repeated inoculation as an strategy for the remediation of low concentrations of phenanthrene in soil . Biodegradation. 12, 201 -207.

Vogel, T. M. 1996. Bioaugmentation as a soil bioremediation approach. Curr. Opin. Biotechnol. 47, 1107-1115.

Wright, S. J., Caunt, P., Carter, D., and Baker, P. B. 1986. Microbial oxidation of alpha-pinene by Serratia marcescens. Appl. Microbiol. Biotechnol. 23, 224 -227.

This page intentionally left blank

CHAPTER 2

BIODEGRADATION OF PAHS IN SOIL BY TWO DEUTEROMYCETE FUNGI

A.R. Clemente; L.R. Durrant
Universidade Estadual de Campinas - UNICAMP

Abstract: The fungal strains used is this work, namely, 984 and 1040, were isolated fr om soil samples collected at the Jureia -Itatins Ecological Reserve, São Paulo-Brazil. Following microscopical examinations these strains were classified as Aspergillus sp. (984) and Verticillium sp. (1040). The degradation of PAHs in soil contaminated with 5 mg naphthalene/g soil; 1,0 mg anthracene/g soil or 0,5 mg pyrene and/or benzo[a]pyrene/g soil, was verified. These strains were grown in wheatbran:water for 3 days, inoculated in sterilized and non- sterilized soil, and cultivated for 2,4, 6, and 8 weeks. The PAHs were then extracted and degradation was determined by HPLC. The best degradation, in sterilized soil was obtained after 8 weeks for the two strains: naphthalene (64.50 - 65.43%), anthracene (77.35 - 85.83%) and pyrene (73.01 - 78.78%). When benzo[a]pyrene was used the best degradation shown by strain 984 was 89.62% in six weeks and 78.06% by strain 1040 in eight weeks. In non-sterilized soil the strains exhibited lower growth and degradation than in sterilized soil, with the exception for benzo[a]pyrene (82.3 - 82.6%). Our results indicate that these two fungal strains have potential for application in the bioremediation of soils contaminated with PAHs.

Keywords: PAHs, biodegradations, fungi, soil.

1. INTRODUCTION

Polycyclic aromatic hydrocarbons (PAHs) represent a class of nonionic, poorly water-soluble, and toxic organic compounds which occur in environmental matrices due to natural causes such as forest fires or anthropogenic processes such as urban and industrial activities (Conte et al., 2001). Commonly used remediation techniques including land removal and incineration or landfilling are now less environmentally acceptable or cost effective than they used to be. Hence, interest in bioremediation as an alternative approach for the cleaning up of contaminated environments has increased (Canet et al., 2001). Research on biodegradation has demonstrated the potential of fungi to degrade PAHs (Canet et al., 1999).

Several factors affect the persistence of PAHs in the environment. Wilcock et al., (1996) reported that low molecular weight PAHs is rapidly lost from sediments, whereas the high molecular weight ones have larger persistence. Ghoshal & Luthy (1996) showed that a mechanism of mass transfer of aromatic hydrocarbons (in particular naphthalene) is responsible for the rate of biomineralization of such contaminants in system containing single coal tar globules. Gustafson & Dickhut (1995) showed that the concentration of hydrophobic PAHs in water depended mainly on their hydrophobicity. Physical-chemical properties of soils have also been considered responsible of the retention of PAHs in soil matrices. The amount of organic carbon (Weissenfelds et al., 1992), and the hydrophobicity of soil organic matter (Murphy et al., 1990), were estimated to be the most significant parameters in decreasing the environmental availability of PAHs.

Several studies have shown that diverse fungi are capable of PAH mineralization, and that rates of mineralization correlate with the production of ligninolytic enzymes (Field et al., 1992; Sack et al., 1997). These microorganisms naturally decompose lignin to obtain the cellulose inside the wood fiber using a non-specific enzymatic complex, which also enables them to degrade a wide range of contaminants if they are introduced to soils. In soil, the fungi release these enzymes to the extracellular medium, allowing the fungi to degrade large molecules that they would otherwise be unable to incorporate across cell walls. Such metabolism has further advantages in so far as the fungi avoid the uptake of potentially toxic substances and the non-specific action of the enzymes involved makes precondition to individual pollutants unnecessary. Furthermore, induction of the extracellular enzyme system is independent of the presence of the contaminants, therefore the fungi can degrade contaminants at extremely low concentrations (Canet et al., 2001).

To date, most investigations of PAHs degradation by fungi have been carried out in vitro by inoculating artificially contaminated liquid media, which do not provide a reliable indication as to their likely success under field conditions (Canet et al., 1999). Here, the degradation of PAHs in soil by two ligninolytic fungi is demonstrated.

2. MATERIAL AND METHODS

2.1 Fungal strains and inoculum preparation.

The fungal strains isolated from soil samples collected at the Jureia/Itatins ecological reserve (São Paulo - Brazil) and classified as deuteromycetes, were a gift from the Tropical Culture Collection - Fundação Tropical André Tosello, Campinas-SP. The strains were maintained on PDA (potato dextrose agar-DIFCO) slants. Petri dishes containing PDA were inoculated with a mycelium portion of each strain and incubated at 30°C for 8-10 days and used as the inocula for the experiments described below. These strains have previously been reported as ligninolytic (Clemente et al., 2001 and Clemente & Durrant, 2003). Following microscopically examinations were classified as Aspergillus sp. (984) and Verticillium sp. (1040), (Clemente & Durrant, 2003).

2.2 Soil characteristics and soil contamination (Andersson & Henrysson, 1996).

A sandy soil having an organic matter content of 3.1% and pH of 6.5, was used throughout this work. This soil type was chosen because of its properties, which made it easier to handle and the extraction of the PAHs more reliable. The soil was divided into two fractions. One fraction was stored in sterile flasks at +4°C, while the other was autoclaved at 121°C for 15 minutes, stored at room temperature for a week, sterilized again and then stored at +4°C. Individual soil fractions were contaminated with naphthalene (5.0%), anthracene (0.75%), pyrene (1.0%) or benzo[a]pyrene (0.5%), which had been previously dissolved in a 250 mL mixture of hexane and acetone 3:1 (v/v), except for anthracene which was dissolved only in acetone. The solutions were added to the soil to give the final concentration of 5.0 mg naphthalene/g soil; 1.0 mg anthracene/g soil or 0.5 mg pyrene or benzo[a]pyrene/g soil, and mixed for a total time of 2 hours. When all the solvents had evaporated the fungi grown in wheatbran:water for 3 days were inoculated into the flasks containing the contaminated soils and were incubated for 2,4, 6 and 8 weeks at 30° C.

2.3 HPLC analyses of the extracted PAHS for the determination of degradation (Andersson & Henrysson, 1996)

At two-week interval samples, consisting of the whole flasks were collected and toluene (100 mL), was added to each flask, which were thoroughly mixed and placed in a water bath at 100°C for 3 hours. When the flasks had reached room temperature, 2 mL of the toluene mixture was removed, filtered with a 0.22 μm membrane using a glass syringe and used for the HPLC analyses.

2.4 High performance liquid chromatography (HPLC).

All HPLC analyses, for the determination of degradation of PAHs, were performed with a Zorbax ODS (0.46 x 15 cm) C_{18} reverse-phase column (SUPELCO Chromatography Products). Separation was achieved by isocratic elution in acetonitrile:water (70:30), with a flow rate of 1.0 mL/min and UV absorbance detector set at 254 nm.

3. RESULTS AND DISCUSSION

In non-stenhzed soil, the strains exhibited lower visual growth and degradation than in sterilized soil, with the exception of benzo[a]pyrene. As shown in figure 1, both strains were able to degrade naphthalene when present in sterile soil. Best degradation rates (~65% for both strains), were obtained from the 6th to 8th week of incubation. Very low degradation was observed following the HPLC analyses of samples from the non-sterile soil. It is possible that the indigenous microorganisms present in the non-sterile soil compete for nutrients or have some kind of antagonism with the fungal strains used here. However, when anthracene was used as the contaminant (figure 2), degradation occurred under both conditions, sterile (~80%) after 8 weeks and non-sterile (~70%) from the 6th to the 8th week. When pyrene was used (figure 3), degradation took place under both conditions, but was superior under sterilized condition for both strains (~70%) than under non-sterile (~50%). Benzo[a]pyrene was the only PAH best degraded by both strains (~82%), under both conditions. Growth of both strains in the sterilized soil may have been stimulated by the effective death of the indigenous microflora but also due to changes in the physicochemical properties of the soil after sterilization, a process that can bring about an increase in the concentration of soluble nutrients and organic matter (Lynch,

1998). Anderson et al., (2000), have observed an increase in the visual growth of *Pleurotus ostreatus* and *Phanerochaete chrysosporium* which showed a significant correlation with the degradation of the PAHs present in the growth medium. The low visual growth and degradation of PAHs in non-sterile soils could be explained by the low ability of fungi to compete with the existent microorganisms, or it may be that their extracellular enzymes are non-active in the soils used or may not be produced (Matens & Zadrazil, 1992, and In der Wiesche et al., 1996). It is also possible that other factors such as soil temperature and moisture could have a negative effect in the growth of the fungi (Canet et al., 2001). In the bioremediation of contaminated soils, it is important that any PAH present be removed by the soil microflora or by any inoculated microorganism or that a consortium can be established among them. When the ligninolytic white-rot fungus *Bjerkandera* sp was inoculated in soils containing high molecular weight PAHs, such as benzo[a]pyrene, it was able to completely degrade this compound following 15 days of incubation (Kotterman et al., 1998). Similarly, the results presented here show that *Aspergillus sp* and *Verticillum sp* were also able to degrade benzo[a]pyrene after 4 weeks. Our results indicate that these two fungal strains have potential for application in the bioremediation of soils contaminated with PAHs.

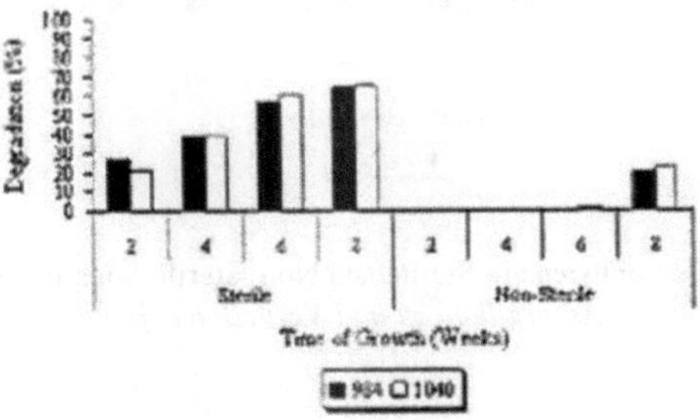

Figure 1. Degradation of Naphthalene in Sterile and Non -sterile Soils following Growth of *Aspergillum sp and Verticillium sp*

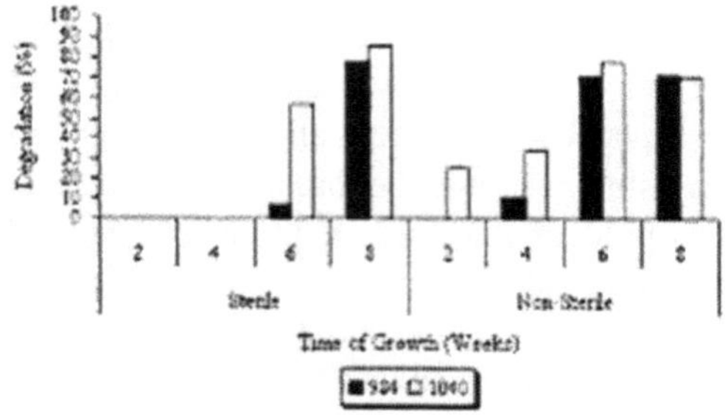

Figure 2. Degradation of Anthracene in Sterile and Non -sterile Soils following Growth of *Aspergillum sp and Verticillium sp*

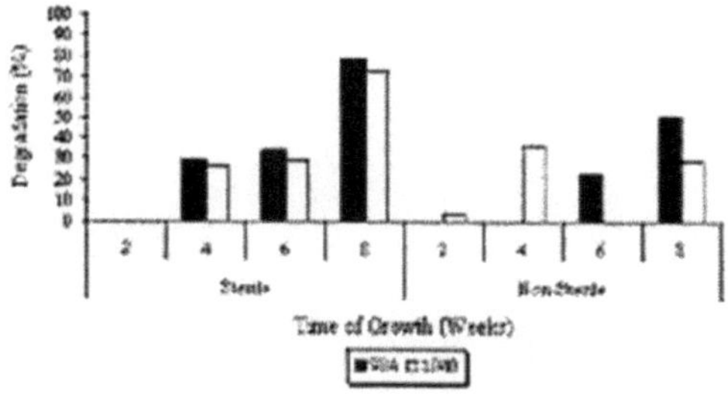

Figure 3. Degradation of Pyrene in Sterile and Non -sterile Soils following Growth of *Aspergillum sp and Verticilium sp*

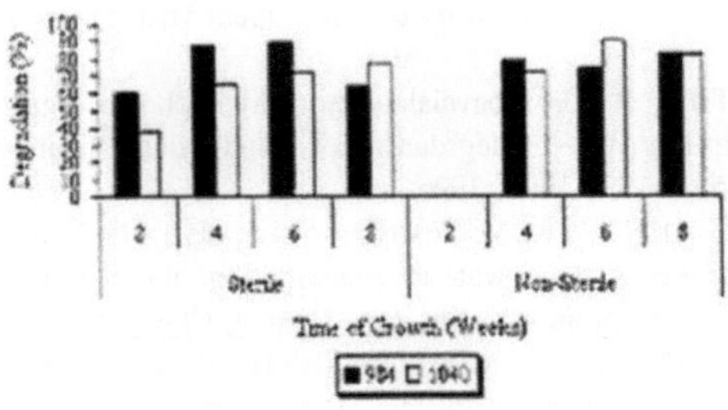

Figure 4. Degradation of Benzo[a]pyrene in Sterile and Non-sterile Soils following Growth of *Aspergillum sp and Verticillium sp*

ACKNOWLEDGEMENTS

The authors wish to Express their thanks to "Fundação de Amparo à Pesquisa do Estado e São Paulo" (FAPESP) for financing the experiment.

REFERENCES

ANDERSON, B. E. & HENRYSSON, T. Accumulation and degradation of dead -end metabolites during treatment of soil contamined with polycyclic aromatic hydrocarbons with five strains of white-rot fungi. Appl. Microbiol. Biotechnol. v. 46, p. 647-652, 1996.

ANDERSSON, B. E.; WELINDER, L.; OLSSON, P. A.; OLSSON, S. & HENRYSSON, T. Growth of inoculated white -rot fungi and their interactions with the bacterial community in soil contaminated with poly cyclic aromatic hydrocarbons, as measured by phospholipid fatty acids. Biores. Technol. v. 73, p. 29-36, 2000.

CANET, R.; LOPEZ-REAL, J. & BECK, A. J. Overview of polycyclic aromatic hydrocarbons biodegradation by white-rot fungi. Land Contam. Reclam. v. 7, p. 191-197, 1999.

CANET, R,; BIRRNSTINGL, J. G.; MALCOM, D. G.; REAL-LOPEZ, J. M. & BECK, A. J. biodegradation of polycyclic aromatic hydrocarbons (PAHs) by native microflora and combinations of white-rot fungi in a coal-tar contamined soil. Biores. Technol. v. 76, p. 113-117,2001.

CLEMENTE, A. R.; ANAZAWA, T. A. and DURRANT, L. R. (2001) Biodegradation of Polycyclic Aromatic Hydrocarbons by Soil Fungi. Brazilian Journal of Microbiology. Brasil, v.32, n.4, pp. 255 -261.

CLEMENTE, A. R. and DURRANT, L. R. (2003) Degradation Of PAHs By Soil Fungi With Potential For Bioremediation. In: Proceedings In Situ and On -Site Bioremediation – The Seventh International Symposium. (Alleman, B. C. & Leeson, A., Eds.). Columbus, Ohio, Battelle Press.

CONTE, P.;ZENA, A.; PILIDIS G. & PICCOLO A. Increased retention of polycyclic aromatic hydrocarbons in soils induced by soil treatment with humic substances. Environ. Pollut. v. 112, p. 27-31,2001.

GHOSHAL, S. & LUTHY, R. G. Bioavaialability of hydrophobic organic compounds from nonaqueus-phase liquids: the biodegrdation of naphthalene from coal tar. Environ. Toxicol. Chem. v. 15, p. 1894-1900, 1996.

GUSTAFSON, K. E. & DIKHUT, R. M. Distribution ofpolycyclic aromatic hydrocarbons in southern Chesapeake Bay surface wate r: evaluation of three methods for determining freely dissolved water concentrations. Environ. Toxicol. Chem. v. 16, p. 452-461, 1995.

IN DER WIESCHE, C.; MARTENS, R. & ZADRAZIL, F. Two-step degradation of pyrene by white-rot fungi and soil microorganisms. Appl. Microbiol. Biotechnol. v. 46, p. 653-659, 1996.

KOTTERMAN, M. J. J.; VIS, E. H. & FIELD, J. A. Successive mineralization and detoxification of benzo[a]pyrene by the white -rot fungus *Bjerkandera* sp. Strain BOS55 and indigenous microflora. Appl. Environ. Microbiol. v. 64, p. 2853-2858, 1998.

LYNCH, J. M. Microorganisms in their natural environments – the terrestrial environment. In: Lynch, J. M., Hobbie, J. E. (Eds), Micro-organisms in action: Comcepts and applications in microbial ecology. Blackwell Scientific Publications, Paolo alto, p. 103 - 131, 1988.

MARTENS, R. & ZADRAZIL, F. Screening of white-rot fungi for their ability to mineralize polycyclic aromatic hydrocarbons in soil. In: Preprints from the international symposium on soil decontamination using biological processes. DECHEMA, Frankfurt am main, p. 505-510, 1992.

MURPHY, E. M.; ZACHARA, J. M. & SMITH, S. C. Influence of mineral-bound humic substances on the sorption of hydrophobic organic compounds. Environ. Sci. Technol. v. 24, p. 1507-1516, 1990.

WEISSENFELS, W. D.; KLEWER, H. J. & LANGHOFF, J. Adsorption of polycyclic aromatic hydrocarbons (PAHs) by soil particles: influence on biodegradability and biotoxicity. Appl. Microbial. Biotechnol. v. 36, p. 689-696, 1992.

WILCOCK, R. J.; CORBAN, G. A.; NORTHCOTT, G. L.; WILKINS, A. L. & LANGDON, A. G. Persistence of polycyclic aromatic compounds f different molecular size and water solubility in surficial sediment of an intertidal sandflat. Environ. Toxicol. Chem. v. 15, p. 670-676, 1996.

CHAPTER 3

BIODEGRADATION OF DIURON AND PYRUTHIOBAC-SODIUM BY WHITE-ROT AND SOIL FUNGI

Rose M.A. Gondim-Tomaz[1], Telma T. Franco[2] and Lucia R. Durrant[3]

Instituto Argronômico (ICA), Campinas, Brazil[1]; Departamento de Processes Químicos, Faculdade de Engenharia Química, Universidade Estadual de Campinas, Campinas, Brazil [2]; Departamento de Ciências de Alimentos, Faculdade de Engenharia de Alimentos, Universidade Estadual de Campinas (SP), Brazil[3]

Abstract: Thirteen strains, namely, *Pleurotus* sp BCCB 507, P. sp CCB 068, *Pleurotus* sp. 016, *Agaricus campestris, Phanerochaete chrysosporium* ATCC 24725 and the soil isolates DP24e, DP24o, DRP02n, SP16a, SRP 17c, SRP17g and SRP20e were selected following their cultivation in solid media containing either the pesticide Diuron or Pyrithibac -sodium (Staple 280CS). These fungi were grown in liquid medium for three days when 25 µg/mL of Diuron or 10 µg/mL of Staple were added and cultivation was carried out for up to a 14 -day period. Ligninolytic activities and also the degradation of the pesticides were determined. When Diuron was used the highest degradation was obtained with *Pleurotus* sp BCCB 507 (60.70% - 7th day), *Pleurotus* sp CCB 068 (80.75% - 10th day), *P.* sp. 016 (58.60% - 7th day), and the soil isolates SRP17g (65.46% - 10th day), SRP17c (67.60% – 14th day) and SRP20e (62.20% - 10th day). When Staple 280 CS was used *Pleurotus* sp BCCB 507 (14.25% - 7th day), *A. campestris* (32.90%-14th day) and *P.* sp. CCB 068 (53.20% - 7th day) showed the highest degradation. MnP was the predominant ligninolytic enzyme produced by all the strains, regardless of the pesticide used.

Key words: biodegradation; ligninolytic enzymes; diuron; pyrithiobac-sodium; fungi.

1. INTRODUCTION

Pesticides are widely used in current agricultural practices. Owing to their toxic effects on non-target organisms, most pesticides may produce serious detrimental effects on ecosystems (Bretaud et al, 2000). Pesticide residues persist in the environment and may be incorporated and accumulated via the food chain, affecting human health (Muñoz de la Peña et al, 2003).

The fate of pesticides in edaphic ecosystems is dependent not only on various abiotic mechanisms such as photochemical degradation, adsorption to soil elements, absorption by plants or leakage, but is also governed to a large extent by the degradation activities of microorganisms (Aislabie et al, 1995). Hence, the knowledge of catabolic pathways of pesticide-degrading microorganisms may help to solve some problems of agricultural pollution (Widehem et al., 2002).

The herbicide Diuron, (3-(3,4-dichlorophenyl)-1,1-dimethylurea is one of the most extensively used in Brazilian agriculture (Tien & Kirk, 1988). It is widely used to selectively control annual weeds species in a variety of crops (cotton, sugar cane, vineyards, etc.), or is also used for total weed control on non-cultivated areas (maintenance of roads, railways, parks, etc.). Diuron is very persistent at the soil surface partly due to its low solubility. However, its degradation has been observed in soil and was attributed to biological activity (Dalton et al.,, 1966). It can be degraded or transformed by fungi *(Fusarium oxysporum)* (Kaufman & Blake, 1973); *Cunninghamela echinulata* (Tilmanns et al., 1978), and by mixed bacterial cultures (Shelton et al., 1996 and Cullington & Walker, 1999). The herbicide Pyrithiobac (sodium 2-chloro-6-(4,6-dimethoxypyrimidin-2-ylthio)benzoate), the active ingredient of Staple (DuPont, 1993), is a new herbicide developed for postemergence control of broad-leaf weeds in cotton. Pyrithiobac degrades relatively slowly in soil, primarily by microbially mediated degradation, with an estimated half-life of 60 days based on laboratory studies (DuPont, 1993).

White-rot fungi are defined by their physiological capacity to degrade lignin, a very complex and heterogeneous polymer (Hataaka, 1994). The white-rot fungi's enzyme system is non-specific, enabling the degradation of compounds which resemble lignin structure to occur, having been implicated in the degradation of a wide variety of aromatic xenobiotics, including polyaromatic hydrocarbons (PAHs), polychlorinated bifenyls, pentachlorophenol and various groups of pesticides (Higson, 1991 and Prado & Airoldi, 2002). The present study was designed to investigate the ability of 13 fungal strains to degrade two herbicides, diuron and pyrithiobac-sodium (Staple 280 CS) and also to determine the production of the ligninolityc enzymes during their growth in these herbicides.

2. MATERIAL AND METHODS

2.1 Chemicals

Diuron and Pyrithiobac-sodium were a gift from Millenia Agro Ciência S.A., Londrina-PR, and Iharabras S.A., Sorocaba-SP, Brazil. The purity of each compound was checked by HPLC prior to use and was higher then 98%.

2.2 Fungal strains and inocula preparation

The 13 fungal strains: 5 ligninolytic white-rot fungi belonging to the culture collection of the Laboratório de Sistemática e Fisiologia Microbiana, UNICAMP, Campinas-SP, Brazil) and 8 strains isolated from soil treated with either Diuron or Pyrithiobac-sodium, were maintained on PDA (potato dextrose agar - DIFCO) slants. Petri dishes containing PDA were inoculated with a mycelium portion of each strain and incubated at 30° C for 8-10 days, after which the agar was cut and either **1 cm^2** or **5 x 1 cm^2** mycelial pieces were removed with a thin spatula and used as the inocula for the experiments described below.

2.3 Culture conditions

Fungal strains were inoculated **(5 x 1 cm^2)**, in liquid medium (50 mL/250 mL Erlenmeyer flasks) containing per liter: **0.5 g $(NH_4)_2$ HPO_4; 0.8 g KH_2PO_4; 0.3 g K_2HPO_4; 0.3g $MgSO_4$ $7H_2O$; 0.055 g $CaCl_2$ H_2O; 4.0 mg Zn SO_4 $6H_2O$;** 0.2 g yeast extract and 1.0 mL of thiamine (2,0 mg/mL). The final pH of the culture media was 5.0 Following inoculation all cultures were incubated under stationary conditions at 30°C. Diuron **(25µg/mL)** or pyrithiobac-sodium **(10µg/mL)** was added on the **3rd** day of cultivation and incubated for a further 3, 7, 10 and 14 days. To correct for possible abiotic losses controls, consisting of uninoculated flasks, were also run.

2.4 Enzyme assays

The assays were carried out as described previously (Clemente et al., 2001). Lignin-peroxidase (LiP), manganese-peroxidase (MnP), Laccase

(Lac.) and peroxidase activities were assayed spectrophotometrically in the extracellular fluid of the culture supernatants. LiP was determined by measuring veratryl alcohol oxidation (Tien, & Kirk, 1984). MnP determinations were based on the oxidation of phenol red, according to Kuwahara *et al.* (1984). Laccase was determined by the oxidation of syringaldazine to its quinone form and peroxidase was determined by the oxidation of syringaldazine to its quinone form in the presence of hydrogen peroxide (Szklarz, 1989). The supernatants were obtained following centrifugation of the content of whole flasks at 17,000 rpm for 15 min at 4°C. Enzyme activities were expressed in units per liter, where one unit is equal to 1 μmol of substrate oxidized per minute.

2.5 High performance liquid chromatography (HPLC)

All HPLC analyses were performed using a C_{18} reverse-phase column (SUPELCO). Separation was achieved by isocratic elution in methanol: water (70:30) as mobile phase (Esposito, 1998), with a flow rate of 0,7 mL/min and UV absorbance detector set at 254 nm for diuron. Pyrithiobac-sodium was eluted using 43% ACN/ 57% pH 3, 30 mM potassium phosphate buffer as the mobile phase (Sumpter, & Peterson, 1994), with a flow rate of 0,8 mL/min, 40°C, and UV absorbance detector set at 254 nm.

3. RESULTS AND DISCUSSION

All the fungal strains used in this work exhibited some ligninolytic activity and degradation after growth in the herbicides, which varied with the time of growth and also with the carbon source. Table 1 shows the greatest values of the enzymes' activities produced following growth of the fungi in diuron.

Table 1. **Maximal Ligninolytic activities and degradation produced following growth o f the fungal strains in Diuron**

Strains	LiP	MnP	Lac.	Perox.	Degradation (%)
BCCB 507	0.19	19.46	ND	0.05	60.74
A.camp.	0.16	7.85	ND	0.49	11.76
P.crysosp.	0.13	13.36	0.06	0.30	3.88
CCB068	0.68	63.94	4.79	3.38	80.75
Pleur.016	0.39	5.96	ND	0.08	58.63
DP24e	0.22	9.33	0.10	0.23	33.80
DP24o	0.13	5.92	ND	0.03	37.58
DRP02n	0.17	19.01	0.07	0.20	5.30
DRP02e	0.21	12.33	0.05	0.20	44.06
SP16a	0.17	10.58	ND	0.10	21.03
SRP17g	ND	14.89	0.02	0.09	65.46
SRP17c	0.16	11.61	ND	0.07	67.03
SRP20e	0.22	12.02	ND	0.03	62.60

ND: Not Detected

Extracellular LiP, MnP, laccase and peroxidase were determined in the supernatant of the cultures grown in diuron and higher levels of MnP were produced by most of the fungal strains, when compared with the other ligninolytic enzymes. MnP catalyzed the **H_2O_2-dependent** oxidation of aromatic compounds via a single-electron oxidation mechanism, followed by a series of non-enzymatic reactions yielding various types of degradation products (Yoshidaet al., 1996). It has been suggested that manganese-peroxidase could be the only enzyme responsible for the oxidation of the aromatic structure of diuron (Esposito et al., 1998).

Pleurotus sp (BCCB 507), *Pleurotus* sp (CCB 068) and the non-identified soil fungi SRP17c showed the highest degradation of diuron (table 1 and fig. 2) among the 13 strains, following 7, 10 and 14 days of growth. *Pleurotus* sp CCB 507 exhibited maximal LiP, MnP and Peroxidase activities on the 3rd, 10th and 7th day, whereas Laccase was not detected. *Pleurotus* sp CCB 068 presented the highest LiP, MnP, Laccase and Peroxidase activities on the 7th, 10th, 10th and 7th day, respectively. The soil fungus SRP17c exhibited maximal LiP, MnP and Peroxidase activities on the 10th, 3rd and 7th day, whereas Laccase was not detected. These results indicate that these three stains expressed their ligninolytic system differently when growing in diuron.

Table 2 shows the highest values for the enzyme activities produced following growth of the 13 strains in Pyrithiobac-sodium. *Agaricus campestris* and *Pleurotus* sp CCB 068 showed highest degradation of this herbicide following 14 and 7 days of growth (table 2 and fig. 4), respectively. Maximal values for MnP were produced on the 10th day by *Pleurotus* sp CCB 068, whereas a low level of MnP was produced by *Ag.campestris* on the 3rd day (table 2). LiP was produced by these two strains on both the 3rd and 7th days of growth. Laccase was not detected in the culture supernatants of *Ag.campestris* and *Pleurotus* sp CCB 068 showed a peak for laccase activity on the 10th day. Maximum values for peroxidase activities by these two strains were on the 7th day.

Table 2. **Maximal ligninolytic activities and degradation produced following growth of the fungal strains in Pyrithiobac -sodium.**

Strains	LiP	MnP	Lac.	Perox.	Degradation (%)
BCCB 507	0.21	6.10	1.09	0.34	14.25
A.camp.	0.17	0.90	ND	0.36	32.94
P.crysosp.	0.33	34.12	0.20	0.46	3.55
CCB068	0.41	96.68	2.38	4.39	53.23
*Pleur.*016	0.13	4.53	0.01	0.44	ND
DP24e	0.32	5.43	0.33	0.41	ND
DP24o	0.68	7.58	0.19	0.54	ND
DRP02n	0.09	19.64	0.27	0.56	ND
DRP02e	0.07	9.06	0.24	0.59	ND
SP16a	1.94	10.90	0.23	0.24	ND
SRP17g	0.23	11.03	0.32	0.52	ND
SRP17c	0.04	8.78	0.14	0.42	ND
SRP20e	ND	7.49	0.11	0.28	ND

ND: Not Detected.

The strains responsible for the greatest degradations of the herbicides, *Pleurotus* sp CCB 507, *Pleurotus* sp (CCB 068) and SRP17c, in diuron, and *Ag.campestris* and *Pleurotus* sp CCB 068 in Pyrithiobac-sodium, were selected and used for more detailed studies on the growth and degradation of these herbicides, as described under culture conditions.

Figure 1 shows the activities of the enzymes produced following the growth of *Pleurotus* sp CCB 507, *Pleurotus* sp CCB 068 and the soil fungi SRP17c in diuron.

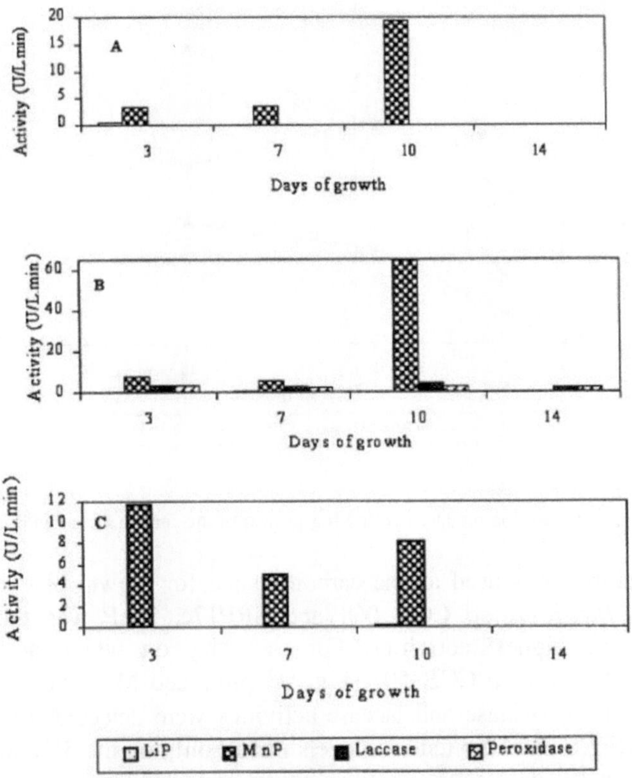

Figure 1. Ligninolytic activities produced following growth of the fungal strains in diuron . A) Pleurotus sp. CCB 507; B) Pleurotus sp. CCB 068; C) Soil fungus SRP17c

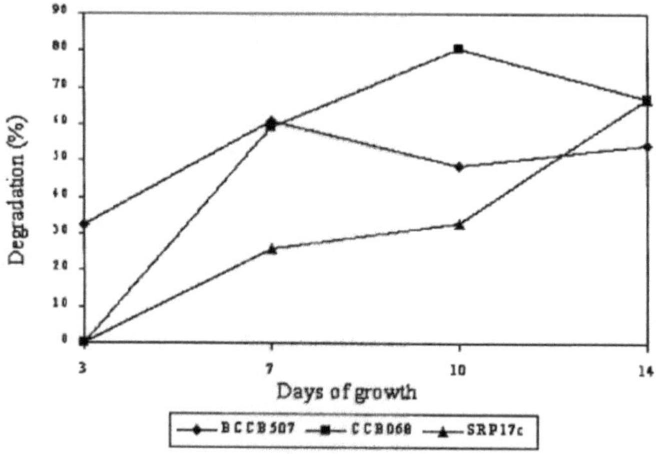

Figure 2. Degradation produced following growth of the fungal strains in diuron

When diuron was used as the carbon source for growth of *Pleurotus* sp CCB 507, *Pleurotus* sp CCB 068 and SRP17c, MnP was the enzyme presenting the highest activities. For each day on which activity was measured, *Pleurotus* sp CCB 507 (Fig.1A) produced MnP, very low levels of LiP and no peroxidase and laccase activities were detected. *Pleurotus* sp CCB 068 (Fig. 1B) exhibited low levels of LiP only on the 3rd and 7th days; laccase and peroxidase activities were always detected when assayed. MnP activity was detected on the 3rd, 7th and 10th days. The soil isolate SRP17c (Fig. C) exhibited MnP on the 3rd, 7th and 10th days, but not on the 14th day. LiP was detected only on the 10th day. Low levels of peroxidase and no laccase activities were detected. These three strains were able to degrade diuron, and maximum levels of degradation 80.75%, 67.03% and 60.74% were produced by the strains CCB 068, SRP17c and BCCB 507, respectively (Fig. 2).

Figure 3 shows the activities of the enzymes produced following growth of *Pleurotus* sp CCB 068 and *Ag.campestris,* in pyrithiobac-sodium. MnP was the enzyme exhibiting the highest activities. *Pleurotus* sp (CCB 068) (Fig. 3A) produced LiP activity on the 3rd, 7th and 10th days. MnP was produced always when assayed. Laccase and peroxidase activities were not detected on the 3rd day. *Ag.campestris* (Fig. 3B) produced LiP and peroxidase on the 3rd and 7th days; MnP on the 3rd and 10th days but no laccase activity was detected. The best degradation rates were obtained on the 7th and 14th days by *Pleurotus* sp CCB 068 (53.23%) and *Ag.campestris* (32.94%), respectively (Fig. 4).

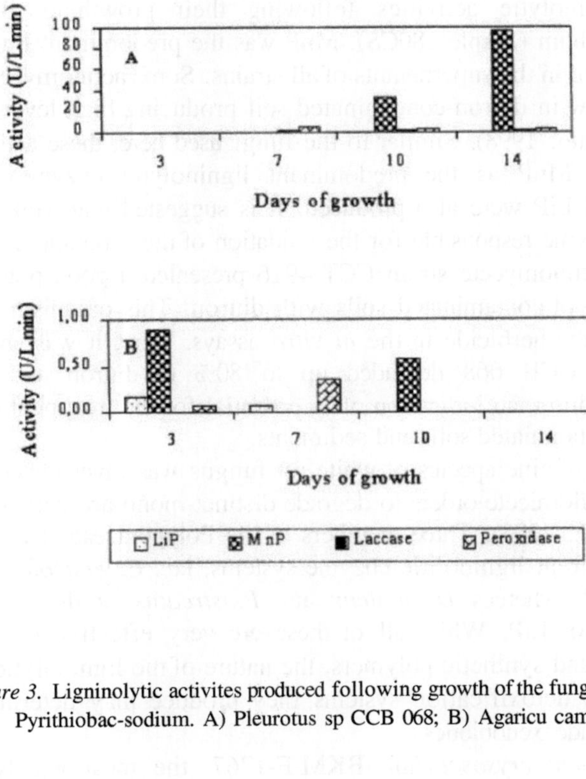

Figure 3. Ligninolytic activites produced following growth of the fungal strains in Pyrithiobac-sodium. A) Pleurotus sp CCB 068; B) Agaricu campestris

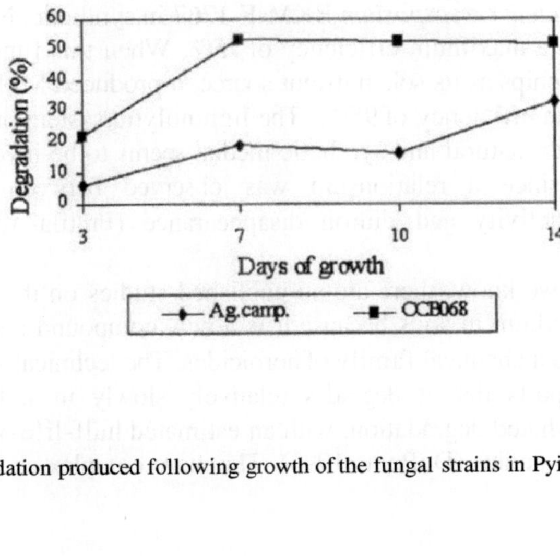

Figure 4. Degradation produced following growth of the fungal strains in Pyithiobac -sodium.

The fungal strains used in this work were able to degrade herbicides and exhibited ligninolytic activities following their growth in diuron or pyrithiobac-sodium (Staple 280CS). MnP was the predominant ligninolytic enzyme detected in the supernatants of all strains. Some actinomycete strains are able to grow in diuron-contaminated soil producing high levels of MnP activity (Esposito, 1998). Similar to the fungi used here, these actinomicete also produced MnP as the predominant ligninolytic enzyme. Laccase, peroxidase and LiP were also produced. It is suggested that MnP could be the unique enzyme responsible for the oxidation of the aromatic structure of diuron. The actinomycete strain CCT 4916 presented a good potential for bioremediation of contaminated soils with diuron. This organism degraded up to 37% of the herbicide in the *in vitro* assays. Here, it was shown that *Pleurotus* sp. CCB 068 degraded up to 80% of diuron and 53% of pyrithiobac-sodium, an indication of its potential for the treatment of diuron and Staple-contaminated soils and sediments.

The ability of nine species of white rot fungus was investigated, from a variety of basidiomicete orders to degrade distinct mono-aromatic pesticides (Bending, et al., 2002). Three members of the Polyporaceae are known to produce a different ligninolitic enzyme systems, i.e., *C.versicolor* produces LiP and MnP, whereas *D.squalens* and *P.ostreatus* produce MnP and Laccase, but not LiP. While all of these are very effective degraders of natural lignin and synthetic polymers, the nature of the ligninolytic enzyme as well as the detoxification systems they produce may determine their ability to degrade xenobiotics.

Phanerochaete crysosporium BKM-F-1767, the most widely studied ligninolytic fungus, when incubated in synthetic and natural media degraded 14 and 10 mg/L diuron, respectively. In this study, diuron could be degraded by *Phanerochaete crysosporium* BKM-F-1767 in synthetic, N-limited liquid medium with a maximum efficiency of 75%. When this fungus was grown on ash wood chips as its sole nutrient source, it produced MnP and degraded diuron with an efficiency of 95%. The ligninolytic system produced by this fungus in both, natural and synthetic media, seems to be involved in diuron degradation since a relationship was observed between the onset of ligninolytic activity and diuron disappearance (Fratila-Apachitel et al., 1999).

As far as we know, there are no published studies on the degradation of pyrithiobac-sodium in soils because it is a new compound and has yet to be classified into a chemical family of herbicides. The technical information on pyritiobac reports that it degrades relatively slowly in soil, primarily by microbial mediated degradation, with an estimated half-life of 60 days based on laboratory studies (DuPont, 1993). The more complex chemical structure of the pesticide Pyrithibac-sodium (Staple 280 CS), when compared to diuron, may be responsible for the lower levels of degradation and of strains able to attack it. Here, two strains of white-rot fungi were able to grow and degrade this herbicide (~53%) in seven days.

Considering both pesticides, *Pleurotus* sp CCB 068 was the best among all the strains tested. It is important to mention that the highest levels of enzymes activities were produced by this strain, regardless of the pesticide. Our results indicate that this fungus has a great potential for application in the bioremediation of soil and sediments contaminated with herbicides.

ACKNOWLEDGEMENTS

We would like to thank the EMBRAPA for financial support.

REFERENCES

Bretaud,S.; Toutant,J.-P.; Saglio,P. Effects of carbofuran, diuron, and nicosulfuron on acetylcholinesterase activity in goldfish (*Carassiues auratus) Ecotoxicology and Environmental safety* v.47,p. 117-124,2000.

Muñoz de la Peña, A.; Mahedero, Bautista -Sánchez, A. Monitoring of phenylu rea and propanil herbicides in river water by solid -phase-extraction high performance liquid chromatography with photoinduced -fluorimetric detection. *Talanta,* v.00, p.1 -7, 2003.

Aislabie, J.; Lloyd-Jones, G. A review of bacterial degradation of pesticides. *Aust.J.Soil Res.* v.33, p.925-942, 1995.

Widehem,P.; Aït-Aïssa, S.; Tixier, C.; Sancelme,H.V.; Truffaut, N. Isolation, characterization and diuron transformation capacities of a bacterial strain *Arthrobacter* sp. N2. *Chemosphere.* v.46, p.527-534, 2002.

Tien,M. & Kirk, T.K. Lignin Preoxidase of *Phanerochete chrysosporium Methods in Enzymology* v.161, p.238-249, 1988.

Dalton, R.L.; Evans, A.W.; Rhodes, R.C. Disappearance of diuron from cotton fields soils. *Weeds.* v.14, p. 14-31, 1966.

Kaufman, D.D.; Blake, J. Microbial degradation of several acetamide, acylanilide, carbamate, toluidine and urea pesticides. *Soil Biol.Biochem.* v. 5, p. 297-308, 1973.

Tilmanns, G.M.; Wallnöffer, P.R.; Engelhardt, G.; Olie, K.; Hutzinger, O. Oxidative dealkylation of five phenylurea herbicides by the fungus *Cunninghamella echinulata* Thaxter. *Chemosphere.* v.1, p.59-74, 1978.

Shelton, D.R.; Khader, S.; Karns, J.S.; Pogell, B.M. Metabolism of twelve herbicides by *Streptomyces. Biodegradation,*v.7, p.129-136, 1996.

Cullington, J.E. & Walker, A. Rapid biodegradation of diuron and other phenylurea herbicides by a soil bacterium. *Soil Biol. Biochem.* v. 31, p. 677-686, 1999.

DuPont Agricultural Products. *Staple herbicide, technical information ;* DuPont Agricultural Products: Wilmington, DE, 1993

Hataaka, A. Lignin-modifying from selected white-rot fungi: production and role in lignin degradatin. *FEM Microbiol.Rev.* v.13, p.125-135, 1994.

Higson, F.K. Degradation of xenobiotics by white-rot fungi. *Rev.Environ.Contam.Toxicol.* v.122,p.111-152, 1991.

Prado, A.G.S. & Airoldi C. The toxic on soil microbial activity by the free immobilized pesticide diuron *Thermochimica Acta* v.394, p.155-162, 2002.

Clemente, A.R.; Anazawa, T.A.; Durrant,L.R. Biodegradation of polycyclic aromatic hydrocarbons by soil fungi. *Brazilian Journal of Microbiology.* v. 32, p.255-261, 2001.

Tien, M. & Kirk, T.K Lignin-degrading enzyme from *Phanerochaete chrysosporium:* purification, characterization and catalytic properties of a unique H_2O_2 requiring oxygenase. *Proceedings of National Academy of Science USA* , v.81, p.2280-2284, 1884.

Kuwahara, M., Glenn, J.K., Morgan, M.A. & Gold, M.H. Separation and Characterization of two extracellular H_2O_2 – dependent oxidases from ligninolytic cultures of *Phanerochaete chrysosporium. FEBS Letters,* 169(2):247-250, 1984.

Szklarz, G.D., Antibus, R.K., Sinsabaugh, R.L. & Linklins, A.E. Production of Phenol oxidases and Peroxidases by Wood -Rotting Fungi. *Mycologia,* 81(2), p.234-240, 1989.

Esposito, E., Paulillo, S.M. & MANFIO, G.P. Biodegradation of the Herbicide Diuron in Soil by Indigenous Actinomicetes *Chemosphere,* Great Britain, **37**(3): 541-548, 1998.

Sumpter, S.R. & Peterson, B.A. *Analytical Method for the Determination of KIH-2031 (DPX-PE350) in Water Using Column-Switching Liquid Chromatography. DuPont Agricultural Products,* Experimental Station Wilmington, DuPont Report **N ^{0}AMR** 2746-93, p.1-58, 1994.

Yoshida, S.; Yonehara, S.; Ha, H -c, Iwahara, K.; Watanabe, T.; Honda, Y., Kuwahara, M. Production and characterization of ligninol ytic of *Bjerkandera adusta* on wood meal/wheat bran culture and production of these enzymes using a rotary-solid fermenter. *Mycoscience* v.37, p. 417-425, 1996.

Bending, G.D.; Friloux, M.; Walker, A. Degradation of contra sting pesticides by white rot fungi and its relationship with ligninolitc potencial. *FEMS Microbilogy Letters,* v. 212, p.59-63, 2002.

Fratila-Apachitel, L.E.; Hirst, J.A.; Siebel, M.A. & Giyzen, M.J.; Diuron Degradation by *Phanerochaete crysosporium* BKMF-F. 1767 in Synthetic and Natural Media *Biotechnology Letters.* v.21, p.147-154, 1999.

CHAPTER 4

DESIGN CHALLENGES FOR LARGE SCALE SULFATE REDUCING BIOREACTORS

James J. Gusek, P.E.
Golder Associates Inc.

Abstract: The first large-scale (1,200 gpm capacity), sulfate -reducing bioreactor (SRBR) was constructed in 1996 to treat water from an underground lead mine in Missouri. Other large-scale SRBR systems have been built elsewhere since then. This technology holds much promise for economically treating heavy metals and has progressed steadily from the laboratory to industrial applications. Scale -up challenges include: designing for seasonal temperature variations, minimizing short circuits, changes in metal loading rate s, storm water impacts, and resistance to vandalism. However, the biggest challenge may be designing for the progressive biological degradation of the organic substrate and its effects on the hydraulics of the SRBR cells.

Due to the wide variability of the organic materials that may be locally available at reasonable costs, the design of organic substrate SRBR systems is not and may never become a "cookbook" approach. Balancing substrate geochemical requirements with intuitive physical resistance to organic decay currently plays a large role in the large -scale system design process.

Key words: Passive Treatment, acid rock drainage, heavy metals, sulfate-reducing bacteria

1. INTRODUCTION

Among the growing list of generic design approaches, there are basically two kinds of *biologically*-driven passive treatment cells for treating mine drainage. *Aerobic Cells* containing cattails and other plants and algae are typically applicable to coal mine drainage where iron and manganese and mild acidity are problematic. *Anaerobic Cells* or *Sulfate-Reducing Bioreactors* are typically applicable to metal mine drainage with low to high acidity and a wide range of metals. Most passive treatment systems employ one or both of these cell types. The track record of aerobic cells in treating coal mine drainage is impressive, especially in the eastern coalfields of the US. Sulfate-reducing bioreactors have tremendous potential at metal mines and coal mines but have not seen as wide an application.

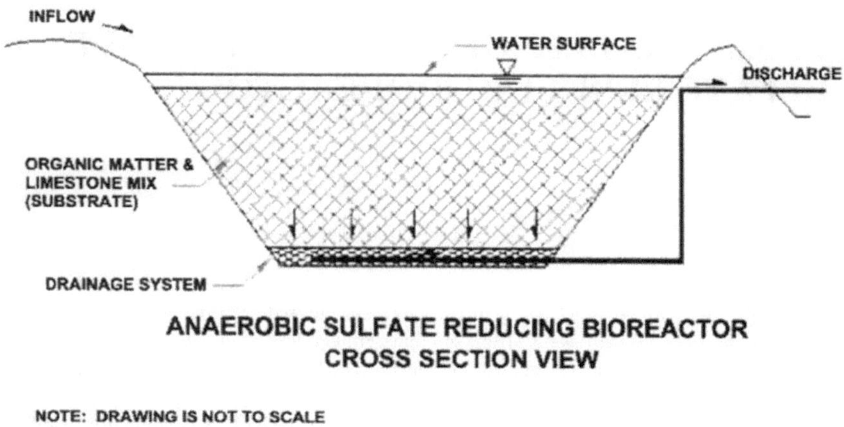

Figure 1. Sulfate-Reducing Bioreactor Schematic Section

Sulfate reduction has been shown to effectively treat mine drainage containing dissolved heavy metals, including aluminum, in a variety of situations (Gusek, 2002). The chemical reactions are facilitated by the bacteria *desulfovibrio* in SRBRs as shown in Figure 1 in schematic form and the photo in Figure 2.

The sulfate-reducing bacterial reactions (equation 1) involve the generation of:

- Sulfide ion (S^{-2}), which combines with dissolved metals to precipitate sulfides (equation 2)

- Bicarbonate (HCO_3^-), which has been shown to raise the pH of the effluent

The sulfate-reducing bacteria produce sulfide ion and bicarbonate in accordance with the following reaction (Wildeman, et al., 1993):

$$SO_4^{-2} + 2\,CH_2O \rightarrow S^{-2} + 2\,HCO_3^{\,-} + 2\,H^+ \qquad 1)$$

Figure 2. A Typical Sulfate-Reducing Bioreactor

The dissolved sulfide ion precipitates metals as sulfides, essentially reversing the reactions that occurred to produce acid mine drainage/acid rock drainage (AMD/ARD). For example, the following reaction occurs for dissolved zinc, forming amorphous zinc sulfide (ZnS):

$$Zn^{+2} + S^{-2} \rightarrow ZnS \qquad 2)$$

Suspected geochemical behavior of aluminum in sulfate-reducing bioreactors has just recently been documented (Thomas and Romanek, 2002). It is suspected that insoluble aluminum sulfate forms in the reducing environments found in sulfate-reducing bioreactors, perhaps in accordance with the following reaction which is one of many that are possible:

$$3Al^{3+} + K^{+} + 6H_2O + 2SO_4^{2-} \rightarrow KAl_3(OH)_6(SO_4)_2 \text{ (Alunite)} + 6H^{+} \qquad 3)$$

The key conditions for sulfate-reducing bacteria health are:

- A pH of 5.0 which can be maintained by the bacteria itself through the bicarbonate reaction in equation 1, and/or the presence of limestone sand

- The presence of a source of sulfate (typically from the water being treated)

- Organic matter (CH_2O) in the substrate.

SRBRs have been successful at substantially reducing metal concentrations and favorably adjusting the pH of metal mine drainages.

There are many advantages to sulfate-reducing bioreactors in treating mine drainage, including the ability to:

- Work in cold, high altitude environments
- Handle high flow rates of mildly affected mine drainage in moderate acreage footprints
- Treat low pH acid drainage with a wide range of metals and anions including uranium, selenium, and sulfate
- Accept acid drainage-containing dissolved aluminum without clogging with hydroxide sludge (Gusek and Wildeman, 2002)
- Provide life-cycle costs on the order of $0.50 per thousand gallons
- Be integrated into "semi-passive" systems that might be powered by liquid organic wastes

Sulfate-reducing bioreactors might not be applicable in every abandoned or active mine situation. A phased design program of laboratory, bench, and pilot scale testing has been shown to increase the likelihood of a successful design (Gusek, 2001).

It is relatively easy to construct and operate small scale SRBRs. Most of the construction materials are available off-the-shelf. For example, a 55-gallon drum can be easily retrofitted into SRBR bench scale cell using less than $10 worth of plumbing materials commonly found at local hardware stores. These sized cells might treat from one to tens of liters of water per day, depending on the drainage chemistry.

Children's swimming pools have been used to construct pilot scale SRBRs, again, using less than $100 worth of common plumbing supplies. These sized cells might treat up to 30,000 liters per day (about five gpm).

Scaling up from pilot scale to full scale is no easy task. Now site-specific conditions become more important in design. This paper addresses some of the more common scale-up design challenges including:

- seasonal temperature variations
- metal loading rate changes

- short circuiting
- storm water impacts
- resistance to vandalism
- changing economics
- organic substrate biodegradation

2. SEASONAL TEMPERATURE VARIATIONS

In most bench and some pilot scale situations, the temperature of the mine drainage may be artificially controlled. This situation is somewhat necessary to minimize the number of experimental variables and, frankly, it is the easiest thing to do, especially given the propensity of small plumbing fixtures to quickly freeze in the winter.

From a microbiological standpoint, it is well understood that depressed temperatures slow bacterial activity. This is why refrigeration is so popular in preserving food. Sulfate-reducing bacteria activity is also sensitive to temperature; however, they have been found to adapt to very harsh environments, including super-cold (-40°C) water beneath the Antarctic ice cap (Postgate, 1979). In designed SRBRs, the limiting activity function may be the rate at which organic materials are decaying – this will be a function of the behavior of that whole suite of microbes that colonize SRBRs and break down the organic matter into simple compounds that become nutrients for the sulfate-reducing bacteria.

Minor temperature fluctuations may not affect SRBR overall activity all that much once the microbial suite has established itself. In a pilot scale SRBR at the Ferris Haggarty Copper Mine/Osceola Tunnel in Wyoming, the mine effluent was typically less than 5°C. During the winter at this high altitude (9,500 ft elev.) site, the treated effluent temperature dropped to as low as 0.5°C. Yet the rate of sulfate reduction reduced to only about 80 percent of the benchmark rate (Gusek, 2000). This observation was incorporated into the design of a full-scale system to treat as much as 600 gpm.

A part of the temperature challenge at the Ferris Haggarty site was mitigated by a number of design features, including:

- collecting the mine water deeper in the mine and delivering it to the SRBR cells in insulated pipes to preserve some of its ambient in-ground temperature, and

- covering the SRBRs with an insulating layer of clean mine waste rock

Covering/burying SRBRs (which do not need plants or sunshine to function) can be used to solve other full-scale design challenges. Other design considerations, such as seasonal changes in metal loading rates, may further mitigate temperature effects on SRBR performance.

3. METAL LOADING RATE CHANGES

SRBR cells are sized based on metal loading and can be resilient to metal loading variations *within reasonable limits.* Those limits are best determined in pilot scale tests where the expected operating ranges of flow and metal concentrations and the reactions of the SRBR cells to those varying conditions can be assessed. For example, a pilot SRBR cell at a lead mine in Missouri was sized for 25 gpm. Once steady state operation was observed for many months, the net alkaline flow was increased to nearly double the design rate. The SRBR cell began to show evidence of stress (i.e., decreased metal removal efficiency) after several months of exposure to the higher flow (Gusek, et al., 1998). Not all SRBR cells might be this resilient, especially if the mine drainage is net acidic, but this observation allowed engineers to include a significant factor of safety in the design of the full-scale system (1,200 gpm capacity) at this site.

Overloading SRBRs with net acidic mine drainage can be catastrophic: sulfate-reducing bacteria populations can be decimated by overexposure to low pH mine water with high concentrations of metal, particularly iron and aluminum. In this situation, the geochemistry of the SRBR might be significantly altered and while some metal removal may still occur, removal efficiencies may suffer and in certain cases even turn negative after the overloading situation has abated. In other words, some particularly unstable metal compounds that had been precipitated during the overloading event might actually be re-dissolved as the geochemistry of the SRBR cell recovers to the extent that the concentrations of metals in the SRBR cell effluent are greater than the influent.

SRBRs are typically sized to deliver treated water with low concentrations of metals and a near neutral pH. However, experience has shown that SRBR cell effluents typically contain excess alkalinity at reasonable concentrations that may be available to ameliorate acidic contributions that might be impacting the receiving stream far removed from the original passive treatment site. This operational by-product of metal removal in an SRBR may be used to lessen the impact of overloading conditions as suggested below.

In the full-scale SRBR designs then, flow management will be a paramount concern to insure that the spikes in flow and/or concentration as defined in pilot testing are attenuated to the most practical extent. Alternatively, overloads may even be temporarily diverted past the system in the hope that excess alkalinity and dilution by system-treated effluent will lessen the downstream impact as temporary as it may be. Overload mitigation may be practically accomplished in a full-scale system design with a variety of methods that include using:

- Properly sized holding/mixing ponds
- Flow-sensitive diversion ditches that are engaged only in overloading conditions, and
- Underground mine workings that are configured as storage reservoirs

4. SHORT CIRCUITING

SRBRs are typically designed as vertical flow systems with mine water traveling downward (or upward) through an organic substrate to be collected at the opposite end (bottom or top) of the cell. Early SRBR designs used organic substrate materials that had very low (on the order of **1×10^{-4} cm/sec)** saturated hydraulic conductivity (K) values. Thus, minor changes in K values introduced by substrate heterogeneity offered a greater opportunity for short circuiting. Recent substrate mixes have measured K values several orders of magnitude less. However, as the capacities of systems have increased the opportunities for intra-cell short circuiting have increased as well. This is primarily due to the increasing cell surface area in larger capacity systems.

Busler, et al. (2002) studied this issue in vertical flow reactors (VFRs) which are similar to SRBRs. They documented significant short circuiting in a VFR that had been designed in accordance with accepted practice. Busler, et al. subsequently plumbed a VFR with multiple adjustable-head discharge points (each connected to a different zone of the VFR) and dramatically reduced short circuiting. They also evaluated the efficacy of variable spacing of solution collection pipe perforations to further minimize short circuiting.

The essence of large-scale SRBR cell design that can be gleaned from field experience and Busler, et al. is to, "divide and conquer." That is, subdivide large flows into smaller flows that are equitably distributed to multiple cells of similar dimensions rather than construct one large SRBR cell. If intra-cell short circuiting is a concern, further subdivide the treated solution collection system into parallel flow nets with virtually identical pipeline headlosses. This approach may increase the cost of the system plumbing, but the design will be less likely to suffer the effects of short circuiting which are neither easily isolated nor rectified.

5. STORM WATER IMPACTS

A properly designed SRBR cell/system will need to survive the physical impacts of storm water flows. This is accomplished with runoff diversion channels and other standard management practices which are not unique to SRBR cells or systems. However, in climates with high precipitation, storm water falling directly on to the SRBR can result in significant operational problems.

For example, an SRBR cell with an open water pond in South Carolina might be exposed to a 100-year, 24-hour rain storm event on the order of 7.5 inches. While the plumbing and hydrologic aspects of the cell can be easily designed to handle the additional influx of water, consideration needs to be given to the effects of the increased flow on the bacterial population and the physics of metal precipitate retention in the organic substrate. Assuming that "solution to pollution is dilution" may be dangerous. It is unlikely that this situation would be naturally encountered during the operation of a pilot cell. Therefore, it may need to be artificially created to evaluate the short-term effects on SRBR cell performance.

In the case of the South Carolina site, pilot test results and a parametric study revealed that the short-term flow increases in response to rainfall events could result in unacceptable system performance. Several alternative SRBR cell designs were considered including adjustments of cell footprint (i.e., watershed area) and increasing the freeboard to temporarily store the 24-hour storm in the SRBR cell. The selected alternative to minimize storm water impacts was the burial of the cell beneath a light-weight fill/geomembrane/soil-vegetation cover. This approach offered significant resistance to long-term vandalism at this site which was part of a permanent mine closure project.

6. RESISTANCE TO VANDALISM

As distasteful as the design consideration may be, any passive treatment system must be designed to resist vandalism by humans as well as animals. The larger the system is, the larger the vandalism target.

Camouflage is probably the best human vandalism deterrent: one cannot harm something that is not easily seen. This can best be accomplished by blending SRBR cells into the landscape by either burying them which can be used also to solve other design issues or by creating visual footprints that look totally natural.

Again, because neither plants nor air are required for SRBRs to function, they can be buried beneath a veneer of rock and soil provided that the feed water plumbing to the cell is not compromised. Settlement of the organic substrate needs to be considered in the design if burial is being considered. However, most organic substrate designs typically include a large component of wood chips or sawdust, which do not readily compress under minor surcharge loads developed by soil/rock covers. This aspect of the design should ideally be evaluated at the pilot stage of the design effort.

The visual aspect of "natural-looking" SRBR cell design with open water ponded on the surface requires a step back from the rectangular or geometric shapes that seem to be the stamp of typical design engineers who may have difficulty drawing lines that are not straight. One way to hide a rectangular-shaped SRBR cell is to extend the ponded water surface beyond the rectangular footprint to a shallow water zone with a more natural-looking irregular shape. This design feature may add a little to the construction cost, but community acceptance and resistance to vandalism of exposed geomembrane liners may make it worth the extra expenditure.

While beavers are known as nature's "engineers," they can do more vandalism damage than humans to SRBR cells and other passive treatment system features on a pound-for-pound basis. The larger the flow rate treated, the larger this problem can be. This is because beavers are naturally attracted to the sound of running water. This acoustical magnet triggers their digging and damming reflex and large-scale SRBR cells become at risk.

To minimize this risk, large-scale SRBR cells near beaver habitat can be designed with internal spillways or water level control structures that muffle the falling water noise. These reasonably-priced pre-fabricated structures are commercially available in a range of sizes. In addition, there are a number of websites that contain conceptual designs for "beaver-proof" spillways.

7. CHANGING ECONOMICS

Procuring the organic and inorganic components for bench and pilot scale SRBRs may involve several hundred kilograms to several thousand kilograms of various materials. Invariably, many businesses or individuals are initially more than willing to provide their waste materials such as manure, sawdust, green waste, or wood chips for free. This is probably done in the hope that they can save money by not having to pay to have their waste disposed in a landfill if it can be put to a beneficial use. Unfortunately, when it is revealed that a particular waste stream source is an integral component of the SRBR substrate, the price of procuring it inevitably goes up. The construction engineer is then faced with a dilemma: change the design and risk more uncertainty in the system's operation, or pay the higher price for the material.

The best solution to the dilemma may be to change the design by substituting similar materials if possible. Fortunately, the uncertainty of substitution can be minimized by a little extra effort in the bench scale phase of the full-scale SRBR cell design. It is recommended that potential substitutions be anticipated in advance and tested on a bench scale. Thus, last-minute changes in material sources can be made with more peace of mind.

8. ORGANIC SUBSTRATE BIODEGRADATION

Organic materials are a key component in the formulation of the substrate of sulfate-reducing bioreactors. Often these materials are considered waste materials and can be obtained for little or no purchase cost. The only expense incurred might be in their transport to the treatment site. In many cases, the site is located in a remote forest environment. In this situation, some of the materials such as wood chips and sawdust might be generated onsite or from local sources. A short list of organic waste materials that might be candidates for use in a sulfate-reducing bioreactor is provided below. The list is not necessarily all inclusive as specialty wastes unique to different locales might be available.

- Wood chips
- Sawdust
- Rice Hulls
- Hay and straw (spoiled)
- Cardboard
- Soy bean hulls

- Yard waste
- Mushroom compost
- Animal manure
- Waste alcohols including antifreeze
- Waste dairy products
- Sugar cane processing residue (Bagasse)

Early SRBR designs included large amounts of compost, manure, and easily degraded organic materials. Unfortunately, these systems were depleted of biologically available carbon in a short time so cell longevities were short – on the order of several years. Subsequently, more biologically durable materials such as sawdust and wood chips were used in substrate design to offer a better balance between short- and long-term sources of organic nutrients. To date, this author and others have relied on intuitive comparisons of the biodegradability of various solid organic materials that might be used in SRBRs. Precisely estimating the progressive biological degradation of the organic substrate and its effects on the hydraulics of the SRBR cells is a significant design challenge.

However, some progress is being made on this front. Seyler et al. (2003) evaluated the effects of solid phase organic substrate characteristics on sulfate reducer activity and metal removal. The identification of a simple standardized analysis method for comparing the biodegradability of various organic materials is probably the most significant advance made by Seyler et al. The analysis method is tiered off of nutrient analysis of agricultural products. Their results conform to the intuitive observations of this author and could be the foundation of a better understanding of organic substrate design to maximize longevity.

9. SUMMARY

Sulfate-reducing bioreactors are not the only type of passive treatment technique available to the design engineer, and they are not applicable in every situation. However, they can handle a wide variety of flows and AMD/ARD chemistries in hostile cold climates, and they can treat aluminum-bearing AMD/ARD without plugging. Furthermore, they can generate excess alkalinity in their effluent that further enhances the quality of the receiving stream.

Applying the results of laboratory, bench, and full-scale tests to full-scale designs is part common sense, part intuition. However, while a standardized design approach may never be realized, advancements in design criteria to support site-specific evaluation as discussed in this paper continue.

REFERENCES

Busler, Shaun, C. Cooper, T. Danehy, D. Peart, and M. Dunn, 2002. "Visual Demonstration of Water Distribution Patterns in Vertical Flow Systems" in Proceedings of the 19th Annual Meeting, ASMR, Lexington, Kentucky, June 9-13.

Gusek, J.J., T.R. Wildeman, A. Miller, and J. Fricke, 1998. "The Challenges of Designing, Permitting and Building a 1,200-GPM Passive Bioreactor for Metal Mine Drainage, West Fork Mine, Missouri," in Proceedings of the 15th Annual Meeting, ASSMR, St. Louis, Missouri, May 17-21.

Gusek, J.J., 2000. "Reality Check: Passive Treatment of Mine Drainage and Emerging Technology or Proven Methodology?" Presented at SME Annual Meeting, Salt Lake City, Utah, February 28.

Gusek, J.J., 2001. "Why Do Some Passive Treatment Systems Fail While Others Work?" in Proceedings of the Nation Association of Abandoned Mine Land Prog rams, Athens, Ohio, August 19-22.

Gusek, J.J., and T.R. Wildeman, 2002. "Passive Treatment of Aluminum-Bearing Acid Rock Drainage" in Proceedings of the 23rd Annual West Virginia Surface Mine Drainage Task Force Symposium, Morgantown, West Virginia, April 16-17.

Gusek, J.J., 2002. "Sulfate-Reducing Bioreactor Design and Operating Issues: Is This the Passive Treatment Technology for Your Mine Drainage?" presented at the Nation Association of Abandoned Mine Land Programs, Park City, Utah, September 15 -18.

Postgate, J.R., 1979. *The Sulfate-Reducing Bacteria,* Cambridge University Press, London, United Kingdom, 151 pp.

Seyler, Jason, L. Figueroa, D. Ahmann, T.R. Wildeman, and M. Robustelli, 2003. "Effect Of Solid Phase Organic Substrate Characteristics On Sulfate Reducer Activity And Metal Removal In Passive Mine Drainage Treatment Systems" in Proceedings of National Meeting of the American Society of Mining and Reclamation and the 9th Billings Land Reclamation Symposium, Billings MT, June 3-6, 2003. Published by ASMR, 3134 Montavesta Rd., Lexington, KY 40502.

Thomas, Robert C. and Christopher S. Romanek, 2002. "Acid Rock Drainage in a Vertical Flow Wetland I: Acidity Neutralization and Alkalinity Generation," in Proceedings of the 19th Annual Meeting, ASMR, Lexington, Kentucky, June 9 -13.

Wildeman, Thomas R., G. A. Brodie, and J. J. Gusek, 1993. *Wetland Design for Mining Operations.* BiTech Publishing Co., Vancouver, BC, Canada.

CHAPTER 5

IN-SITU BIOREMEDIATION OF A CHLORINATED SOLVENT RESIDUAL SOURCE IN UNCONSOLIDATED SEDIMENTS AND BEDROCK USING BIOAUGMENTATION

Allen Kane[1], Jim Vidumsky[2], David W. Major[3], Nicholas B. Bauer[4]
[1] Golder Associates. Cherry Hill, NJ, [2]E.I. DuPont, Wilmington, DE, [3]GeoSyntec Consultants, Guelph, Ontario, [4]Bauer & Associates, Inc., Reston, VA

Abstract: The Caldwell Trucking Superfund Site is located in Essex County, NJ, and covers approximately 15 acres. Groundwater in glacial deposits as well as fractured basalt bedrock is contaminated with chlorinated ethenes, ethanes, and methanes up to approximately 4,000 ft (1,200 m) downgradient of the site. Trichloroethene (TCE) concentrations in the source area were as high as 700,000 μg/L (about 60% of TCE solubility). Natural biodegradation is present over much of the site; however some areas, particularly near the source, appeared to be substrate limited. Following microcosm studies which demonstrated that complete degradation of the contaminants could be achieved, a comprehensive groundwater remedy was proposed that included bioremediation of the source area. A field pilot test of in-situ enhanced bioremediation in the source area was initiated in 2001. The layout included six nutrient injection wells screened in glacial deposits and fractured bedrock, and seven monitoring wells. Injection wells were bioaugmented with a culture of naturally occurring microorganisms (KB-1 Culture including *Dehalococcoides ethenogenes*) in March 2001. In over 30 months of operation, the system was optimized by adjustment of the amendment composition and the injection frequency. Gene probe techniques were used to verify initial and continued survival and propagation of the *Dehalococcoides ethenogenes* organisms.

Keywords: Bioaugmentation, biostimulation, bedrock, chlorinated solvents, Dehalococcoides

1. INTRODUCTION

The Caldwell Trucking Company Superfund Site Trust (Trust) is currently undertaking the remediation of the Caldwell Trucking Company Site (Site) located in Fairfield Township, Essex County, New Jersey. Beginning in the early 1950s, the site was used for disposal of septic waste from residential and industrial sources, in unlined lagoons. Previous remedial actions have included vadose zone treatment utilizing soil vapor extraction and solidification/stabilization. A significant groundwater source area remains associated with a former lagoon area, and a dissolved-phase plume of chlorinated solvents extends 4,000 feet (1,200 m) downgradient. After undertaking laboratory microcosm studies that showed complete degradation of TCE, the Trust proposed to perform a field Pilot Test of Accelerated Biological Treatment with bioaugmentation in the source zone. Accelerated Biological Treatment is being considered as part of an alternative to the groundwater remedy proposed in 1993, which comprised pumping and treating of groundwater.

2. FIELD PILOT STUDY DESIGN

The Pilot Test of Accelerated Biological Treatment System is designed to drive rapid conversion of dissolved PCE and TCE to DCE, VC, and ethene, and thus to accelerate the dissolution of source material. The goal is not to achieve specific concentrations of parent and daughter compounds in the dissolved phase, but rather to accelerate the dissolution of source material, and so reduce the overall lifetime and impact of the source.

2.1 Pilot Study Area Layout

The Pilot Test configuration included three nested pairs of injection wells and three nested pairs of monitoring wells, 25 to 40 feet (7.6 to 12.2 m) downgradient, screened in the corresponding units. Each nested pair comprises one well completed in the deep overburden (referred to as the 'B zone') and one well in the upper bedrock ('C zone'). An existing nearby sidegradient well (C-15) is also monitored as a control well.

2.2 Hydraulic Testing

Hydraulic conductivity values were estimated for each well by conducting rising head slug tests in all the wells. Using the Hvorslev as well as Bouwer and Rice methods, geometric mean values for hydraulic conductivity of the overburden and upper bedrock were calculated as $1x10^{-3}$ centimeters per second (cm/sec) and $2.4x10^{-5}$ cm/sec, respectively.

2.3 Tracer Testing

In order to assess groundwater velocity and connectivity between wells in the pilot test zone, potassium bromide tracer tests were conducted. In summary, the bromide testing demonstrated that wells throughout the pilot test zone are hydraulically connected and groundwater flow velocities range between 1.4 and 2.5 ft/day (0.43 and 0.76 m/day) in the deep overburden (B-zone) and 1.8 and 3.0 ft/day (0.55 and 0.91 m/day) in the upper bedrock (C-zone).

2.4 Baseline Sampling

Baseline sampling of the pilot test wells showed high concentrations and dominance of TCE and tetrachloroethene (PCE) in the C-zone wells, along with the virtual absence of cis-1,2-dichloroethene (cis-DCE) and vinyl chloride (VC), indicating that biodegradation was negligible in the C-zone in the pilot test area. In the B-zone, some conversion to cis-DCE had occurred but production of vinyl chloride was very limited.

2.5 Nutrient Addition and Bioaugmentation

Monthly nutrient injections were initiated in February 2001 using approximately 100 gallons (380 L) of nutrient solution for each B-zone well and approximately 50 gallons for each C-zone well. Nutrient solutions were prepared under anaerobic conditions using site groundwater and each injection was followed by a "wash" consisting of 50 gallons of groundwater without added nutrients. All upgradient injection wells were inoculated in March 2001 with 10 to 15 liters of the KB-1 consortia of microorganisms while downgradient well C-22 was inoculated on May 30, 2001 to address the higher concentrations of constituents present in this well. The nutrient concentrations and injection frequency were modified throughout the pilot test as shown in the table:

Table 1.

Period	Injection Frequency	Molar Ratio	Total Concentration
Feb 01 – April 01	Monthly	1 M: 1 L: 1 A	4,500 mg/L
May 01 – Jan 02	Weekly	1 M: 1 L: 1 A	4,500 mg/L
Feb 02 – Dec 02	5x per week	1 M: 2 L	5,300 mg/L
Dec 02 – Date	5x per week	1 M: 1 L: 1 E*	13,000 mg/L

M=methanol; L=lactate; E=ethanol; *ratio of electron -equivalents

The injection frequency was increased (while decreasing the amount per injection proportionally) in order to provide a more constant flow of nutrients to the microbes, (which yielded more consistent results in the field, e.g., accelerated and permanent reductions in TCE and accelerated production of daughter products). Additionally, the total potential electron acceptor loading was calculated from all known compounds including VOCs and inorganics, and the nutrient loading was increased to equal the potential loading.

3. RESULTS AND DISCUSSION

3.1 DHE Trends

PCR-based tests for DHE (PCR assay) allow the detection and relative quantification of the *Dehalococcoides ethenogenes* microorganisms (DHE) in soil and groundwater samples. Standard control tests are undertaken together with each set of assays for quality assurance purposes. The results of the DHE testing generally show widespread presence and high "intensity" (a measure of the relative abundance) of the dechlorinating microorganism, DHE. In particular, (1) DHE is present and has been sustained in *all* of the injection wells and *all* of the monitoring wells; and (2) the signal strength is high in 9 of the 12 wells and comparable between injection and monitoring wells, indicating that a treatment zone is established in both the B and C aquifer zones between the injection and monitoring wells.

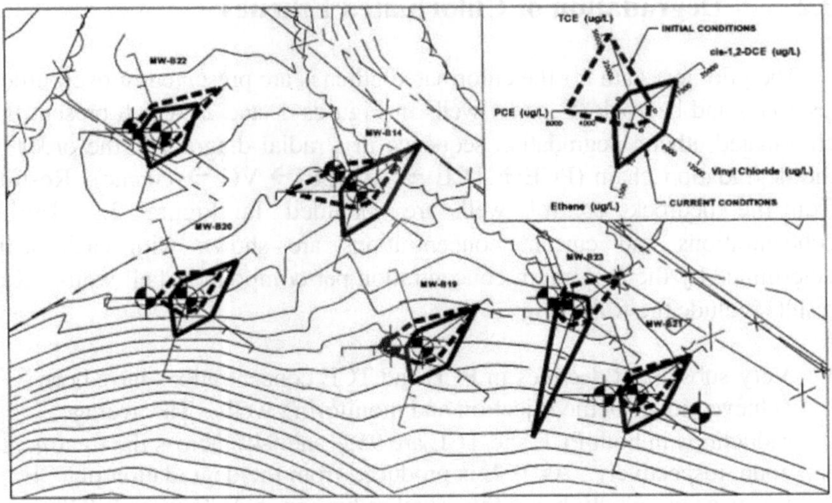

Figure 1. Degradation sequence of chlorinated ethenes in pilot overburden wells

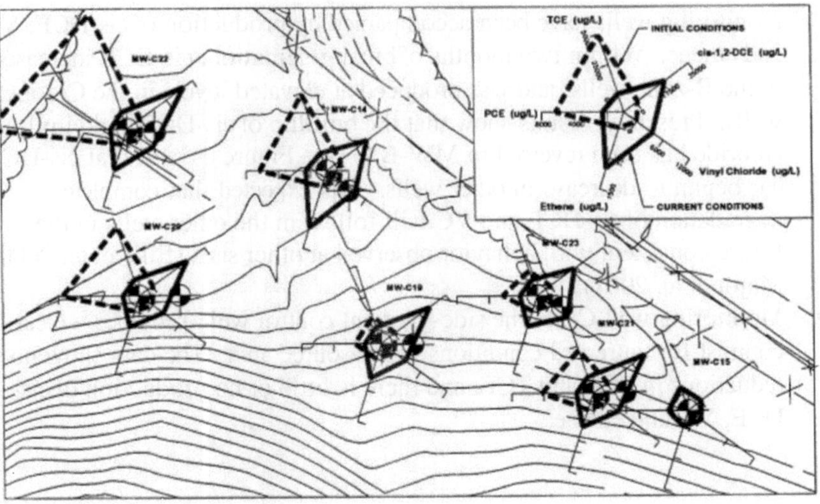

Figure 2. Degradation sequence of chlorinated ethenes in pilot bedrock wells

3.2 Degradation of Chlorinated Ethenes

The pilot test data for the chlorinated ethenes are presented in overburden (B zone) and bedrock (C zone) wells in Figures 1 and 2, which present the chlorinated ethene degradation sequence in a radial diagram in the order of the degradation chain (PCE→ TCE → cisDCE → VC → ethene). Results from the bedrock control well are included in Figure 2. Initial concentrations and current concentrations are shown with each scale determined by the maximum concentration per compound of all wells. Key results include the following:

- Very substantial declines in PCE and TCE concentrations have been achieved in all of the injection and monitoring wells. The average reductions in both PCE and TCE are 95% and 94% across the treatment zone, respectively. As TCE is produced from the degradation of PCE, this is a net reduction and the actual reduction in TCE exceeds 94%. These reductions have occurred equally in the injection wells and monitoring wells indicating that a continuous treatment zone is present across the pilot test area;
- The reductions in PCE and TCE in the injection wells and the monitoring wells have been accompanied by production of cis-DCE, VC and ethene. Within two months of bioaugmentation, cis-DCE increased in the B-zone wells, and was produced at elevated levels in the C-zone wells. Presently, results show that the buildup of cis-DCE and vinyl chloride has been reversed in MW-B23 (see Figure 3), and that cis-DCE has begun to decrease in other wells. It is expected that complete degradation of cis-DCE and VC will follow in the other wells in the future consistent with behavior observed at other sites (Ellis et al, 2000; Major et al, 2001);
- Monitoring well C-15, the side-gradient control well, provides a clear contrast for untreated conditions in the source area. The data show no reductions in PCE and TCE, and there is little or no production of cis-DCE, VC and ethene.

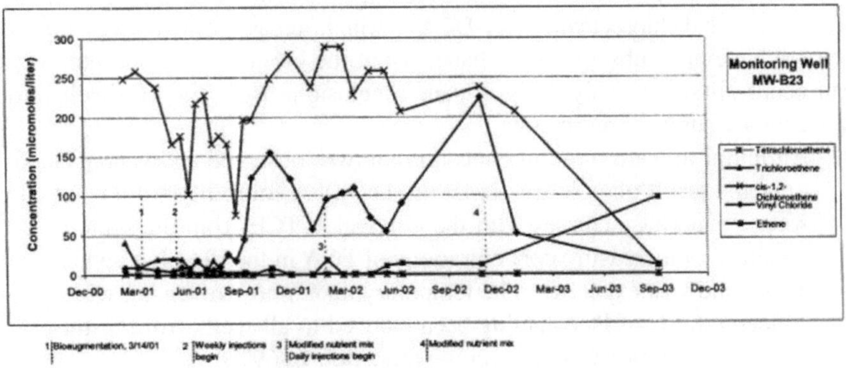

Figure 3. Chlorinated ethenes in pilot test overburden monitoring well MW -B23 showing transformation of cis-DCE and vinyl chloride to ethene

3.3 Well C-22 Performance

Downgradient well C-22 contained baseline concentrations an order of magnitude higher than other wells for several chlorinated VOCs, and so direct nutrient addition and bioaugmentation was initiated in this well on May 30, 2001 as part of the pilot study program. The results obtained include the following:

- PCE reduced by 99.8% from 27,000 ug/L to 50 ug/L;
- TCE, once as high as 680,000 ug/L reduced by 99.8% to 1,400 ug/L;
- cis-DCE concentration increased from non-detect up to 36,000 ug/L and now declined to 29,000 ug/L; and,
- Production of VC and ethene sustained with VC at 2,800 ug/L and ethene at 620 ug/L. Concentrations of VC expected to decline in the future, consistent with well MW-B23.

These improvements are attributable to biodegradation as evidenced by the production of daughter products and confirmed by the PCR assay results, which indicate that healthy levels of DHE are being maintained in this well.

3.4 Degradation of Chlorinated Ethanes

1,1,1-TCA can degrade in the environment by both abiotic and biotic mechanisms. The abiotic mechanism (specifically hydrolysis) results in production of acetic acid (80%) and 1,1 dichloroethene (20%), and published half-lives are on the order of 1 year (Montgomery, 2000; Schwarzenbach et al, 1993; Howard et al, 1991). 1,1-dichloroethene (1,1-DCE) degrades biotically to vinyl chloride, which may be oxidized or further degraded to ethene, depending upon the conditions. The biotic degradation pathway produces 1,1-dichloroethane (1,1-DCA), which breaks down biotically to chloroethane and subsequently ethane. Both the biotic and abiotic pathways can occur simultaneously. Significant conclusions from the Caldwell pilot test data are summarized below:

- An initial upward trend in concentration was observed and is concluded to be the result of mole fractionation, i.e., TCA solubility from residual source material increases with the removal of TCE from the source resulting in increasing concentrations of TCA in the dissolved phase. This trend was expected to be temporary and is now reversing with concentrations of TCA having been reduced in all wells from historic high concentrations by an average reduction of 69%.
- The abiotic pathway of degradation (1,1-DCE) has been shown to be consistently active in all wells.
- The biotic pathway of degradation (via 1,1 -DCA), previously not evident in many wells (notably in the C-zone) has been stimulated by the bioaugmentation and nutrient additions with all wells showing significant levels (1,000 ug/L to 1,900 ug/L).
- Complete biotic degradation of TCA, represented by ethane, has been observed in all wells throughout the test zone, and recent data shows consistent presence at levels well above those in the control well.
- The control well, C-15, has shown little evidence of biotic degradation of TCA. Abiotic degradation, via 1,1-DCE, is present.

3.5 Degradation of Chlorinated Methanes

The chlorinated methanes, in particular chloroform (CF or trichloromethane) degrade biotically, primarily to methylene chloride (DCM). Significant conclusions from the Caldwell pilot test data are summarized below:

- Concentrations of CF were initially much lower than other constituents, with concentrations generally less than about 1,200 ug/L. An initial upward trend in concentration in some wells, possibly due to mole fractionation in a similar manner to the chlorinated ethanes, has been reversed, with all wells lower than their historic highs. Specifically, concentrations of CF have been reduced from historic high concentrations by 70% to 100%, with an average concentration reduction of 87%. The average concentration of all wells is reduced from 783 ug/L to 91 ug/L.
- The biotic pathway of degradation (via DCM), has been stimulated in all wells. Recently, DCM concentrations have been reducing with the average concentration of all wells now at 170 ug/L.

- Control well C-15 has shown little reduction in CF concentration, and no production of DCM, further confirming that bioaugmentation and nutrient addition are responsible for the observed concentration reductions in the pilot test zone.
- Reductions in chloroform may be associated with the complete reduction of cis-DCE to ethene, as has been reported elsewhere (Maymo-Gatell, et al., 2001). Well MW-B23, in which cis-DCE and vinyl chloride have been nearly eliminated, has had less than 50 ug/L of chloroform for over one year and was at non-detect levels in the most recent sampling.

3.6 Geochemical Conditions

Geochemical parameters have been monitored throughout the test and key conclusions are as follows:

- Dissolved oxygen levels have been maintained at near zero levels;
- Reducing conditions exist in the pilot test zone as evidenced by iron- and sulfate-reduction in all wells, throughout the duration of the test. Complete sulfate depletion from levels of 40-80 mg/L has occurred in some wells and ferrous iron production is also substantial in all wells, with a maximum of over 6.0 mg/L. Control well C-15 has shown steady levels of sulfate of approximately 35 mg/L, and ferrous iron concentrations around zero, indicating no occurrence of sulfate or iron reduction.
- Levels of Biological Oxygen Demand (BOD) indicate the effectiveness of the distribution of nutrients throughout the pilot test zone. BOD levels have generally increased throughout the pilot test zone including both B and C zone monitoring wells, indicating nutrients have been distributed throughout the system.
- Methane production is occurring and levels have recently risen to 500 ug/L.
- All of the above geochemical trends are consistent with the sustained occurrence of biologically mediated reductive dechlorination.

3.7 Measured Field Half-Lives

Pilot study data have been interpreted with an analytical degradation model for estimating the half-lives of PCE and TCE. The spreadsheet-based model uses a formulation based on first-order decay kinetics, with the additional contribution of the decay of the parent compound included. The basic formulation is as follows:

$$C_{n,t} = [C_{n,t-1} + \varepsilon(C_{m,t-1} - C_{m,t})]\exp(-\lambda_n \Delta t) \quad (1)$$

where:
C_n = concentration of compound n (daughter) (μg/L);
C_m = concentration of compound m (parent) (μg/L);
t, t-1 = current and previous time step (years);
Δt = time step value (years);
ε = (molecular weight of compound n)/(molecular weight of compound m); and,
λ_n = decay constant of compound n
For PCE, the uppermost parent compound, equation (1) reduces to

$$C_{n,t} = C_{n,t-1} \exp(-\lambda_n \Delta t)$$

The model solves equation (1) simultaneously for PCE and TCE. The resulting average half-lives are 25 days for PCE and 21 days for TCE. These values are approaching those measured in the laboratory microcosm studies and compare vary favorably to natural (unaccelerated) half-lives in the downgradient plume that are on the order of 500 days.

4. CONCLUSION

In summary, a continuous treatment zone has been established across the pilot test area that has achieved average reductions of PCE and TCE concentrations of 94-95% in a period of 30 months; continued concentration reductions are expected. Where complete degradation to ethene has become predominant, it is accompanied by elimination of chlorinated methanes (including chloroform) and substantial reductions in 1,1,1 -TCA. Sulfate and ferric iron reduction are significant and widespread and methane production has increased over the duration of the test. The results indicate that in-situ bioremediation is a viable source treatment/control technology that is synergistic with monitored natural attenuation of a downgradient dissolved phase plume. The system achieved rapid conversion of dissolved PCE and TCE to DCE, VC, and ethene, which accelerates the dissolution of source material, and so will reduce the overall lifetime and impact of the source. An expansion of the pilot test to a full-scale source area treatment is anticipated.

REFERENCES

Ellis et. al. 2000. Bioaugmentation for Accelerated In-Situ Anearobic Bioremediation. Environ. Sci. Techno. 34:2254-2260.

Howard, et al., Handbook of Environmental Degradation Rates, CRC Press, 1991.

Major et al. 2001. Successful Field Demonstration of Bioaugmentation to Degrade PCE and TCE to Ethene. In: Bioaugmentation, Bioba rriers and Biogeochemistry, pp. 27 -34 E.A. Leeson, B. Alleman, P. Alvarez and V. Major. Battelle Press

Maymo-Gatell et al. 2001. Reductive Dechlorinatio of cis -1,2-dichlorethene and vinyl chloride by "Dehalococcoides Ethenogenes". Environ. Sci. Techno. 35:516 -521.

Montgomery J.H., Groundwater Chemicals, Desk Reference, Third Edition. CRC Press, 2000.

Schwarzenbach R.P., et al., Environmental Organic Chemistry, Wiley & Sons, 1993.

This page intentionally left blank

CHAPTER 6

BIOREMEDIATION OF A RAILROAD DIESEL FUEL SPILL IN PALMER, MASSACHUSETTS

Todd D. Kirton; Paul G. Beaulieu
Tighe & Bond Consulting Engineers

Abstract: The nature of the railroad industry nearly ensures the release of petroleum hydrocarbons (in the form of diesel fuel, motor oil, lubricating oils, etc.) to the soils and subsoils along the thousands of miles of rail tracks throughout the United States. The operation of a locomotive, for obvious reasons, is dependent upon the use of these forms of petroleum hydrocarbons, and leaks, spills and accidents are unavoidable in many cases. Nevertheless, these releases may be regulated by state environmental agencies. In Massachusetts, for example, the regulations governing the cleanup of releases of oil and hazardous materials -- the so-called Massachusetts Contingency Plan, 310 CMR 40.000 -- do not exempt the railroad industry from remediating spills in excess of the 10 -gallon 'Reportable Quantity'. Cleanup of spills along active rail lines, however, can be quite challenging. Rail traffic presents significant dangers to remediation personnel. Moreover, although excavation of contaminated soils is often the most expedient method for remediating a spill, this option is not always available at sites along active lines, particularly those with no alternative routes, as interruption of rail traffic is not economically acceptable for the rail company or its clientele. At the 2001 International Conference on Contaminated Soils, Sediments and Water, the authors presented the preliminary results of a bioremediation program that was implemented at the Palmer Railroad Yard to address the release of 700 gallons of diesel fuel along 1,000 linear feet of railroad tracks. The initial results of the project were inconclusive; however, in the intervening period additional inoculation, fertilization and wetting of the affected soils has produced a reduction of soil petroleum hydrocarbon concentrations to levels that are less than the regulatory cleanup standards, thereby allowing site closure. In this paper we will present the details of the additional remediation activities undertaken at the site since 2001 and report on the final analytical results to make the case for using this remediation approach successfully at other active railroad sites.

Key words: rail traffic, inoculation, cleanup timeframe, background conditions

1. SITE RELEASE REVIEW:

A description of the site release, initial response actions and the selection of bioremediation as the remedial alternative to be used to complete site cleanup were presented by Paul G. Beaulieu and Francis J. Hopcroft at the 2001 International Conference on Contaminated Soils, Sediments and Water (Beaulieu and Hopcroft, 2001). As cited in the 2001 paper, the Palmer Railroad Yard facility, which is located in the commercial center of the Town of Palmer, has been used for railroad operations since the mid-19th century. The site facility, which is currently owned by New England Central Railroad (NECR), includes a parking area and a freight house building that is used for managing rail yard operations. An "Old Depot" building is located to the immediate northwest of the site facility. One set of main and secondary tracks run from northwest to southeast at the rear of the site facility. Another set of active tracks, which run west/northwest to east/southeast, intersect the first set of tracks at the immediate northwest of the Old Depot building. There are several other secondary tracks, used primarily for switching, located adjacent to the site facility.

The release of approximately 700 gallons of diesel fuel occurred at the Palmer Railroad Yard facility on December 20, 1999 when a collision occurred between a train engine and a separate train car. The accident occurred to the northwest of where the two sets of tracks intersect. The impact punctured the saddle tank on the train engine leaving a 24-inch gash along the tank. Because of impending rail traffic, the train engine was immediately mobilized to a secondary track located to the immediate southeast of the site facility. As a consequence, diesel fuel was released from the saddle tank at the initial point of damage and the final parked location, and the approximately 1,000 feet of track in between these areas. The release area is depicted on Figure 1.

The site release met the criteria requiring notification to the Massachusetts Department of Environmental Protection (DEP) because the volume of the spill, estimated later at approximately 700 gallons, exceeded the reportable quantity of 10 gallons established under Massachusetts Contingency Plan or "MCP" (310 CMR 40.0000 et seq.).

Figure 1. Site Plan

2. REVIEW OF INITIAL RESPONSE ACTIONS:

As cited in the 2001 paper, initial response actions at the Palmer Railroad Yard facility included the removal of contaminated soil and railroad ballast along the impacted railroad track using a "vactor" truck (Beaulieu and Hopcroft, 2001). More conventional excavation equipment was not used due to the spacing between rail ties and because the client (NECR) wanted to avoid any disruption of rail service caused by the removal of the train tracks. Furthermore, the use of the vactor truck allowed for easier mobility in working around the tracks to accommodate the active train schedule.

As part of excavation activities, a total of approximately 100 tons (or 65 cubic yards) of visually impacted soil and ballast were removed and stockpiled on-site for ultimate off-site disposal. Our initial observations were that the diesel fuel had migrated through the porous ballast, which was of varying total thickness throughout the entire release area, and become trapped in the underlying fill material of tight fine sand to silt with little gravel. In order not to risk destabilizing the structural integrity of the rail track, the depth of excavation on any portion of the release area did not exceed approximately 15 inches below grade. Also, the excavation areas were backfilled with clean rail ballast following excavation.

Prior to backfilling, a total of 19 confirmatory soil samples were collected along the length of the release area. Each sample was collected from the base of the excavation to approximately 3 or 6 inches further below grade. The soil samples were submitted for extractable petroleum hydrocarbons (EPH) analysis and diesel range organics (DRO) were also quantified. In addition, two of the samples were also submitted for volatile petroleum hydrocarbons (VPH) analysis. Laboratory results indicated that "elevated" concentrations of EPH carbon ranges related to diesel fuel were detected in samples collected in the areas where the initial release or accident occurred and where the train engine was ultimately parked.

Laboratory results for confirmatory soil samples collected from the 1,000 feet of area between the Initial and Ending release areas were similar to laboratory results for "background" samples that were randomly collected from areas outside of the release area. Based on these findings, it was determined that further remediation was warranted only in the Initial and Ending portions of the release area. In general, both of these remaining impacted areas measured approximately 1,500 square feet, respectively. Monitoring well sampling indicated that groundwater was not impacted by the site release.

3. IN-SITU BIOREMEDIATION OF SITE SOILS:

As cited in the 2001 paper, a cost-benefit analysis of potential remediation approaches was conducted to determine how to complete site cleanup in the two remaining impacted areas (Beaulieu and Hopcroft, 2001). The client was opposed to the removal of any rail tracks or the disruption of rail service unless absolutely necessary. Therefore, an evaluation of other remedial methods that were capable of reducing petroleum hydrocarbons concentrations in-situ was conducted. Cost projections for the use of bioremediation versus the use of hydrogen peroxide indicated a potential cost savings of 25% to 50%. In addition, above ground injection points and underground piping were necessary to disperse the hydrogen peroxide through the release area. This could have created a walking hazard for railroad personnel who needed to traverse the tracks at night.

CJH Environmental, Inc. (CJH), the selected remedial contractor, conducted a "Treatability Test" to determine whether in-situ conditions at the site were conducive to bioremediation treatment and to identify the bacteria mixture that would be most effective. A complete description of the test process and the effectiveness of the bacteria mixtures at other Massachusetts release sites were reviewed in the 2001 paper. In summary, the bacteria utilized with this technology are naturally occurring, non-engineered, non-pathogenic organisms that are carefully selected and concentrated for their ability to degrade petroleum hydrocarbons to carbon dioxide and water (Beaulieu and Hopcroft, 2001). The exact chemical composition of the nutrient mixture used in any specific instance is proprietary.

As part of this project's inoculation process, the surface areas at the two areas of concern were first moistened through surface watering. The bacteria mixture was then applied, as evenly as possible, across the surface areas of the impacted zones and allowed to penetrate through the ballast and into the underlying soils. The nutrient loading instructions were one gallon of diluted bacteria mix per every four cubic yards of soil to be treated.

Two representative soil samples were collected after each inoculation event from both areas of concern to monitor the progress of bioremediation. The samples were collected from the impacted soils situated below the "clean" ballast in each area. As cited in the 2001 paper, two sampling "ports" (4-inch in diameter) were installed in the Ending release area to allow for easier access through the ballast, which was up 30 inches in total thickness in some areas, to the underlying soils (Beaulieu and Hopcroft, 2001). The monitoring samples were analyzed for total petroleum hydrocarbons (TPH – by GC-FID) nitrates, nitrites and total phosphorous. These results were compared to baseline samples collected before inoculation and were further used by

CJH to determine if additional fertilizing nutrients were necessary to further enhance the bioremediation process.

Based on the results of the initial monitoring event, an "enhancement" program of watering the impacted areas once every week during the growing season, unless a significant rain event occurred, was initiated to further stimulate bacterial growth. In addition, a diluted fertilizer mix was also applied to the impacted areas generally on a monthly basis during the growing season.

4. TIMELINE TO ACHIEVE SITE CLEANUP:

The release of approximately 700 gallons of diesel fuel occurred at the Palmer Railroad Yard facility on December 20, 1999. Initial response actions involving the excavation of approximately 65 cubic yards of impacted soil and ballast occurred between December 20 and December 30, 1999.

After a lengthy review of potential remedial options and costs with the client, the Treatability Test for the use of bioremediation was initiated in August 2000. The test findings were completed in October 2000 and site conditions were deemed favorable for treatment. However, CJH advised that it would be more effective and cost efficient if inoculations begin in the early part of the growing season rather than that fall. In January 2001, the test results and remedial description were forwarded to DEP for their review of the process.

Baseline samples were collected from the two remaining impacted areas in April 2001, and the initial inoculation event was conducted in May 2001. In August 2001, the first round of soil monitoring was conducted. Two additional inoculation events were conducted in December 2001 and April 2002, respectively, with follow-up soil monitoring events conducted in December 2001 and May 2001.

The results of the post-inoculation monitoring event of May 2001 indicated that TPH concentrations in the representative samples from both areas of concern were approaching regulatory cleanup standards. An extensive confirmatory sampling event was then conducted in September 2001, and the findings from that event indicated that the site release could be closed out under Massachusetts regulations.

5. FINAL RESPONSE ACTIONS:

TPH concentrations for soil samples collected in April 2001, prior to beginning bioremediation, were 7,200 milligrams per kilogram (mg/kg) and 1,700 mg/kg for the Initial and Ending release areas, respectively. As cited in the 2001 paper, the monitoring results of the first post-inoculation sampling event of August 2001 indicated a minimal decrease in TPH concentrations. However, subsequent monitoring events results indicated that bioremediation was occurring as TPH concentrations continually decreased in soil samples. For the final monitoring event of May 2002, TPH concentrations were 464 mg/kg and 835 mg/kg in the representative samples collected from the Initial release area and 179 mg/kg and 20.8 mg/kg in the representative samples collected from the Ending release area. The soil monitoring results are summarized in Table 1.

In September 2002, an extensive confirmatory sampling event was conducted which involved the collection five confirmatory soil samples from the Initial release area and four confirmatory soil samples from the Ending release area. In general, the soil samples were collected along the length of the respective release areas. At each of the impacted areas, the clean ballast was temporarily removed in order to expose the underlying soils for sampling. Visual and olfactory observations indicated that there was little evidence of contamination left in these soils.

Each of the nine confirmatory soil samples was submitted for EPH analysis. EPH carbon ranges that were detected above laboratory detection limits in the samples were, on average, less than background concentrations from earlier samples. The confirmatory soil sample results are summarized in Table 2.

Table 1
Soil Monitoring Results (mg/kg)
New England Central Railroad, Palmer, MA
RTN: 1-13246

Sample ID	Initial Area	Initial Release Area 1			Initial Release Area 2		Ending Area	Ending Release Area 1			Ending Release Area 2	
Data Collected	4-19-01	8-30-01	12-21-01	5-21-02	8-30-01	5-21-02	4-19-01	8-30-01	12-21-01	5-21-02	8-30-01	5-21-02
Nitrate/ Nitrogen (as N)	1.3	5.4U	7.4	5U	4.8U	-	1.3	2.8U	12	6U	4.6U	6U
Phosphor-ous (total as P)	210	250	270	198	280	-	330	280	510	462	320	541
Total Petroleum Hydrocarbons (TPH)												
Fuel Oil #2 (C_9 - C_{25})	7200 (1)	8100 (2)	1400 (2)	-	4800 (1)	-	1700 (1)	U	U	-	1700 (2)	-
Unmatched	-	U	U	-	U	-	-	44(3)	U	-	U	-
Diesel Range Organics	-	-	-	462	-	835	-	-	-	179	-	20.8

Notes:
"U" indicates that the analyte was analyzed for but not detected
"(1)" indicates an identical or nearly identical GC pattern
"(2)" indicates a similar GC pattern showing moderate differences
"(3)" indicates "no agreement" with GC patterns in the target list

Table 2
Confirmation Soil Results in (mg/kg)
IRA Completion (Samples collected September 18, 2002)
New England Central Railroad, Palmer MA
RTN: 1-13246

Sample Area	Initial Release Area				Ending Release Area					MCP Method 1 Cleanup Standards	
Sample ID	SS-6	SS-7	SS-8	SS-9	SS-1	SS-2	SS-3	SS-4	SS-5	S-1/GW-1	S-1/GW-3
EPH											
C_9-C_{18} Aliphatic	<3.4	25.3	170	213	<3.4	<3.4	18.4	<3.9	<3.9	1,000	1,000
C_{19}-C_{36} Aliphatic	23.3	28.3	110	117	<3.4	<3.4	12.8	<3.9	<3.9	2,500	2,500
C_{11}-C_{22} Aromatic	13.3	18.1	69.4	58.6	<3.4	<3.4	9.5	<3.9	<3.9	200	800
Targeted PAH compounds											
Acenaphthene	<0.34	<0.34	<0.34	<0.35	<0.34	<0.34	<0.41	<0.39	<0.39	20	1,000
Acenaphthylene	<0.34	<0.34	<0.34	<0.35	<0.34	<0.34	<0.41	<0.39	<0.39	100	100
Anthracene	<0.34	<0.34	<0.34	<0.35	<0.34	<0.34	<0.41	<0.39	<0.39	1,000	1,000
Benzo(a)anthracene	<0.34	<0.34	<0.34	<0.35	<0.34	<0.34	<0.41	<0.39	<0.39	0.7	0.7
Benzo(a)pyrene	<0.34	<0.34	<0.34	<0.35	<0.34	<0.34	<0.41	<0.39	<0.39	0.7	0.7
Benzo(b)fluoranthene	<0.34	<0.34	<0.34	<0.35	<0.34	<0.34	<0.41	<0.39	<0.39	0.7	0.7
Benzo(g,h,i)perylene	<0.34	<0.34	<0.34	<0.35	<0.34	<0.34	<0.41	<0.39	<0.39	1,000	1,000
Benzo(k)fluoranthene	0.35	<0.34	<0.34	<0.35	<0.34	<0.34	<0.41	<0.39	<0.39	7	7
Chrysene	<0.34	<0.34	<0.34	<0.35	<0.34	<0.34	<0.41	<0.39	<0.39	7	7
Dibenzo(a,h)anthracene	<0.34	<0.34	<0.34	<0.35	<0.34	<0.34	<0.41	<0.39	<0.39	0.7	0.7
Fluoranthene	0.45	<0.34	<0.34	<0.35	<0.34	<0.34	<0.41	<0.39	<0.39	1,000	1,000
Fluorene	<0.34	<0.34	<0.34	<0.35	<0.34	<0.34	<0.41	<0.39	<0.39	400	1,000

Ideno(1,2,3-cd)pyrene	<0.34	<0.34	<0.34	<0.35	<0.34	<0.34	<0.41	<0.39	<0.39	0.7	0.7
2-Methylnaphthalene	<0.34	<0.34	<0.34	<0.35	<0.34	<0.34	<0.41	<0.39	<0.39	4	500
Naphthalene	<0.34	<0.34	<0.34	<0.35	<0.34	<0.34	<0.41	<0.39	<0.39	4	100
Phenanthrene	<0.34	<0.34	<0.34	<0.35	<0.34	<0.34	<0.41	<0.39	<0.39	700	100
Pyrene	0.45	<0.34	<0.34	<0.35	<0.34	<0.34	<0.41	<0.39	<0.39	700	700

Bold indicates exceedence of S-1/GW-1 Standard

<5 indicates not detected above the referenced laboratory detection limit

6. COMPARISON TO CLEANUP STANDARDS AND SITE CLOSURE

A risk characterization for the site was conducted in accordance with the requirements of the MCP to determine the risk of harm to health, safety, public welfare and the environment posed by the residual contamination in site soils caused by the site release. Three methods are described in the MCP for evaluating site risk. The Method 1 approach uses a set of soil and groundwater standards that are compared directly to site Exposure Point Concentrations (EPCs). The Method 1 standards were developed by the DEP according to defined site of exposure scenarios that represent a conservative estimate of potential exposures that may occur at most sites.

EPH results for confirmatory samples collected from the two areas where bioremediation occurred were all below applicable Method 1 standards, these results were also below conservative Method 1 standards, which are also used for comparison to be protective of potential unrestricted future uses. The calculated EPCs for soils along the other portion of the release area, where only earlier soil excavation occurred, exceeded conservative Method 1 standards for TPH, which was qualified as related to motor oil. However, the site release involved the release of diesel fuel. Also, the calculated EPCs for background samples also exceeded Method 1 standards for TPH, which was also qualified as related to motor oil. Therefore, the site release was closed out in accordance with Massachusetts regulations and no property deed restriction for future site uses was required.

7. CONCLUSIONS

Although leaks or releases of diesel fuel, motor oil, lubricating oils, etc. along active rail lines is a common occurrence for the railroad industry, the release of oil in quantities greater than 10 gallons in Massachusetts requires that the release be reported and ultimately cleaned up in accordance with State regulations. Cleanup along active rail lines can be a challenging experience in that traditional excavation methods may not be feasible to achieve site closure due to structural concerns associated with the rail lines or rail traffic. The release of 700 gallons of diesel fuel along active rail lines in Palmer, Massachusetts was effectively cleaned up using limited soil excavation methods followed by bioremediation. Due to the shallow nature of remaining contamination in soils after excavation, the bacteria mixture could be effectively applied at the surface. From beginning inoculations through to when the diesel-related contaminant concentrations in soils were reduced to State cleanup standards took approximately 1.5 years or two growing seasons. Therefore, even though bioremediation is more time-consuming than traditional soil excavation methods in achieving site closure, it can be used successfully at active railroad facilities where other ex-situ methods may be infeasible.

REFERENCES

Beaulieu, P.G and Hopcroft, F.J., "Emergency Response actions and Bioremediation of a 700-Gallon Diesel Fuel Spill, Palmer Railroad Yard Facility, Palmer Massachusetts", *Proceedings of the 17th Annual International Conference on Contaminated Soils, Sediments and Water,* University of Massachusetts, Amherst, Ma, October 2001.

Commonwealth of Massachusetts, 1997, The Massachusetts Contingency Plan 310 CMR 40.0000 et seq

Tighe & Bond, Inc. *Immediate Response Action Completion and Response action Outcome Statement Report, Palmer Rail Yard, Depot Street, Palmer, Massachusetts, DEP RTN 1 - 13246.* Submitted to DEP Western Regional Office in December, 2002.

CHAPTER 7

ENHANCED REDUCTIVE DECHLORINATION – A BROADER PERSPECTIVE

Denice K. Nelson, Frederick C. Payne, Suthan S. Suthersan
ARCADIS G&M

Abstract: Enhanced reductive dechlorination of chlorinated compounds has been applied via *in situ* engineered anaerobic systems in increasing numbers over the past decade. Knowledge gained from these field applications, combined with academic literature, allows us to have a broader understanding of the technology. Microbial communities can be engineered through manipulation of aquifer inputs, creating *in-situ* reactive zones. Over a short segment of the groundwater flow path, the microbial biomass is shifted to a succession of facultative degraders and fermenters, sulfate reducers, acetogens and methanogens. At low - to-moderate electron donor loading rates, the aquifer microbial continuum is dominated by species that produce C-1 compounds and acetate as the fermentation products. When electron donor loading increases to high levels a significant portion of the electron donor flows into pathways that generate fatty acids. The network of anaerobic metabolic processes is introduced as an electron donor cascade. Bacteria occupying several points in the electron donor metabolic cascade have been shown to reductively dechlorinate halogenated compounds. Biostimulation using late -cascade compounds such as lactate and met hanol by-pass much of the productive dechlorinating communities. These observations were developed from the authors' research and published reports of numerous other field applications and lab research.

Enhanced reductive dechlorination of chlorinated compounds has been applied via in -situ engineered anaerobic systems in increasing numbers over the past decade. Knowledge gained from these field applications, combined with academic literature, allows us to have a broader understanding of the technology.

Plumes of contaminated groundwater historically were managed by conventional pump -and-treat programs. These programs often require long -term system operation, are expensive to operate and maintain, and rarely are effective in meeting stringent groundwater cleanup objectives. Other technologies, such as conventional air sparging, lack the ability to address sorbed-phase contaminants, often resulting in a rebound effect of contaminant concentrations upon system shutdown. The propagation of in-situ reactive zones through reagent injection is an emerging remedial strategy that has been evaluated, developed, and implemented only within the past decade. The strategy is gaining attention due to the increasing recognition of the limitations of pump-and-treat and air sparging systems, and the ability to implement, in - situ, most of the treatment processes otherwise used in above -ground systems.

ARCADIS pioneered the in-situ reactive zone strategy and has implemented more than 100 in-situ reactive zones for enhanced reductive dechlorination, at a broad range of sites, since 1993. These sites have included a variety of constituents to be treated, including tetrachloroethene (PCE), trichloroethene (TCE), 1,1,1-trichloroethane (1,1,1-TCA), carbon tetrachloride, pentachlorophenol, and chlorinated pesticides in various groundwater concentration ranges with numerous hydrogeologic settings (including shale and karst limestone bedrock, low permeability glacial tills and saprolite, and high permeability alluvium and glacial outwash environments).

To successfully apply this technology, it is crucial to understand the processes that occur in the reactive zones and under what conditions reductive dechlorination will occur. The following sections detail these processes, and provide case studies to illustrate this technology.

Key words: reductive dechlorination , in-situ reactive zones, biostimulation

1. MICROBIAL ECOLOGY

The composition of aquifer microbial communities shifts continually in response to the quantity and quality of dissolved organic matter supply, electron acceptors, and the availability of essential nutrients. Living organisms require four basic commodities for survival: electron donor, electron acceptor, a carbon source and mineral nutrients. Organisms use reactions between electron donors and electron acceptors to obtain energy that can be used to build proteins and other molecules. A carbon source is required to build proteins and other molecules, and in many cases the electron donor and carbon source are the same. However, certain bacterial species can gain energy from inorganic electron donor compounds, such as hydrogen gas, and consequently require a separate carbon source in order to build biomass. The supply of chemicals that serve as terminal electron acceptors controls the aquifer microbial community structure by determining the viable microbial respiratory pathways. Oxygen is the dominant terminal electron acceptor in most natural aquifer systems and respiratory processes that utilize oxygen as the terminal electron acceptor are termed aerobic. Nitrogen, sulfate, dissolved metals and carbon dioxide can also serve as terminal electron acceptors when oxygen is unavailable. Metabolic processes that utilize these compounds as electron acceptors are termed anoxic or anaerobic.

The microbial communities that thrive under very reduced communities utilize sulfates and carbon as terminal electron acceptors, generating sulfides, elemental sulfur, carbon dioxide and methane as by-products. These communities are comprised of bacteria known as sulfate reducers and methanogens. It is unusual for open aquifers to receive sufficient loadings of carbon to sustain sulfate reducing and methanogenic conditions, but redox potentials often fall to moderately reducing levels. These are referred to as "transition" redox conditions and in contaminated aquifers they are recognized by the presence of *cis*-dichloroethene (cis-DCE) and vinyl chloride (when chlorinated solvents PCE or TCE have been released), and redox potentials in the 0 to –150 mV range. Similarly, transition conditions in aquifers impacted with chlorinated alkanes are marked by the appearance of degradation products such as 1,1-dichloroethane (1,1-DCA). Other markers for transition conditions are absence of dissolved oxygen and nitrate, and elevated levels of reduced metals such as iron (Fe^{2+}) and manganese (Mn^{2+}).

Transition-stage microbial communities can be managed, through increased carbon loadings, to generate fully reducing conditions, including sulfate reduction and methanogenesis. This process is termed "reductive biostimulation", and has been shown to be capable of inducing dramatic reductions of redox potential in contaminated aquifers. Reductive biostimulation can be used to induce abiotic precipitation of dissolved transition metals such as Cr^{6+}, and to support enhancement of reductive dechlorination reactions to achieve complete dechlorination.

2. ELECTRON ACCEPTORS AND THERMODYNAMICS

Reductive dechlorination occurs by two basic mechanisms: metabolic – bacteria utilize chlorinated compounds as electron acceptors, and cometabolic – where enzymes and cofactors produced by bacteria can dechlorinate solvents.

All organisms live by incorporating carbon into physical structures necessary for growth and movement, and also obtaining energy for use in growth and movement. The energy for growth and movement is captured through chemical reactions at the cellular level. These reactions are broadly described as the organisms' metabolism, and the principal metabolic reactions are oxidation-reduction reactions.

In reduction-oxidation reactions, two compounds are transformed simultaneously - one is reduced (gains electrons) and the other is oxidized (loses electrons). The compound that is reduced is called the electron acceptor (or oxidant) and the compound that is oxidized is termed the electron donor (or reductant). In aerobic systems, oxygen is typically the electron acceptor. However, in anaerobic systems, chlorinated organic compounds such as tetrachloroethene can function as an electron acceptor in the process termed reductive dechlorination.

Two important questions can be asked about any prospective chemical reaction; Does the reaction yield energy or consume energy?, and How fast does the reaction proceed? The first question is one of chemical thermodynamics. Calculations can be performed to determine whether a particular reaction will be profitable (by yielding energy) for an organism. Organisms gaining the most energy from the available electron donors and electron acceptors will dominate microbial community. When oxygen is available, aerobic bacteria dominate. When oxygen becomes unavailable, opportunities arise for anaerobic bacteria that survive on less energy-effective reduction-oxidation reactions. The second question is one of chemical kinetics and for organisms, the kinetics of interest are enzyme kinetics. Enzymes facilitate metabolic reactions by lowering energy thresholds for reactions to occur. These enzymes determine how fast a reaction proceeds.

Analysis of chemical thermodynamics shows energy gained from chlorinated ethenes as electron acceptors is sufficient to support growth under a wide range of pH and temperature conditions (Table 1). Although some dechlorinating species do not degrade vinyl chloride at meaningful rates, this is not due to thermodynamic limitation. Under non-standard conditions, including low H_2 levels, low temperatures, and low pH, the reactions remain strongly exergonic showing that the differences in utilization of these compounds as electron acceptors is more likely related to enzyme kinetics than to chemical thermodynamics.

Reaction	ΔG°' kJ/mol	ΔG[a] kJ/mol	ΔG[b] kJ/mol	ΔG[c] kJ/mol	ΔG[d] kJ/mol
Reductive Dechlorination Reactions					
PCE to TCE $C_2Cl_4 + H_2 \rightarrow C_2HCl_3 + H^+ + Cl^-$	-173	-162	-156	-171	-150
TCE to cis-DCE $C_2HCl_3 + H_2 \rightarrow C_2H_2Cl_2 + H^+ + Cl^-$	-169	-157	-152	-167	-145
cis-DCE to vinyl chloride $C_2H_2Cl_2 + H_2 \rightarrow C_2H_3Cl + H^+ + Cl^-$	-139	-128	-122	-137	-115
vinyl chloride to ethene $C_2H_3Cl + H_2 \rightarrow C_2H_4 + H^+ + Cl^-$	-149	-138	-132	-148	-126
Native Microbial Metabolism					
Hydrogenotrophic methanogenesis $CO_2 + 4H_2 \rightarrow CH_4 + 2H_2O$	-130	-130	-62	-130	-130
Acetoclastic methanogenesis $CH_3COO^- + H_2O \rightarrow CH_4 + HCO_3^-$	-31	-31	-31	-31	-31
Lithotrophic sulfate reduction $SO_4^{2-} + 4H_2 + H^+ \rightarrow HS^- + 4H_2O$	-152	-163	-83	-154	-94
Acetate-oxidizing sulfate reduction $CH_3COO^- + SO_4^{2-} \rightarrow HS^- + 2HCO_3^-$	-48	-48	-48	-48	-48
pH	7	5	7	7	6
H_2 (atm)	1	1	0.001	1	0.001
Temp (°C)	25	25	25	10	10
PCE (M)	1	1	1	1	1
TCE (M)	1	1	1	1	1
cis-DCE (M)	1	1	1	1	1
Vinyl Chloride (M)	1	1	1	1	1

Table 1. Gibbs free energy yield fo r reactions in standard biochemical conditions (298 K, 1 atm, pH 7) and at 4 non-standard conditions of interest in aquifer formations

3. ENHANCED REDUCTIVE DECHLORINATION

Enhanced Reductive Dechlorination (ERD) technology is intended to facilitate and expedite the biological reductive dechlorination of chlorinated compounds through reductive biostimulation. Molasses, milk whey and corn syrup are commonly available carbon-rich materials that can be used to stimulate aquifer microbial communities. Naturally occurring aquifer bacteria can be managed to very high rates of dehalogenation through the injection of these low-cost food-grade electron donors by driving the aquifer into sulfate reduction, acetogenesis and methanogenesis, which can provide broad-spectrum cometabolic dehalogenation.

Cometabolic degradation occurs when enzymes created by bacteria for other purposes destroy the target contaminant. Metabolic degradation is a natural process whereby the bacteria populations can utilize chlorinated compounds such as PCE as an energy-yielding electron acceptor. Degradation of these compounds dominates the metabolism among bacteria termed "dehalorespirers", which include bacterial species from several genera, including *Desulfuromonas, Dehalospirillum, Dehalococcoides, Dehalobacter,* and *Desulfomonile.*

4. ENGINEERING THE MICROBIAL CONTINUUM - THE ANAEROBIC METABOLIC CASCADE

Aquifer microbial communities adapt quickly and predictably to changes in the electron donor and acceptor loadings, giving us an opportunity to engineer the continuum to create an in-situ reactive zone for the destruction of contaminants. When highly degradable organic carbon is injected into the groundwater flow, bacterial biomass increases and is sustained as long as the loading continues.

Small additions of degradable carbon can be accommodated by the aerobic microbial continuum with minimal impact. An increase in aerobic biomass is observed over a short segment of the continuum, and the associated increase in metabolic oxygen demand is accommodated easily by diffusion of oxygen from the overlying vadose zone. However, the aerobic biomass that can develop in response to electron donor loading is limited by several factors, including space where bacteria occupy space on the aquifer soil matrix, and it's possible for populations to reach space-limited levels; dissolved oxygen – when the bacterial biomass reaches high levels, the diffusion of oxygen from above cannot sustain the electron acceptor demand of the biomass, and the population turns to alternative electron acceptors to support metabolism of the abundant electron donors; and by-product buildup – by-products of metabolism, including carbon dioxide and organic acids, can increase to inhibitory levels.

When electron donor loadings are sufficient to induce consumption of alternative electron acceptors, the composition of the aquifer microbial community undergoes a fundamental shift. Microbial consumption of dissolved organic carbon drives a succession of habitat changes that lead to a continuum of decreasing redox potential and the associated progressive shift of terminal electron acceptors. The sequence of electron acceptors, ranked in approximate order of decreasing energy yield per reaction is:

$$O_2 >> NO_3^- > Mn^{4+} > Fe^{3-} > SO_4^{2-} > CO_2$$

The microbial communities that develop under high electron donor loads are described in Figure 1. These are anaerobic metabolic processes, and the several steps that occur in the processing of electron donors are carried out by distinct groups of organisms (unlike aerobic degradation, which can be completed with competitive energy yields by a single organism).

Many of the metabolic steps in anaerobic respiration are not the classic reduction-oxidation half-cell reactions in which one compound (the electron donor) is oxidized and a second compound (the electron acceptor) is reduced. In anaerobic systems, oxidation and reduction often occurs simultaneously, through the action of enzymes on a single compound in a process known as fermentation. The anaerobic metabolism that can be sustained by fermentation and by oxidation-reduction reactions that consume alternative electron acceptors yields much less energy than oxidative metabolism. The anaerobic metabolic cascade depicted in Figure 1 provides numerous pathways for the decomposition of electron donors along the aquifer microbial continuum. Reactions in the upper portion of the cascade consume electron acceptors, leaving reactions in the lower parts of the cascade to be supported by other, potentially less energy-yielding, electron acceptors.

The metabolic process occurs in several steps and, at any point in the cascade, the most likely next reaction is determined by the concentration of donors and acceptors, and the energy that competing organisms can gain from the reaction. The reaction that gains the most energy at any point in the cascade is thought to be the most competitive, therefore most likely to occur.

For example, in Figure 1, hydrogenotrophic methanogens compete with lithotrophic sulfate reducers for hydrogen as an electron donor (these organisms will also require a source of carbon for growth). When hydrogen supplies are limited, the sulfate reducers are expected to out-compete the methanogens, because they gain more energy from the oxidation of hydrogen (they use a more energetic electron acceptor). It is important to note that when hydrogen is abundant, sulfate reducers and methanogens can co-exist – they may compete for other resources, but the energy disadvantage of CO_2 as an electron acceptor is not a major factor.

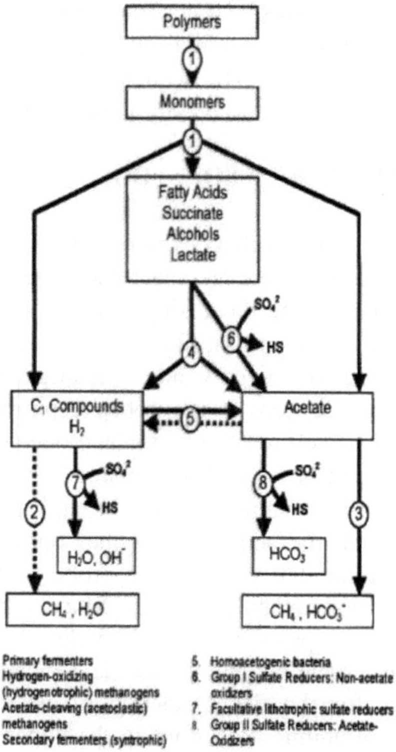

Figure 1. Major metabolic pathways in the aquifer microbial continuum, when it is engineered to load electron donors at a rate that exceeds electron acceptor recharge. Dashed - line pathways (2 and 5) decrease in importance at temperatures below 20 °C. Adapted from Schink (1997).

A zonation of dominant metabolic processes forms in the continuum, when electron donor loading is sustained a high levels with each zone progressively reduced until the electron donor supply is exhausted. This is shown in Figure 2.

The most critical aspect of engineering the microbial community is achieving reductive dechlorination of the chlorinated compounds present in the aquifer. This is accomplished by encouraging microbial population shifts that favor organisms capable of reductive dechlorination. Fortunately, there are many organisms that can accomplish this task.

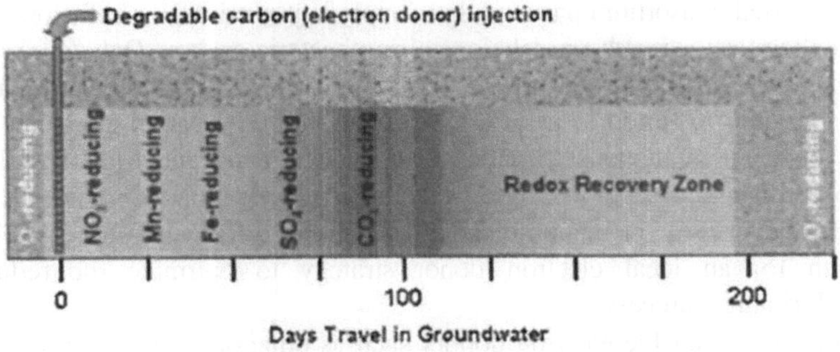

Figure 2. Zonation of dominant metabolic processes in the aquifer microbial continuum, when electron donor loading is sustained at levels that exceed electron acceptor recharge

5. DECHLORINATING BACTERIA

Organisms that can function with oxygen or alternative electron acceptors populate the early portion of the in-situ reactive zones (see Figure 2). These bacterial communities are termed "facultative" since they can grow either in the presence or absence of oxygen. Sharma and McCarty (1996) documented the dechlorination capability of facultative species *Enterobacter* capable of dechlorinating PCE to cis-DCE. PCE dehalogenation required the absence of oxygen, the presence of nitrate, and high concentrations of fermentable compounds such as glucose.

Reductive dechlorination reactions have also been observed in *Clostridium bifermentans. Clostridium* is a well-known fermentative bacterial genus, many times producing ketones as a product of fermentation. *Clostridium* has been shown capable of degrading chlorinated ethanes, chlorinated ethenes, dichloromethane and dichloropropane (Chang, et al., 2000). Case study data indicate there may be a correlation between high rates of enhanced reductive dechlorination and elevation of ketone concentrations from baseline levels. This may be an indication of the participation of *Clostridium* in the most successful dechlorination sites.

The most highly studied dechlorinating species occur late in the reactive zone. Methanogens, acetogens and sulfate-reducing species all have been shown to participate in dechlorination of chlorinated compounds (Fathepure and Boyd, 1988; He *et al.,* 2002; Ndon *et al.,* 2000). Highly specialized bacterial species (dehalorespirers) have also been observed in this portion of the reactive zone, which solely use chlorinated compounds as electron acceptors (Holliger, 1993; Magnuson *et al.,* 1998; Neumann *et al.,* 1994).

These bacteria are completely reliant on these compounds as well as unidentified consortium partners that supply key nutrients or cofactors.

Laboratory research on dehalorespiring bacteria such as *Dehalococcoides ethenogenes* indicates that they have a high affinity for molecular hydrogen and are able to function at lower hydrogen gas (H_2) partial pressures than any of the methanogens (Smatlak et al, 1996; Yang and McCarty, 1998). The methanogens achieve higher rates of hydrogen consumption when the H_2 levels exceed 11 nanomolar (nM). These differences have fueled a search for an ideal electron donor strategy to optimize the reductive dechlorination process.

Easily degradable electron donors such as ethanol yield a very high rate of H_2 production, thought to stimulate methanogenesis. Slowly degradable donors, such as propionic and butyric acids, and slow-release carbon sources are expected to yield lower H_2 propagation rates, thereby minimizing aquifer H_2 levels, and theoretically giving a competitive advantage to dehalorespirers. This strategy is designed to give *Dehalococcoides ethenogenes* a competitive advantage, relative to methanogens, sulfate reducers, and other consumers of molecular hydrogen. However, the total rate of dechlorination (dehalorespiration + methanogenic co-metabolism + dechlorination contributions of sulfate reducers and others) may be largely independent of the contribution of dehalorespirers. Field evidence indicates PCE and TCE degradation proceeding under sulfate-reducing conditions. Full dechlorination, including direct, cis-DCE to ethene conversion is observed under methanogenic conditions.

At low carbon loading rates, primary fermenters produce C1 compounds and acetate as the main products. Dehalorespiring bacteria can rapidly degrade PCE often time with the formation of a strong vinyl chloride peak as dechlorination nears completion. However case studies have shown that low carbon loading rates cannot sustain both the recharge of electron acceptors and the sulfate reducing and methanogenic zones where reductive dechlorination occurs, leading to incomplete dechlorination of compounds.

At high carbon loading rates, the C1 and acetate pathways are saturated, forcing a significant portion of the carbon through the fatty acid synthesis pathways (cf. Schink, 1997). Under these conditions, additional dechlorinators are encouraged, such as *Clostridium* species, and vinyl chloride dechlorination can proceed at a rate that suppresses peak development. Ketone formation (acetone and 2-butanone) often accompanies the high-rate dechlorination processes that suppress the vinyl chloride peak. These compounds will degrade over time in the highly anaerobic environment, and most often will not appear in downgradient wells.

To optimize degradable carbon usage and time for aquifer treatment, it is critical to manage the carbon loadings into the reactive zone to deliver a steady carbon loading to create a continuous zonation of redox processes. There are many practical constraints on the engineering of the aquifer microbial continuum. Among these are:

Groundwater flow velocity – The zonation depicted in Figure 2 depends on the movement of groundwater to spread the microbial succession over a long distance in the aquifer. At low groundwater flow velocities, the succession must occur in a fixed space, one species replacing another in time, at any particular location.

Aquifer alkalinity – Many of the fermentation products are organic acids that can lower groundwater pH if the aquifer's buffering capacity is too low.

Aquifer microbial populations – The aquifer must contain a sufficient bacteria population to perform the needed metabolic transformations, including groundwork reactions among the fermentative bacteria, and functional, dechlorinating reactions among the primary fermenters, sulfate reducers, methanogens and acetogens. Microbial populations in near-surface aquifer formations have been sufficient to drive commercial-scale, enhanced reductive dechlorination at each of more than 100 sites undertaken, to-date. The risk of inadequate bacterial populations is more likely at locations with low background microbial counts, such as might occur at great depths in aquifers with low-organic recharge.

Case studies in tetrachloroethene and trichloroethene-contaminated aquifers show that the aquifer microbial can be manipulated, through controlled injection of highly-degradable carbon, to carry out dechlorination reactions in a predictable and highly-effective manner. The following case studies were chosen to demonstrate the effect of carbon loading on anaerobic biostimulation of reductive dechlorination.

6. CASE STUDIES

Three case studies have been drawn to illustrate the range of site conditions in which the technology has been applied, and the range of dechlorination patterns we have observed.

6.1 Case Study 1

The Case Study 1 site lies in the central United States and was selected to demonstrate the reductive dechlorination of TCA. The injection area of this site was concentrated in the middle of a former underground storage tank basin. Enhanced reductive dechlorination was applied to a plume consisting of TCE and 1,1,1-TCA and their respective degradation products, with a horizontal groundwater velocity approximately 0.4 feet per day. The carbon injection system consisted of 3 wells, spaced at 15-foot intervals. Figure C1-1 shows the injection and monitoring well layout for Case Study 1.

Molasses was selected as the most cost-effective electron donor for the site, and injections were performed at 2-week intervals for five months. The dechlorination process began on Day 0, with molasses injections that raised the average injection zone total organic carbon (TOC) concentration to 4,000 mg/L. The monitoring location shown in Figure C1-1 is approximately 120 days downgradient from the injection zone, and TOC concentrations ranged between 85 and 300 mg/L at the monitoring location.

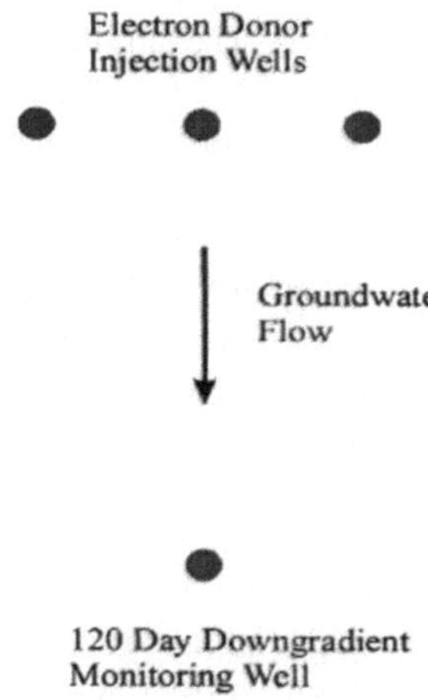

Figure C1-1. . Reactive zone layout for Case Study 1. The injection system consists of 3 injection wells, spaced at 15 -foot intervals. Groundwater velocity is approximately 0.4 feet per day

Within the five-month injection period, TCE and 1,1,1-TCA concentrations decreased over 99% at a monitoring well located 120 days downgradient from the injection zone, as shown on Figures C1-2 and C1-3, respectively. Both ethane and ethene concentrations had peaked and declined by the end of the five-month study. Vinyl chloride degradation rates exceeded those of cis-DCE by such a margin that vinyl chloride concentrations never rose above low levels (0.05 to 0.5 umol/L). Figure C1-4 contains ethane data plotted on a log scale to more clearly demonstrate the fate of 1,1,1 -TCA and daughter products.

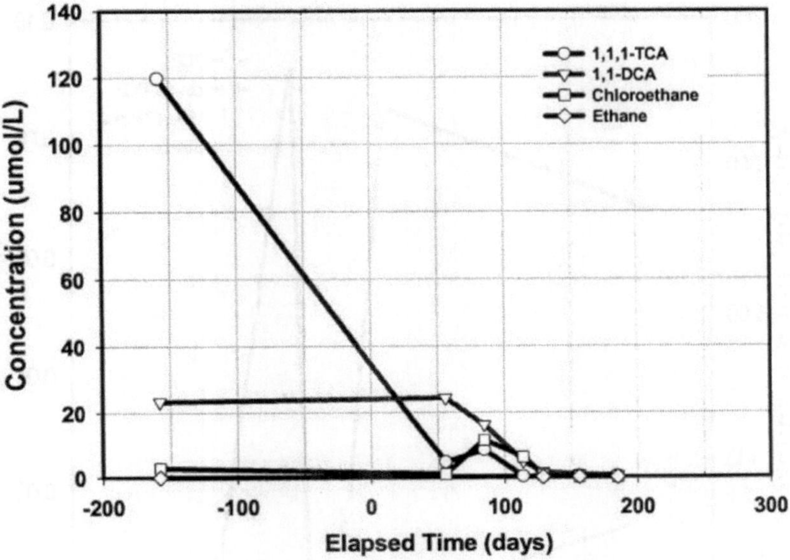

C1-2. Chlorinated alkane concentrations in groundwater at Case Study 1 site. Over 99% reduction was observed concentrations over the five -month injection period. Injections started at time = 0

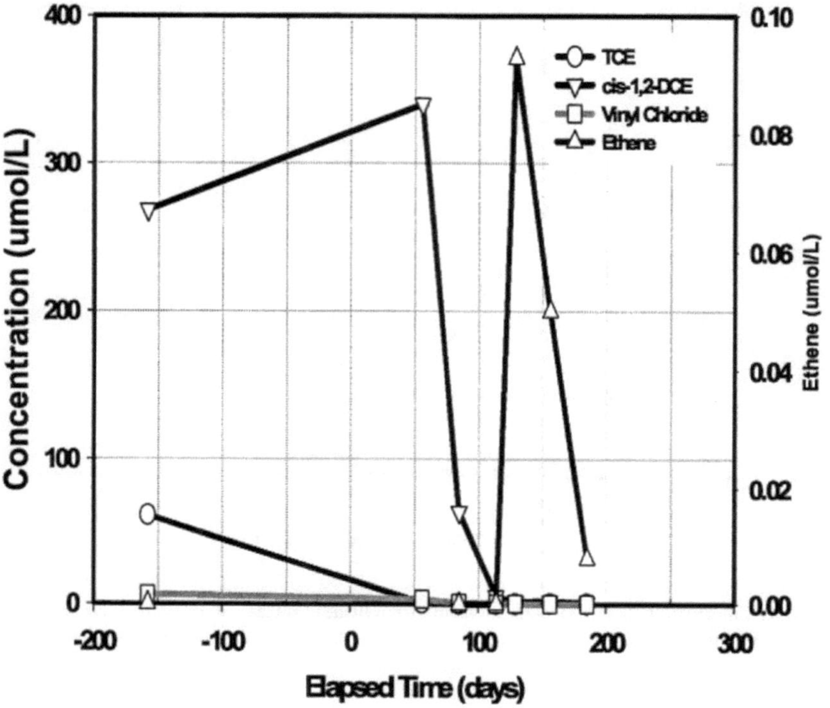

C1-3. Chlorinated alkene concentrations in groundwater at Case Study 1 site. Over 99% reduction was observed in concentrations over the five -month injection period. Injections started at time = 0

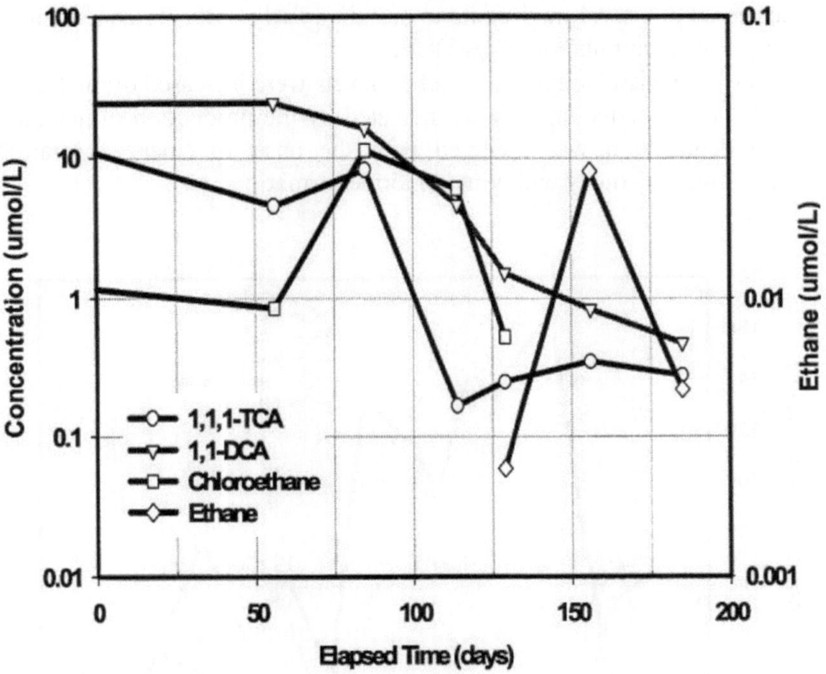

C1-4. Chlorinated alkene concentrations in groundwater at Case Study 1 site (log scale). Over 99% reduction was observed in concentrations over the five -month injection period. Injections started at time = 0

6.2 Case Study 2

This second case study was selected to show the effects of lower carbon loadings on the dechlorination of cis-DCE and vinyl chloride. Although the site is achieving complete dechlorination, as demonstrated by the formation of ethene, there is a noticeable vinyl chloride peak. The rate of vinyl chloride degradation appears to be sensitive to carbon loading rates. The Case Study 2 site is located in the western United States, with TCE as the contaminant of concern. Prior to implementation of remedial action at this site, some cis-DCE was observed as a degradation product.

An in-situ reactive zone was constructed to employ enhanced reductive dechlorination through biostimulation. Molasses was selected as the most cost-effective electron donor for the site. Figure C2-1 describes concentrations of each compound in the TCE-to-ethene sequence for a well located within the Case Study 2 injection zone. As shown in Figure C2-2, the initial carbon loading was sufficient to rapidly dechlorinate TCE, but then decreased to a level inadequate to sustain a fully reducing zone thereby resulting in an accumulation of cis-DCE.

On Day 270, dissolved organic carbon levels were increased dramatically to reestablish the reducing zone as indicated by the production of methane. Cis-DCE degradation was observed with the onset of ethene formation occurring simultaneously with vinyl chloride formation.

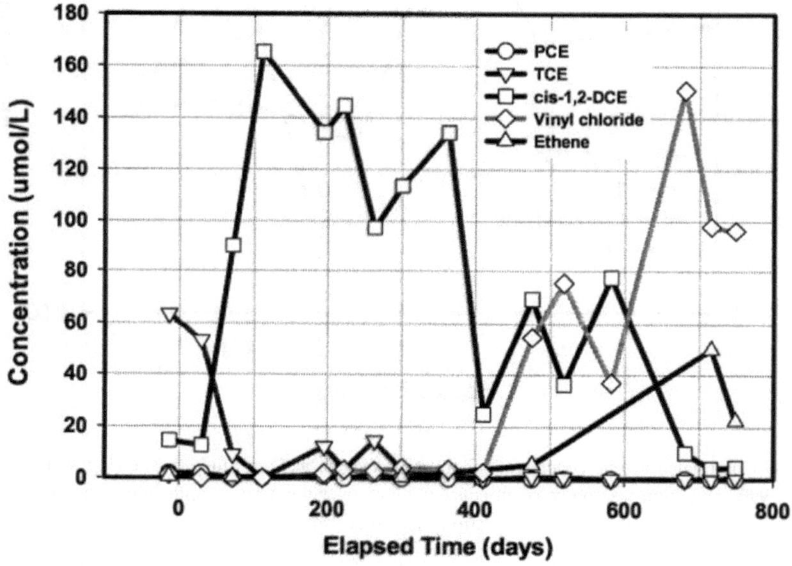

Figure C2-1. Decreases in chlorinated alkene concentrations in groundwater at the Case Study 2 site.

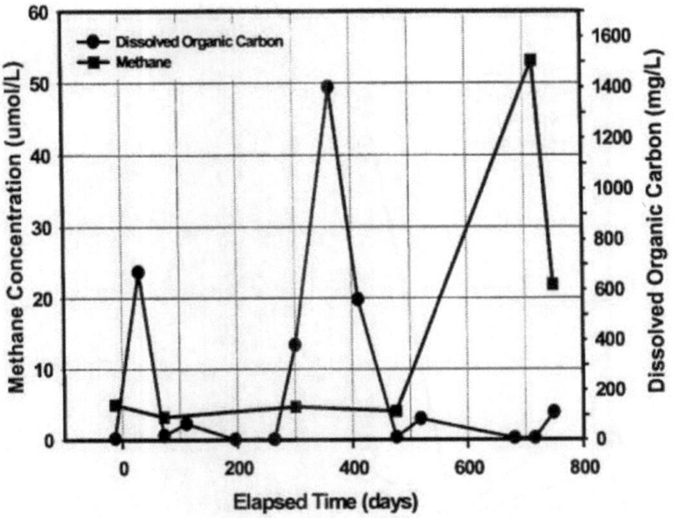

C2-2. Dissolved organic carbon and methane observed within a biostimulation injection zone, where readily degradable organic carbon (molasses) was injected into the aquifer

6.3 Case Study 3

Two sites were selected to examine the relationship of ketone production and degradation of chlorinated compounds. The first site is located in the south-central United States and contained tetrachloroethene-impacted groundwater. Molasses was chosen as the electron donor at this site. A high level of carbon loading was maintained at the site with total organic carbon ranging from 8,000 to 16,000 mg/L in the injection zone.

Chlorinated ethene concentrations decreased to low levels at Day 200 and increased to high levels at day 300, indicating a significant sorbed mass was present at the site (Figure C3-1). As with Case Study 2, vinyl chloride degradation was slow enough to accumulate to more than half the observed cis-DCE level, with the onset of ethene formation occurring simultaneously with vinyl chloride formation. The reduction of all ethene compounds occurred during the same period where the production of ketones was observed.

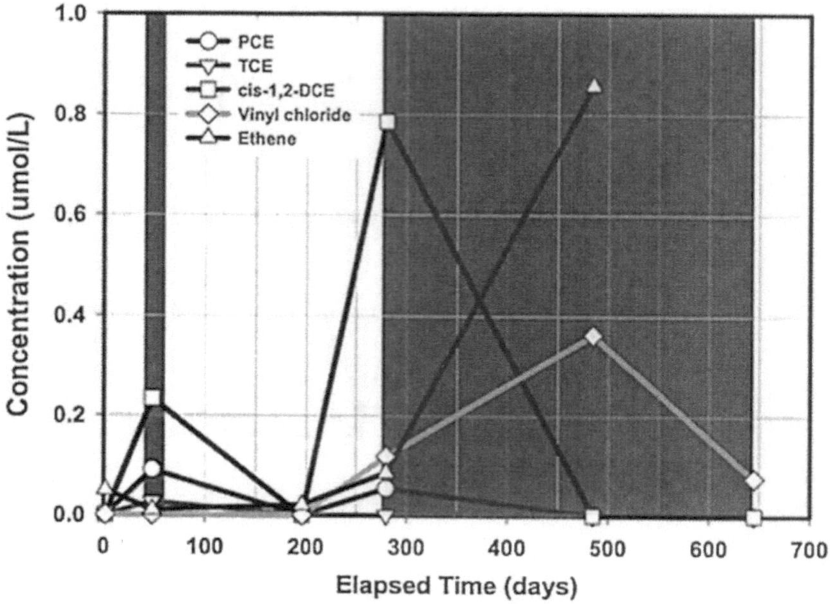

Figure C3-1. Cholorinated alkene concentrations in groundwater at a Case Study located in the south-central United States. Injections started at time = 0. Periods of ketone production shown as colored bands.

The second site is located in the mid-Atlantic United States, where groundwater was impacted with carbon tetrachloride. Molasses was applied at the site to enhance reductive dechlorination with a high level of carbon loading (total organic carbon ranging from 16,000 to 24,000 mg/L).

Chlorinated ethane concentrations decreased to low levels by Day 300 with no noticeable rebound in chlorinated compound concentrations, indicating there was not a significant sorbed phase present at the site (Figure C3-2). The reduction of ethane compounds, with the exception of carbon tetrachloride, occurred during the period of ketone production. As noted earlier, *Clostridium* species has been shown capable of degrading dichloromethane, consistent with this finding, ketone production was observed during degradation of dichloromethane.

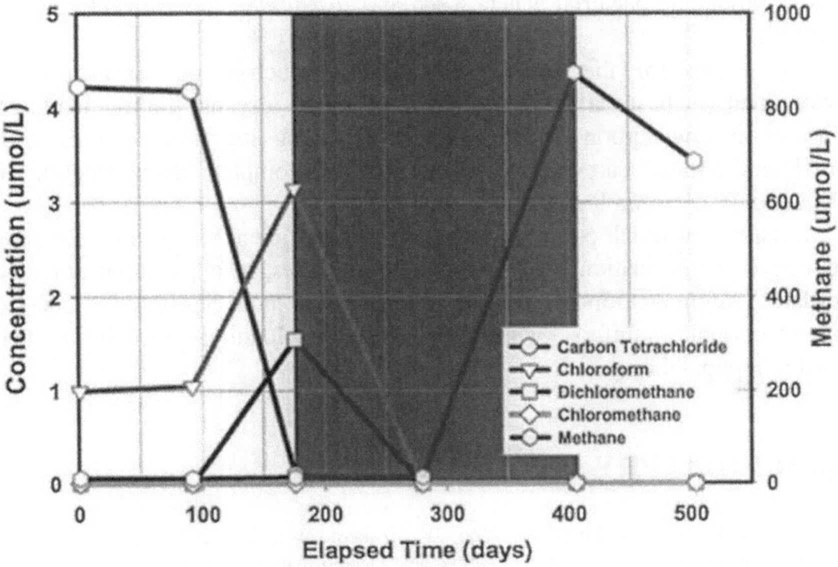

Figure C3-2. Chlorinated alkane concentrations in groundwater at a Case Study located in the mid-Atlantic United States. Injections started at time = 0. Periods of ketone production shown as colored bands.

7. SUMMARY AND CONCLUSIONS

Enhanced reductive dechlorination through biostimulation has been successfully deployed by the authors at more than 100 sites in the United States, Europe and South America. A number of important observations have emerged from the operation of these systems that can be used by designers of in-situ reactive zone systems.

Organic carbon dosing is the critical driving force in enhanced reductive dechlorination. The dose mass and injection frequency must be sustained at a level that fends off electron acceptor recharge and loss of reducing equivalents to aquifer minerals.

Competition from sulfate reducers and methanogens does not limit the enhanced reductive dechlorination process. Sulfate reducers can coexist with dechlorinating bacteria and methanogens when dissolved organic carbon supplies are abundant.

Ketone production has been observed to accompany the highest rates of reductive dechlorination. This may be an indication of the participation of ketone-producing bacterial genera, such as *Clostridium,* in the dechlorination process.

Lag times for the onset of complete reductive dechlorination are determined by bacteria population levels and by organic carbon loadings. Sites with low carbon dosing rates exhibit long lag times. Sites with sufficient organic carbon loading can achieve complete dechlorination of PCE and TCE within 9 months.

Existing bacterial populations have been sufficient to achieve complete reductive dechlorination, with acceptable lag times, at all sites encountered to-date. Augmentation of bacterial populations is more likely to be required in oligotrophic aquifers, where sparse carbon loading severely restricts bacterial populations.

BIBLIOGRAPHY

Chang, Y., Hatsu, M., Jung, K., Yoo, Y., Takamizawa, K. (2000). "Isolation and Characterization of a Tetrachlorethylene Dechlorinating Bacterium *Clostridium Bifermentans* DPH-1." *Journal of Bioscience and Bioeng ineering,* 89(5), 489-491

Fathepure, B. Z., Boyd, S. A. (1988). "Dependence of tetrachloroethylene dechlorination on methanogenic substrate consumption by Methanosarcina sp. strain DCM." *Appl. Environ. Microbiol.* 54, 2976-2980.

He, J., Youlboong, S., Dollhopf, M., Fathepure, B., Tiedje, J., Loffler, F. (2002). "Acetate versus Hydrogen as Direct Electron Donors to Stimulate the Microbial Reductive Dechlorination Process at Chloroethene -Contaminated Sites" *Environ. Sci. Technol. 36(18):3945-3952, 2002 Sep 15*

Holliger, C. (1993). "A highly purified enrichment culture that couples the reductive dechlorination of tetrachloroethene to growth." *Appl. Environ. Microbiol,* 59(9), 2991-2997.

Magnuson, J., Stern, R., Gossett, J., Zinder, S., Burris, D. (1998). "Reductive Dechlorination of Tetrachloroethene to Ethene by a Two -Component Enzyme Pathway." *Appl. Environ. Microbiol.* 64 (4), 1270-1275.

Ndon, J. J., Randall, A. A., Khouri, T. Z. (2000). "Reductive dechlorination of tetrachloroethylene by soil sulfate-reducing microbes under various electron donor conditions." *Environmental Monitoring and Assessment,* 60, 329-336.

Neumann, A., Scholz-Muramatsu, H., Diekert, G. (1994). "Tetrachlorethene metabolism of Dehalospirillum multivorans."*Arch. Microbiol,* 162, 295-301.

Sharma, P., McCarty, P. (1996). "Isolation and Characterization of a Facultatively Aerobic Bacterium that Reductively Dehalogenates Tetrachloroethene to *cis*-1,2-Dichlorethene." *Appl. Environ. Microbiol.,* 62 (3), 761 -765.

Schink, B. (1997). "Energetics of Syntrophic Cooperation in Methanogenic Degradation." *Microbiology and Molecular Biology Reviews* , 61 (2), 262-280.

Smatlack, C., Gossett, J., (1996). "Comparative Kinetics of Hydrogen Utilization for Reductive Dechlorination of Tetrachloroethene and Methanogenesis in an Anaerobic Enrichment Culture." *Environ. Sci. Technol.* 30(9), 2850-2858.

Suthersan, S. (2002). *Natural and enhanced remediation systems.* ARCADIS Environmental Science and Engineering Series. Boca Raton, FL: CRC Press. ISBN 1 -56670-282-8.

Suthersan, S., Lutes, C., Palmer, P., Lenzo, F., Payne, F., Liles, D, Burdick, J. (2002). FINAL: Technical Protocol for Using Soluble Carbohydrates to Enhance Reductive Dechlorination of Chlorinated A liphatic Hydrocarbons. ARCADIS AFCEE Protocol.

Yang, Y, McCarty, P. (1998). "Competition for Hydrogen within a Chlorinated Solvent Dehalogenating Anaerobic Mixed Culture ." *Environ. Sci. Technol.* 32(22), 3591 -3597.

This page intentionally left blank

CHAPTER 8

COMPARISON OF DINITROTOLUENE DEGRADATION BY A MIXED CULTURE IN AQUEOUS BATCH SYSTEM

Jan Paca[1], Jiri Barta[1], and Rakesh Bajpai[2]
[1]Institute of Chemical Technology, Prague, Czech Republic, [2]University of Missouri, Columbia, MO

Abstract: A mixed culture, enriched from nitrotoluenes -contaminated aged -soil from an ammunition plant in the Czech Republic, was used to study biodegradation of 2,4-dinitrotoluene (2,4-DNT) and 2,6-dinitrotoluene (2,6-DNT) in aqueous media under aerobic conditions. The study was conducted in shake flasks with mineral salt medium with or without additional carbon and energy source. The mineral salt medium was supplemented with either glycerol or succinate as t he primary carbon and energy source. When glycerol supplementation was used with **5.2 mg.L** $^{-1}$ initial concentrations of DNT, the results showed total disappearance of 2,4 -DNT in 6 days and that of 2,6 -DNT in 10 days. With succinate as the carbon and energy source, total disappearance of 2,4 -DNT was observed in 10 days, but the disappearance of 2,6 -DNT was poor. One or more metabolites also accumulated in the medium. After six months of a selection pressure by DNTs, the mixed culture was still able to degrade 2,4-DNT almost completely; its ability to degrade 2,6 -DNT in basal salt medium also increased. Also no intermediates accumulated in the medium. When the dinitrotoluenes were used as the sole source of C -, N- and energy, ~ 40 % of each DNT was removed within 4 days. However, uptake of 2,4 -DNT showed a lag phase and one unidentified intermediate accumulated in the medium. 2,6-DNT uptake did not show any lag-phase and no intermediate accumulated in the medium.

Key Words: Dinitrotoluene, Biodegradation, Mixed Culture, Glycerol, Succinate

1. INTRODUCTION

Dinitrotoluenes (DNTs) have been synthesized by successive nitration of toluene for use as explosives (Cooper, 1996) and as raw materials for manufacture of diisocyanate and dyestuffs (Dunlap, 1981). Mainly two isomeric forms of DNT (2,4-DNT and 2,6-DNT) are produced due to the tendency of methyl group to direct substitutions at ortho- and para-positions and of nitro-group to meta-position. The production and usage practices, discharges, and accidental spills have resulted in contamination of large tracks of surface and subsurface soil and groundwater by dinitrotoluenes along with other (mono- and tri-) nitrotoluenes (Broder and Westmoreland, 1998). Since nitrotoluenes are categorized as priority pollutants (Kieth and Telliard, 1979) with strict limits on concentrations permitted in soil and water, several attempts have been made to clean up the soil, groundwater, and wastewater contaminated with dinitrotoluenes (Nishino, et al.,2000; Razo-Flores et al, 1999; Davel et al., 2000; Mueller et al., 1993). Microorganisms with capability to degrade dinitrotoluenes have been isolated from contaminated sites. Spain and coworkers (Nishino, et al., 2000) isolated pure cultures capable of degrading 2,4- and 2,6-DNT. These cultures were able to metabolize the DNTs as sole carbon, energy, and nitrogen sources. Presence of nitrates in the environment was found to be inhibitory to the cells. 2,6-DNT inhibited 2,4-DNT degraders, and vice-versa. 2,6-DNT was degraded considerably more slowly compared to 2,4-DNT. In a fluidized-bed reactor in which the suspended sand (Smets et al., 1999) or granular activated carbon particles (Malcolm Pirnie, 1999) were colonized by the 2,4- and 2,6- degraders, degradation of 2,6-DNT was determined to control the throughput from the reactor for effective clean-up of liquid streams. Given the known general problems of bioaugmentation and the difficulties reported in degradation of nitrotoluenes (Lindeman et al., Former et al., 2003), it is desirable to characterize the nitrotoluene-degrading microorganisms associated with specific sites of contamination. Hence, this effort involved isolating a mixed culture from an explosives-contaminated site in Czech Republic and investigations of its capability to degrade the different nitrotoluenes. The results dealing with degradations of 2,4-DNT and 2,6-DNT by this mixed culture are presented in this paper.

2. MATERIALS AND METHODS

2.1 Organisms

Grab-samples of contaminated surface soil were collected from several different locations on the grounds of Synthesia Company, Pardubice (Czech Rep.). These samples were used to enrich nitrotoluene degraders at 26 °C using a medium containing basal salts (BSM) and sodium succinate, followed by subculturing in the BSM with a mixture of nitrotoluenes. The mixed culture thus obtained was stored in BSM supplemented with nitrotoluenes and 25% glycerol at –80 °C. For propagation purposes, the cells were maintained at 4 °C on Bacto-agar (DIFCO) slants and subcultured every 6-8 weeks by transferring to BSM liquid medium (supplemented with sodium succinate and mixture of nitrotoluenes). Inocula were prepared from the Bacto-agar slants by inoculating BSM liquid medium (with sodium succinate and mixture of nitrotoluenes) and incubating the cells for seven days.

2.2 Culture medium

The basal salt medium (BSM) consisted of (in $g{\cdot}L^{-1}$): KH_2PO_4, 4.30; K_2HPO_4, 3.40; KNO_3, 0.80; $MgCl_2{\cdot}6H_2O$, 0.34; and 1 $mL{\cdot}L^{-1}$ of trace-element solution. The composition of trace element solution was ($g{\cdot}L^{-1}$) $FeSO_4{\cdot}7H_2O$, 5.0; $ZnSO_4{\cdot}7H_2O$, 5.0; $MnSO_4{\cdot}H_2O$, 5.0; $CuSO_4{\cdot}5H_2O$, 5.0; $CoCl_2{\cdot}6H_2O$, 0.1; $Na_2B_4O_7{\cdot}10H_2O$, 0.1; $Na_2MoO_4{\cdot}2H_2O$, 0.1. pH of the basal salt medium was 7.2. In experiments involving cometabolic degradation of nitrotoluenes, 1.0 $g{\cdot}L^{-1}$ glycerol or sodium succinate was added to the flasks along with the dinitrotoluenes (DNT). Medium containing sodium succinate was supplied with 50 $mg{\cdot}L^{-1}$ yeast extract also. 2,4-DNT and 2,6-DNT were obtained as high purity chemicals from Sigma-Aldrich, St. Louis.

2.3 Biodegradation experiments

All the experiments were conducted in 500 mL shake flasks (100 mL working volume) incubated at 26 °C on an orbital shaker at 120 rpm. Samples were collected either once or twice a day, and were analyzed for cell density, DNTs, and intermediate metabolites. Cell density was determined by measuring light absorption at 400 nm using a spectrophotometer model Spekol 11, Carl Zeiss, Germany. The concentrations of the DNTs and metabolites were determined by isocratic reversed-phase HPLC (Waters 510 HPLC pump, 25 cm x 4 mm Nucleosil

120-5 C18 column, Spectromonitor 3200 UV detector, and model 'Basic Marathon' type 816 autosampler). The solvent phase was a 50:50 mixture of water-methanol that was pumped through the column at a rate of 1 mL min^{-1}. Peak detection was carried out at 230 nm. Peak integration was conducted by using CSW (Chromatography Station for Windows) version 3. Samples for HPLC were prepared by centrifugation at 10,000 g for five minutes. Comparing the retention times of peaks with retention time of pure species standards identified nitrotoluene peaks. Concentrations were determined by using calibration curves. Unknown metabolites were identified from their peak retention times as metabolites A through F, and their areas are reported as such. Attempts are underway to identify the unknown metabolites. The pH of the medium was measured at the start and the end of each experiment using a pH meter model MPH 44L, INSA. All the experiments were conducted in triplicate and the data reported are average values from the three flasks.

3. RESULTS AND DISCUSSION

3.1 Biodegradation of 2,4-DNT

When the mixed culture was inoculated in basal salt medium without any additional carbon source, the cells grew slowly but steadily as shown in Figure 1(a), but consumed 2,4-DNT quite rapidly. Starting with a 2,4-DNT concentration of 12 mg·L^{-1}, all of 2,4-DNT was consumed in six days even though there was a lag period of three days in DNT consumption. Accumulation of only one metabolite (metabolite A) was observed in association with the consumption of DNT. Cell growth and DNT consumption were concurrent.

When the cells were inoculated in basal salt medium with 1 g·L^{-1} sodium succinate (Figure 1b) or with 1 g·L^{-1} glycerol (Figure 1c), cell growth was rapid. Cell growth on glycerol was more abundant than on succinate. In both the cases, the cell concentration as suggested by optical density, decreased after peaking at the end of the first day. Consumption of 2,4-DNT was rapid during the period of cell growth and it too slowed down considerably after cell growth ceased. In presence of succinate (Figure 1b), metabolite A accumulated again concurrently with the consumption of 2,4-DNT. No reduction in the concentration of this metabolite was observed during the 10-day operation of the batch. Small quantities of another metabolite (metabolite B) were also observed, but it was completely reconsumed after the second day of operation. At the end of day 10, all the 2,4-DNT was gone from the broth. In the presence of glycerol (Figure 1c), however, accumulation of metabolite A was not as high as in presence of succinate and it was reconsumed. Small amounts of a third metabolite (metabolite C) were observed in the broth at the end of day one but it was rapidly consumed and none was observed again after day two.

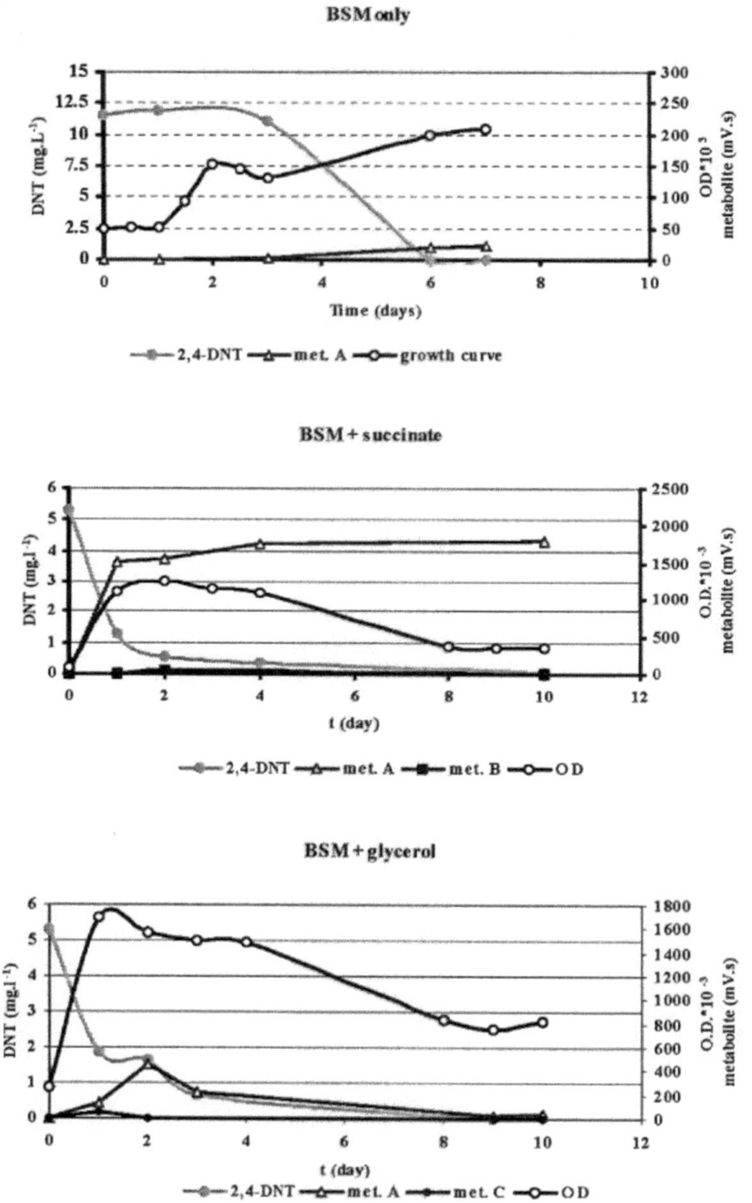

Figure 1a, 1b, 1c. Biodegradation of 2,4-DNT by the Mixed Culture

3.2 Biodegradation of 2,6-DNT

Figure 2 shows the experimental results dealing with biodegradation of 2,6-DNT by the mixed culture. When only the basal salt medium containing 2,6-DNT was inoculated with the mixed culture, there is no cell growth and no change in the concentration of 2,6-DNT in broth over seven days (Figure 2a). However, in presence of succinate or glycerol, cell growth occurred and disappearance of DNT took place.

With the addition of 1 $g \cdot L^{-1}$ sodium succinate to the basal salt medium, cell growth was rapid and cell concentration peaked at the end of day 1 (Figure 2b). The cell concentration then decreased and remained steady from day five onwards. The concentration of 2,6-DNT decreased almost linearly from 5.5 $mg \cdot L^{-1}$ initially to 2.5 $mg \cdot L^{-1}$ on day 10. Significant amounts of only one metabolite (metabolite A) were produced in basal salt medium supplemented with succinate.

In the presence of 1 $g \cdot L^{-1}$ glycerol in the basal medium, cell growth was even more prolific compared to growth on succinate (Figure 2c). Maximum cell concentration was observed at the end of day one. Degradation of 2,6-DNT was rapid in presence of glycerol and all the 5.5 $mg \cdot L^{-1}$ disappeared by day seven. In this case, two metabolites (metabolites A and E) were observed during HPLC analysis of the samples, but their concentrations, as measured by area under the curve, were very small.

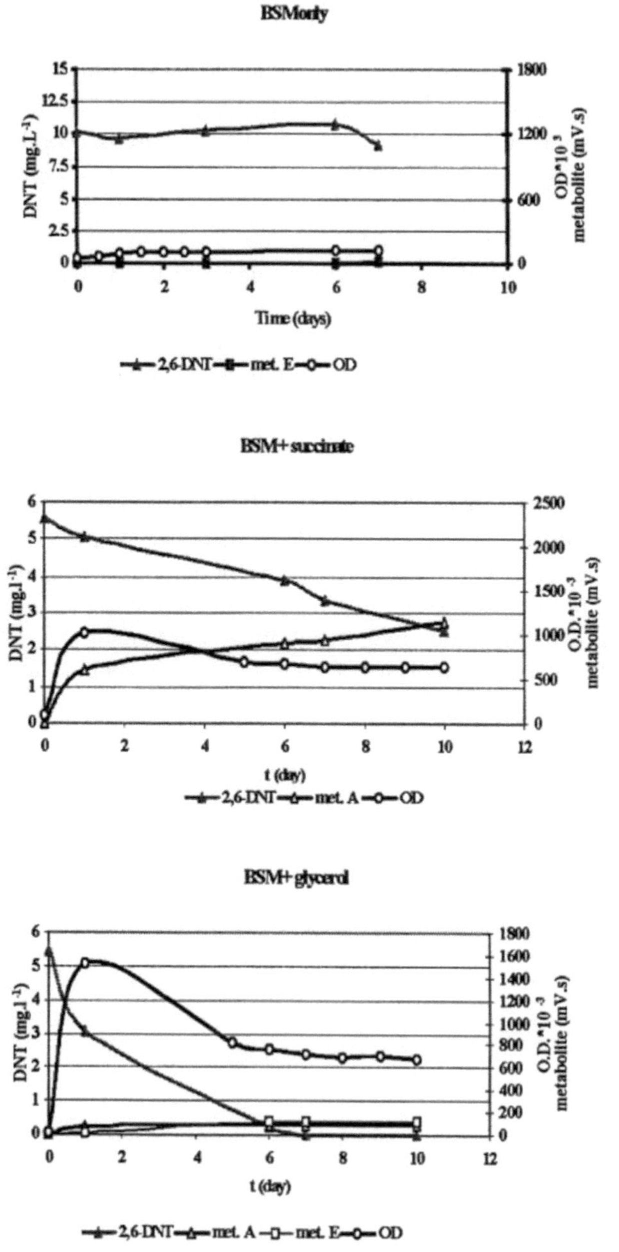

Figure 2a, 2b, 2c. Biodegradation of 2,6-DNT by the Mixed Culture

3.3 Biodegradation of nitrotoluenes in a mixture

Data dealing with the biodegradation of a mixture of mono- and dinitrotoluenes are presented in Figure 3. The mixture contained 3-nitrotoluene and 4-nitrotoluene also, in addition to 2,4-DNT and 2,6-DNT. The experiments were conducted with BSM medium supplemented either with succinate or with glycerol. As observed earlier with pure dinitrotoluenes, the cells grew well on these media. More growth was observed in presence of glycerol medium than on succinate medium. In medium supplemented with glycerol, no drop in optical density was observed.

In medium supplemented with succinate (Figure 3a), the mononitrotoluenes and 2,4-DNT were transformed simultaneously and rapidly to exhaustion. But 2,6-DNT was not metabolized to any significant extent even after 10 days. Two metabolites (metabolites A and D) accumulated in the medium. The concentration of metabolite A increased through the end of sixth day and then stabilized around 1100 (mV.s). Metabolite D, on the other hand, did not accumulate as extensively; it was metabolized also and completely disappeared by day eight.

In glycerol -supplemented medium (Figure 3b), cell growth was extensive. 2,4-DNT and 3-NT were rapidly degraded in the first two days. The concentration of 2,6-DNT decreased initially but then leveled off. Approximately 60% 2,6-DNT remained in the broth at the end of 10th day; all the other nitrotoluenes disappeared from broth by end of day 6. Metabolite A accumulated again and its concentration increased only to level off. Metabolite D was produced in this broth as well and it was metabolized completely by the end of day 10. In presence of glycerol, another metabolite (metabolite C) also accumulated in broth. It too disappeared by the 10th day.

Several authors have reported that 2,6-DNT is an inhibitor of cell growth and biodegradation of 2,4-DNT and vice versa. This mixed culture also appeared to show such a tendency. The biodegradation rate of 2.6-DNT was adversely affected by the presence of mononitrotoluenes and 2,4-DNT (Figures 3a and 3b). The effect of 2,6-DNT on biodegradation of 2,4-DNT is not clear.

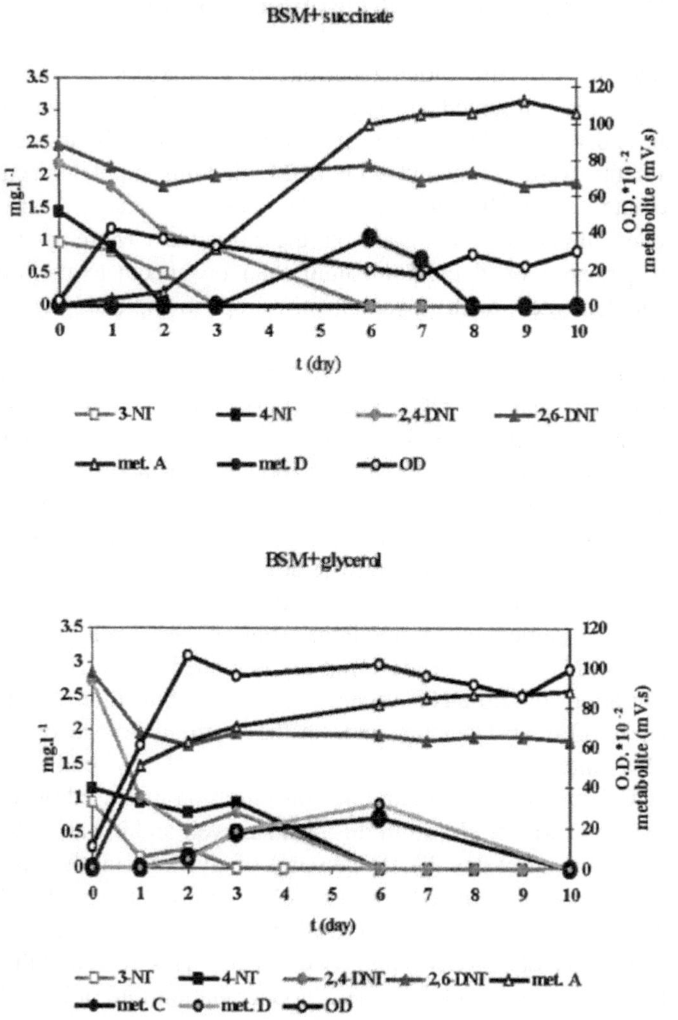

Figure 3a, 3b. Biodegradation of a Mixture of Nitrotoluenes

3.4 Effect of storing cells in a deep freezer on their ability to biodegrade nitrotoluenes

Figure 4 shows the results of experiments conducted with mixed culture stored in a deep freezer (-80°C) for six months. The experiments were conducted in BSM medium without any supplementations with succinate or glycerol, and the results were compared with those obtained with fresh enriched mixed culture (Figures 1a and 2a). As shown in Figure 4(a), degradation of 2,4-DNT does not appear to be affected by storing the cells. There was a small lag period in degradation of 2,4-DNT and metabolite A accumulated in the broth. The levels of accumulation of metabolite A were significantly higher than those observed with fresh mixed culture. With respect to degradation of 2,6-DNT, the stored cells were able to degrade the dinitrotoluenes (Figure 4b) even in absence of any supplementation with carbon source and without accumulating any metabolite in the broth over a period of 10 hours.

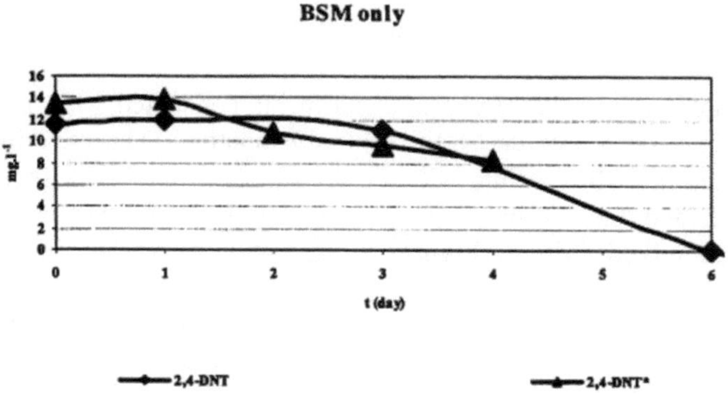

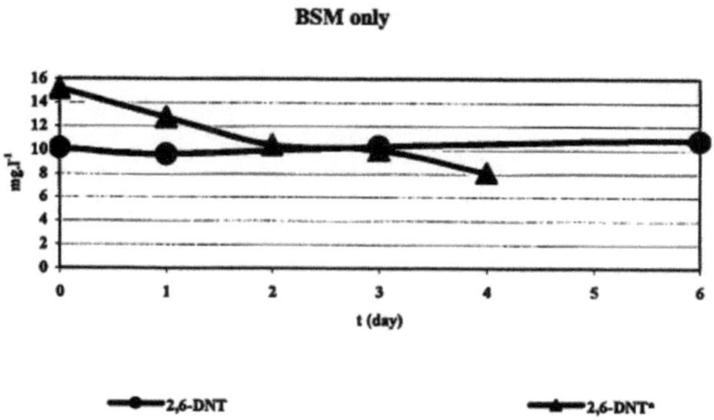

Figure 4a, 4b. Effect of Storage of the Mixed Culture on its ability to Biodegrade DNTs

3.5 Discussion

The mixed culture isolated from the Synthesia site showed an ability to degrade both the dinitrotoluenes in shake flasks. 2,4-DNT was degraded with or without supplementation with a carbon source. On the other hand, 2,6-DNT was degraded by freshly enriched mixed culture only when supplemented with a carbon source. In other words, 2,4-DNT could be metabolized by the mixed culture as carbon and energy source, but the cells were not able to use 2,6-DNT as carbon and energy source. When the cells were stored in presence of a mixture of nitrotoluenes (including 2,6-DNT), the cells were able to use 2,6-DNT as carbon and energy source. These observations are in agreement with the reports published in literature by

Spain and coworkers, (Nishino et al, 2000.; Lindeman et al., 1998; Fortner et al., 2003; Nishino et al. 1999; Smets and Mueller, 2001; Zhang et al., 2000). Nasr et al., (2001), have also reported results concerning the effect of nutrients on biodegradation of dinitrotoluenes. These authors found that biodegradation of 2,4-DNT by pure culture of *Burkholderia* sp. Strain DNT and its genetically modified variants was enhanced by addition of carbon and nitrogen sources in the medium. The strain used by Nasr *et al.* (2001) was isolated by Nishino et al.(2000). But it must be a different strain than the one used by Spain and coworkers, (Nishino et al., 2000; Smets et al., 1999) as Spain reported that no additional carbon and nitrogen sources were required for the biodegradation activity of 2,4-DNT. Several metabolites have been observed in this research and efforts to identify these are in progress.

4. CONCLUSIONS

A mixed culture isolated from a nitrotoluene-contaminated site, efficiently degraded 2,4 DNT and 2,6-DNT when present singly or as a mixture together with mononitrotoluenes. Cell growth could be improved by addition of supplemental carbon source in the basal salt medium, but the supplemental carbon was not necessary for biodegradation of 2,4-DNT. Biodegradation of 2,6-DNT by freshly isolated mixed culture did require supplemental carbon, but this requirement eased in after storing the culture in freezer for several months.

REFERENCES

Broder, M. F. and Westmoreland, R. A. 1998. An estimate of soils contaminated with secondary explosives. Final Report. SFTM -AEC-ET-CR-98002. US Army Environmental Center, Aberdeen Proving Grounds, Maryland.

Cooper, P. W. 1996. Explosives Engineering. VCH (Publisher). New York, N. Y.

Davel, J. L.; Suidan, M. T.; Moteleb, M. A.; Macomber, J. R.; Kim, J.; Adrian, N. R. 2000. Biodegradation of energetic compounds in anaerobic fluidized bed reactors. WEFTEC 2000, Annual Conference & Exposition on Water Quality and Wastewater Treatment, 73rd, Anaheim, CA, United States, Oct. 14-18, 2000, p.1404-1420.

Dunlap, K. L. 1981. 'Nitrobenzenes and Nitrotoluenes' in Encyclopedia of Chemical Technology, **3**rd Edition, vol. 15, p. 916 -932.

Fortner, J. D., Zhang, C., Spain, J. C., and Hughes, J. B. 2003. *Env. Sci. Technol.,* 37(15), 3382-3391.

Kieth, L. H. and Telliard, W. A. 1979. *Env. Sci. Technol.,* 13, 416-423.

Lindemann, U., Smith, J. C., and Smets, B. F. 1998. *Env. Sci. Eng.,* 32(1), 82-87.

Malcolm Pirnie, Inc. 1999. Aerobic Biological Fluid Bed Bioreactor Study for the Mineralization of DNT: Volunteer Army Ammunition Plant. Final Report submitted to the Federal Integrated Biotreatment Research Consortium, USAERDC, Vicksburg, MS.

Mueller, W. F., Bedell, G. W., and Jackson, P. J. 1993. 'Biodegradation of Explosives'. Final Report to the US DOE and New Mexico Waste-Management Education and Research Consortium, Project number WERC -89-059. p. 998-1008.

Nasr, M.A., Hwang, K.-W., Akbas, M., Webster, D.A., and Stark, B.C. 2001. *Biotechnical Prog.,* 17 (2), 359-361

Nishino, S. F., Spain, J. C., Lenke, H., and Knackmuss, H. -J. 1999. *Env. Sci. Technol.,* 33 (7), 1060-1064.

Nishino, S. F., Paoli, G. C., and Spain, J. C. 2000. *Appl. Env. Microbiol.,* 66(5), 2139-2147.

Razo-Flores, Elias; Lettinga, Gatze; Field, Jim A. 1999. *Biotechnol. Prog.,* 15(3), 358-365.

Smets, B. F., Riefler, R. G., Lendenmann, U., and Spain, J. C. 1999. *Biotechnol. Bioeng.,* 63 (6), 642-653.

Smets, B. F. and Mueller, R. J. 2001. *Biodegradation.,* 12, 209-217.

Zhang, C., Hughes, J. B., Nishino, S. F., and Spain, J. C. 2000. *Env. Sci. Technol.,* 34 (13), 2810-2815.

CHAPTER 9

SUBSTRATE VERSATILITY STUDIES ON THE AEROBIC DEGRADATION OF BTX COMPOUNDS

Eldon R. Rene, Shihabudheen M. Maliyekkal, Ligy Philip, and T. Swaminathan
Indian Institute of Technology Madras, Chennai, India

Abstract: Biodegradation studies were carried out in aerobic batch systems using Benzene, Toluene and Xylene (BTX) as the sole carbon source to evaluate the versatility of a bacterial consortium taken from a domestic sewage treatment plant. This mixed microbial culture was acclimatized using individual BTX compounds and the potential capability of microbes precultured on a single substrate to degrade other compounds was studied over a wide range of BTX concentrations (~50-500mg/L). The biodegradation pattern and growth rates ascertained experimentally in these versatility studies were compared with results obtained from individual degradation experiments studied earlier. Similar growth pattern was observed for all the substrate with inhibition occurring at higher concentrations (~200 mg/L for Benzene and Xylene, ~250mg/L for Toluene). Toluene was degraded more than the other substrates followed by benzene and xylene. Adaptation to a more toxic compound like benzene and xylene improved the utilization of toluene. On the other hand microbes grown on a less toxic compound (toluene), grew at a lower rate in the presence of more toxic compounds. Kinetic parameters such as μ_{max}, K_I, and K_S were determined using the Haldane's and Levenspiel's substrate inhibition models. Results from Haldane's model were more illustrative of the experimental observations and were found to satisfactorily explain the system behavior.

Key words: Substrate versatility, Biodegradation, Mixed culture, BTX, Acclimatization.

1. INTRODUCTION

Widespread application of hazardous pollutants such as Benzene, Toluene and Xylene (commonly called as BTX) in fine chemical and petrochemical industries has caused considerable concern over their potential adverse effects to human health and the environment. These compounds frequently enter into the air, water and soil environments through improper handling and disposal practices, refining activities, leakage during storage and transportation and disposal of petroleum by products. BTX compounds are either confirmed, or are suspected, carcinogens that primarily pollute the ground water and pose significant threats to human health. The USEPA has included these compounds among the 188 Hazardous Air Pollutants (HAP's) of potential significance. The fate of these compounds is naturally controlled by microbial activity at the remediation site (Schwarzenbach et al., 1993), and their solubility in water makes them travel hundreds of meters downstream with ground water from the contaminated site (Cozzareli et al., 1990). The biodegradation kinetics and reaction rates are kinetically controlled and influenced by parameters such as pH, temperature, initial substrate concentration, type of inoculum added, nutrient concentration, substrate inhibition and the presence of other complex carbon sources (Alexander, 1977). These compounds have been found to be degraded effectively by naturally occurring soil microorganisms, and other mixed aerobic bacterial populations (Smith, 1990; Alvarez and Vogel, 1991; Chang et al., 1993; Oh et al., 1994). Most of the studies on the biodegradation of BTX compounds have primarily been carried out with mixed cultures acclimatized to specific compounds. A few studies on the biodegradation of mixtures of BTX compounds have also been reported.

Chang et al., 1993 experimented the degradation of mixtures of BTX compounds, and the results from their study revealed three different substrate utilization pattern; no interaction, competitive inhibition and cometabolism. Bielefeldt and Stensel (1999) have used three different cultures isolated from a contaminated vadose zone soil at a gas manufacturing plant site to degrade BTEX compounds. Their study revealed competitive inhibition, which significantly reduced the biodegradation rates of individual BTEX compounds in mixtures when compared to their degradation rates studied individually. Shim and Yang (1999) worked with relatively high concentration of BTEX compounds using co-cultures of *Pseudomonas putida* and *Pseudomonas fluorescens* acclimatized with toluene as the carbon source. Their study showed substrate inhibition at certain threshold concentrations for all the substrates studied. Biodegradation of homologous compounds by mixed cultures may involve similar pathways. Most of the environmental contamination situations involve mixtures of chemicals. Under such conditions, the biodegradation patterns of different substrates using the same microorganisms (precultured on a specific carbon source) can reveal very useful information. The effect of preculturing on a single carbon source (B, T or X) on the biodegradation of other compounds or their mixtures has not been systematically carried out so far. This paper presents the study on the qualitative comparison of the biodegradation patterns of individual BTX compounds with biomass precultured on any one of them.

2. MATERIALS AND METHODS

2.1 Chemicals

All chemicals used in this study were of laboratory grade: Benzene (>99%), Toluene and Xylene (>99%, sulphur free) were purchased from Ranbaxy Fine Chemicals Limited, India

2.2 Microbial culture and media composition

The microbial mixed consortia was obtained from a municipal sewage treatment plant. The final inoculum was obtained by series of repeated inoculation in a mineral salt medium (MSM) that had the following composition: **K_2HPO_4 – 0.8 g/L, KH_2PO_4- 0.2 g/L, $CaSO_4.2H_2O$- 0.05 g/L, $MgSO_4.7H_2O$ – 0.5 g/L, $(NH_4)_2SO_4$- 1.0 g/L and $FeSO_4$ - 0.01 g/L.** The pH of the mineral salt media was adjusted to 6.85.

2.3 Experimental

The mixed microbial consortia were precultured in 100 ml of the MSM containing 150 mg/L of either benzene, toluene or xylene as the carbon source for about 40-50 hours. The batch culture biodegradation of BTX compounds was studied over a concentration range of ~50-500 mg/L of BTX individually. 100 ml of MSM containing the required concentration of B, T or X was inoculated with 5 ml of the precultured suspension and incubated on a rotary shaker at 120 rpm and ambient temperature (28-30°C). Flasks were sealed with stoppers to minimize volatile losses. Samples collected at regular intervals were analyzed for pH, biomass and residual B, T or X concentration.

2.4 Analytical methods

Cell concentration was determined by spectrophotometric method (OD_{540}) using a UV-visible spectrophotometer (Shimadzu UV-1601 PC, Japan).

Residual concentrations of B, T or X in the aqueous phase were measured in cell free samples by Gas Chromatography (Nucon gas chromatograph, Model 5765, Aimil Engg. India) fitted with a flame ionization detector using a Chromatopak (10% FFAP on CH-WIHP) packed column. Nitrogen was used as the carrier gas at a flow rate of 25 ml/min. The temperatures of injector, oven and detector were 150, 120 and 250°C respectively.

3. RESULTS AND DISCUSSION

Laboratory studies were performed in aerobic batch experiments to evaluate the growth and degradation pattern of mixed cultures precultured on benzene, toluene and xylene on each of the BTX compounds (~ 50–500 mg/L). The biomass growth profile with different substrates is shown in Figure 1. Though biomass growth was observed in all cases, there were different patterns of growth for different inoculums.

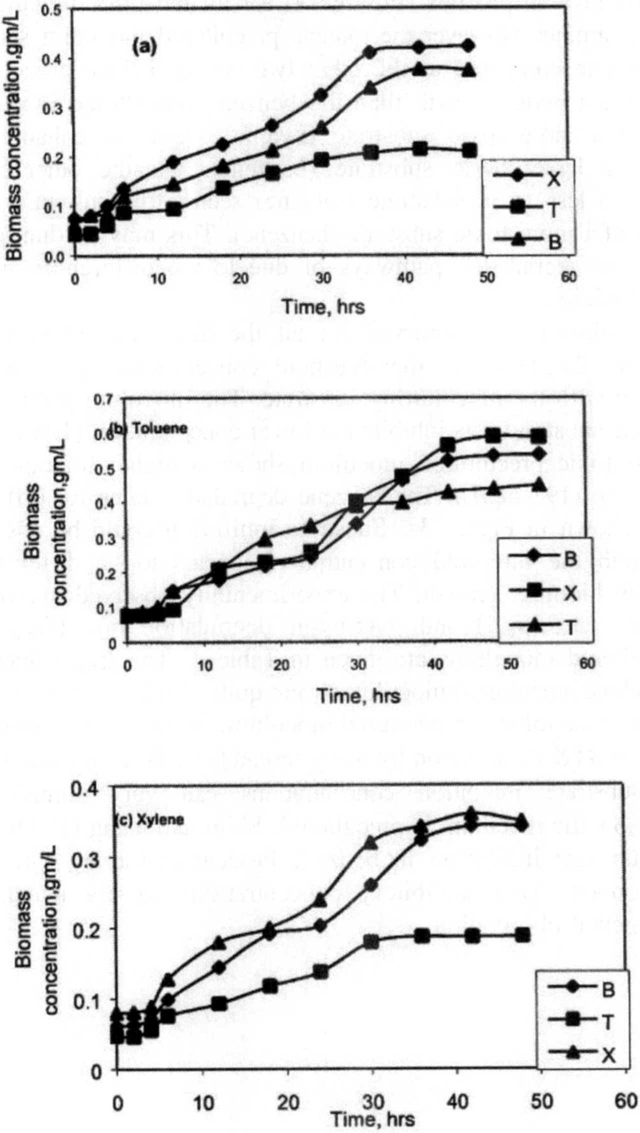

Figure 1. Biomass growth profile for different substrates using inoculums precultured with different carbon resources

3.1 Benzene biodegradation

The biomass growth profiles (Figure 1a) for all the three precultured inoculums were similar. However the toluene precultured inoculum showed a lesser growth rate compared to the other two, while xylene precultured inoculum showed a better growth than the benzene precultured inoculum. Preculturing on a more toxic substrate (xylene) seems to enhance the degradation of a lesser toxic substrate (benzene). On the other hand, preculturing on a less toxic substrate (toluene) seems to result in a poor biodegradation of a more toxic substrate (benzene). This may be due to the differences in the degradation pathways or due to the differences in the mechanism of toxicity.

Substrate inhibition was observed for all the three inoculums and is shown in Figure 2a. However the threshold concentration for substrate inhibition varied with the preculturing substrate. The inoculum precultured on the least toxic substrate gets inhibited at lower concentration (108 mg/L), while the most toxic precultured inoculum shows a higher tolerance for substrate inhibition (197 mg/L). The benzene degradation rates for different inoculums are shown in Figure 3a. Substrate inhibition could be observed here also, though the threshold concentration appears to be different in comparison with biomass growth. The experimentally observed maximum specific growth rate (μ_{max}) and maximum degradation rate (D_{max}) for different precultured inoculums are given in Table 1. The μ_{max} values for benzene and xylene precultured inoculums were quite similar and marginally higher than that of the toluene precultured inoculum. Substrate inhibition has been reported for BTX degradation by many researchers. However this study proves that substrate inhibition concentrations can vary significantly depending on how the inoculum is precultured. Shim and Yang (1999) have reported the substrate inhibition in benzene biodegradation by a toluene precultured biomass. Their inhibitory concentrations closely match the present experimental observation.

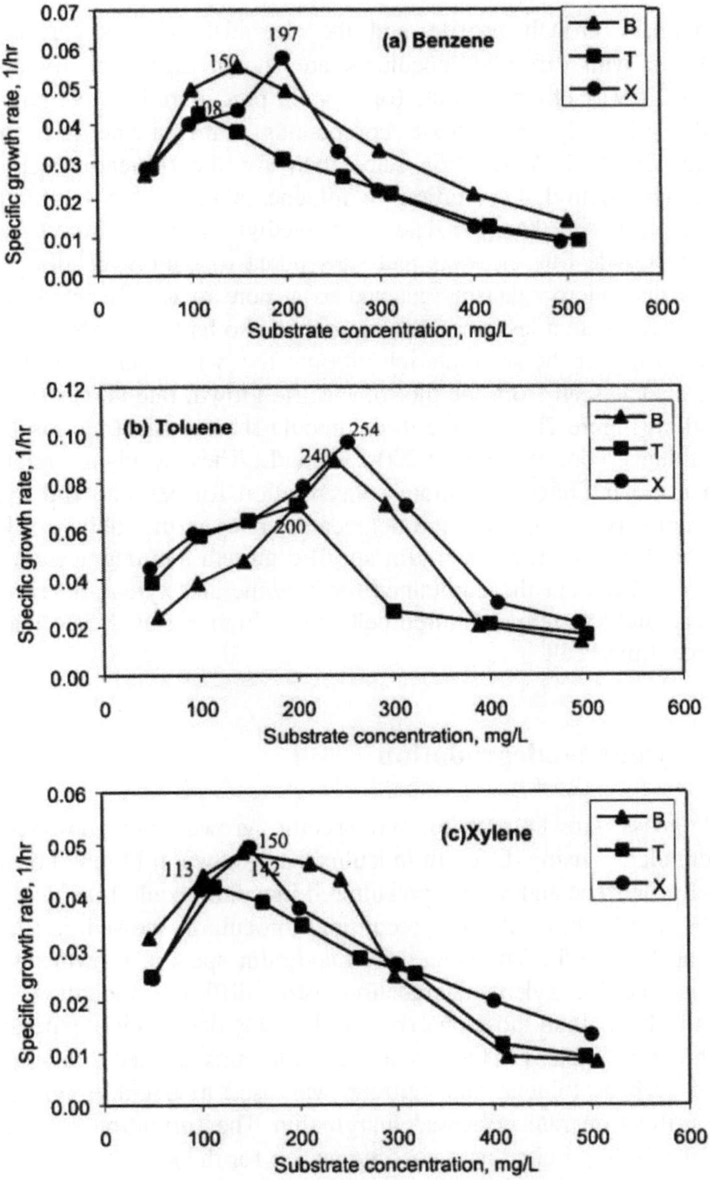

Figure 2. Effect of substrate concentration on the degradation rate of mixed cultures in the biodegradation of B, T, or X

3.2 Toluene biodegradation

The biomass growth profile and the degradation rates for toluene biodegradation with different inoculums are shown in Figure 1b and 3b respectively. It was observed that, for inocula precultured on benzene and xylene, the relatively more toxic compounds than toluene gave better biomass growth and degradation rates than for the toluene precultured inoculum. The methyl substitution in toluene is known to facilitate its biodegradation faster than benzene. The methyl group hydroxylation is believed to precede ring cleavage and subsequent oxidation reaction. This also proves that microorganisms adapted to a more toxic substrate can, not only easily assimilate a less toxic substrate, but also have increased substrate utilization. However the substrate inhibition effect was observed in the case of toluene also as seen from the plot of specific growth rate versus substrate concentration (Figure 2b). All the three inocula showed substrate inhibition around a toluene concentration of 200-250 mg/L. These levels are relatively higher than the inhibitory substrate concentration for benzene and xylene. The maximum specific growth and degradation rates at the inhibitory levels are given in Table 1. The maximum specific growth and degradation rates were much higher than those obtained for benzene and xylene degradation. The benzene and xylene precultured cells had a higher growth rate than the toluene precultured cell.

3.3 Xylene biodegradation

The biomass growth profiles and specific growth rates observed for xylene degradation using different inoculums are shown in Figure 1 c and 2c respectively. Benzene and xylene precultured inoculum could tolerate xylene up to 150 mg/L, but toluene precultured inoculum showed much less tolerance at 113 mg/L. The values for maximum specific growth rate and degradation rate for xylene degradation using different inoculums were substantially lower than those observed for benzene degradation experiments and are shown in Table 1. Thus when microorganisms adapted to a less toxic compound such as toluene and benzene was used to degrade xylene, the substrate utilization rate decreased marginally. The formation of complex intermediates is often cited as a possible reason for the slow biodegradation of Xylene and its isomers. The benzene and xylene precultured inocula gave similar profiles for the degradation rate as illustrated in Figure 3c. Chang et al., (1993) have used a BTX mixture enriched consortia that could mineralize only benzene and toluene whereas *p*-xylene was cometabolized to inhibitory end products. Oh et al., (1994) studied the degradation of individual and mixtures of BTX compounds by a mixed microbial consortium. In mixtures, benzene and toluene were removed by competitive inhibition kinetics, but *p*-xylene was cometabolized in the presence of benzene and toluene. These individual studies carried out, significantly explain the importance of preculturing while studying the degradation kinetics of toxic substrates such as B, T or X.

Table 1. **Specific growth rate and degradation rates observed from Experiments**

Inoculum Precultured with:	Substrate	μ_{max}, 1/hr	D_{max}, mg/L.hr	Inhibition Observed, mg/L
Benzene	Benzene	0.0553	2.55	~150
	Toluene	0.0892	3.66	~250
	Xylene	0.0486	2.43	~150
Toluene	Benzene	0.0427	1.95	~100
	Toluene	0.0707	2.9	~200
	Xylene	0.0420	1.9	~100
Xylene	Benzene	0.0574	3.05	~200
	Toluene	0.0966	4.25	~250
	Xylene	0.0495	2.3	~150

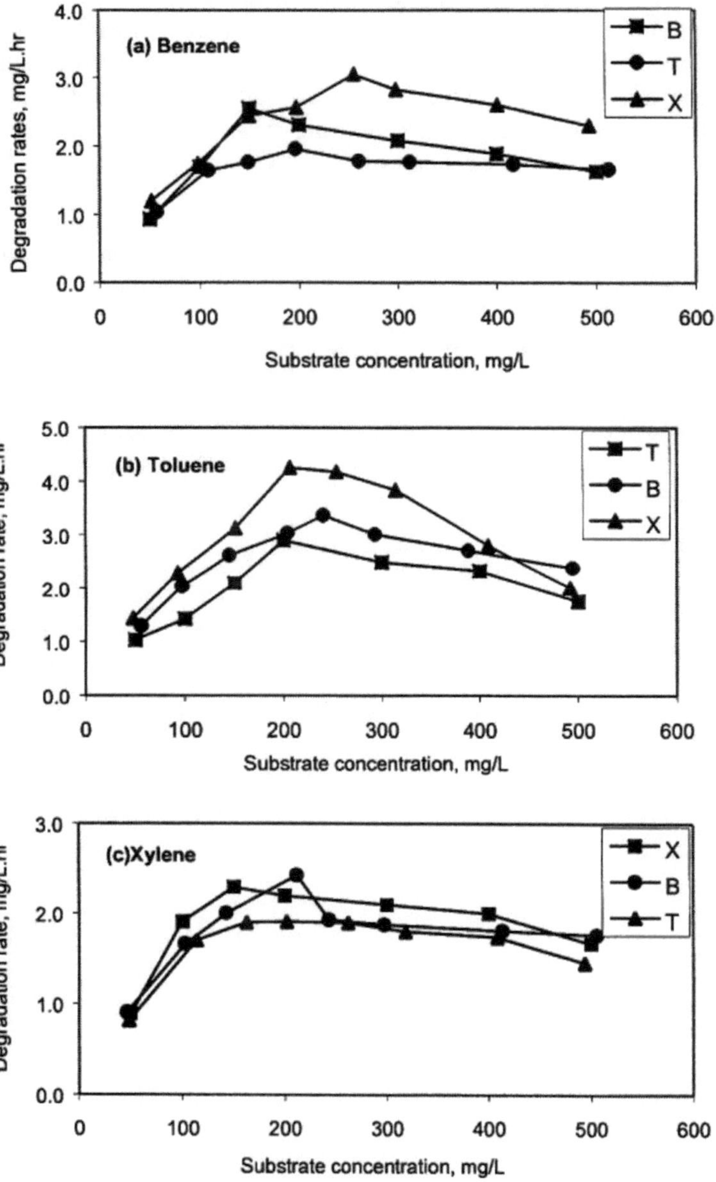

Figure 3. Effect of substrate concentration on the degradation rate of mixed cultures in the biodegradation of B, T, or X

3.4 Effect of preculturing on removal efficiencies

The removal efficiency of individual compounds B, T or X was determined as a function of time with different inocula. These experiments were done at three different concentrations of the specific substrate, i.e., at the inhibitory concentration observed for the selected precultured inoculum and at below and above the inhibitory concentration. Some typical examples of the results are shown in Figure 4a, 4b and 4c respectively. The pattern was similar in all the cases with concentration higher than the inhibitory concentration significantly reducing the removal efficiency. It was observed that the toluene removal efficiencies were quite high for all the three inocula reaching almost as high as 95%. Benzene and xylene removal efficiencies were also above 70% if the substrate concentration was maintained below the inhibitory levels. Toluene precultured cells gave relatively lower removals for all the three compounds as with the biomass growth.

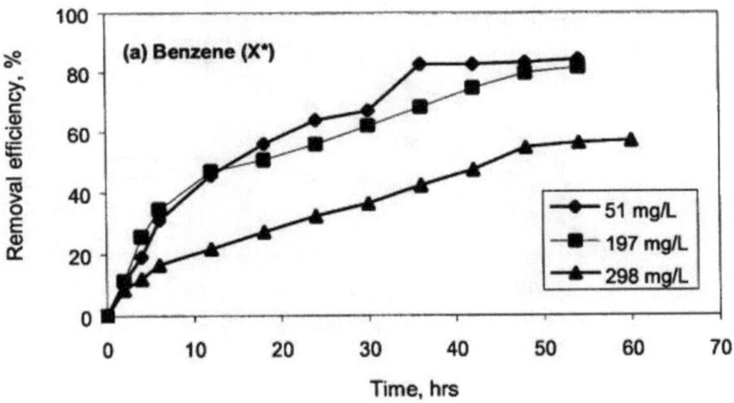

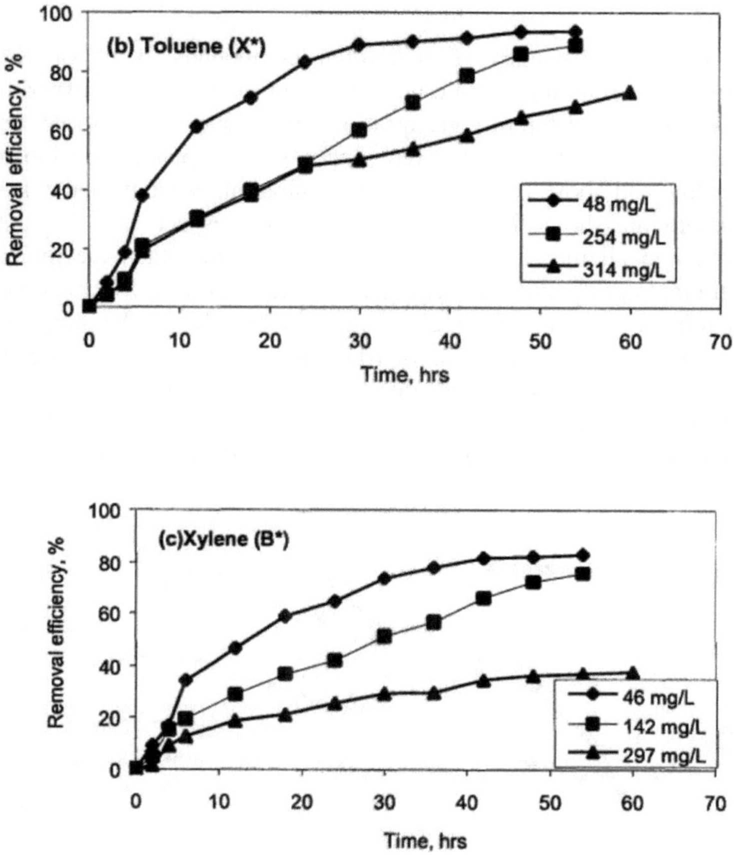

Figure 4. Removal Efficiency profiles of B, T, or X at three critical concentrations studied *-B and X refer to benzene and xylene pre-cultured inoculums

3.5 Quantification of substrate inhibition kinetics

To develop more predictive capability of kinetic parameters, the rate data observed experimentally has to be fitted to appropriate kinetic models (Simkins and Alexander, 1984). The Monod growth model does not take into account the phenomenon of substrate inhibition, which is more frequently observed while studying the degradation of toxic compound like BTX. Having experimentally observed substrate inhibition, BTX kinetics data were suitably modeled by Haldane and Levenspiel's inhibition models. Mathematical representation of the these models are given below.

Model proposed by Haldane (1930):

$$\mu = \frac{\mu_{\max} S}{K_S + S + \frac{S^2}{K_I}}$$

Model proposed by Levenspiel (1980):

$$\mu = \mu_{\max}\left(1 - \frac{S}{K_I}\right)\left(\frac{S}{K_S + S}\right)$$

Where μ is the specific growth rate, 1/hr
S is the initial substrate concentration, mg/L
μ_{max} is the maximum specific growth rate, 1/hr
K_S is the half saturation constant, mg/L
K_I is the threshold substrate inhibition constant, mg/L.

Values of K_S indicate the ability of microbes to grow at low substrate levels (Reardon et al., 2000) and K_I values indicate the sensitivity of the culture to substrate inhibition (Shim and Yang, 1999). The higher K_I value physically means that the culture is less sensitive to substrate inhibition and vice versa. The parameters μ_{max}, K_S and K_I from various models were evaluated by the method of least squares and implemented in the MATLAB 5.1 package. A comparison of values of μ_{max}, K_S and K_I for BTX degradation using precultured cells of B, T and X could be seen in Table 2. Values of K_I from Haldane's model using three different culture vary between 142.1- 211.8 mg/L, 209.4 - 269.5 mg/L and 134.3 -175.6 mg/L for benzene, toluene and xylene degradation respectively. As seen from Table 2, K_I values obtained from the Levenspiel model were found to be on the higher side i.e., greater than 500 mg/L, which was well above the experimental range. K_S values varied between 30.3 mg/L - 72.3 mg/L for various experimental runs. On close observation it was seen that the μ_{max}, K_S and K_I value predicted from these models were slightly higher than the experimental observations. But on the other hand Haldane's model predicted parameters concurred with the experimental value and also simulated the experimental data well. A plot of various experimental and model predicted specific growth rates for BTX compounds grown with B, T and X precultured cells is shown in Figure 5. Even though the Levenspiel model fitted the specific growth rate profile well, the predicted $\mathbf{K_I}$ is about three times higher than the experimental value. The results from Haldane's models were hence found to be more illustrative of the experimental behavior for mixed culture degradation studies.

Table 2. **Kinetic parameters evaluated from substrate inhibition models**

Inoculum precultured with:	Substrate	μ_{max}	K_s	K_I	Standard Deviation
Haldane's Model		1/hr	mg/L	mg/L	
Benzene	Benzene	0.0973	64.3	170.2	0.0093
	Toluene	0.1311	49.2	269.5	0.0250
	Xylene	0.0976	51.1	154.8	0.0113
Toluene	Benzene	0.0826	43.8	142.1	0.0058
	Toluene	0.1117	56.6	209.4	0.0152
	Xylene	0.0879	52.0	134.3	0.0065
Xylene	Benzene	0.0629	30.3	211.8	0.0119
	Toluene	0.1229	72.3	252.3	0.0252
	Xylene	0.0811	58.2	175.6	0.0065
Levenspiel's Model					
Benzene	Benzene	0.0953	86	555	0.0059
	Toluene	0.1624	77.4	582	0.0186
	Xylene	0.0861	40.3	536.2	0.0061
Toluene	Benzene	0.0684	70.9	592.1	0.0043
	Toluene	0.1166	125	570	0.0069
	Xylene	0.0743	54.8	543.7	0.0029
Xylene	Benzene	0.0860	88.0	515.9	0.0086
	Toluene	0.1358	98.2	636.6	0.0208
	Xylene	0.0865	85	561.7	0.0048

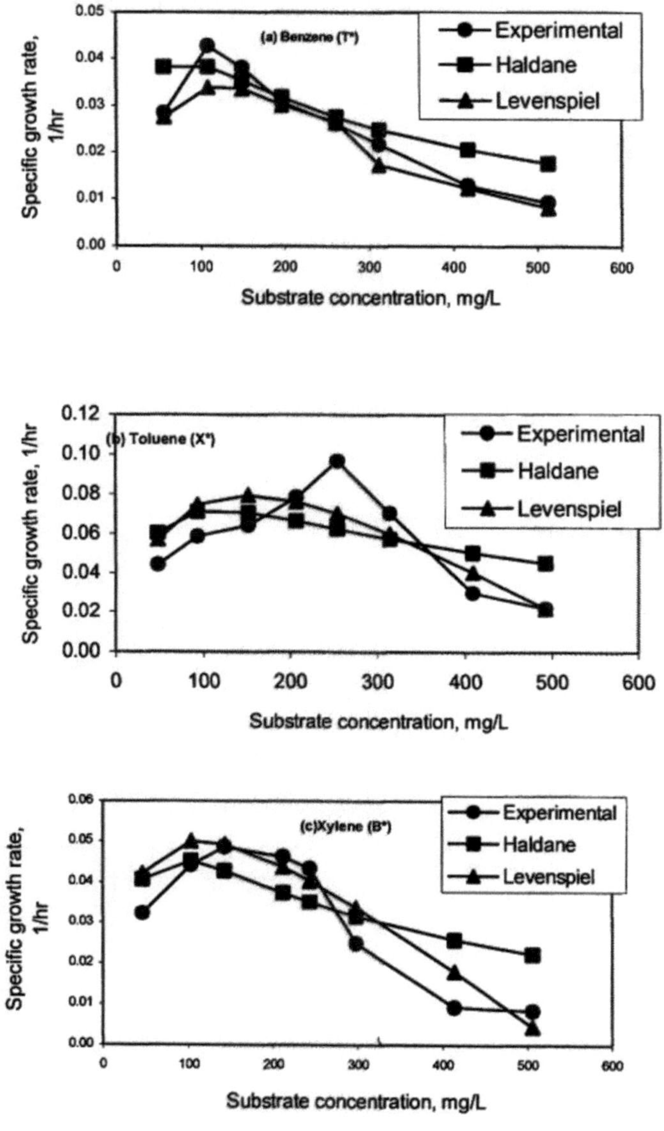

Figure 5. Experimental and model predicted profiles of specific growth rate for benzene , toluene and xylene degraded with different pre -cultured inoculums.* - B, T and X refer to benzene, toluene and xylene pre-cultured inoculums

4. CONCLUSION

Substrate versatility and inhibition kinetics of a mixed culture capable of degrading individual BTX compounds was studied in free cell systems with B, T and X precultured cells. All cultures were able to grow well in the MSM with B, T or X as the sole carbon source. In general, cells precultured on more toxic compounds had a better growth in the presence of a less toxic substrate. The inhibitory concentration of the substrate also varied as a function of the preculturing substrate. These studies also prove that toluene is a less toxic and more easily biodegradable substrate than either benzene or xylene. Substrate inhibition kinetics is well explained by the Haldane's and Levenspiel's model, but the Haldane model gave more realistic constants.

REFERENCES

Alexander, M. 1977. In: Introduction to soil microbiology, Wiley, Newyork, NY, 2 nd ed. pp. 467.

Alvarez, P.J., and Vogel, T.M. 1991. Substrate interact ions of benzene, toluene and para - xylene during microbial degradation by pure cultures and mixed culture aquifer slurries. Appl. Environ. Microbiol. 57, 2981 –2985.

Bielefeldt, A.R., and Stensel, H.D. 1999. Modeling competitive inhibition effects during biodegradation of BTEX mixtures. Water Res. 33(3), 707–714.

Chang, M.K., Voice, T.C., and Criddle, C.S. 1993. Kinetics of competitive inhibition of cometabolism in the biodegradation of benzene, toluene and para xylene by two *Pseudomonas* isolates. Biotechnol. Bioeng. 41, 1057 -1065.

Cozzarelli, I., Eganhous, R.P., and Baedecker, M.J. 1990. Transformation of mono aromatic hydrocarbons to organic acids in anoxic grou ndwater environment. Environ. Geol. Wat. Sci. 16, 135–141.

Haldane, J.B.S. 1930. In: Enzymes, Longmans Green and Co. Ltd. London, Great Britain.

Levenspiel, O. 1980. The Monod equation: A revisit and a generalization to product inhibition situation. Biotechnol. Bioengg. 22, 1671 -1687.

Oh, Y., Shareefdeen, Z., Baltziz, B.C., and Bartha, R. 1994. Interactions between benzene, toluene and *p*–xylene during their biodegradation . Biotechnol. Bioeng. 44, 533–538.

Reardon, K.F., Mosteller, D.C., and Rogers, J.D.B. 2000. Biodegradation kinetics of benzene, toluene and phenol as single and mixed substrates for *Pseudomonas putida* F1. Biotechnol. Bioeng. 69(4), 385–400.

Schwarzenbach, R.P., Gschwend, P.M., and Imboden, D.M.1993. In: Environmental organic chemistry, Wiley, Newyork, NY, pp. 681.

Shim, H., and Yang, S.T., 1999. Biodegradation of benzene, toluene, ethylbenzene and o - xylene by a coculture of *Pseudomonas putida* and *Pseudomonas fluoresens* immobilized in a fibrous - bed bioreactor. J. Biotechnol. 67, 99 -112.

Simkins, S., and Alexander, M. 1984. Models for mineralization kinetics with the variables of substrate concentration and population density. Appl. Environ. Microbiol. 47, 1299 -1306.

Smith, M.R. 1990. The biodegradation of aromatic hydrocarbons by bacteria. Biodegradation . 1, 191–206.

CHAPTER 10

BIOREMEDIATION OF PERCHLORATE AND EXPLOSIVES IN GROUNDWATER

Katherine R. Weeks[1], Scott C. Veenstra[2], Mr. David L. Hill[3]
[1]AMEC Earth & Environmental, Inc., Westford, MA, [2] AMEC Earth & Environmental, Inc. Greensboro, NC, [3]US Army/National Guard Bureau, Camp Edwards, MA

Abstract: Potential remediation processes for explosives - and perchlorate-impacted soil and groundwater are being evaluated on a fast -track schedule at the Camp Edwards Training Area on the Massachusetts Military Reservation (MMR). Laboratory and field treatability studies have been performed since 2000. The most recent round of studies focuses on ex situ remediation of both perchlorate and explosives in groundwater using biological fluidized bed reactor (BFBR), with other processes to be evaluated in the coming year. Historically, range training operations at MMR resulted in the deposition onto soil of propellants, explosives, and pyrotechnic compounds, some of which have been detected in groundwater. Perchlorate and the explosive RDX are present at concentrations up to 300 μg/L and 200 μg/L, respectively.

The BFBR process has been used to degrade relatively high perchlorate concentrations at other sites (milligrams/liter); however Camp Edwards presents challenges for treatment at significantly lower concentrations and for groundwater commingled with explosives such as RDX. The challenges for the BFBR treatability studies were to remediate groundwater in the operable units to project goals of less than 1.5 μg/L for perchlorate and to show significant degradation of RDX. The results of these studies have provided information to help design and implement full -scale treatment systems at MMR.

Key words: Perchlorate, Explosives, Groundwater, Remediation, Bioremediation

1. INTRODUCTION

The Massachusetts Military Reservation (MMR) is a 21,000-acre facility located on Cape Cod, Massachusetts. Approximately 15,000 acres of MMR constitute the Camp Edwards Training Ranges and Impact Area. Historic ordnance disposal, as well as range training activities resulted in the deposition of propellants, explosives, and pyrotechnic (PEP) compounds on the surface soils within the training ranges and impact area. Leaching of rainwater through the soils has transported some of these contaminants to the groundwater underlying Camp Edwards.

The Army National Guard initiated an Innovative Technology Evaluation (ITE) Program in March 2000 to identify alternatives for remediation of the contaminated soils and groundwater. Studies performed in 2002 evaluated the biological fluidized bed reactor (BFBR) process for the destruction of a) commingled perchlorate and explosives and b) of low concentrations of perchlorate only. The process was selected for evaluation based upon its demonstrated success in destruction of perchlorate at other sites and the potential that the technology could simultaneously degrade RDX.

The goal of Study #1 was to determine the ability of BFBR systems to destroy explosives as well as perchlorate in groundwater. Both perchlorate and RDX are present in groundwater from one of the operable units at the site, at concentrations of approximately 100 and 190 μg/L, respectively. Traditionally, BFBRs have proven successful in degrading perchlorate, but few BFBR studies have been performed on explosives, and most of these studies have addressed trinitrotoluene (TNT) but not RDX. Because RDX and perchlorate both degrade anaerobically, it was anticipated that BFBRs would be able to treat groundwater containing both contaminants. The results of the study would determine whether the BFBR system could function as a stand-alone treatment system instead of a lead-lag style system of BFBR to treat perchlorate and granular activated carbon to remove explosives.

At the toe of the Demo 1 plume, groundwater concentrations of perchlorate have been detected in the range of 2 to 6 μg/L. Study #2 was therefore conducted to determine whether the BFBR process would be successful in degrading these low concentrations of perchlorate under conditions where insufficient electron acceptors (such as perchlorate and nitrate) are present to stimulate the biological processes necessary for such degradation.

2. METHODS AND PROCEDURES

Initial concentrations of the target compounds from groundwater samples collected from Camp Edwards and their corresponding ITE performance goals and analytical detection limits are shown in Table 1. Also shown are geochemical parameters that can contribute to the biological process or compete as electron acceptors with perchlorate.

Compound	ITE Performance Goal	Detection Limit	Study #1 Initial Concentration	Study #2 Initial Concentration
Perchlorate (ClO_4^-) (μg/L)	<1.5	0.35	100	3.0
RDX (μg/L)	Evidence of Destruction	0.25	190	<0.25
HMX (μg/L)	Evidence of Destruction	0.25	18	<0.25
Dissolved Oxygen (mg/L)			9.8	9.4
Nitrate as N (mg/L)			2.2	<0.12
Sulfate (mg/L)			4.6	6.1
Total Organic Carbon (mg/L)			<1.0	<1.0
Orthophosphate as P (mg/L)			<0.2	<0.2
Ammonia as N (mg/L)			<0.5	<1.0
pH (S.U.)			5.8	6.3

Table 1. ITE Fluidized Bed Reactor Study Initial Concentrations

A BFBR system consists of a reactor vessel containing a granular medium that is colonized with active bacterial biofilm. The medium is fluidized by the upward flow of groundwater through the vessel, and provides support for bacteria to attach and grow. A schematic of a typical BFBR is presented in Figure 1.

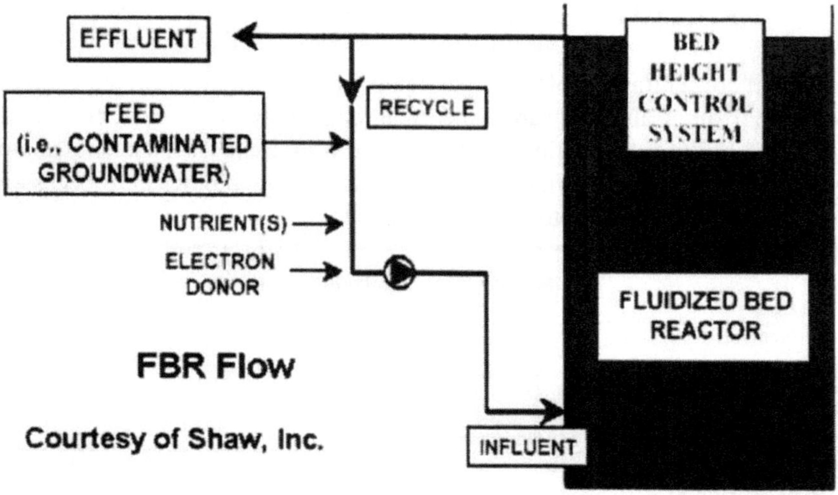

Figure 1. General BFBR System Process Schematic

BFBR systems typically include the following features:

An influent stream of impacted groundwater, which contains the contaminants of interest. At Camp Edwards, both perchlorate and explosives such as RDX act as electron acceptors that are critical to the growth of the biofilm;

A granular medium (typically sand or granular activated carbon - GAC) that is colonized by active bacterial biomass. The MMR studies used GAC as the bed medium;

Controlled addition of a nutrient substrate, such as acetic acid (vinegar), denatured alcohol (ethanol), or molasses to provide an electron donor for the biofilm to interact with explosives and perchlorate;

Controlled addition of growth nutrients (nitrogen, phosphorous), and pH control chemicals such as sulfuric acids and sodium hydroxide;

Hydraulic control to maintain fluidization of the system, by suspending the GAC, and provide enough hydraulic retention time to treat the influent water to desired performance goals; and

Treated water exiting the reactor, which is recycled or discharged.

Envirogen, Inc. (recently acquired by Shaw Environmental and Infrastructure, Inc.) was selected to perform bench-scale studies of the BFBR technology based on their experience at the Longhorn Army Ammunitions Plant in Karnack, Texas (Polk et al, 2001), as well as other sites. Samples were collected and analyzed following AMEC requirements and quality assurance/quality control (QA/QC) procedures (Ogden, 2000a). Envirogen's field laboratory at Camp Edwards analyzed geochemical parameters and field-quality RDX concentrations. Severn Trent Laboratories (STL) of Colchester, Vermont performed confirmatory explosives analyses using EPA Method 8330. Applied Research Associates (ARA) of South Royalton, Vermont performed field-quality analyses of perchlorate using a colorimetric technique that was concurrently being tested elsewhere at Camp Edwards. Envirogen's New Jersey laboratory performed perchlorate analyses using EPA Method 314.0 and RDX analyses on the GAC medium using EPA Method 8330. CEIMIC Laboratories of Narragansett, Rhode Island performed confirmatory perchlorate analyses using the same method.

Study #1: Three laboratory-scale BFBR columns were operated in parallel. In the first study, one (BFBR #1) was fed acetic acid, a simple organic substrate. A second reactor (BFBR #2) was fed molasses, a complex organic substrate. The third column (BFBR #3) was a control, operated in an identical manner to the other two, but without an organic substrate or nutrient (nitrogen and phosphorus) feed. Figure 2 shows a photograph of the laboratory BFBR systems in operation.

Figure 2. Laboratory Scale BFBR

Each BFBR column was fed groundwater from Camp Edwards until perchlorate effluent concentrations approached influent concentrations. It could then be inferred that the GAC sites in the BFBRs were saturated with perchlorate. Subsequent phases of operation and effluent measurements could then be viewed as true measurements of biodegradation within the BFBR systems. The temperature of the reactors was kept at approximately 17°C to simulate field-scale operations.

In Phase 1 of Study #1, once perchlorate breakthrough was seen, BFBRs #1 and #2 were both inoculated with naturally occurring bacteria that were already acclimated to the contaminants of concern. Although the BFBR study could have been performed with naturally occurring bacteria alone, inoculation significantly reduced the time required for the study. The reactors were then operated until they were acclimated for the destruction of perchlorate, with effluent concentrations of perchlorate less than 5 µg/L.

In Phases 2 and 3 of Study #1, when perchlorate effluent concentrations reached 5 µg/L, the reactors were operated at an excess organic substrate feed rate (2 times the theoretical requirement for biological growth) with an effective hydraulic retention time (HRT) of 80 minutes to determine whether the BFBR systems could degrade both perchlorate and explosives. When RDX effluent concentrations were consistently below 2.0 µg/L, an attempt was made to decrease the effective HRT to 35 minutes. Although perchlorate effluent concentrations remained below the 1.0 µg/L laboratory reporting limit, RDX concentrations rose to approximately 5 µg/L. It was therefore determined that an HRT of 80 minutes was more effective for both perchlorate and RDX destruction. Figure 3 presents a graphic depiction of the study influent and effluent perchlorate concentrations vs. time for the acetic acid BFBR (BFBR #1) in Study #1.

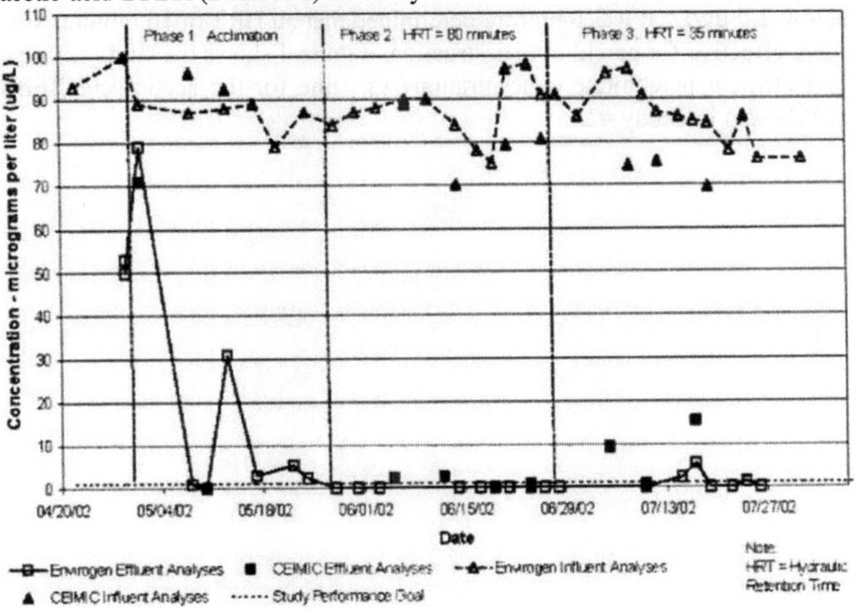

Figure 3. Study #1: Reactor One Effluent Perchlorate Concentrations vs. Time.

Study #2: Again, three laboratory-scale BFBR columns were operated in parallel, and operating parameters such as flow, temperature, and pH were controlled in the same manner as in Study #1. BFBR #1 was again fed acetic acid, BFBR #2 was fed ethanol, and BFBR #3 was again operated as a control. Each reactor was also fed phosphorus and nitrogen for the growth of the bacterial mass.

For this study, it was recognized that like perchlorate, natural constituents of groundwater such as oxygen, sulfates, and nitrates act as electron acceptors critical to biological growth. If the number of electron acceptors is insufficient for biological growth, as in the second study where perchlorate and nitrate concentrations are very low, additional electron acceptors must be added to the system to maintain growth of the bacteria. Therefore, sodium nitrate was added at a rate of 2.5 mg/L (as nitrogen) in the Study #2 as an additional electron acceptor.

The reactors in Study #2 were operated at an excess organic substrate feed rate with an effective hydraulic-retention time (HRT) of 16 minutes to determine whether the BFBR systems could degrade perchlorate (Phase 1). When perchlorate effluent concentrations were consistently below 1.0 μg/L in the acetic acid BFBR, an attempt was made to decrease the effective HRT to 11 minutes (Phase 2). Perchlorate effluent concentrations did not remain below 1.0 μg/L. It was therefore determined that an HRT of 16 minutes was more effective for perchlorate destruction. Figure 4 shows the study influent and effluent perchlorate concentrations vs. time for the acetic acid BFBR (BFBR #1) in Study #2.

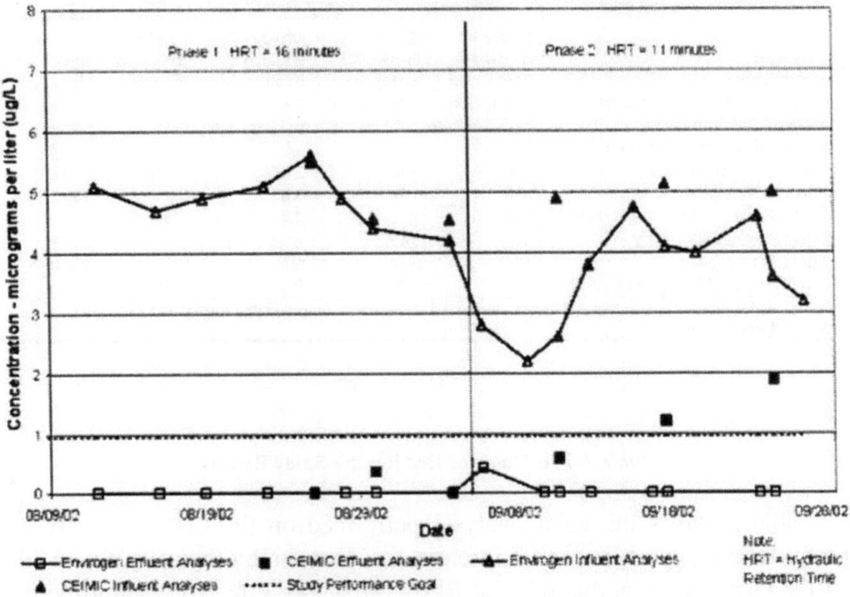

Figure 4. Study #2: Reactor One Effluent Perchlorate Concentrations vs. Time

3. RESULTS AND CONCLUSION

Table 2 displays results under the optimal conditions for each study. Neither the molasses BFBR in Study #1 nor the ethanol BFBR in Study #2 was successful in destroying perchlorate in comparison with the acetic acid BFBR. The acetic acid reactors degraded perchlorate to below 1.0 µg/L in groundwater. The acetic acid reactor degraded RDX to below 2.0 µg/L, a 99% reduction of the initial RDX concentration.

Parameter	Study #1 BFBR #1	Study #1 BFBR #2	Study #2 BFBR #1	Study #2 BFBR #2	Studies #1 / #2 BFBR #3
Electron Donor	Acetic acid	Molasses	Acetic acid	Ethanol	None
Dosage of Sodium Nitrate (mg/L)	Not needed	Not needed	2.5	2.5	None
Dosage of Electron Donor	2 - 5 X theoretical maximum for biological growth				None
Hydraulic Retention Time (min)	80	80	16	16	80 / 16
Influent Perchlorate (µg/L)	100	100	2 to 6	2 to 6	100 / 2-6
Influent RDX (µg/L)	190	190	0.0	0.0	190 / 0
Influent Sodium Nitrate (mg/L as N)	0	0	2.5	2.5	0
Effluent Perchlorate (µg/L)	<1.0	13.5	<1.0	2.8	39 / 6
Effluent RDX (µg/L)	<2.0	4.6	NA	NA	3.0 / NA
Effluent Sodium Nitrate (mg/L as N)	NA	NA	0.52	0.91	NA / 0.13

Notes: Effluent concentrations averaged over last 5 samples collected under specified HRT
NA = Not applicable

Table 2. ITE Fluidized Bed Reactor Study Results

Table 3 shows the RDX analyses performed on the GAC in Study #1. From these analyses, it can be inferred that that the biologically active film on the GAC in the acetic acid-fed BFBR was more effective at destroying a significant amount of sorbed RDX than in the molasses-fed BFBR (BFBR #2) or control BFBR (BFBR #3).

Sample Location within BFBR	End of Phase 1 RDX (mg/kg of GAC)	End of Phase 2 RDX (mg/kg of GAC)	End of Phase 3 RDX (mg/kg of GAC)
BFBR #1 - Top	309	4	4
BFBR #1 - Bottom	330	3	4
BFBR #2 - Top	590	626	784
BFBR #2 - Bottom	728	558	545
BFBR #3 - Top	591	558	1019
BFBR #3 - Bottom	641	718	888

Notes: Data represent average concentrations. Duplicate samples were taken at the end of Phase 1, and triplicate samples were taken at the end of Phases 2 and 3.

Table 3. Study #1: ITE Fluidized Bed Reactor GAC Analyses for RDX

The following conclusions can be drawn from the studies.

A single BFBR fed with acetic acid can successfully degrade perchlorate from 100 µg/L to concentrations below 1.0 µg/l, and RDX from 190 µg/l to below 2 µg/l, using a field-equivalent hydraulic retention time (HRT) of 80 minutes.

A single BFBR fed with acetic acid can successfully degrade perchlorate from 2 to 6 µg/L to concentrations below 1.0 µg/L, using a field-equivalent hydraulic retention time (HRT) of 16 minutes, and with the addition of sodium nitrate as an additional electron acceptor.

BFBRs fed with molasses or ethanol as a substrate degraded perchlorate, but not to the studies' performance goals.

Based on the results of the first study, if reduction of RDX is not a primary factor in treatment, and the reactor design is based on an HRT for perchlorate destruction alone, the equivalent HRT might be reduced to 35 minutes.

The biologically active film on the GAC in an acetic acid-fed BFBR is more effective in the removal of sorbed RDX than in a molasses-fed BFBR or a control reactor.

The above results were validated by the effluent concentrations from the control reactor (BFBR #3), which were consistently higher than effluent concentrations from the other two reactors. No attempt was made to quantify the amount of perchlorate sorbed to the GAC during the test because the adsorption capacity of GAC for perchlorate is very low.

Based on the success of the studies, the BFBR process was selected as the fundamental component for treatment at MMR in 2002. The system is currently in the final design stages. The ITE program is continuing to evaluate other technologies to treat low concentrations of perchlorate, such as granular activated carbon (GAC), GAC that has been tailored with ionic monomers or polymers, and ion exchange (IX) resins.

ACKNOWLEDGEMENTS

The authors would like to thank Mr. Ben Gregson of the Impact Area Groundwater Study Office, and Ms. Heather Sullivan and Mr. Ian Osgerby of the U.S. Army Corps of Engineers for their support of the Innovative Technology Evaluation Team.

REFERENCES

AMEC, 2003. *Innovative Technology Evaluation Groundwater Treatability Study Summary: Fluidized Bed Reactor Study #2. Camp Edwards Impact Area Groundwater Study Program Massachusetts Military Reservation Cape Cod, Massachusetts*. Prepared for The National Guard Bureau and U.S. Army Corps of Engineers by AMEC Earth & Environmental, Westford Massachusetts. January, 2003.

AMEC, 2002. *Innovative Technology Evaluation Groundwater Treatability Study Summary: Fluidized Bed Reactor. Camp Edwards Impact Area Groundwater Study Program Massachusetts Military Reservation Cape Cod, Massachusetts*. Prepared for The National Guard Bureau and U.S. Army Corps of Engineers by AMEC Earth & Environmental, Westford Massachusetts. November, 2002.

Ogden, 2000a. *Quality Assurance/Quality Control Plan Revision 7 of the Camp Edwards Impact Area Groundwater Quality Study, Massachusetts Military Reservation Cape Cod, Massachusetts.* Prepared for the National Guard Bureau by Ogden Environmental and Energy Services, Westford, Massachusetts (Ogden Environmental and Energy Services became AMEC Earth & Environmental in December 2000). April, 2000.

Polk et al., 2001. Polk, J., C. Murray, C. Onowokae, D.E. Tolbert, A.P. Togna, W.J. Guarini, S. Frisch, and M. Del Vecchio. "Case Study of Ex -Situ Biological Treatment of Perchlorate-Contaminated Groundwater." *Proceedings of the Fourth Tri-Services Environmental Technology Symposium.* San Diego, CA. June, 2001.

Weeks, K.R., S. C. Veenstra, and B.P. Gregson, 2001. "Comparison of Innovative Technologies for Soil Cleanup at Camp Edwards, Massachusetts Military Reservation". *Proceedings of the Fourth Annual Tri -Services Environmental Technology Symposium,* San Diego, CA. June, 2001.

Weeks, K.R., S. C. Veenstra, Hill, B.P. Gregson, and D. Taege, 2002. "Comparison of Innovative Technologies for Groundwater Remediation at Camp Edwards, Massachusetts". *Proceedings of the Third International Conference on Remediation of Chlorinated and Recalcitrant Compounds ,* Monterey, CA. May, 2002.

PART II: CHEMICAL OXIDATION

CHAPTER 11

APPLICATION OF IN-SITU OXIDATION (ISCO) AT A SITE WITH MULTIPLE DIVERSE HYDROGEOLOGIC SETTINGS

Raymond Cadorette; David Walker, PG ;Lawrence Nesbitt, PE
Shaw Enviornmental, Hopkington, MA

Abstract: As part of an effort to reach state cleanup goals and alleviate the continued need for source area hydraulic control, the largest known *in-situ* oxidation (ISCO) effort was implemented at this manufacturing facility site located in New England. TCE and PCE are the primary contaminants at the site. Groundwater extraction and treatment was conducted in the primary contaminant source areas at the site for a period of eleven years. Over the course of the pumping and treatment program, di ssolved concentrations of total chlorinated organic compounds (CVOCs) in the most impacted recovery well averaged 457 mg/L. An ISCO pilot test was conducted in the primary contaminant source area, which reduced the average (over 12 months) dissolved concentration of total CVOCs to 63 mg/L in the most impacted recovery well, a sustained decrease of over 85%. Based on this initial successful application of ISCO technology, a full-scale ISCO system was designed and implemented. One key to successful ISCO is to achieve intimate contact between the contaminant and the oxidant. To achieve this contact and account for the varying depths of the targeted CVOCs and the multiple diverse hydrogeologic settings at the site, various permanganate application methods were employed at different depths in the full -scale ISCO system. Application methods utilized at this site included gravity addition through vertical and horizontal wells and pressurized application through vertical wells to a maximum depth of 160 feet. Th is ISCO full-scale treatment program used a 20% sodium permanganate solution in 28 injection locations in the primary contaminant source area and downgradient locations. This paper summarizes the various application methods employed at this site, the safety measures employed and the results of the first year's application of over 75,000 gallons of sodium permanganate solution. This dosage is part of the largest ISCO application of permanganate to be applied to date.

1. INTRODUCTION

A large *in-situ* chemical oxidation (ISCO) effort was implemented by Shaw Environmental, Inc. (Shaw) at a manufacturing facility site located in New England as part of an effort to reach state cleanup goals in groundwater and alleviate the continued need for source area hydraulic control. The chlorinated volatile organic compounds (CVOCs) trichloroethene (TCE) and tetrachloroethene (PCE) are the primary contaminants at the site. An ISCO pilot test was conducted in one of the contaminant source areas, which, over the course of 12 months, reduced the average dissolved concentration of total CVOCs by over 85%.

Based on this initial and successful application of ISCO technology, a full-scale system was designed and implemented to address CVOC impacts in the shallow and deep overburden and bedrock aquifers. Treatment was applied at the two main source areas and two impacted locations downgradient of the source areas (Figure 1).

This paper summarizes the various application methods employed at this site, the safety measures in use, and the results of the first year application of over 75,000 gallons of a 20% sodium permanganate solution (over 175,000 pounds of oxidant) in different hydrogeologic settings.

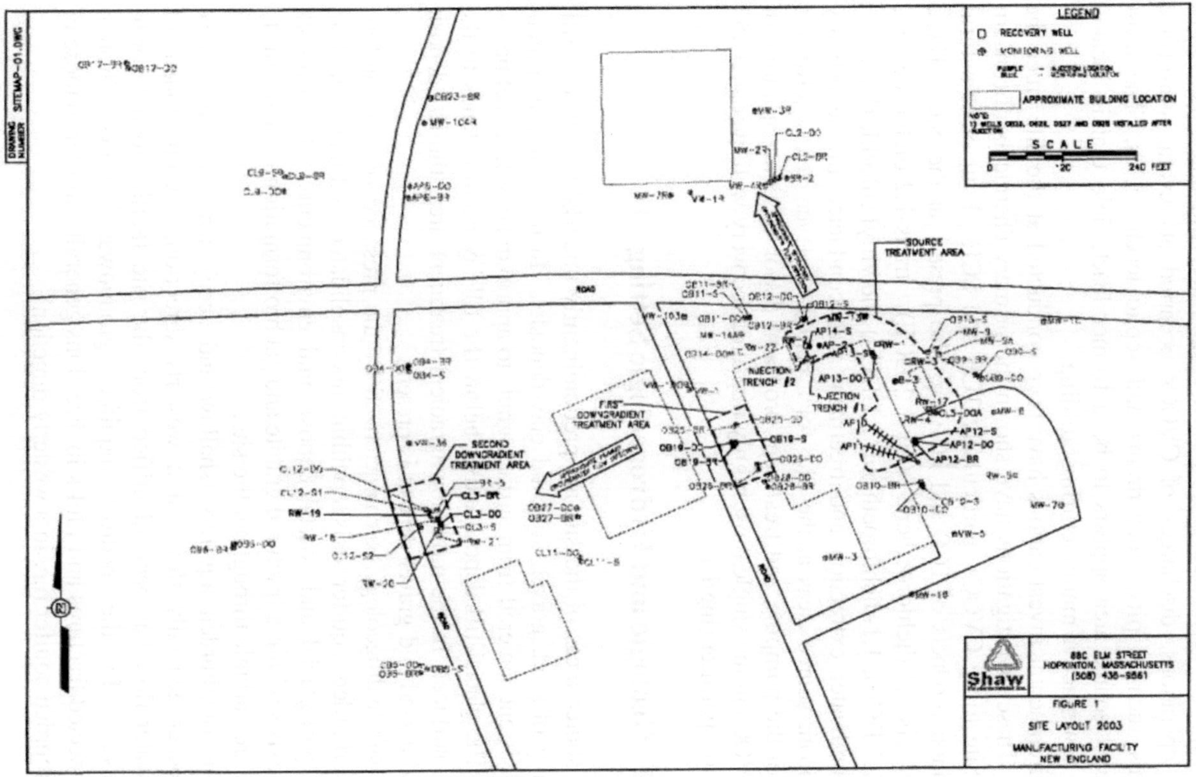

Figure 1. Site Layout 2003

1.1 Site History

The site has been used for industrial manufacturing since the 1940s. During operations, discharges of CVOCs occurred at several subsurface structures such as pits, floor drains and septic leach fields. The most significant source area appears to be a former leach field that is located under a newer active manufacturing building.

Environmental investigations have been conducted at the site since the 1980s. These investigations have shown the primary chemicals of concern at the Site to be CVOCs, particularly TCE and PCE. Four other natural degradation products of TCE and PCE are also present at the Site. These include cis-1,2-dichloroethene (cis-1,2-DCE), trans-1,2-dichloroethene (trans-1,2-DCE), 1,1-dichloroethene (1,1-DCE) and vinyl chloride (VC).

Groundwater extraction and treatment was conducted in the primary contaminant source areas at the Site for a period of eleven years. Over the course of the pumping and treatment program, dissolved concentrations of total CVOCs in the single most highly impacted recovery well averaged 457 milligrams per liter (mg/L).

1.2 Geologic and Hydrogeologic Setting

The source areas of the site are predominantly underlain by glacial till, consisting of sand, gravel, silt, and clay deposits. In the source treatment area, fill is encountered from surface grade to approximately five feet below grade. The glacial till deposit exists below the fill layer and extends to the bedrock surface. In the most downgradient treatment area, the till deposit ends and a sand and gravel deposit is present.

The site hydrogeology is characterized by the presence of two aquifers. The overburden aquifer exists within unconsolidated sediments laid in glacial, post-glacial and coastal depositional environments. The bedrock aquifer exists within a fractured granite and gabbro formation. The depths to bedrock are variable throughout the site.

In the overburden aquifers (shallow and deep) near the source area, groundwater generally flows in a westerly direction. A component of groundwater flow has been identified from the site facility to the north-northwest. In the bedrock aquifer, groundwater flows through interconnected fractures and faults in directions generally similar to those of the overburden aquifer (i.e. in a westerly direction).

2. PERMANGANATE APPLICATION POINTS AND INJECTION METHODS

2.1 Application Points

Shaw utilized both new and existing wells and new recharge trenches for the addition of permanganate into the subsurface. Existing monitoring wells were chosen as addition points based on their location and their screened interval relative to the source areas. New injection points were installed in areas where permanganate additions would provide additional and more complete coverage in and around the contaminant source areas, as well as providing oxidant to different lithologies in the subsurface.

A total of 12 new injection wells were installed across the Site prior to conducting the remedial action. Nine of the injection wells were completed in the shallow overburden, two injection wells were installed in the deep overburden and one injection well was completed in bedrock. The locations of these new wells were based on the historic reports of releases at the Site and the results of previous assessment activities. In addition, 14 existing wells were used as injection points. These included one shallow overburden well, eight deep overburden wells and five bedrock wells.

Two of the new wells used for shallow overburden injection (AP-10 and AP-11) were horizontal wells installed beneath an actively used building located in one of the potential source areas. The wells were installed slightly above the water table extending approximately 60 feet from the wellhead. The horizontal wells were installed to allow nearly continuous injection for treatment of potential vadose zone contamination and shallow aquifer impacts. The wellheads of the horizontal wells are located outside the building, and allow permanganate injection to occur without disrupting the on-site workers activities.

The wells were generally installed with typical monitoring well construction, including PVC well materials in the shallow and deep overburden wells, and steel casing with open boreholes in the bedrock wells.

Two horizontal trenches were installed in a source area of low permeability soils (Injection Trench #1 Trench #2). These trenches were constructed of horizontal perforated PVC pipe, surrounded by stone, at a depth below adjacent utility lines. A vertical riser provided access for chemical injection. This design allowed for a useful volume of permanganate to be injected and stored in the stone fill for slow infiltration over time, and at a depth that would not impact utilities.

2.2 Application Methods

A liquid permanganate storage, mixing, injection and delivery system was constructed in an available permanent building on the subject property. Storage and mixing tanks were used to contain the oxidant prior to its injection. Sodium permanganate was delivered to the site as a 40% permanganate solution in bulk tanker trucks. This 40% solution was mixed with an equal amount of water, creating an approximate 20% sodium permanganate solution by volume, prior to injection.

The 20% permanganate solution was delivered to the injection points through either flexible vinyl hose directly from the permanganate storage piping system, or the solution was transferred into a portable permanganate storage container skid for injection at remote locations on the Site. The permanganate additions were conducted by either gravity flow, or in limited circumstances, under pressure (using chemically resistant positive displacement pumps) to increase infiltration rates. Flow totalizers were utilized during the additions to quantify the volumes of permanganate solution injected to each well on a daily basis.

3. SAFETY AND COMPLIANCE MEASURES

As previously mentioned, sodium permanganate is delivered to the site as a liquid 40% concentrated solution and stored in a dedicated tank. It is incompatible with acids, peroxides, metal powders, oil and grease, and combustible organics. Combustible organics of concern include, but are not limited to, paper towels, wooden pallets and cotton clothes such as blue jeans and tee shirts. The permanganate is diluted with an equal volume of water prior to handling for well injection to reduce, but not eliminate, safety concerns. Overhead sprinkler systems in the permanganate storage building were originally charged with ethylene glycol to prevent freezing. This is incompatible with permanganate and was purged from the permanganate storage areas.

The bulk tank truck driver is required to have adequate training and spill response equipment, notify Shaw three days in advance of deliveries, fax his employer ID, and phone with arrival time the day before delivery. Shaw oversees the truck offloading delivery activity.

Procurement of Permanganate is subject to requirements of the United States Drug Enforcement Diversion and Trafficking Act. The concern is its potential diversion to make illicit drugs.

Eye protection with safety glasses and side shields, along with goggles or a face shield is recommended. Coated Tyvek suits and PVC gloves are required to protect skin and clothing. An emergency shower and an eye wash station are available in the permanganate storage area. (The water in the shower and in the eye wash station is a tepid temperature in order to avoid hypothermia during extended dousing). Portable eyewash equipment is also stationed at remote injection points.

All permanganate handling activities are performed using the "buddy system", with two technicians at the point of activity. This provides someone to assist immediately in an emergency. In the case of a two-man crew workday, one technician injecting at a remote well while the other transfers permanganate at the treatment plant is not permitted.

Paper towels or cotton rags were not used to clean up any incidental permanganate spills or drips as these materials have the potential to spontaneously combust with the permanganate solution. All minor drips or spills were neutralized with small amounts of acetic acid, water and hydrogen peroxide.

Large-scale permanganate injection projects also must consider building code issues related to the storage of oxidants and the SARA reporting requirements for the large quantity of permanganate.

4. RESULTS OF APPLICATION

4.1 Calculated Addition Volumes

The proposed volume of permanganate to be injected at each location was based upon the estimated mass of CVOCs to be treated, along with the soil matrix demand. The soil matrix demand is that presented by natural organics in the soil, and other demands such us unoxidized iron. Field tests were conducted during the injection well installations to confirm or revise proposed injection volumes. The calculated volumes were split in order to provide a controlled multi-year program.

4.2 Addition Flow Rates

The permanganate addition rates were generally dependant upon the aquifer of which the solution was being added. Shallow overburden injection proved to be the easiest due to the loose unconsolidated sand and gravel fill and the limited hydrostatic pressure head. All of the additions to the shallow injection wells were conducted under gravity flow either directly from the permanganate dilution tank from within the treatment building, or from the portable permanganate storage tank. When using this gravity flow method and a ½-inch inside diameter transfer hose, an approximate two gallons per minute (gpm) injection rate was achieved. In some instances, the injection to certain wells would need to be cycled if the injection rate was greater than the individual well capacity to disperse the permanganate. Once a well was filled with the permanganate solution, the injection would need to be stopped. A manually operated ball valve near the wellhead was used to shut the injection flow when necessary. Once the permanganate level in the well dropped, the process was repeated.

Permanganate flow to the deep overburden aquifer wells was conducted with variable injection rates. The multiple depositional environments identified at the site resulted in the expected variability in well capacity to disperse the permanganate. Deep overburden wells in the source area were installed within dense glacial till that limited the wells ability to receive permanganate. The deep overburden wells located in the downgradient sand and gravel deposit revealed a high well yield, which accepted continuous gravity permanganate additions.

Permanganate additions to the bedrock aquifer wells also indicated variable injection rates, however, there did not seem to be any relevance to the well location and seemed to be determined mainly on whether the well was installed within major bedrock fractures.

When necessary, and where possible, the permanganate solution was pressure injected into the subsurface utilizing a portable, variable speed transfer pump. Permanganate solution was transferred from the portable permanganate tank to the transfer pump through vinyl hose. The injection hose was connected directly to the injection wellhead utilizing a pipe union or like fitting. Generally, the permanganate solution was injected at pressures less than 30 pounds per square inch (psi), at a rate of approximately two gpm, however, the addition rates and pressures varied based on subsurface conditions. The option to inject to wells under pressure was limited based on nearby utilities, buildings, and the shallow groundwater areas.

4.3 Field Monitoring Data

Groundwater monitoring data from field instruments was collected prior to, during, and following the permanganate additions are included pH, and oxidation reduction potential (ORP).

4.3.1 Oxidation Reduction Potential

Measurements of ORP were used to track the presence and effect of permanganate during the injection. Prior to permanganate additions at the

Site, groundwater generally displayed ORP values from 200 millivolts (mv) to negative values approaching –200 mv. The negative ORP values generally indicate reducing conditions. The oxidizing nature of permanganate however produces a positive ORP and its presence will often result in measurements in excess of 500 mv. As the permanganate is consumed and its concentrations are reduced, ORP values typically return to a lower value.

The ORP readings collected at nearby, non-injection monitoring points during the additions indicated variable results, but generally indicated an increase following the permanganate injections. Before injection was initiated at the OB19 cluster, ORP changes of over 400 mv were observed. The only permanganate injection completed prior to this observed ORP increase was at the two horizontal wells (AP-10 and AP-11), located approximately 200 feet upgradient of the OB19 cluster. This suggests a beneficial downgradient influence in the deep overburden and bedrock aquifer from injection activities. In several wells, the ORP values indicated a discernible return to pre-injection values following cessation of the additions.

4.3.2 pH

Measurements of pH were made to evaluate whether there were any measurable effects following the addition of permanganate and/or any dechlorination processes. Basic chemistry, along with previous studies by Shaw on other sites have revealed that the breakdown of PCE and TCE by permanganate can produce acids (resulting in a lower pH), and that the breakdown of DCE compounds and vinyl chloride consumes acid (resulting in a more basic, higher pH). Typical groundwater monitoring may not detect these changes however since the breakdown of other organics at the site and the buffering capacity of the soils also have an effect on the pH.

A site-wide data review of groundwater pH measurements revealed fluctuating data. However, at a few wells the trends discussed above were observed in the field monitoring data collected at the Site. For example, at well OB12-BR where pre-treatment DCE levels were high and TCE/PCE levels were low, the pH increased approximately two points after injection. Conversely, at well OB14-DO where pre treatment TCE levels were significantly higher than the DCE levels, an approximate two point pH decrease was observed after treatment began. The pH effects remained inconsistent across the site but generally remained near neutral.

4.4 Post Addition Analytical Results

Groundwater samples were collected approximately six months following the cessation of permanganate injections and submitted for laboratory analysis of VOCs. The results indicated total VOC concentrations were significantly reduced in the source area in both the overburden and the bedrock aquifers.

For presentation, isoconcentration maps were prepared to illustrate the distribution of total VOC impacts both prior to, and following the permanganate injections. The isoconcentration contours were generated using groundwater VOC data from all the sampled wells. In compiling the isoconcentration maps, the highest concentration detected within each well is presented where more than one sample was analyzed from a well. The exact location of each isoconcentration contour is interpolated based on the limited number of control points available for contouring, and manual interpretation.

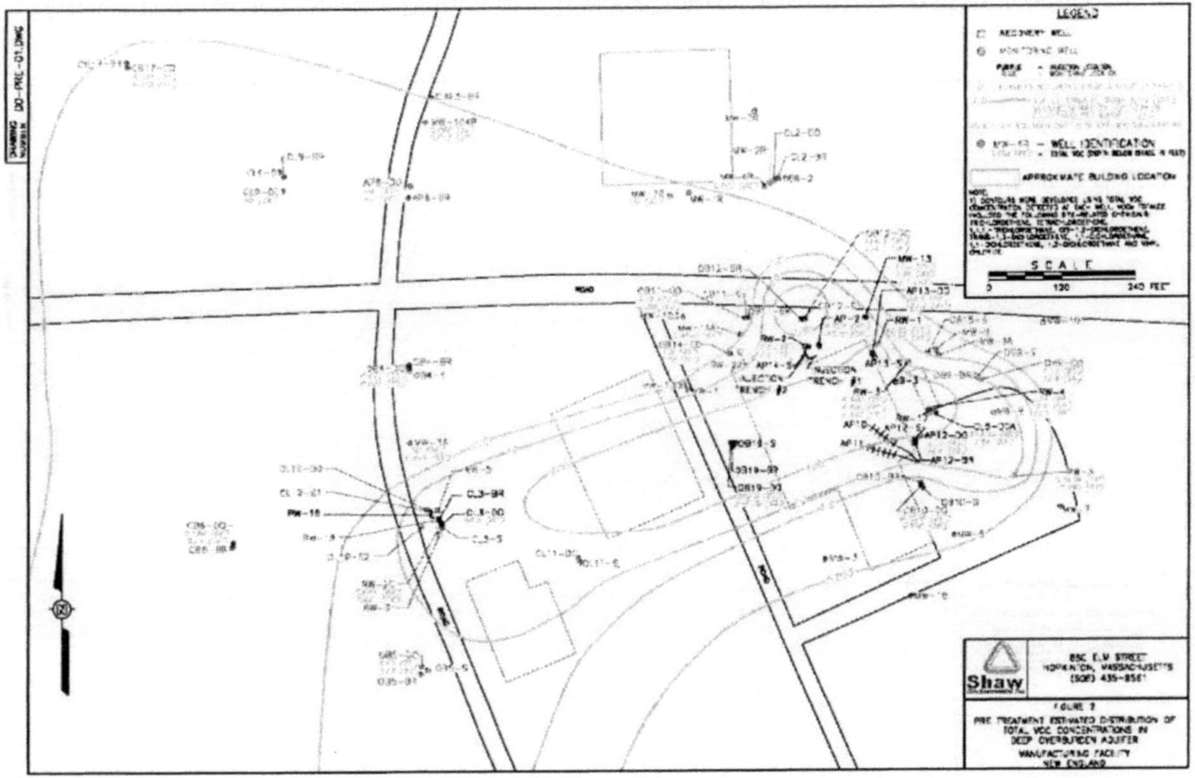

Figure 2. Pre Treatment Estimated Distribution of Total VOC Concentrations in Deep Overburden Aquifer

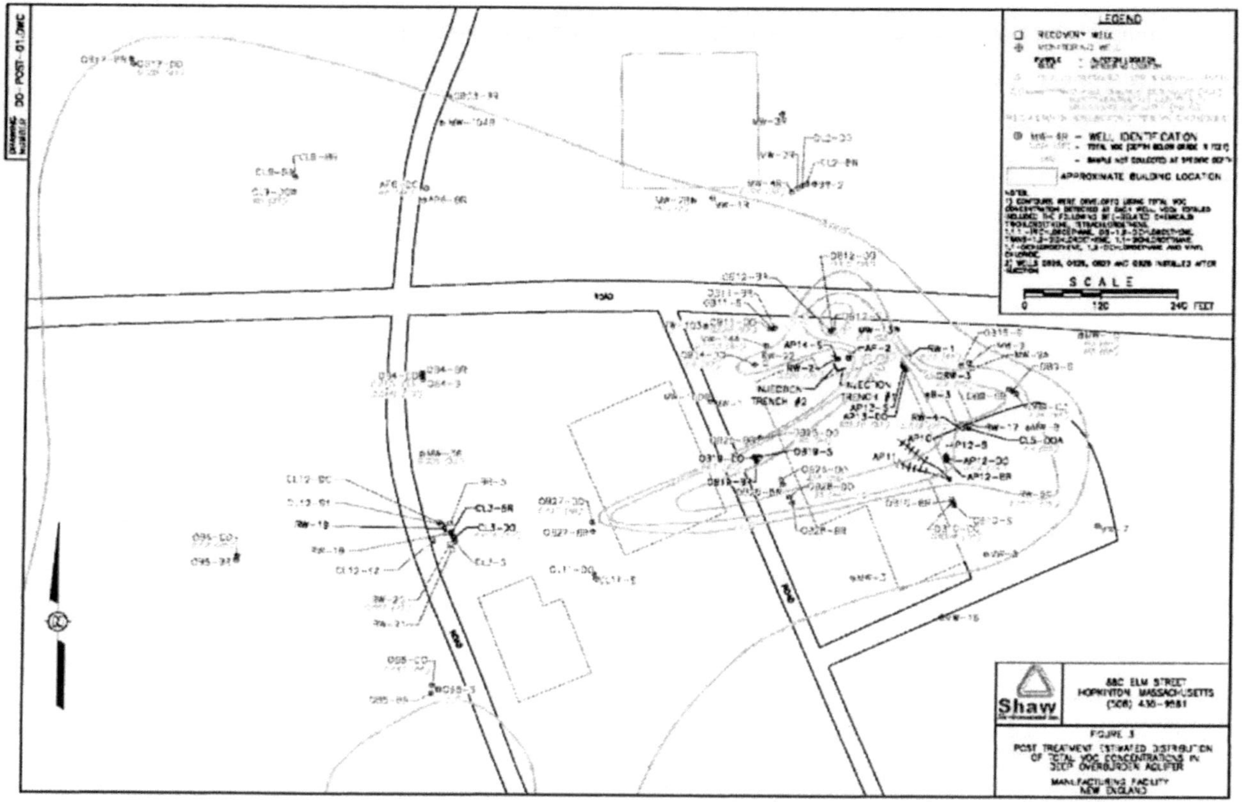

Figure 3. Post Treatment Estimated Distribution of Total VOC Concentrations in Deep Overburden Aquifer

The deep overburden well data shows remarkable response in downgradient injection areas with sand and gravel deposits (Figures 2 and 3). For example, a reduction of approximately 72% was observed at monitoring well OB5-DO and a reduction of 99% was noted at injection well CL3-DO (Table 1). Source areas indicate significant progress towards reduction, generally consistent with the multi-year injection program expectations. Notable is the 99%reduction atRW-2; which was part of the pilot test and has had two years of permanganate treatment. Other significant reductions were seen at injection locations CL5-DOA (approximately 98%) and RW-4 (99%).

Table 1: Comparison of Pre and Post Treatment Total VOC Results

Monitoring Wells (1)	Pre Treatment Total VOCs (2) (mg/L)	Post Treatment Total VOCs (2) (mg/L)	Percent Reduction
Deep Overburden			
OB5-DO	3.05	0.857	71.9%
OB9-DO	13.6	2.76	79.7%
OB12-DO	146	119	18.5%
OB14-DO	1.82	1.8	1.1%
OB19-DO	**240.7**	**84.7**	**64.8%**
MW-13	**13**	**0.8**	**93.8%**
AP12-DO	**49.4**	**41.4**	**16.2%**
AP13-DO	**121.2**	**625.1**	**-415.8%**
CL3-DO	**46.5**	**0.026**	**99.9%**
CL5-DOA	**212.3**	**4.4**	**97.9%**
RW-1	**67.1**	**0.23**	**99.7%**
RW-2	**45.1**	**0.098**	**99.8%**
RW-3	**8.34**	**2.6**	**68.8%**
RW-4	**33.8**	**0.019**	**99.9%**
RW-5	10.09	3.383	66.5%
Bedrock			
OB5-BR	13	4.35	66.5%
OB6-BR	3.3	3.27	0.9%
OB9-BR	15	12.6	16.0%
OB10-BR	2.1	0.735	65.0%
OB12-BR	1.6	20.127	-1157.9%
OB17-BR	1.2	0.782	34.8%
OB19-BR	**65**	**0.018**	**100.0%**
AP12-BR	**60**	**0.019**	**100.0%**
CL3-BR	**49**	**ND**	**100.0%**

CL2-BR	1.6	0.77	51.9%
CL9-BR zone 2	5.3	8.4	-58.5%
CL9-BR zone 3	8.3	3.95	52.4%
BR-5 zone 2	4.7	1.999	57.5%
BR-5 zone 3	4.7	0.76	83.8%
RW-17	**37**	**0.004**	**100.0%**
RW-19	**69**	**0.014**	**100.0%**
RW-21	8.5	0.007	99.9%
RW-22	3.5	2.15	38.6%

Notes:

(1) = includes wells with pretreatment concentration greater than 1 mg/L

(2) = highest concentration detected in well used for comparison

mg/L = milligrams per liter.

ND = non detect

OB19-DO = indicates well used for injection

Overall reductions were more pronounced in the bedrock aquifer, which has post-injection contours much tighter to the source areas (Figures 4 and 5). Notable reduction were observed at monitoring locations OB5-BR (approximately 67%), OB10-BR (65%), BR-5 zone 3 (approximately 84%) and at RW-21 (99%). Reductions of over 99% were noted at all of the bedrock injection locations.

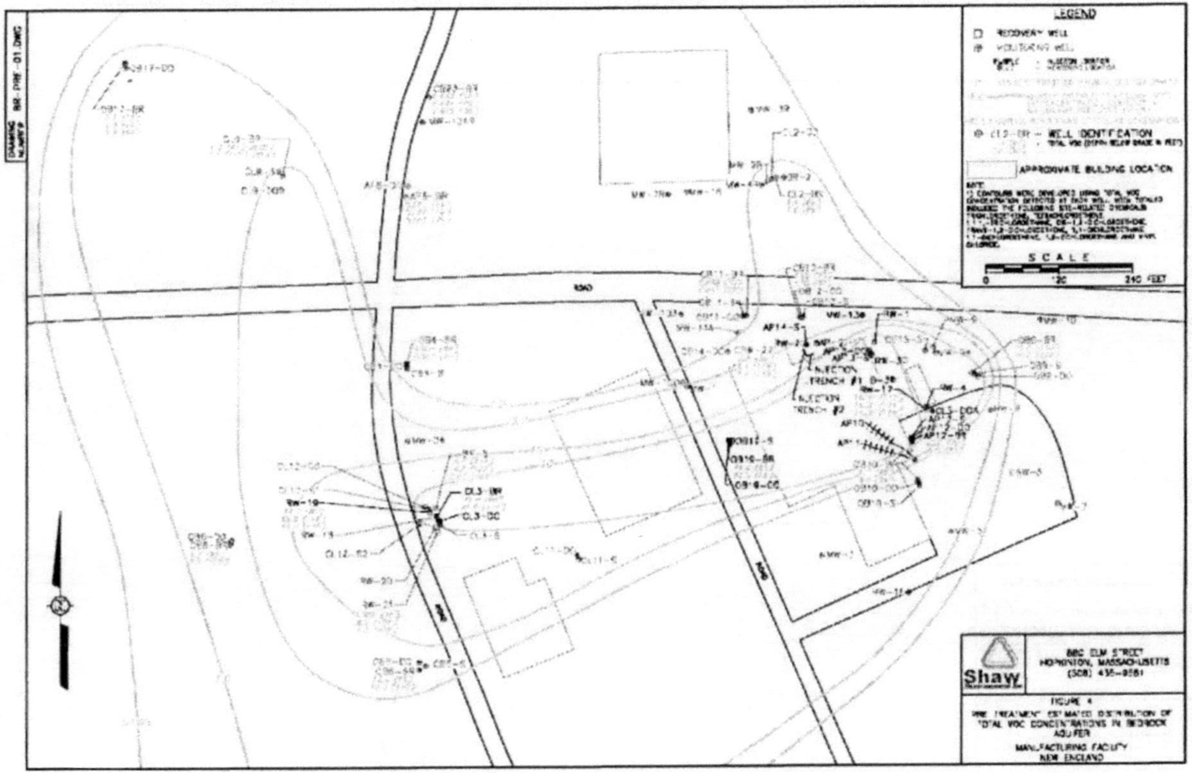

Figure 4. Pre Treatment Estimated Distribution of Total VOC Concentrations in Bedrock Aquifer

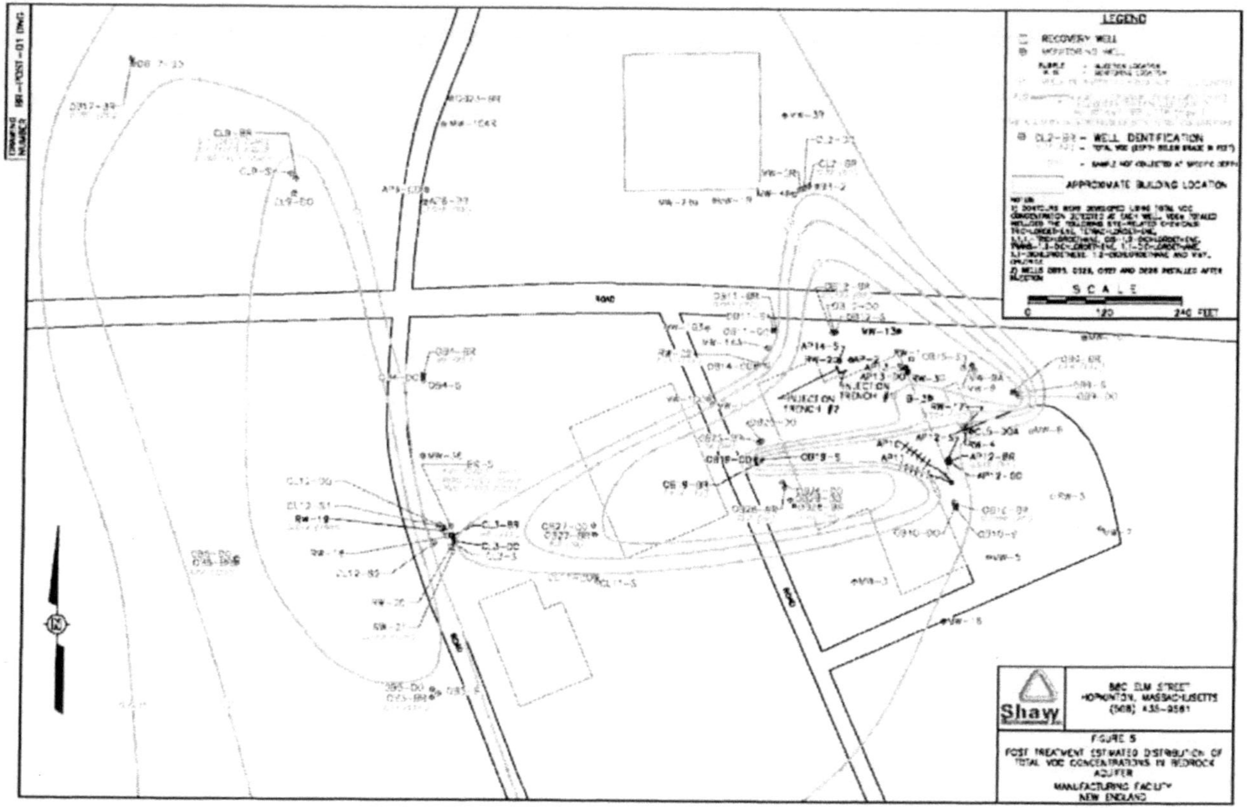

Figure 5. Post Treatment Estimated Distribution of Total VOC Concentrations in Bedrock Aquifer

Increases were observed at a few wells. Generally the increases were minor; however, significant increases were observed in a limited number of wells (Table 1). Specifically, at deep overburden injection well OB13-DO and at bedrock monitoring well OB12-BR. These increases appear to be related to the desorbing of VOC from the soil matrix (as a result of the interaction between permanganate and the natural organic materials) and the shutdown of the former groundwater pumping system. However, where historic data exist, these increased concentrations remain less than historic maximum concentrations from samples collected from this vicinity before and during the operation of the pumping system.

Concentrations of VOC at the outer extent of both the bedrock and deep overburden plumes were not significantly changed during the six month post-injection period.

5. LESSONS LEARNED

The unique scale and duration of this injection program provided some insight useful for future project planning. They are as follows:

Material incompatibilities: Removal of the pump and treat system resulted in a concern with well sampling purge water disposal. Shaw decided to use passive diffusive bag samplers for VOC sample events. The bags exclude the permanganate ion from entering the diffusive bag sampler when sampling the wells with permanganate presence, so there is no need to submit concentrated permanganate groundwater to the laboratory. The bags are deployed for a minimum two-week period for sample equilibrium. Post-injection sampling in certain wells with high residual permanganate was problematic due to the two-week immersion of nylon cord in permanganate. Since the nylon cord broke down too easily, subsequent use of polyethylene cord solved this problem.

Contaminant Desorption: As Shaw has seen at other sites, permanganate appears to have the potential to desorb CVOCs adhering to other organic material in the subsurface by oxidizing the organic content of the soil in the treatment area. The permanganate reaction with target chlorinated solvent compounds is quick, but some of these compounds are less readily oxidizable than others. In addition, some of the target compounds such as cis-1,2-DCE desorb more readily from soil particles. The result can be an increase in some of the compounds within and immediately downgradient of a treatment area. These local increases are treated by residual permanganate in the formation and subsequent treatment events.

Residual permanganate: Monitoring residual permanganate during post-injection monitoring proved to be very valuable. Its presence in non-injection monitoring points may indicate favorable pathways, dispersion, or lack of remaining contaminants. Its absence in injection points indicates permanganate consumption and the potential need for additional injection. Of course this data needs to be considered in conjunction with adjacent well data, groundwater contours and CVOC data.

CHAPTER 12

A COST-EFFECTIVE DECISION: ACCELERATED CLEANUP USING PERMANGANATE

Elizabeth Rasmussen[1], Richard W. Lewis[2], Robert Luhrs[3], Timothy Pac[2], and Ronald C. Slager[1]
[1]Raytheon Company, Sudbury, MA, [2]ERM, Sharon, MA, and [3]Raytheon Company, Lexington, MA

Abstract: Raytheon Company has a former manufacturing facility in Quincy, Massachusetts that had on-going environmental restoration work to treat a release of chlorinated volatile organic compounds (VOCs) to soil and groundwater. The site soil consists of 8 to 10 feet of a granular fill with some peat still present intermittently. Underlying the fill to depths of 40 to 50 feet are interbedded fine sands, silts and clays. Groundwater is shallow at a depth of 2 to 8 feet. The highest VOC groundwater concentrations are generally immediately on top of the silt layer.

Raytheon installed a treatment system at the facility consisting of a groundwater /soil vapor extraction system with air stripping of groundwater and thermal oxidation of stripper and soil vapor off-gasses from 1997-1999. Consistent long-term operation of the thermal oxidizer was expected to be expensive with an anticipated utility bill of $150,000 per year.

Raytheon sold the facility in 2001. The new owner has redeveloped the site as a retail store. Pending imminent site redevelopment, Raytheon considered the use of innovative remedial technologies that would save the company time and money. As a result of this analysis and successful pilot testing, Raytheon used chemical oxidation at the site.

Applications of both sodium and potassium permanganate were selected for four target areas of the site (approximately 1 -acre). Permanganate was added into a network of temporary injection points, completed 12-14 feet below ground surface. A total of 44,680 pounds of permanganate was applied through 97 points during the 10 weeks of addition in 2001.

Post-application monitoring indicates significantly decreased VOC concentrations indicating about an order of magnitude reduction in total VOCs at most locations. Several small programs of treatment for residual areas have resulted in application of an additional 7,000 pounds of permanganate. Raytheon anticipates continued monitoring of the groundwater until site closure can be achieved.

Key words: Permanganate, case study, chlorinated solvents, groundwater, oxidation

1. INTRODUCTION/OVERVIEW

Raytheon Company has a former manufacturing facility in Quincy, Massachusetts that had on-going environmental restoration work to treat a release of chlorinated volatile organic compounds (VOCs) to soil and groundwater. Raytheon operated an industrial manufacturing plant on the property between 1951 and 2001. VOCs were found at the site in five primary areas of concern and include trichloroethylene (TCE), tetrachloroethylene (PCE) and associated daughter products. Remedial actions were initiated at the site in 1997 when Raytheon still owned the site, and consisted of groundwater and soil vapor extraction and treatment coupled with downgradient groundwater containment. The treatment system consisted of a shallow tray air stripper for groundwater treatment and thermal oxidation of stripper and soil off-gasses. Raytheon sold the property in December 2000 to a national retail chain. The sale of the property and pending site redevelopment prompted Raytheon to reevaluate remedial options for the site. The redevelopment of the site included the demolition of the existing structures, allowing accessibility to all areas of the site. Raytheon completed a cost-benefit analysis that considered in-situ chemical oxidation versus long-term operation of the treatment system. The results of the cost-benefit analysis and successful pilot testing of the technology led Raytheon to implement a full-scale chemical oxidation treatment program which is described in more detail herein.

2. SITE DESCRIPTION

The site is located in Quincy, Massachusetts in a commercial/industrial area, northeast of a wetland area. The site is a former low-lying marsh area selected for development in the 1950s. The property is approximately 16 acres of land on six parcels with VOC concentrations in groundwater over about 6 acres of the site. The former Raytheon buildings consisted of two separate structures known as Buildings 2 and 3. Figure 1 shows a site map. After site redevelopment, the property currently contains one primary 120,000 square-foot retail store.

The upper soil consists of 8 to 10 feet of a granular fill with some peat still present intermittently. Underlying the fill to depths of 40 to 50 feet below ground surface (bgs) is a glacial lacustrine deposit of interbedded fine sands, silts and clays. Groundwater is shallow, at a depth of approximately 2 to 8 feet bgs. Wetlands surround the site to the south and east. The highest VOC groundwater concentrations are generally immediately on top of the silt layer. Groundwater is the primary medium of concern at the site, and groundwater flow is in a northeasterly direction toward the front of former Raytheon Building 2.

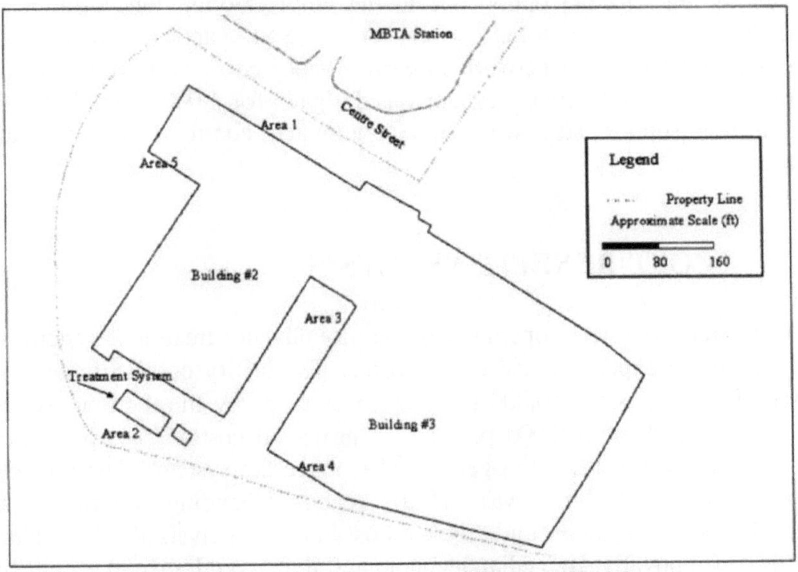

Figure 1: Site Map

Figure 1. Site Map

3. SITE BACKGROUND

The five primary areas of concern for VOCs in groundwater are shown on Figure 1. Area 1 is on the northern side of the property and is not considered a source area for VOCs. Area 2 is on the southwestern corner of the former Raytheon Building 2. Area 3 is found in the former courtyard area between Buildings 2 and 3. Area 4 is near the southwest corner of former Raytheon Building 3. Area 5 is on the northwest corner of former Raytheon Building 2 and was discovered in 2000. The 1998 risk characterization for the site identified average concentrations of PCE of 17,600 ug/L and TCE of 17,200 ug/L in groundwater at the site (GZA, 1998).

From 1997 to 1999 Raytheon installed a groundwater and soil vapor extraction and treatment system. The system consisted of a ShallowTray ™ air stripper with a design groundwater flow of forty-five gallons per minute. Twenty-one groundwater extraction wells were distributed through four known areas of concern to extract contaminated groundwater for conveyance to the treatment system. Off-gasses from the air stripper and contaminated soil vapor were treated by the thermal oxidizer. Thermal oxidation was selected as the remedial option due to the vinyl chloride concentrations in groundwater in some areas of concern. Soil vapor extraction was accomplished through a network of eleven vapor extraction wells and five air inlet wells. The thermal oxidizer was designed for 1,000 scfm flow. The system was sophisticated with some automated controls and a security system.

4. COST-BENEFIT ANALYSIS

Consistent long-term operation of the groundwater treatment system was expected to be expensive and cumbersome. The utility cost estimate alone was projected to be $150,000 per year. Gas usage by the thermal oxidizer accounted for about $80,000 per year in projected costs. These projections did not include the labor to operate and manage the system. The projected system operational time was 15-20 years. Pending imminent site redevelopment, Raytheon undertook a cost-benefit analysis that considered the use of innovative remedial technologies that would save the company time and money. The cost-benefit analysis included an engineering estimate of the current treatment system costs and benefits under the new property ownership structure, compared to the costs and benefits for implementation of a full-scale chemical oxidation program. The cost-benefit analysis included pilot testing of the proposed alternate/supplemental remedy as well as laboratory testing to estimate full-scale treatment costs.

Between December 2000 and May 2001, Raytheon undertook several phases of pilot testing in two areas of concern at the site. The first pilot program included an application of a total of 4,100 pounds of permanganate into five injection wells. Baseline monitoring was conducted in each pilot area to pre-characterize the site conditions in November and December 2000 prior to permanganate addition. Groundwater samples were analyzed for VOCs, chemical oxygen demand (COD), chloride, and selected dissolved metals (copper [Cu], iron [Fe], manganese [Mn], and sodium [Na]). Liquid sodium permanganate was applied into several points, and wells surrounding the application point were monitored at various frequencies for 12 weeks. Injection rates achieved in the field during the pilot program were less than 1.5 gallons per minute per point. The radii of influence (ROI) achieved in the short-term program were variable from about ten to twenty linear feet.

A second pilot application was conducted to test the addition of a small quantity of sodium permanganate (400 pounds of oxidant) into a second area of the site. Injections in this area utilized a small diameter (¾-inch) driven well point and a drilled monitoring well. Composite soils were prepared and analyzed for soil oxidant demand (SOD) using a field test kit. The results of the second application indicated an approximate 15-foot radius of influence and confirmed that an addition into small-diameter points was a viable technique for distributing permanganate at the site.

As part of the cost-benefit analysis, samples were collected to estimate the SOD. SOD measures the amount of oxidant required to oxidize all available compounds within the geologic matrix where the oxidant is applied. Non-contaminant compounds may exert a demand for oxidant that is greater than the stoichiometric demand for the contaminants. Table 1 shows the results of the SOD testing for the Quincy site. Results of the SOD testing indicated that a full-scale program would be cost-effective and that permanganate costs would not be prohibitively high.

Table 1: SOD Sampling Results

Area of Concern	Soil Oxidant Demand (SOD) (pounds/cubic yard)
Area 2	21-36
Area 3	10.5
Area 5	4.5-10.5

In summary, pilot-testing application activities indicated that permanganate significantly reduced VOCs in local groundwater within treatment areas. The cost-benefit analysis concluded that chemical oxidation could potentially be effective as a full-scale remedy at the site.

5. CHEMICAL OXIDATION PROGRAM

5.1 Pre-Characterization Program

Based on the results of the cost-benefit analysis, chemical oxidation activities were expanded to a full-scale application program. Raytheon undertook an extensive pre-characterization design program starting in June 2001. The purpose of the program was to gather enough information regarding contaminant distribution to design a full-scale program. This program included installation of 114 carbon steel injection/monitoring wells from June to September 2001. Results from the pilot testing had indicated that temporary points could be used for both monitoring and permanganate injection.

Injection points were ½-inch inner diameter carbon steel points installed by direct push drilling methods using both an ATV-mounted and dolly-mounted hydraulic hammer drill. The wells were generally installed to depths of about 12 to 14 feet bgs with a 5-foot well screen. Figure 2 shows the location of the injection/monitoring points, generally within Areas 2, 3, 4 and 5. In conjunction with a site-wide sampling event, groundwater samples were collected from each of the newly installed wells to provide an extensive baseline sampling program. Figure 3 shows an interpretation of the baseline groundwater sampling results from 2001. The baseline sampling program indicated total levels of VOCs in groundwater at the site from about 1-100 mg/l in areas of concern in 2001.

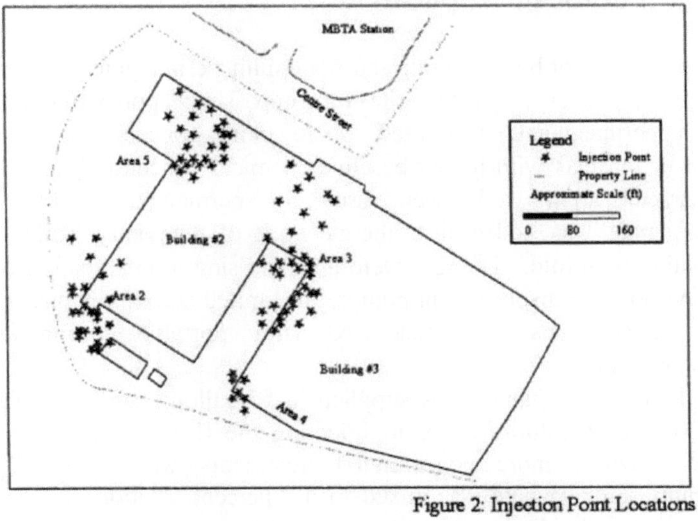

Figure 2. Injection Point Locations

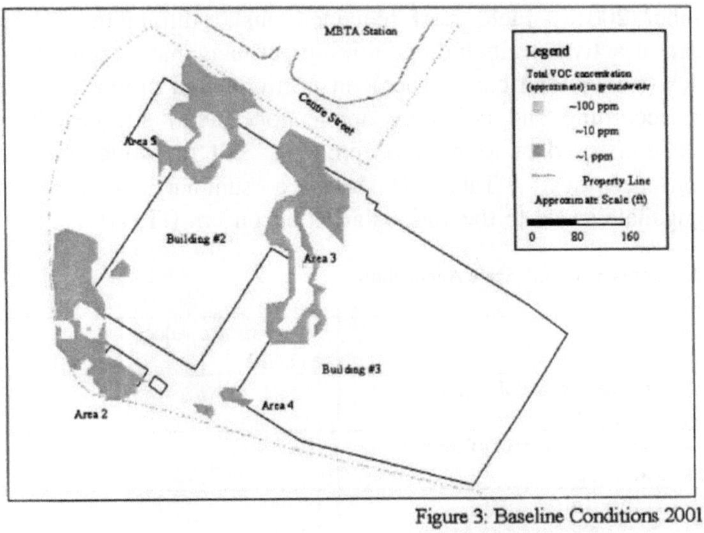

Figure 3. Baseline Conditions

5.2 Full-Scale Application

Applications of both sodium and potassium permanganate were selected for four target areas of the site (encompassing approximately 1-acre). Sodium permanganate was used where more concentrated dosing was needed or in areas with higher baseline chemical concentrations (IT, 2001). Pre-characterization wells were used for permanganate injection and permanganate was added into the network of temporary points using a distribution manifold. Dosages were applied using a combination of gravity feed and positive displacement pumps. In limited access, remote, or specific locations, additions were conducted using portable equipment and/or temporary piping.

Sodium permanganate was supplied in 55-gallon drums and was mixed to a 20-percent solution prior to injecting the fluid directly into selected locations where more concentrated treatments were required. Solid potassium permanganate was mixed to a 2-percent solution on site and was conveyed to individual application points through a series of pvc pipes that were connected above-ground to the various treatment areas. In Area 3, permanganate was also added through a shallow trench. A total of 44,680 pounds of permanganate (about 250,000 gallons of fluid) was applied through 97 points during the 10 weeks of addition from September to November 2001. Field staff remained onsite throughout the full-scale application activities, including: mixing permanganate; coordination of site logistics (storage and conveyance); inspection of all piping and application points; recording and adjusting application rate(s) as needed; collecting water chemistry data to monitor progress; and real-time optimization of addition activities. Table 2 shows a summary of the pounds of permanganate added to the various areas of concern (IT, 2001).

Table 2. Summary of Full-Scale Application

Application Area	Permanganate Added (pounds)
Area 5 – Northwest corner of former Raytheon Building 2	11,920
Area 2 – Southwestern corner of former Raytheon Building 2	10,970
Area 3 – Courtyard area Area 3D – North-Northwest extension of courtyard area	13,690 7,430
Area 4 – Southwest corner of former Raytheon Building 3	670
Subtotal Full-Scale Program 2001	44,680

5.3 Construction Considerations

The construction aspects of the full-scale implementation program were coordinated with the site redevelopment team. Scheduling and logistical considerations were important to the overall success of the program. Site redevelopment included utility construction, building demolition and building construction. The permanganate injections were coordinated with the site redevelopment schedule to minimize costs and increase the long-term benefit to the site of the full-scale program. Staging of the injection program with phasing of work in various areas of concern allowed the site redevelopment to continue unimpeded by the injection program.

6. POST-APPLICATION MONITORING

Periodic groundwater-monitoring events were conducted post-application from 2001 to 2003. These events included groundwater sampling in October/November 2001, December 2001, January 2002, March 2002, June 2002, November 2002, and June 2003. Groundwater monitoring events included selected field parameters from key wells in the vicinity of injection locations including pH, dissolved oxygen (DO), specific conductance, oxidation-reduction potential (ORP), temperature and color. Depth to water was also monitored during the program. Analytical measurements included VOC analysis by EPA Methods 8021B, 8260 and Gas Chromatograph (GC) Screen methods. When permanganate was still present in groundwater samples, quenching techniques were often employed prior to submitting samples for laboratory analysis. Quenching was accomplished through the use of sodium thiosulfate. Where applicable, following the permanganate application, groundwater samples indicating the presence of permanganate were submitted for colorimetric analysis of permanganate concentration. Raytheon has found the quantification of residual permanganate levels to be especially useful in evaluating the amount of permanganate remaining in the treatment area.

Residual areas identified during post-application monitoring have been selected for additional treatment using chemical oxidation. Several small programs of treatment for residual areas have resulted in application of an additional 7,000 pounds of permanganate through various application methods. Figure 4 shows an aerial photograph taken during a supplemental injection program from April 2002. This particular program consisted of a "test pit" injection program in two areas of concern at the site. In conjunction with the test pit permanganate application, this program also involved limited soil removal.

Figure 4. Test-Pit Application of Sodium Permanganate

7. DISCUSSION OF RESULTS

Figure 5 shows the post-application total VOC concentrations at the site measured in November 2002 at all accessible locations. Comparison of Figures 3 and 5 illustrates the decrease in measurable concentrations of total VOCs between the 2001 and 2002. The 2003 risk assessment for the site identifies the mean concentration of PCE as 153 ug/L compared to 17,600 ug/L in 1998. Also, the mean concentration of TCE is identified in the 2003 risk assessment as 513 ug/L compared to 17,200 ug/L in 1998 (GZA, 2003). (The data sets considered to develop site averages and means are hybrid sets due to the site redevelopment, well destruction and data quality issues, etc.) Table 3 provides data from select key monitoring wells at the site and includes both baseline data from June to July 2001 and recent data from June 2003.

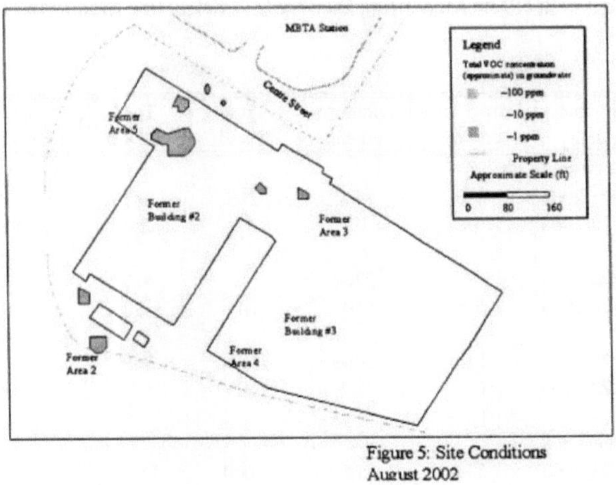

Figure 5. Site Conditions, August 2002

Table 3. Select Site Monitoring Data*Table 3.* Select Site Monitoring Data

Location	Sample Date	Tetrachloro-ethene	Trichloro-ethene	Vinyl chloride	Total VOCs	Percent Decrease
Select Locations Area 1						
B-405S	6/5/2001	810	4500	210	8835	
	5/22/2003	19	71	13	258	97
GZ-801	6/5/2001	45	2700	94	4239	
	5/22/2003	7.5	10	ND	30	99
EW-102	6/20/2001	110	2400	ND	3490	
	5/22/2003	ND	5.7	ND	7	100
Select Locations Area 2						
EW-202	7/23/2003	530	75	5000	13605	
	5/27/2003	19	11	2.5	513	96
EW-205	6/5/2001	4600	3700	1600	13600	
	5/23/2003	ND	2.1	ND	15	100
SV-211	6/23/2001	2	29	13	77	
	5/23/2003	ND	ND	28	128	-66
Select Location Area 4						
B-104	6/5/2001	ND	4900	ND	6610	
	5/27/2003	2.5	45	ND	70	99

Notes:
Reporting Units are ug/L
Laboratory Method used was 8021B
ND = Analyte not detected above reporting limits

8. CONCLUSIONS

To date, Raytheon estimates that savings of about $2.25 million dollars have been achieved due to the extensive mass removal successfully completed over a short duration from 2001 to the present. The need to run the groundwater and soil vapor extraction system as a primary remedial solution has been eliminated by the mass reduction achieved through the aggressive oxidation program.

Implementation of this technology resulted in removal of significant mass of VOCs at the site within a short time frame. Post-application monitoring results indicate decreases in total VOC concentrations at the site of about one to two orders of magnitude in many locations. Raytheon anticipates continued monitoring of the groundwater until site closure can be achieved.

REFERENCES

ERM, Post-Application VOC Concentrations, Raytheon Systems Company, 465 Centre Street, Quincy, Massachusetts, August 2002.

IT Corporation, "Remedial Work Plan, Raytheon Systems Company, 465 Centre Street, Quincy, Massachusetts," September 10, 2001.

IT Corporation, "Remedial Status Report, Raytheon Systems Company, 465 Centre Street, Quincy, Massachusetts," January 15, 2001.

IT Corporation, Baseline VOC Concentrations, Raytheon Systems Company, 465 Centre Street, Quincy, Massachusetts, June -July 2001.

GZA GeoEnvironmental, Inc. " Release Abatement Measure Completion Report, Modified Response Action Outcome and Periodic Evaluation RTN 3 -0872, Former Raytheon Quincy Site, 465 Centre Street, Quincy, Massachusetts," September 2003.

GZA GeoEnvironmental, Inc. "Response Action Outcome RTN 3 -0872, Former Raytheon Quincy Site, 465 Centre Street, Quincy, Massachusetts," September 1998

This page intentionally left blank

PART III: HEAVY METALS

CHAPTER 13

HEAVY METAL CONTENT OF SOILS IN THE KARSTIC AREA OF NORTH HUNGARY

Rita Kaszala; Prof. Dr. Ilona Barany-Kevei
University of Szeged, Szeged, Hungary

Abstract: Research on Aggtelek karsts includes examinations of physical and chemical parameters of the soils, especially the acid soluble and the extractable (with EDTA solution) heavy metal content. The assessment of anthropogenic influent has come to the forefront during the last decades. Research of heavy metal contamination has become increasingly more significant among investigations. The article presents the relationship between two forms of heavy metal content in the karstic soil .

Key words: acid soluble heavy metal content

1. INTRODUCTION

Karst areas have a specific karstecology system and the soil plays an important role in that system in the relationship of the clime-host rock and the greenery. Knowing this, the state of the soil is indispensable.

The interferences in the carbonated rocks and water result in karst development and karstic frame at the surface and close to the surface. The procession takes place mostly at the surface of the rocks and the aggregation soil of the burst, which in turn effects the shallow soil layer. This process involves the properties of the soil claim on the process of the karst development and settles the whole dynamic of the run.

1.1 Property of the sampled area

The area is 10 sq km, the catchment basin of the Béke-cave. It is situated in Middle Europe in the Northern part of Hungary (see map 1), in the Southwestern part of the National Park of Aggtelek. In the northern part of the area Trias limestone comes up to the surface, for this reason it is called the uncovered karstic area. Going south, this Trias limestone goes deeper, and on this rock, sediment from the age of pannon settles. We can observe this duality on the developed soil at the parent material. In the northern part of the area one finds mainly reddish-tone soils remaining, which are rich in clay minerals, and brown forest soils (these soil sample numbers are lower than 30). On the other hand in the covered karstic area we find bright, yellowish-brown colored soils, which contain loam and sand (like terra fusca): soil samples which come from this part of the area have numbers higher than 30.

The greenery on the open karstic area is hornbeam-oak forest mixed with beech (Carpinus Quercetum Fagetosum) with scattered undergrowth. In some places, there is the more preferable warm cornel-oak forest (Como Quercetum Pubescenti Petrea). In the treeless places, juniper (Juniperus Communis) and blackthorn build up impenetrable scrub. Most of the southern terrain is covered with slope steppe-grassland (Salvio festucetum rupicole).

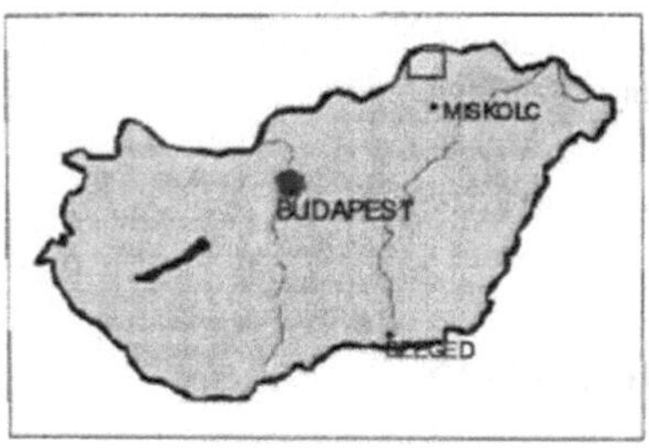

Map 1. Map of Hungary, with Aggtelek Mountain

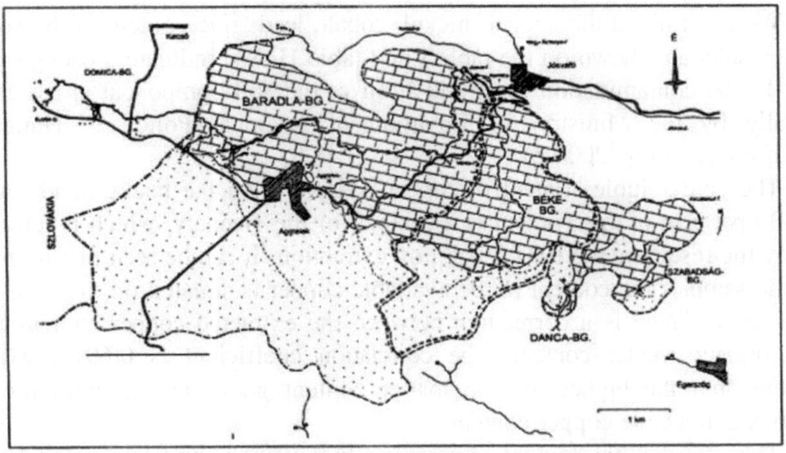

Map 2. The geological Map of Aggtelek Mountain

2. METHODS

Soil samples were collected during the summer of 2002. These samples came from two depths: one from the surface (0-10 cm) and the other from 20-30 cm deep.

Determination of the pH value took place according to the pedological practice (2,5:1 range of distilled water and soil, and with a 24-hour wait before measuring). The organic matter-content was calculated from the loss of the mass after heating (700 °C). The acid-soluble heavy metal content was measured by ICP-OES techniques after digestion with an acid mixture **(HNO_3-H_2O_2-$HClO_4$)**. The complexed metal content of the soils was determined by ICP-OES after extraction with 0,02M concentrate EDTA solution which has pH = 4,67±0,05. (Rowell, D.L. 1994). The measurements took place in the University of Veszprém.

3. RESULTS

The heavy metal content of limestone is not too high. According to (Merian E. 1984) the average metal content of limestone is: Cu: 4ppm, Co: 2ppm, Ni: 15ppm, Zn: 23ppm. In accordance with Kabata -Pendias H. (1984) Cu: 2-10ppm, Co: 0,1-30, Ni: 7-20ppm.

We determined the copper, nickel, cobalt, lead, zinc content of the soils. The results are shown on the table I and table II. We indicate a background level. The contamination threshold limit of the soil component is assigned legally by the Ministry for Environment Regional Policy of Hungary (Magyar Közlöny, 2000 b).

The acid-soluble content of copper is lower than the background level. 20-30 per cent of this content is available for the greenery, which is referred to by the results of the EDTA solution extraction. It can be seen that in most of the samples the content of the available copper is situated on the 0-10 cm soil layer. There is a correlation between the extracted copper content and the organic matter content. The correlation coefficient is 0,6911, which means that the higher organic matter content goes together with higher EDTA extractable copper content.

The acid-soluble content of nickel is higher than the background level. This concentration is higher in the northern part in 20-30 cm layer, while in the southern part it is higher in 0-10 cm layer. The extractable nickel content is higher in the 0-10 cm soil layer. The extractable nickel concentration is about 4-6 percent of the acid soluble nickel concentration but in some samples it goes up to 12-14 percent.

Cobalt concentration is about the background level. The acid soluble cobalt concentration is higher in layer of 0-10 cm. However the extractable concentration is higher in the lower layer. The average extractable cobalt content is 30-40 percent.

In the case of lead, the acid-soluble metal content is near the contamination threshold limit, and in some samples it exceeds this limit. 15-30 percent of the acid-soluble metal content is extractable, so it is a potential danger source for the vegetation. (We would like to prove this statement with further analysis of the greenery).

The concentration of the acid-soluble form of zinc is near the background. The extractable part is very low: 5-7 per cent in layer of 0-10cm, and 2-4 per cent in layer 20-30cm. In case of zinc, there is a correlation between the organic matter content and the extractable metal concentration in the layer of 20-30 cm.

Table 1. The heavy metal content of soils from 0-10 cm depth, pH value and the organic matter content

Cu Acid Soluble 0-10cm (ppm)	Cu EDTA extraction 0-10cm (ppm)	Ni Acid soluble 0-10cm (ppm)	Ni EDTA extraction 0-10cm (ppm)	Co Acid soluble 0-10cm (ppm)	Co EDTA extraction 0-10cm (ppm)	Pb Acid soluble 0-10cm (ppm)	Pb EDTA extraction 0-10com (ppm)	Zn Acid soluble 0-10cm (ppm)	Zn EDTA extraction 0-10cm (ppm)	pH (H20)	Organice matter content %
25,74	4,15	45,41	2,02	17,97	4,55	62,67	14,50	91,68	4,64	6,08	15,30
25,71	4,41	45,91	2,05	13,87	3,84	46,83	12,16	92,76	6,20	6,35	16,60
25,62	5,55	36,49	3,70	13,15	6,73	68,08	7,38	88,88	56,11	6,57	12,00
19,26	3,83	37,50	1,83	14,47	5,19	58,47	12,63	99,59	6,49	6,90	12,30
19,44	4,99	33,34	2,07	18,18	5,30	63,57	21,33	103,81	3,07	6,88	10,40
19,31	3,90	28,02	1,38	9,85	3,48	65,76	14,22	93,41	1,90	5,87	11,80
16,50	3,30	31,82	1,91	9,52	3,27	71,87	17,03	105,10	4,31	5,33	12,90
11,68	3,78	29,89	0,98	14,87	4,16	135,14	37,96	110,57	8,16	5,76	14,70
11,91	4,66	33,16	3,60	16,37	7,24	65,07	21,61	80,07	6,23	5,56	12,50
8,20	4,07	21,40	2,17	11,21	4,73	46,72	24,03	76,07	8,63	4,50	11,20
8,22	4,19	25,02	3,88	16,50	6,25	37,89	19,69	70,15	6,13	5,60	9,60
18,12	3,94	35,74	1,90	11,42	4,62	83,23	19,89	98,34	4,78	5,41	12,20
17,92	5,93	38,53	3,17	7,64	4,68	114,52	24,19	100,63	7,13	5,90	22,30
12,50	4,21	14,64	0,04	8,88	4,45	65,41	19,53	42,98	3,37	5,66	8,80
15,35	4,84	32,88	5,12	17,22	6,67	69,11	20,90	78,57	6,37	5,69	6,70
9,62	3,08	11,45	1,39	8,64	1,86	30,23	11,27	25,83	1,11	5,18	6,80
7,18	2,26	2,02	1,59	1,12	0,84	38,34	11,28	38,31	2,52	4,70	7,60
17,32	2,66	20,52	0,81	1,59	2,89	41,22	5,27	42,50	1,44	5,60	8,40

Cu Acid Soluble 0-10cm (ppm)	Cu EDTA extraction 0-10cm (ppm)	Ni Acid soluble 0-10cm (ppm)	Ni EDTA extraction 0-10cm (ppm)	Co Acid soluble 0-10cm (ppm)	Co EDTA extraction 0-10cm (ppm)	Pb Acid soluble 0-10cm (ppm)	Pb EDTA extraction 0-10com (ppm)	Zn Acid soluble 0-10cm (ppm)	Zn EDTA extraction 0-10cm (ppm)	pH (H20)	Organice matter content %
30,00		25,00		15,00		30,00		100,00			
100,00		40,00		30,00		75,00		200,00			

Table 2. The heavy metal content of soils from 20-30 cm depth, pH value and the organic matter content

Number of the sample places	Cu Acid soluble 20-30cm (ppm)	Cu EDTA extraction 20-30 cm (ppm)	Ni Acid soluble 20-30cm (ppm)	Ni EDTA extraction 20-30 cm (ppm)	Co Acid soluble 20-30cm (ppm)	Co EDTA extraction 20-30 cm (ppm)	Pb Acid soluble 20-30cm (ppm)	Pb EDTA extraction 20-30 cm (ppm)	Zn Acid soluble 20-30cm (ppm)	Zn EDTA extraction 20-30 cm (ppm)	pH(H2O)	organic matter content %
1	28,42	3,64	44,23	1,66	16,13	3,94	61,37	9,70	101,67	1,14	6,51	13,90
2	25,78	5,23	43,15	1,68	10,95	3,79	57,73	9,72	75,54	9,83	6,57	18,10
3	25,46	5,54	38,64	3,75	16,44	7,32	63,22	16,43	85,15	3,51	6,68	10,20
4	18,91	3,89	38,95	1,43	15,09	5,13	63,89	9,50	98,54	1,82	6,81	9,60
5	19,77	3,52	37,56	1,36	19,03	6,05	92,20	17,69	111,14	1,41	6,91	9,20
6	16,51	3,71	28,31	1,51	10,73	4,78	57,20	15,11	83,39	6,37	6,00	7,40
15	16,77	2,82	36,11	0,50	31,25	2,81	68,34	13,67	96,65	1,26	5,60	9,98
19	13,31	3,87	37,06	0,00	14,51	4,75	119,36	31,98	114,02	0,39	5,33	9,40
21	13,73	5,00	40,60	0,57	14,32	9,71	60,08	13,80	81,52	0,00	5,80	8,20
22	10,90	4,21	29,28	2,22	18,08	6,51	39,83	14,83	81,13	1,15	4,73	6,70
23	11,00	3,99	22,77	1,00	13,46	4,58	24,52	11,66	69,34	1,46	5,45	6,00
24	12,30	3,56	38,27	0,74	12,57	4,24	72,50	14,69	93,87	1,20	6,01	9,40
25	23,66	6,66	42,16	1,76	8,73	4,62	100,44	21,28	109,46	5,49	6,60	18,80
32	9,71	2,44	14,41	0,76	8,72	3,59	68,71	12,37	36,91	0,78	5,58	4,90
34	15,31	3,24	32,76	4,16	17,31	6,47	47,82	15,81	66,93	2,67	5,95	9,80
35	7,23	2,60	7,77	0,80	8,80	1,94	23,85	6,26	25,66	0,69	5,02	1,20
36	7,20	3,37	1,88	0,88	0,89	1,04	18,25	3,51	26,16	0,77	4,96	3,70
39	15,05	2,51	18,58	0,70	1,05	2,57	37,65	5,09	35,18	0,92	5,83	6,85
background level	30,00		25,00		15,00		30,00		100,00			
threshold limit	100,00		40,00		30,00		75,00		200,00			

4. CONCLUSION

The examined area is presently a protected area (National Park). That is the reason why the soils are not contaminated by heavy metals. In some samples we find that the concentration values are above the contamination threshold limit (e.g. Pb).

Those metals which have high concentrations of the extractable form also present potential dangers for the greenery (Co, Pb).

We have observed that all of the examined metals have lower concentrations in the southern part soils. This condition comes from the different structures of the soils, the different organic matter and clay mineral content. In the future, heavy metal contamination can mean a potential danger in the whole area. For that reason we need further and regular investigations to generate information about the heavy metal content of the soils and the greenery, and to make observations about the relationship between them.

REFERENCES

A felsz n alatti vizek min seget erint tevekenysegekkel sszefugg egyes feladatokr l sz l 33/2000 (II. 17.) Kormányrendelet. – Magyar Közlöny 2000/23. szám, p. 1078 -1098

A felsz n alatti v z es a f ldtani k zeg min segi vedelmehez szukseges hatarertekekr l sz l 10/2000. (VI.2.) KöM -EüM-FVM-KHVM együttes rendelete. – Magyar Közlöny 2000/53. szám, p. 3156-3467.

Füleky Gy. (1999): Tápanyag-gazdálkodás - Mez gazda Kiad p. 70-80

Kabata-Pendias A. –Pendias H. (1984).: Trace elements in soil and plants. – CRC Press, Boca Raton, p. 315

Kádár I. (1995): Környezet- és Természetvédelmi Kutatások. - MTA Agrokémiai és Talajtani Kutató Intézet, Budapest, p. 131 -132

Rowell, D.L. (1994): Soil Science: Methods and Applications. – Longmann Group UK Limited, p. 350.

Talaj és agrokémiai vizsgálati módszerkönyv2. A talajok fizikai -kémiai és kémiai vizsgálati módszerei. – (szerk. BUZAS I.), Mez gazdasagi Kiad Budapest, 1988. p. 243

CHAPTER 14

MITIGATION OF ACID MINE DRAINAGE AND OTHER CONTAMINATION IN THE PRINCE WILLIAM FOREST NATIONAL PARK IN NORTHERN VIRGINIA

Michael Komalowski, Douglas Mose and George Mushrush
George Mason University, Chemistry Department, Fairfax, VA

1. INTRODUCTION

Quantico Creek is located in Prince William Forest Park in the southern end of Prince William County, Virginia. The creek meanders through 26.5 square kilometers of parkland. Quantico Creek and its tributaries have a horizontal "Y" shape, flowing eastward. The southwest tributary of Quantico Creek is called South Fork and the northwest tributary is called North Fork of Quantico Creek. North Fork and South Fork merge near the eastern margin of the Park. This study is a comparison between previous and new data on metal concentrations in the river sediment to determine metal concentration trends. The metals analyzed in this study include cadmium, copper, lead and zinc. These metals were selected because the United States Environmental Protection Agency (USEPA) has determined that these metals can cause serious damage to the environment.

The USEPA, in conjunction with the Agency for Toxic Substances and Disease Registry and the Center for Disease Control has established an Internet site (www.atsdr.cdc.gov) which provides useful information on many substances, including the chemical elements which are used in this study.

Cadmium is an element found in all soil, but in higher concentrations in areas where ore deposits of copper, lead, and zinc exist. Cadmium is found as a compound, such as cadmium oxide, cadmium sulfate, and cadmium chloride. Cadmium is odorless and tasteless, and hazardous to human health because it can accumulate in the body, and damage the lungs, cause kidney disease and irritate the digestive tract. The Department of Health and Human Services has determined that cadmium and cadmium compounds may reasonably be anticipated to be carcinogens. Animals, which have been exposed to low levels of this metal, have shown signs of high blood pressure, iron-poor blood, liver disease, and nerve or brain damage. The USEPA allowable limit for cadmium in drinking water is 5 parts per billion (ppb).

Copper compounds adhere to soil particles and only wash into streams under acidic conditions, but remain in soil or water for a long time. The most common compound is copper sulfate, but many compounds exist naturally. Copper is a naturally occurring substance and is odorless. The average level of copper in rivers and streams is 4 ppb. For humans, copper is an essential dietary element, but too much can be dangerous. It can cause headaches, dizziness, vomiting, nausea, diarrhea, and stomach cramps. High levels of copper can cause liver and kidney damage, and even death. The USEPA allowable limit for copper in drinking water is 1.3 parts per million (ppm).

Lead compounds in soil wash into streams under acidic conditions, but remain in the soil or water for a long time. Lead is odorless and tasteless, and can affect almost every organ and system in the body. The most sensitive area is the central nervous system, particularly in children. Lead can also damage kidneys and the immune system. Exposure to lead is more dangerous for young and unborn children. Harmful effects include premature births, smaller babies, decreased mental ability in infants, learning difficulties, and reduced growth in young children. In adults, lead may decrease reaction time, cause weakness in fingers, wrists, or ankles, possible memory loss, anemia, abortions and damage to male reproductive systems. The USEPA allowable limit for lead in drinking water is 15 ppb.

Zinc compounds, such as zinc chloride, zinc oxide, zinc sulfate, and zinc sulfide, adhere to soil particles, wash into streams under acidic conditions, and remain in the soil or water for a long time. Zinc accumulates in fish and other organisms, but not in plants. Zinc is a naturally occurring substance and is odorless. For humans, zinc is an essential dietary element, but too much can be dangerous. It can cause anemia, pancreas damage, and lower levels of high-density lipoprotein cholesterol. In animals, zinc has been found to cause infertility or smaller offspring. The USEPA allowable limit for zinc in drinking water is 5 ppm.

Several studies show that excess cadmium, copper, lead and zinc are present in Quantico Creek. In fact, three major problems areas have been cited in Prince William National Forest Park. The first is in Quantico Creek, on the east side of the Park, where earlier studies found metal pollution from an abandoned pyrite mining area. The second problem is in Quantico Creek, along the west side of the Park, which had metal pollution from the adjacent Quantico Marine Corps Base. Subsequent reports concluded that these two pollution problems have been remediated, but the permanence of the remediations is unsure. The third pollution problem is runoff from roadways and parking areas. Just north of the park and within the Quantico Creek watershed, two single-family and one townhouse communities have recently been constructed which could add runoff and metal pollution to North Fork and Quantico Creek. Similarly, development inside the Park (roads, parking lots, campsites, etc.) could add runoff and metal pollution.

1.1 Pyrite Mining

Prince William Forest National Park lies in a geological region where the Piedmont and Coastal Plains provinces meet (Hopkins, 1984). The relief of the area topography is about one-hundred meters, and is characterized by narrow ridges and steep-sided valleys, mainly along Quantico Creek, the primary waterway in the Park. Since these ridges and valleys were formed by surficial water movement, they trend west to east through the Park.

The western three-fourths of the Park is in the Piedmont Province and is underlain by metamorphic rocks ranging in age from Precambrian and Early Cambrian to Late Ordovician (i.e., they formed by geological events during the interval between about 700 and 450 million years ago). The rocks, which include gneiss, greenstone, phyllite, schist and slate, have been folded and faulted and the layering they exhibit dips nearly vertically. Outcrops are found along the streambeds of South Fork Quantico Creek and some of its tributaries. Along the ridge tops and valley walls, the rocks are overlain by saprolite (thoroughly decomposed rock formed in place by chemical weathering of bedrock) ranging in thickness from less than 5 feet to more than 150 feet. In the eastern third of the Park, younger Coastal Plain sediments consisting of sand, clay, silt and gravel of Cretaceous age (about 50-30 million years old) unconformably overlie the saprolite and bedrock, and while generally thin or absent, they are up to about 150 feet thick in the eastern past of the Park, filling an ancient river valley that formed prior to the deposition of the Coastal Plain sediments. Locally, recent (up to about 5000 years old) alluvial deposits are found along some of the streams throughout the Park (Hopkins, 1984).

In 1889 a pyrite mine was opened along North Fork, just upstream of its confluence with South Fork, near the eastern park boundary. Iron sulfide (pyrite) was the primary ore, but other metal compounds were mined. These included mainly copper-iron sulfide (chalcopyrite), zinc sulfide (sphalerite), gold, and lead sulfide (galena). In these mining operations, sulfuric acid was used to wash ore removed from the ground. The sulfuric acid wash, rich in dissolved metals, was allowed to soak into the soil outside of the mining operation.

Until 1994, the mine tailings were exposed to precipitation that washed residual acid and metals into the Quantico Creek. A reclamation project on the mine site was initiated in 1994. This included the removal of many tons of contaminated topsoil from and around the tailing piles, the establishment of runoff control ponds, laying down of caustic soda to neutralize the acidic soil, the grading of the area to reduce topographic irregularities, and the planting of trees.

In the more than 40 years of data compiled in NRWRD (1994), one third of the over 600 measurements along Quantico Creek showed pH readings that are above the USEPA standard range of 6.5 to 9.0. Downstream of the pyrite mine, the cadmium concentration was measured 30 times between 1972 and 1993. The USEPA's acute freshwater criteria value of 3.9 ug/L (micrograms of metal per liter of water, or ppb) for cadmium was exceeded two times, and the drinking water criteria value of 5.0 ug/L was exceeded three times. Similarly, the copper concentration was measured 40 times from 1972 to 1993, and the acute freshwater criteria value for copper (18 ug/L) was exceeded sixteen times at three locations on Quantico Creek. The lead concentration was measured 60 times from 1972 to 1993, and the drinking water criteria value of 30 ug/L for lead was exceeded 9 times at two locations. The zinc concentration was measured 59 times from 1972 to 1993, and the acute freshwater criteria value of 120 ug/L for zinc was exceeded 15 times.

Scientists at George Mason University, with funding from the National Park Service Water Resources Division, performed an assessment to determine whether water quality and aquatic habitat improvement objectives of a reclamation and restoration program were achieved in Quantico Creek (Hamblin-Katnik, 2001). Biotic life diversity and abundance were measured as well as various factors of water chemistry at five locations across the pyrite mine reclaimed area. The biological data and water chemistry data were collected monthly over a two-year period, from September 1997 to August 1999. The study collected more than 25 measurements of each metal during the period of the study, and while this study did not conclude that the abandoned mine reclamation was completely successful, it did conclude that there was an improvement in water quality based on comparisons to earlier measurements compiled in NRWRD (1994). In particular, the post-reclamation metal concentration was less than the pre-reclamation measurements for copper, zinc and iron.

The sediment transport from the pyrite mine area in the Park is significantly influenced by storm action. Three trends where found to be consistent when the sediment loads were measured during storms. The sediment load downstream of the pyrite mine site is known to be doubled during storms but after the storm-facilitated increase in sediment load, the sediment load returns to its pre-storm level. Also, most of the transport of sediment through the Park occurred during storm events. In one study, it was found that 56% of sediment load in 1985 was moved during a two-day storm in February of that year (Prugh, 1988).

1.2 Military Activities

The Marine Corps Development and Education Command training area, formerly known as the Marine Corps Combat Development Command (MCCDC), is on heavily wooded land immediately west of the Park. The Marines have used this area for track vehicle and helicopter assault training for many years.

In 1993 and 1995, the U. S. Marine Corps published Natural Resources Conservation Reports for the command facilities and areas (U. S. Marine Corps 1993; U. S. Marine Corps 1995). The reports describe work that was performed to protect the Quantico Creek watershed. Also during this period, permanent water-monitoring stations were established throughout the training area on all streams and creeks.

More recently, studies used George Mason University personnel to monitor South Fork on the Marine Corps Base (west of the park) for the health of biological activity. Although no indication of stream degradation was found (Kelso, 2000), riverbed sediment samples from South Fork near the Quantico Marine Corps Base were included in the present study to examine the possibility of renewed metal pollution.

1.3 Urban Sprawl and Park Development

Prince William County is about 60 km from the District of Columbia. Over the past few decades, the number of people needed to operate and interact with the Federal Government has been growing steadily. Area roads have rapidly improved since the 1960's, and now many Federal employees live in Prince William County. This community growth is due in large part to the construction of the National Transportation System of Interstate Highways. Part of this system is Interstate Highway 95 (I-95), which was built from Maine to Florida. A heavily traveled part of I-95 that connects

Washington to a large number of housing developments between Washington and Fredericksburg passes along the eastern side of Prince William Forest Park.

The current comprehensive plan for county development is focussed on attempts to restrict growth and maintain a rural area west and north of Manassas in Prince William County. The plan anticipates that 50,000 new dwelling units will be approved in the county, and that the majority of these will be in currently undeveloped land to the north of the Park.

Three multi-home developments border the Park on the north side. Nottingham Forest is a single-family community that is only 100 meters away from where the North Fork begins. Brittany is a much larger single-family community that reaches to within 400 meters of North Fork. This community is completely surrounded by parkland, and its surface runoff enters the Park. A third development called Forest Park is now under construction: it is a townhouse development located on the northeast corner of the Park. Its runoff will enter Quantico Creek near the northeast corner of the Park.

In addition, there are several construction projects on property adjacent to the Park (new asphalt roadways and bridges), and there are projects inside the Park (asphalt roadways, bridges, camping areas and maintenance facilities). All of these projects have the potential of having an adverse effect on streams in the Quantico Creek watershed.

2. ANALYTICAL METHOD

2.1 Sampling Sites

The sites along North Fork and South Fork extend west to east across the National Park property, to the east side of the Park where the North and South Forks of Quantico Creek converge. The site selection was made to bracket or to be near tributaries of North Fork and South Fork where storm water runoff occurred. Most tributaries carried water from potential contamination sources such as the pyrite mining area along the east side of the Park, the military area along the west side of the Park, housing developments along the north side of the Park, and developments inside the park (roadways, parking lots, camping areas). Also, some additional tributaries were sampled which did not appear to be from potential contamination sites.

Data collection sites for this research in the Park are at some of the sites used in an earlier compilation study (NRWRD, 1994), and others are near some of the sites used a recent study. (Hamblin-Katnik, 2001). This was done so as to facilitate comparisons between historic data and the new metal measurements collected during this research.

2.2 Method Overview

To achieve the objective of determining the concentration of cadmium, copper, lead and zinc in the sediment of North and South Fork of Quantico Creek, stream sediment samples were collected at various locations during a short interval of time. The sediment samples were dehydrated, sieved and then the mass of the small particles was measured. Each sample was digested in nitric acid and the volume diluted to make a 50-ml sample. The concentration of each metal was measured for each of these samples using an atomic absorption spectrometer, and from these measurements the amount of each metal was calculated for each mass of dehydrated, sieved sample (micrograms of metal per gram of sediment).

2.3 Sediment Sampling Strategy

At most of the North Fork and South Fork sites, the sample strategy was to collect a set of ten closely spaced samples of sediment from upstream and another set downstream from where a tributary is located whose water flow might (or might not) be expected to bring contamination to the site. To summarize:

(a) South Fork collection site 1 was used to measure metal measure contamination from the Quantico Marine Corps Base; no sites were used for North Fork.

(b) South Fork collection sites 2 and 6 were used to measure the contamination from roadways; site 13 was used on North Fork.

(c) South Fork collection sites 3 and 9 were used to measure the contamination from camping areas; site 14 was used on North Fork.

(d) No South Fork collection sites were used to measure the contamination from housing developments; sites 12, 17 and 18 were used for North Fork

(e) South Fork collection sites 7 and 8 were used to measure the contamination from a maintenance area parking lot; no sites were used on North Fork.

(f) No collection sites on South Fork were used to measure contamination from the abandoned pyrite mine site; sites 19 and 20 were used on North Fork.

(g) South Fork collection sites 0, 4 and 5 were used to measure sites with no visible source of contamination; sites 10, 11, 15 and 16 were used on North Fork.

For the sediment sampling, collections were obtained by using a one-meter-long, 2-cm-diameter PVC pipe that was pushed a few centimenters into the stream sediment. The PVC pipe was cleaned and rinsed thoroughly before each sample was collected. New zip-lock bags were purchased for the data collection and only used once. Each time a cleaned wooden dowel was used to extract the sediment from the PVC pipe, and the samples were placed directly into the zip-lock bags and sealed. An "A" sampling spot was located approximately one meter upstream from where the tributary entered the stream and where the bottom sediment appeared to contain fine clay. The same method was followed to collect at a downstream "B" sample spot. This procedure was repeated until a total of ten sediment samples from each "A" sampling spot and ten from each "B" sampling spot were obtained. Each sample had a mass of between 5 and 10 grams, about 20 - 50% of which was normally of clay-size particles. The sets of ten samples of sediment were each subsequently analyzed and used to determine the mean sediment concentration of each of the metals at the "A" spot and at the "B" spot at each sampling site.

2.4 Laboratory Metal Measurement Protocols

Cleaning of the laboratory equipment was necessary to reduce the chance that contamination could alter the trace metal measurements. The oven used for the drying was cleaned before the samples were placed inside the oven. The sieve that was used to separate the larger particles of sand from the smaller clay particles of each sediment sample was thoroughly cleaned and inspected before each sieving to insure that no sediment was present in the sieve.

Prior to the chemical analysis procedures, all glassware and sampler cups were washed. A soap and hot water solution cleaned the glassware, which was then rinsed with distilled water to remove the soap. A 0.2% nitric acid solution was used as a final rinse and the glassware was allowed to dry, during which the openings to the glassware and sampler cups were covered.

2.5 Stream Sediment Sample Preparation Procedure

Since the large number of sediment samples could not all be analyzed in a reasonable period of time, all samples of sediment were immediately frozen upon their removal from the park. When the analytical equipment was ready, a portion of the sediment samples were unfrozen and brought into the laboratory for preparation. Each sediment sample was transferred to a 150 mL beaker and placed in an 80°C oven and heated long enough to remove water from the sediment. The dried samples were then placed in a sieve to separate out the clay-size and smaller particles in the sediment.

The sieve was an USA Standard Testing Sieve, #120, Brass, with a mesh opening of 125 micrometers. This meant that the sediment samples to be analyzed contained both silt-size and clay-size particles. In addition, inorganic material, mineral grains and metal oxides may have been included in the sample along with organic materials such as organic colloids. None of the samples was tested for these materials even though both of these materials adsorb metals more than clay sediment particles. The assumption in this research is that because the make-up of the samples from the various sites do not visually have a wide change in composition of organic and inorganic components, any changes in composition will not account for large concentration variations.

A brass-sieve method was used because of the availability of the equipment. The USGS uses a wet-sieve method to separate fine grain sediments. This method insures that all the fine grain particles are passed through the cloth mesh filter and no contamination occurs as it might with metal sieves. During the course of the research, it was hypothesized that the metal in the brass sieve (copper and zinc) might have impacted the results, particularly in sediment samples with low metal concentrations. However, tests showed that any additions of metal from the sieve were at concentrations below the analysis threshold. That is, if sieve-derived contamination was present, it was very small and did not change the concentrations of these metals in the river sediments.

The mass of the sediment sample to be used for analysis was measured and recorded. Each sample was digested in concentrated nitric acid, and Whatman filter paper Number 1 was used in a funnel and to separate the dissolved sample from any solid residue using gravity. The Whatman filter paper Number 1 is considered to have a medium porosity with a particle size retention of 11 micrometers. The nitric acid solutions containing the dissolved metals were placed in 50 mL volumetric flasks and distilled water was used to form a solution of 50 mL.

2.6 Metal Concentration Measurement Procedure

The analyses of the dissolved sediment samples were performed on a Perkin–Elmer, model HGA-600 Graphite Furnace Atomic Absorption Spectrometer (AAS). The AAS is an analysis tool used for measuring metal concentrations in solutions, and the graphite furnace is a special application of the AAS for metal concentration measurements at the parts-per-billion order of magnitude.

The AAS graphite furnace has a light source specific for each metal to be analyzed. In order for the AAS to determine the concentration of unknown solutions, a calibration curve is developed for each metal. Placing solutions of known concentration into the AAS and recording the associated light absorption produces a calibration curve. That is, the AAS automatically records the concentration versus absorption data and produces a calibration curve. The AAS then compares the absorption of a field sample to the calibration curve and then determines a specific metal concentration. To perform this function, the absorption verses time profile generated during a sample analysis is integrated and the total mass is determined; this is then converted to concentration using sample size and dilution factors.

Standard and matrix modifier solutions were purchased form Perkin–Elmer. The Perkin–Elmer Standard Solutions were 1000 milligrams per liter concentrations of cadmium, copper, lead, and zinc in nitric acid solution. The matrix modifier was a mixture of palladium nitrate and magnesium nitrate solutions which was required to obtain consistent measurements. The water was distilled and de-ionized. The nitric acid used in all solutions was research grade purity.

Dilutions of the 1000 mg/L Standard Cadmium, Standard Copper, Standard Lead, and Standard Zinc Solutions were prepared using nitric acid mixed with distilled water to make 0.2% nitric acid solutions. The zinc, lead, copper and cadmium were diluted to concentrations of 10 mg/L and stored for no more than six months. From these secondary standard solutions, more dilute standards of 250 micrograms per liter (ug/L) were prepared on a monthly basis and stored in FEP bottles. When AAS analyses were performed for each metal, these 250 ug/L solutions were diluted even further to obtain the concentrations used to make the calibration curves (typically 10, 20, and 50 ug/L).

The sediment samples had been digested in a wet chemistry laboratory for 1 – 2 days prior to the AAS analysis. These digested samples were in labeled vials and brought to the AAS laboratory. Using a micro-pipette, 80 microliters of liquid was extracted from a labeled vial and placed in an analysis cup to be analyzed using the AAS. The concentrations of the digested samples were sequentially measured automatically and the order of measurement was cadmium, copper, lead and zinc. The cadmium was the first measured because it was found to have the lowest measured concentrations, and no dilution was generally needed to keep the cadmium concentration within the range of the AAS cadmium calibration curve. It was possible to use the magnitude of the cadmium concentrations as an indicator of the amount of dilution that might be necessary for the other metals. By experience, it was found that the concentrations of the copper, lead and zinc were about 100 times more concentrated than the cadmium.

Copper and lead measurements were performed next, and zinc was the last because it sometimes required the largest amount of sample dilution in order to be analyzed using the AAS calibration curve for zinc.

3. DISCUSSION

3.1 South Fork Results

1. All cadmium measurements are similar (about 0.01 – 0.10 ug/g cadmium, that is about 0.01-0.10 micrograms of cadmium per gram of sediment), except site 2B (about 50 ug/g Cd), which was downstream of a tributary off a roadway.
2. All copper measurements are similar (about 10–20 ug/g Cu), except site 2B (about 300 ug/g Cu).
3. All lead measurements are similar (about 5-15 ug/g Pb), except site 2B (about 900 ug/g Pb).
4. All zinc measurements are similar (about 10-40 ug/g Zn, except site 2B (about 600 ug/g Zn).

3.2 North Fork Results

1. All cadmium measurements are similar (about 0.02-0.05 ug/g Cd), except near the pyrite mine. This includes site 19A (about 0.2 ug/g Cd), site 19B (about 5 micrograms/g Cd) and site 20 (about 0.5 ug/g Cd). With these exceptions, the North Fork cadmium measurements are similar to the South Fork measurements.
2. All copper measurements are similar (about 15-30 ug/g Cu), except the pyrite mining area sites 19A (about 100 ug/g Cu), 19B (about 150 ug/g Cu) and 20 (about 250 ug/g Cu). With these exceptions, the North Fork copper measurements are similar to the South Fork copper measurements.
3. All lead measurements are similar (about 5-15 ug/g Pb), except the pyrite mining area sites 19A (about 20 ug/g Pb), 19B (about 70 ug/g Pb) and 20 (about 50 ug/g Pb). With these exceptions, the North Fork lead measurements are similar to the South Fork lead measurements.
4. All zinc measurements are similar (about 10-25 ug/g Zn, except the no-visible-contamination site 15 (about 180 ug/g Zn), and the pyrite-mine-area sites 19A (about 70 ug/g Zn), 19B (about 2100 ug/g Zn) and 20 (about 500 ug/g Zn). With these exceptions, the North Fork zinc measurements are similar to the South Fork zinc measurements.

3.3 USEPA Allowable Maximum Stream Water Metal Concentrations

The USEPA has established Guidelines for deriving Numerical National Water Quality Criteria (NNWQC) for the Protection of Aquatic Organisms and their uses. The USEPA guidelines for deriving numerical national water quality criteria for the protection of aquatic organisms and their use is given in the form of a formula based on the hardness value of the storm water sample. For each trace metal, there is a formula to calculate the USEPA's metal concentration limit (CFR 40, 2001).

The metal concentrations in North Fork and South Fork waters can be used to determine if these metal-in-water concentrations are an indication of pollution problems based on USEPA criteria. To make this determination, estimates of the metal-in-water concentrations (based on metal-in-sediment concentrations) can be compared to the environmental limits established by the USEPA. The sites' numbers at which the EPA limits for stream water were exceeded include site 2B (all metals), site 15 (cadmium only), site 16A (copper only), site 19B (cadmium and zinc), and site 20 (cadmium, copper and zinc). Sites 19 and 20 were visually anticipated to be contaminated, but sites 15 and 16 were not.

3.4 Comparison to Previous Data

The Natural Resources and Water Resources Division of the National Park Service in 1994 included (within its Baseline Water Quality Data Inventory and Analysis Report for Prince William Forest Park) some specific data that were gathered during the period 1975 through 1979 and also in 1983 (NRWRD, 1994). These data were trace-metal concentration measurements from Quantico Creek at several locations within Prince William Forest Park. Several of the data collection sites that were chosen for this research (sites 1, 10, 18-20) are close to the NRWRD data collection sites. Comparison of five sites of the NRWRD (1994) study on the North Fork and South Fork with similar sites in the present research (sites 1, 10,18 and 20) shows that copper and lead metal concentrations decreased by 50-90%, and zinc increased by 50-100% at sites 1, 10, 18. However, zinc increased by a factor of 10 at site 19. Data are not available to make cadmium comparisons.

A comparison of the results between Hamblin-Katnik (2001) and this research also cannot include cadmium comparisons because cadmium was not measured in the Hamblin-Katnik (2001) research. However, comparisons using the other metals are possible for four sites. Sites 10 and 18 showed a decrease in copper, lead and zinc of 5-50%. Conversely, metal at site 19 remained about the same, except for an increase in zinc, and site 20 showed an increase in copper, lead and zinc.

4. CONCLUSION

The areas that were investigated as sources of metal contamination to Quantico Creek sediment were a pyrite mining area along the eastern margin of the Park, a military training area along the western margin of the Park, housing developments along the northern side of the Park, and construction projects inside the Park. The measurements reveal that the abandoned pyrite mining area (now a "reclaimed" area) still contributes metal contamination to North Fork of Quantico Creek. Evidently, the restoration of this abandoned site did not stop trace metals from entering the North Fork sediment from storm-water runoff at the pyrite mine area.

The Quantico Marine Corps Base is presently not causing metal-concentration contamination in South Fork stream sediment. Evidently the environmental measures performed by the Marine facility on the South Fork were successful.

Water coming out of areas of urban sprawl (housing developments) was thought to have caused some metal contamination in North Fork sediments, but this is not happening, probably because the communities have storm-water holding ponds, so contamination in surface runoff is not reaching the National Park stream sediment. At least one of the developments (a road) within the Park has caused metal contamination in part of the South Fork.

In short, a comparison between historical data and the new data in this report show that of the 37 new metal measurement locations, only four had metal concentrations greater than the previous mean concentration data collected many years ago (sites 2B, 15, 19 and 20). Site 2 is a small tributary collecting runoff from a roadway roadway inside the Park along South Fork, and its metal concentrations exceeded the historical measurements for all four metals (cadmium, copper, lead and zinc) by a large amount. Site 15 is a South Fork site with no visible source of contamination, which had not been sampled earlier, but now is known to be contaminated. Sites 19 and 20 are near the pyrite mining area and have high metal concentrations, some of which exceed their historical concentrations.

5. SUMMARY

Measurements of riverbed sediments near the pyrite mining area along the east side of the Prince William Forest National Park show that the stream sediment is contaminated with metals derived from the century-old mining operation. The Quanitico Marine Corps Base along the west side of the Park no longer contributes metal contamination. The three housing developments along the north side of the Park are not causing metal contamination in the Park stream sediment. Some contamination was found within the Park, originating from a road.

REFERENCES

Code of Federal Regulations, CFR 40, July 1, 2001, U.S. Government Printing Office, Washington, D.C.

Comber, S. D. W., A. M. Gunn and C. Whalley, 1995. Comparison of the p artitioning of trace metals in the Humber and Mersey estuaries. Marine Pollution Bulletin. 30(12): 851 - 860.

Hamblin-Katnik, 2001. Post-Reclamation Water Quality Monitoring of Cabin Branch Pyrite Mine: Final Report. George Mason University, Fairfax, VA

Hopkins, H.T, 1984. Water resources reconnaissance of Prince William Forest Park. U.S. Geological Survey Water Resources Investigation Report. OFR 84 -4009.

Kelso, D., 2000. Quantico Marine Corps base stream monitoring 1998
-1999, Draft Report, George Mason University, Fairfax, Va., March 10.

National Park Service (NPS), 1994. Cabin Branch pyrite mine in Prince William Forest Park, Phase II, U.S. Department of Interior, April 5.

Natural Resources and Water Resources Division (NRWRD), NPS, 1994. Baseline
Water Quality Data Inventory and Analysis for Prince William Forest Park, U.S. Department of Interior, NPS/NRWRD/NRTR-94/26

Prugh, B. J. Jr., 1988. Flow variability, water quality, and sedimentation studies in Prince William Forest Park, Virginia, U.S. Geological Survey Draft Report

U.S. Marine Corps Combat Development Command. 1993. 1991 National resources 1993 conservation report. U.S. Department of Defense

U.S. Marine Corps Combat Development Command. 1995. 1993 National resources 1995 conservation report. U.S. Department of Defense

CHAPTER 15

SORPTION, DESORPTION AND LEACHING TRANSPORT OF HEAVY METALS IN SOILS COMMON TO NEW ENGLAND

Alton Day Stone[1]; James C. O'Shaughnessy, Ph.D.[2]
[1] *Alton Engineering, Sterling, MA,* [2] *Worcester Polytechnic Institue, Worcester, MA*

Abstract: Heavy metals are recognized as human health and environmental contaminants of concern. Non-industrial human exposures typically involve metals dissolved in water, sorbed in the soil, or contained in foods. Leaching of heavy metals to groundwater supplies is of particular concern. Heavy metals bioaccumulate in biota, and bioconcentrate in animals at higher trophic levels in the food chain. Animal exposures typically include ingestion of water, plant material and contact with soil. Risk characterization is a formal part of the process for remediation of properties contaminated with hazardous materials. The risk posed by heavy metals in the environment is a function of toxicity, physical parameters, and the mobility of the metal in the soil. To date, there is limited quantitative data regarding the sorption, leaching, and transport of metals in the vadose zone. This paper presents the results of bench scale experiments to evaluate these phenomena. Sixteen study soil columns were prepared using soils common to New England. Soils contained varying amounts of plastic and non-plastic fines, and included outwash sands, clay and till deposits. Physical parameters of each column were determined, including cation exchange capacity, particle size parameters and plasticity. Each column was filled with one liter of solution contain ing known concentrations of cadmium, chromium, copper, nickel and zinc. After a four -day contact period, columns were gravity drained for two days, and the volume of effluent solution and associated metals concentrations determined. Columns were then leached by application of 20 tap water rinses of approximately one pore volume each, and the concentration of each metal in the rinse lixiviant determined. Data were used to investigate the following questions: the capacity of soils to sorb metals, whether soil physical parameters are predictive of sorption, the ability of tap water to leach the metals, and the kinetics of the leaching process.

Key Words: heavy-metals, soil, leaching, adsorption, transport

1. INTRODUCTION

1.1 Metals and the Environment

Man has used and been dependent upon heavy metals since the emergence of metallurgy in the copper and bronze ages. Metals such as arsenic and lead have long been known to be poisons and the effects of acute high-concentration exposures are well documented. Heavy metals are intrinsic constituents of the geologic environment, and many are essential to the enzymatic biological processes upon which life depends. However, society recently has come to recognize that chronic low concentration exposure to many heavy metals represents an environmental and human health risk. In particular, society's increasing dependence upon ground water for drinking water supplies has focused interest on preventing or reducing contamination of aquifers used for potable supplies.

Environmental contamination by heavy metals is relatively common. The Environmental Protection Agency (EPA) (USEPA, 1987) estimates that approximately 10% of the Superfund sites involve heavy metal contamination. In recognition of the hazard posed by exposures to heavy metals, Federal and State agencies have established regulations to limit environmental contamination and reduce human and environmental exposures. The EPA currently lists 13 heavy metals as "priority pollutants" under the Comprehensive Environmental Recovery, Compensation and Liability Act (CERCLA) commonly known as "Superfund": antimony, arsenic, beryllium, cadmium, chromium, copper, lead, mercury, nickel, selenium, silver, thallium, zinc. Clean up of heavy metals contaminated soil has proved difficult. Unlike organic substances, metals cannot be mineralized or easily detoxified in-situ by biological or other processes. Excavation and landfill disposal, with or without stabilization, is the most common remedial alternative. However this is an expensive process and other alternatives have been investigated. Increasingly, remedial action alternatives, including the "no action" alternative as well as clean up target levels, are being based on risk assessment which is conducted to evaluate the site-specific level of risk to human health and the environment posed by a given metal or other hazardous material in soil and groundwater. Characterization of contaminant transport processes and migration are an important consideration of the risk assessment process.

The fate of heavy metals in the soil is a function of the complex physical and chemical interactions between the soil, water, metal and biota. In particular, the leaching and transport of metals through the soil column is controlled by two interdependent but separate categories regarding the chemistry of metals in soil (Sims & Wagner, 1983): (1) solution chemistry and (2) interfacial chemistry. Although there is a large body of theoretical literature describing and modeling cation soil chemistry, primarily in regard to agriculture, until recently there was little applied information regarding soil sorption and desorption of heavy metals.

1.2 Overview of Objectives and Research

In response to the need for information on the behavior of heavy metals in soil, bench-scale research was conducted at Worcester Polytechnic Institute (WPI) in Worcester, MA under a research grant from the EPA, Contract No. R 818627-01-0 "Enhancement of 'In-Situ' Soil Flushing for Heavy Metals Removal". The overall objectives of the research were to investigate sorption and desorption of heavy metals from soils common to Superfund sites and the suitability of various solvents to enhance the extraction of metals from contaminated soil using a leaching (flushing) process. The experiments employed small columns packed with soil that were contaminated with heavy metals. Tap water, and acid and/or chelating solutions were then applied to the soil columns to evaluate the extraction process. Research results have been presented by Caniff (1993), St. Germain (1993), Macaulay (1994), Venkataramanappa (1994), O'Shaughnessy and D'Andrea (1994), and O'Shaughnessy and Stone (1997).

The above-listed studies focused on evaluating the ability of the acid and/or chelating solutions to leach and flush sorbed-phase metals from the soil column, as a potential remedial action method. However, as described below, each test was initiated with an application of tap water. This focus of this paper is to present the results of the tap water application portion of the experiments, as the results provide insight into the migration of heavy metals in vadose zone soils.

Five heavy metals, cadmium (Cd), chromium (Cr), copper (Cu), nickel (Ni) and zinc (Zn), were chosen for the WPI research studies. Each of the metals is an EPA Priority Pollutant and each metal except Ni is present in high concentrations at sites investigated by the EPA Field Investigation Team. Environmental contamination by these metals may result from dispersion of material during industrial or manufacturing processes, e.g. fugitive emissions from smelting facilities, industrial byproducts, e.g. mine tailings, spills of concentrated substances, or the improper disposal of waste products. Heavy metal wastes are generated by numerous industries including mining, ore refining, metals production, electroplating, automotive, tanning, pesticide production, lithography, engraving, soldering, welding, tire wear and rubber products, combustion of coal and manufacturing. All five of the study metals are normally present in soil; Copper, Zn and possibly Ni are essential plant micronutrients.

The objectives of the research described in the subject paper were to investigate:

1. The sorption of cadmium, chromium, copper, nickel and zinc by soils common to New England, under laboratory conditions which approximated those of an actual chemical spill.
2. Whether physical soil properties, including - unit mass of fines (**uM_F**), coefficient of uniformity (**C_U**), coefficient of concavity (**C_C**) and bulk density (**D_B**) could be used to predict metals sorption in soil.
3. Whether the presence of plastic (clay mineral) fines affected the sorption of heavy metals.
4. The effectiveness of tap water to desorb, transport and leach heavy metals from the soil column.

The experiments were intended to duplicate a mixed spill (release) to vadose zone soil, which is then subject to rainfall leaching. The following general model applies to the experiments: (1) a liquid spill occurs and heavy metals enter the top of the soil column through unsaturated advective flow in response to gravity and capillary forces; (2) metal mass becomes partitioned between sorbed-phase on soil particles and dissolved-phase in liquid contained in the inter-pore space *soil solution (pedicular water);* (3) infiltration and percolation of rainfall flush dissolved-phase mass deeper into the soil column, and leach (desorb) sorbed mass which is also is transported deeper into the soil column and resorbed; (4) continued rainfall, in a series of unsteady flow events, distributes the metal throughout the soil column through a sequence of leaching and sorption processes and eventually transports the metal to the saturated zone. The same sorption, leaching and transport processes may then occur in the saturated zone.

2. STUDY METHODS AND MATERIALS

2.1 Apparatus and Procedures

Two plexiglass columns were packed with each of the eight soil types described below, for a total of 16 soil columns. Two empty columns were used as equipment blanks. To prepare each column, approximately 3,316 grams of soil were moistened to achieve a water content of approximately 10%. Samples of each soil were collected, weighed to determine the wet weight and oven dried and reweighed to determine the dry weight. Soil was then packed into a column in approximately 5 cm thick lifts, to achieve a total depth of 45.7 cm and a total soil column volume of approximately

1,752 cm. Soil was packed using a custom-made, controlled-drop hammer which fit inside the column. Each lift was tamped five times by dropping the hammer from a height of 30.5 cm above the soil surface to achieve a desired compaction density of 105 pcf. Upon completion of packing, each column was weighed to determine the soil column weight. The physical and chemical parameters of each soil type and column were determined and are summarized in Tables 1, 2, and 3.

Table 1. Parent Soils Particle Size and Chemical Parameters

Soil	% Fines	% Silt	% Clay	D_{10}	Cu	Cc	pH	% TOC	CEC m-eq / 100g
Outwash Sand	10.3	10.3	0	0.05	8.6	1.5	7.4	0.09	1.92
Fine Sand	<1	<1	0	0.18	1.6	1.3	5.9	0.00	0.51
Moulton Sand	88	44	44	<0.001	-	-	8.0	0.1	8.97
Cambridge Clay	80	75	5	0.004	5	1.0	8.3	0.22	29.13
Fuller Till	17.3	12.7	4.6	0.013	42.3	4.5	8.1	0.14	0.14

% Fines = -0.075 mm
% Silt = -0.075 mm, + 0.002 mm
% Clay = -0.002 mm
CEC = cation exchange capacity
D_{10} = Effective Size
Cu = Uniformity Coefficient = D_{60}/D_{10}
Cc = Coefficient of Concavity = $D_{30}^2/D_{10} \times D_{60}$
TOC = total organic carbon

Table 2. Column Soils Particle Size Parameters

Soil #	Type	% Fines	% Silt	% Clay	D_{10}	D_{30}	D_{60}	Cu	Cc
1	Outwash Sand + 10% NPF	10.3	10.3	0	0.05	0.18	0.43	8.6	1.51
2	Fuller Till	17.3	12.7	4.6	0.013	0.18	0.55	42.30	4.50
3	Outwash Sand + 4% NPF	4.1	4.1	0	0.11	0.22	0.48	4.36	0.46
4	Outwash								

Soil #	Type	% Fines	% Silt	% Clay	D_{10}	D_{30}	D_{60}	Cu	Cc
	Sand + 16% NPF	16.0	13.5	3.5	0.04	0.16	0.38	9.50	1.68
5	Outwash Sand + 21% NPF	21.0	17.0	6.0	0.004	0.13	0.36	90.0	11.74
6	Fine Sand	<0.1	<0.1	0	0.18	0.26	0.29	1.61	1.30
7	Outwash Sand + 5% PF	5.0	5.0	<0.01	0.10	0.22	0.43	1.95	1.23
8	Outwash Sand + 4.5% PF + 4.5% NPF	9.0	8.5	0.5	0.08	0.22	0.4	5.0	1.5

% Fines = -0.075 mm
% Silt = -0.075 mm, + 0.002 mm
% Clay = -0.002 mm
D_{10} = Effective Size
D_{30} = Size for which 30% of soil particles are smaller
D_{60} = Size for which 60% of soil particles are smaller
Cc = Coefficient of Concavity = $D_{30}^2/D_{10} \times D_{60}$
Cu = Uniformity Coefficient = D_{60}/D_{10}
NPF = non-plastic fines, obtained by adding Moulton Silt
PF = plastic fines, obtained by adding Cambridge Clay
Moulton Silt is 44% silt- and 44% clay-sized particles
Cambridge Clay contains 4% clay-sized particles

Soil columns were contaminated by saturating the soil column with solution containing established concentrations of each of the study metals (Cd, Cr, Cu, Ni, Zn). The mixed metal solution was prepared by dissolving known weights of cadmium sulfate ($3CdSO_4*8H_2O$), chromic acid, cupric sulfate ($CuSO_4*5H_2O$), nickelous nitrate ($Ni(NO_3)_2*6H_2O$) and zinc sulfate ($ZnSO_4*7H_2O$) in a known volume of distilled water. The pH of the mixed metal solution was recorded. Samples of the solution were analyzed to determine metals concentrations. Samples were collected in 50 ml vials, acidified to pH of 2 with HNO_3 for preservation, and subsequently analyzed at the WPI Laboratory. The concentrations of metals in the solutions used to contaminate each column are shown in Table 4. Each column was contaminated by adding one liter of mixed metal solution to the column, and then allowing the solution to stand in the column for a 4-day contact period. This time period was considered sufficient for each contaminant to achieve equilibrium between the dissolved- and sorbed-phases in soil. Each column was then drained by gravity for a 2-day period (field capacity), the volume of effluent solution measured, and the effluent analyzed to determine the concentration of each metal.

Table 3. Physical Characteristics of Soil Columns

Column #	Soil #	Type	Unit Weight ¥ (g/cm^3)	Bulk (dry) Density Db (g/cm^3)	Dry Column Soil Mass dMsoil (g)	Water Content Ø (%)	Void Ratio e	Porosity n	Void Volume Vv (cm^3)	Hyd. Cond. K (cm/s)
1	1	OWS + 10% NPF	1.89	1.72	3013.44	9.80	0.60	0.38	666	1.04E-03
2	1	OWS + 10% NPF	1.89	1.74	3048.48	8.90	0.58	0.37	648	9.10E-04
3	2	Fuller Till	1.58	1.44	2522.88	10.10	0.90	0.47	823	8.52E-04
4	2	Fuller Till	1.58	1.43	2505.36	10.20	0.92	0.48	841	7.29E-04
5	3	OWS + 4% NPF	1.76	1.63	2855.76	8.30	0.69	0.41	718	4.55E-03
6	3	OWS + 4% NPF	1.76	1.59	2785.68	10.40	0.73	0.42	736	2.61E-03
7	4	OWF + 16% NPF	1.89	1.73	3030.96	9.20	0.59	0.37	648	5.12E-04
8	4	OWF + 16% NPF	1.89	1.70	2978.40	10.10	0.62	0.38	666	6.61E-04
9	5	OWF + 21% NPF	1.89	1.72	3013.44	9.40	0.60	0.37	648	4.31E-04
10	5	OWF + 21% NPF	1.91	1.74	3048.48	9.90	0.58	0.37	648	5.19E-04
11	6	Fine Sand	1.60	1.47	2575.44	9.00	0.87	0.47	823	1.24E-02
12	6	Fine Sand	1.55	1.42	2487.84	9.20	0.94	0.49	858	8.90E-03
13	7	OWS + 5% PF	1.95	1.65	2890.80	9.30	0.67	0.40	701	4.12E-03
14	7	OWS + 5% PF	1.58	1.76	3083.52	10.50	0.56	0.36	631	1.29E-04
15	8	OWS + 5% PF + 5% NPF	1.58	1.45	2540.40	9.00	0.90	0.47	823	3.91E-03
16	8	OWS + 5% PF + 5% NPF	1.79	1.64	2873.28	9.40	0.68	0.40	701	5.17E-03

OWS = Outwash Sand
NPF = Non Plastic Fines
PF = Plastic Fines

¥ = Wet column soil mass/column soil volume
Db = Dry column soil mass/column soil volume
Ø = Volume water in wet soil column/column soil volume
e = Volume voids in soil column/volume solids in soil column

n = Volume voids in soil column/ column soil volume
Vv = n * soil column volume
Column soil volume = 1752 cm^3

Table 4. Contamination Solutions - Metals Concentrations (mg/L)

Columns	Cadmium	Chromium	Copper	Nickel	Zinc
1 to 10	1318.0	1055.5	52.76	294.5	555.5
11, 12, 14	1349.3	1075.5	53.09	299.5	370.3
13	1700.0	1224.3	41.55	610.2	731.5
15	1349.3	1075.5	53.09	299.5	731.5
16	1318.0	1224.3	41.55	610.2	731.5

Each column was then subjected to 14 to 20 tap water rinses of one soil column pore volume each to promote desorption and transport, duplicating the natural processes of infiltration and downward percolation of precipitation through a natural soil column. The volume of effluent from each rinse and metals concentrations in the effluent were determined. Mass balance calculations were used to determine the mass remaining in the soil column and mass removed at the completion of each rinse. The resulting data were used to quantitatively describe sorption and desorption-leaching processes.

2.2 Study Soils

Eight soils or soil types, numbered 1 through 8, were used in the column studies. Three of the test soils were "parent" soils and five were "manufactured" soils by combining two or three of the parent soils. The soils are common to New England and were all collected from the Worcester MA area.

2.2.1 Parent Soils

Bulk samples of the parent soils were collected from field locations without attempting to preserve the in-situ structure. Thus, all soils used in the column studies were disturbed and were reconstituted in laboratory into unit weights which resembled the natural in-situ conditions. All parent soils were collected from locations where they were unlikely to be contaminated by heavy metals at concentrations above normal background levels. In the laboratory, the bulk parent soil samples (except Cambridge Clay) were air dried and sieved to remove all particles coarser than 4.76 mm (US Standard No. 4 sieve) and thoroughly mixed. The Cambridge Clay was stored in barrels in a humid room and was not dried, sieved or mixed. Samples of each parent soil were subjected to mechanical (sieve and hydrometer) analysis to determine the particle (grain) size distribution. Parent soils were classified using the Burmister Classification System. Soil descriptions provided below are based on the results of the mechanical analysis and observations made at the collection locations.

2.2.1.1 Outwash Sand [crs-med to fn SAND, little Silt, little fn Gravel]

The Outwash Sand appeared to have been deposited by glacial meltwater in an outwash environment. The soil is moderately poorly graded (i.e. uniform or well sorted), with a coefficient of uniformity C_U = **8.6** and a coefficient of concavity C_C = **1.5**. Uniform soils are indicative of stratified drift deposits. However, stratification was not observed at the collection location, which suggests the soil may be derived from kame or kame terrace deposits. The relatively high uniformity of the soil indicates fluvial sedimentation. Plastic fines (clay < 0.002 mm) were absent from the soil.

2.2.1.2 Fine Sand [Fn SAND]

The Fine Sand also appears to be an outwash deposit, as evidenced by the high degree of uniformity C_U = **1.3**. Both non-plastic and plastic fines were absent.

2.2.1.3 Moulton Silt [Silt and Clay, some med-to fn Sand]

The Moulton Silt was comprised of equal amounts (approximately 44% each) of silt and clay-sized particles, and approximately 12% fine sand. However, testing for the Atterberg Limits determined that the soil was non-plastic as moistened samples would not roll. Therefore, it was assumed that the clay-sized particles were non-plastic and probably consisted of rock flour rather than clay minerals. The fact that the soil is relatively well graded over its particle size range and that coarse fragments were absent from the original soil suggest a fluvial or lacustrine depositional environment of moderately low energy and seasonal variation in flow velocity.

2.2.1.4 Cambridge Clay [SILT, little Clay, some fn Sand]

Based on particle size, the Cambridge Clay was actually about 75% silt, 20% fine sand and only 5% clay-sized particles. However, testing for Atterberg Limits determined that the soil was plastic; plasticity index I_P 14.07, liquid limit W_L of 34.5 and plastic limit W_P of 20.4.

2.2.1.5 Fuller Till [fn-med to crs SAND, some Silt, some Gravel]

The location from which the soil was collected and its well-graded nature indicated the soil was a glacial till. Tills are comprised of unstratified sediments which were directly deposited by glacial ice and in general, were largely unaltered by fluvial processes. The soil contained approximately 17% fines, and 4.6% clay-sized particles. This soil was the most well graded of the parent soils, with $Ç_U$ = **42.3** and $Ç_C$ = **4.5**. The soil was highly plastic, having a I_P = **5.22**, with W_L = **21%** and W_P = **15.78%**.

2.2.2 Study Soils

Study soils were either a parent soil or a mixture of two parent soils.

2.2.2.1 Soil 1 Outwash Sand with 10% non-plastic fines

This soil was the parent Outwash Sand.

2.2.2.2 Soil 2 Fuller Till

This soil was the parent Fuller Till.

2.2.2.3 Soil 3 Outwash Sand with 4% non-plastic fines

This soil was the parent Outwash Sand with approximately 6% of fines removed.

2.2.2.4 Soil 4 Outwash Sand with 16% non-plastic fines

This soil was the parent Outwash Sand with the addition of Moulton Silt to raise the non-plastic fines content to 16%. The clay-sized fraction was estimated to be 3.5%.

2.2.2.5 Soil 5 Outwash Sand with 21% non-plastic fines

This soil was the parent Outwash Sand with the addition of Moulton Silt to raise the non-plastic fines content to 21%. The clay-sized fraction was estimated to be 6%.

2.2.2.6 Soil 6 Fine Sand

This soil was the parent Fine Sand.

2.2.2.7 Soil 7 Outwash Sand with 5% plastic fines

This soil was the parent Outwash Sand with the removal of the original fines and the addition of Cambridge Clay to provide a plastic fines content of 5%. The actual clay-sized fraction was estimated to be less than 1%.

2.2.2.8 Soil 8 Outwash Sand with 5% plastic fines and 5% non-plastic fines

This soil was the parent Outwash Sand with the reduction of the original fines content to 4.5% and the addition of Cambridge Clay to provide a plastic fines content of 4.5%. The actual clay-sized fraction was estimated to be 0.5%.

3. METALS SORPTION – DATA ANALYSIS AND RESULTS

3.1 Overview

In the soil column sorption study, the contaminating process was intended to be representative of conditions under which sorption would occur following an actual spill of liquid waste in the field. Solutions containing known masses of the test metals were poured (spilled) into the test soil columns and the fully saturated soil column allowed to achieve (assumed) phase equilibrium without agitation for a 4-day period. The columns were then gravity drained and mass balance calculations used to determine sorbed concentrations [Mols] (moles metal/kg-soil) for individual metals in the soil column, and the total sorbed concentration of all metals in the column T[Mols].

In the soil, organic humus micelle, mineral clays, particularly in colloidal form (ASTM, 1994), and hydrous oxides of manganese and iron (Brady, 1990) are considered to be the three primary sites for sorption of heavy metals. The organic content for the parent soils, measured as total organic carbon TOC, ranged from 0.014% to.22%. Because the TOC values were very low, and the TOC of the manufactured column study soils was not determined, the effect of TOC on sorption could not be quantitatively evaluated. Similarly, testing for hydrous manganese and iron was not conducted, and the effect of these substances on sorption could not be evaluated. In effect, the quantitative investigation of sorption was conducted under the defacto assumption that TOC and hydrous manganese and iron were not significant factors in the sorptive capacity of the column soils.

Sorption is proportional to surface area. The specific surface $\mathbf{S_S}$ (surface area/unit mass or surface area/unit volume) of a soil is inversely proportional to grain size of a soil[3]. The $\mathbf{S_S}$ of mineral soil particles increases dramatically with decreasing grain size. Typical values of $\mathbf{S_S}$ for various soil particle sizes range in orders of magnitude from $\mathbf{10^3\ cm^2/g}$ for silt, $10^4\ \mathbf{cm^2/g}$ for silt for clay, to $\mathbf{10^5\ cm^2/g}$ for silt for colloidal clay (Overcash, 1979). Cation exchange capacity (CEC) also increases with $\mathbf{S_S}$ as particle size decreases. CEC values for various particle sizes range from 3 meq/100 g soil for silt, to 22 meq/100 g for clay, to 63 meq/100 g for very fine colloidal clay (Brady, 1990). In particular, clay colloids are negatively charged and have extensive external and internal surfaces, making the particles ideal site for both physical, chemical and ion replacement sorption.

Because sorption and CEC are proportional to $\mathbf{S_S}$, soil physical properties which express or are a function of the soil particle size distribution, were considered as possible parameters which could be used to predict the sorptive capacity of soils. The total sorbed moles concentration T[Mols] (dependent variable) for each soil column was plotted as a function of soil column mass fines M_{FINES}, coefficient of concavity C_C, coefficient of uniformity C_U, and bulk density D_B (independent variables). Regression analysis was performed to evaluate the correlation between variables.

Because only Soils 2, 7 and 8 contained plastic fines (i.e. clay minerals), and the actual mass of fines in each column was not determined, a specific quantitative evaluation between [Mols] or T[Mols] and mass of plastic fines could not be performed. However, soil columns were ranked by T[Mols], to evaluate the importance of the mass of plastic fines in sorption of heavy metals. Histograms showing [Mols] for individual metals and soils were also prepared for the purpose of identifying the preference of metals for sorption on specific soil types.

3.2 Data Analysis

Data for the following parameters are presented in Table 5. Mass values were converted to sorbed moles concentrations $[Mol_S]$. This was performed because the mass of each metal in the contaminating solutions was not always equal, and the mass of soil in each column was not always equal. Under the assumption that one metallic ion occupies one sorption site, converting mass to mole concentrations would provide for a better comparison of the sorptive capacities of the various soils for the different metals. Sorbed moles concentrations were calculated using the procedures outlined below.

- The initial mass retained iM_R of each metal in each soil column was the mass remaining in the column after the contamination period and the initial 2-day drainage period. This was determined by subtracting the metal mass in the drainage effluent from the metal mass in the contaminating solution.
- The mass of each metal removed in rinse pore volume 1 M_{pv1} was calculated from the volume of rinse elutriate and the metal concentration in the lixiviant (extraction solution). These data were used to calculate the percent of iM_R which was removed in the first pore volume rinse.
- The mass sorbed M_S was defined as the mass of each metal retained in the soil column after the first pore volume rinse (PV1) of tap water; $M_S = iM_R - M_{pv1}$. At the completion of the 2-day drainage period, water content in the columns was at field capacity, and water was retained under capillary suction (pressure) in the smaller soil pores. Tap water rinse PV1 typically removed 30% to 60% of the iM_R of Cd, Cr, Ni and

Zn. Mass removed in PV1 was considered to primarily be dissolved-phase mass present in the soil capillary pore that was flushed from the column by the initial rinse. Thus, M_S was considered to be representative of the actual sorbed mass of each metal. The iM_R and M_S values were obtained by mass balance calculations.

- M_S for each metal was converted to moles sorbed (Mol_S) by dividing M_S by the respective gram molar weights GMW.
- The total moles sorbed ($tMol_S$) for each column was calculated by summing the Mol_S values for the five metals.
- The sorbed moles concentration [Mol_S] for each metal and the total sorbed moles concentration T[Mol_S] for each column were then calculated by dividing the Mol_S and $tMol_S$ values by the dry mass of the soil column; i.e. **T[Mols] = $tMol_S$/dMsoil = moles metal/kg soil.**

Table 5a. Summary of Column Chemical Parameters after Contamination and Rinse PV1

Column #	Soil #	Description	Dmsoil (kg)	Total iMr (mg)	Total Ms (mg)	Total Mol-s (moles)	Total [Mol-s] (moles/kg soil)	% iMr Removed in PV1 (%)				
								Cd	Cu	Cr	Ni	Zn
1	1	OW sand with 10% NPF	3.013	2329	1476	0.0192	0.0064	33	<1	50	51	28
2	1	OW sand with 10% NPF	3.048	2249	1406	0.0181	0.0059	36	<1	56	40	27
3	2	Filler Till	2.532	2133	1387	0.0189	0.0075	34	<1	39	45	29
5	3	OW sand with 4% NPF	2.856	1954	1208	0.0163	0.0057	37	<1	46	48	32
6	3	OW sand with 4% NPF	2.786	1954	1400	0.0189	0.0068	26	<1	37	36	22
7	4	OW sand with 16% NPF	3.031	2376	1405	0.0182	0.0060	39	<1	60	39	31
8	4	OW sand with 16% NPF	2.978	2379	1667	0.0216	0.0072	27	<1	46	31	24
9	5	OW sand with 21% NPF	3.013	2486	1469	0.0189	0.0063	38	<1	61	46	29
10	5	OW sand with 21% NPF	3.048	2475	1497	0.0186	0.0061	35	<1	60	42	31
11	6	Fine Sand	2.575	1644	746	0.0103	0.0040	50	48	49	91	73
12	6	Fine Sand	2.488	1592	759	0.0104	0.0042	47	46	48	90	70
13	7	OW sand with 5% PF	2.891	3233	2430	0.0324	0.0112	24	<1	47	15	16
14	7	OW sand with 5% PF	3.084	2497	1676	0.0211	0.0068	27	<1	58	32	20
15	8	OW sand with 5% NPF + 5% PF	2.540	2217	1707	0.0218	0.0086	20	<1	46	30	12
16	8	OW sand with 5% NPF + 5% PF	2.873	2926	2116	0.0295	0.0103	30	<1	50	17	17

DMsoil = Dry mass of soil column

NPF = non-plastic fines, PF = plastic fines

Total iMr = Total initial mass retained, after contact period and two day drainage period

Total Ms = Total mass sorbed, equals iMr – mass in first pore volume of rinse

Total Mol-s = Total moles sorbed, equals sum of moles sorbed for all five metals

Totals [Mol-s] = Total sorbed moesl concentration equals Total Mol-s/dMsoil

Gram Molecular Weights (GMW): Cd = 112.40; Cr = 52.00; Cu = 63.55; Ni = 58.71; Zn = 65.37

Table 5b. Summary of Column Chemical Parameters after Contamination and Rinse PV1

Column	Soil	Description	Sorbed Moles Concentration (milli-moles/kg soil)				
#	#		Cd	Cu	Cr	Ni	Zn
1	1	OW sand with 10% NPF	2.10	0.28	1.15	0.60	1.75
2	1	OW sand with 10% NPF	1.93	0.27	1.52	0.78	1.65
3	2	Filler Till	3.86	0.33	4.75	1.52	2.01
5	3	OW sand with 4% NPF	1.66	0.29	1.73	1.39	5.68
6	3	OW sand with 4% NPF	1.99	0.30	1.85	0.82	1.64
7	4	OW sand with 16% NPF	5.96	0.83	3.83	2.36	5.25
8	4	OW sand with 16% NPF	7.37	0.83	5.08	2.76	5.79
9	5	OW sand with 21% NPF	6.49	0.83	3.66	2.39	5.71
10	5	OW sand with 21% NPF	7.42	0.83	3.87	2.68	3.80
11	6	Fine Sand	3.25	0.25	5.83	0.16	0.69
12	6	Fine Sand	3.42	0.27	5.84	0.17	0.68
13	7	OW sand with 5% PF	10.77	0.72	7.27	8.81	9.43
14	7	OW sand with 5% PF	8.18	0.83	4.49	3.26	4.42
15	8	OW sand with 5% NPF + 5% PF	10.19	1.09	6.37	3.34	10.06
16	8	OW sand with 5% NPF + 5% PF	7.86	0.72	7.43	8.71	9.90

3.3 Mass Removal in Rinse Pore Volume 1

Rinse PV1 removed a significant mass of each of the metals except Cu. Similar results were observed by Caniff, (1993), Macaulay (1994), and Venkataramanappa (1994). This result indicates that the initial mass of each metal, other than Cu, in the contaminating solution was in excess of that required to achieve equilibrium between sorption and solution. Due to the comparatively low mass of Cu in the contaminating solution, a very high percentage of mass was sorbed. The Freundlich and Langmuir adsorption models predict that as the mass of sorbate in solution increases, the total mass adsorbed increases, but the percentage of the original dissolved phase mass that is adsorbed decreases. Because, with the exception of Fine Sand (Columns 11 and 12), the mass of copper probably was not in excess of the equilibrium concentration, the copper sorption data cannot be compared directly with the other metals.

These results are pertinent to the containment and abatement of environmental spills of high-concentration solutions. As indicated by the column experiments, significant dissolved- or lightly sorbed-phase metal mass can be retained in the micopore spaces of a soil. However, percolation of storm water may rapidly flush the mass downward through the soil column, resulting in contamination of deeper soils through re-sorption, or transmission of dissolved-phase mass directly to the saturated zone.

3.4 Sorption and Soil Physical Parameters

The total moles sorbed concentrations $T[Mol_S]$ for each study soil column were considered representative of the total capacities of the soil columns for sorption of heavy metals. $T[Mol_S]$ values were plotted as a function of several soil column physical parameters, and regression analysis was performed using the least squares method to evaluate whether a correlation existed which could be used to predict sorption.

Mass of Fines Data for $T[Mol_S]$ and soil column mass of fines M_{fines} (particle size <0.075 mm, -#200 sieve) are presented in Table 6. $T[Mol_S]$ is plotted as a function of soil column mass of fines in Figures 1a (odd-numbered columns) and 1b (even-numbered columns). The regression coefficients R of 0.0018 and 0.1966 indicate there is not a strong correlation between increased M_{fines} and sorption.

Table 6. Total Sorbed Moles Concentration T[Mol-s], Soil Column Mass Fines, Bulk Density Db, and Coefficients of Uniformity Cu and Concavity Cc.

Col #	Rank	Soil # and Description	% Fines (-#200)	DMsoil (g)	Mass Fines (g)	Total Ms (mg)	Total Mol-s (moles)	Total [Mol-s] (moles/kg soil)	Db (g/cm^3)	Cc	Cu
11	A	6 Fine sand	0.10	2575.44	2.58	746	0.0103	0.0040	1.47	1.30	1.61
5	B	3 OW sand with 4% NPF	4.10	2855.76	117.09	1208	0.0163	0.0057	1.63	0.46	4.36
13	H	7 OW sand with 5%PF	5.00	2890.80	144.54	2430	0.0324	0.0112	1.65	1.23	1.95
15	G	8 OWS with 5% NPF + 5% PF	9.00	2540.40	228.64	1707	0.0218	0.0086	1.45	1.50	5.00
1	E	1 OW sand with 10% NPF	10.30	3013.44	310.38	1476	0.0192	0.0064	1.72	1.51	5.15
3	F	2 Fuller Till	17.30	2522.88	436.46	1387	0.0189	0.0075	1.44	4.50	9.50
7	C	4 OW sand with 16% NPF	16.00	3030.96	484.95	1405	0.0182	0.0060	1.73	1.68	42.30
9	D	5 OW sand with 21% NPF	21.00	3013.44	632.82	1469	0.0189	0.0063	1.72	11.74	90.00
12	I	6 Fine sand	0.10	2487.84	2.49	759	0.0104	0.0042	1.42	1.30	1.61
6	L	3 OW sand with 4% NPF	4.10	2785.66	114.21	1400	0.0189	0.0068	1.59	0.46	4.36

Col #	Rank	Soil # and Description	% Fines (-#200)	DMsoil (g)	Mass Fines (g)	Total Ms (mg)	Total Mol-s (moles)	Total [Mol-s] (moles/kg soil)	Db (g/cm^3)	Cc	Cu
14	M	7 OW sand with 5% PF	5.00	3083.52	154.18	1676	0.0211	0.0068	1.76	1.23	1.95
16	O	8 OWS with 5% NPF + 5% PF	9.00	2873.28	258.60	2116	0.0295	0.0103	1.64	1.50	5.00
2	J	1 OW sand with 10% NPF	10.30	3048.48	313.99	1406	0.0181	0.0059	1.74	1.51	5.15
8	N	4 OW sand with 16% NPF	16.00	2978.40	476.54	1667	0.0216	0.0072	1.70	1.68	42.30
10	K	5 OW sand with 21% NPF	20.80	3048.48	634.08	1497	0.0186	0.0061	1.74	11.74	90.00

Table 7. Sorbed Moles Concentration [Mol-s] for Each Metal, Soil Type and Column

	Columns		Metal [mol-s]=mol-s/kg soil				
Soil #	**ODD**	**Even**	**Cd ODD**	**Cd EVEN**	**Cu ODD**	**Cu EVEN**	**Cr ODD**
1 Outwash Sand + 10% NPF	1	2	0.0021	0.0020	0.0003	0.0003	0.0016
2 Fuller Till	3	-	0.0021	-	0.0003	-	0.0026
3 Outwash Sand + 4% NPF	5	6	0.0017	0.0020	0.0003	0.0003	0.0017
4 Outwash Sand +16% NPF	7	8	0.0019	0.0024	0.0003	0.0003	0.0013
5 Outwash Sand +21% NPF	9	10	0.0021	0.0024	0.0003	0.0003	0.0012
6 Fine Sand	11	12	0.0013	0.0014	0.0001	0.0001	0.0023
7 Outwash Sand + 5% PF	13	14	0.0033	0.0026	0.0002	0.0003	0.0021
8 Outwash Sand + 5% NPF + 5% PF	15	16	0.0029	0.0024	0.0003	0.0002	0.0016

NPF = *non-plastic fines*, PF= *plastic fines*

Table 7. Sorbed Moles Concentration [Mol-s] for Each Metal, Soil Type and Column Cont.

Soil #	Columns		Metal [mol-s]=mol-s/kg soil				
	ODD	**Even**	**Cr EVEN**	**Ni ODD**	**Ni EVEN**	**Zn ODD**	**Zn Even**
1 Outwash Sand + 10% NPF	1	2	0.0012	0.0006	0.0007	0.0017	0.0017
2 Fuller Till	3	-	-	0.0008	-	0.0017	-
3 Outwash Sand + 4% NPF	5	6	0.0021	0.0006	0.0008	0.0014	0.0016
4 Outwash Sand +16% NPF	7	8	0.0017	0.0008	0.0009	0.0017	0.0019
5 Outwash Sand +21% NPF	9	10	0.0013	0.0008	0.0009	0.0019	0.0012
6 Fine Sand	11	12	0.0023	0.0001	0.0001	0.0003	0.0003
7 Outwash Sand + 5% PF	13	14	0.0014	0.0027	0.0011	0.0029	0.0014
8 Outwash Sand + 5% NPF + 5% PF	15	16	0.0020	0.0029	0.0027	0.0029	0.0030

NPF = non-plastic fines, PF= plastic fine

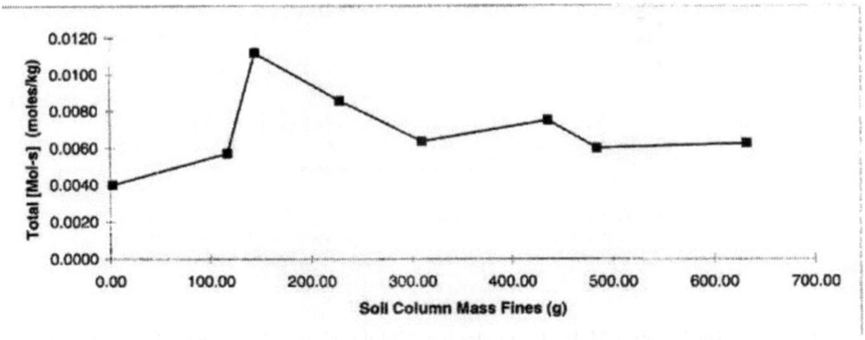

Figure 1a. Total Sorbed Moles Concentration T[Mols -s] vs. Mass of Fines (-0.75mm particle size) in Soil Column - Odd Numbered Columns; R = 0.0018

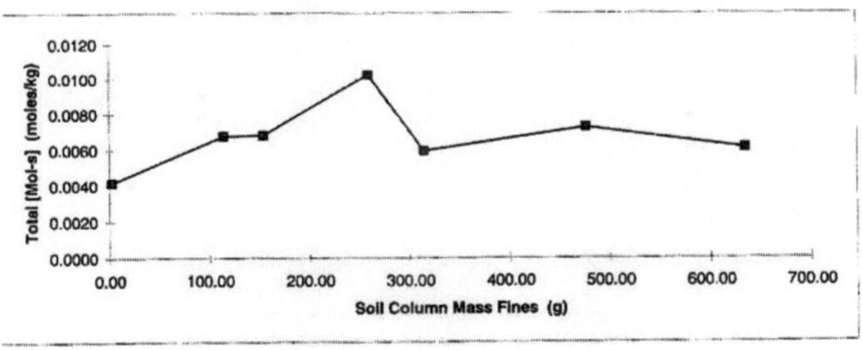

Figure 1b. Total Sorbed Moles Concentration T[Mols -s] vs. Mass of Fines (-0.75mm particle size) in Soil Column - Even Numbered Columns; R = 0.1966

Coefficient of Concavity Data for **T[Mol$_S$]** and Cc are presented in Table 6. T[Mol$_S$] is plotted as a function of C_C in Figure 3. The regression coefficients R of 0.0859 and 0.1107 indicate there is not a strong correlation between C_C and sorption.

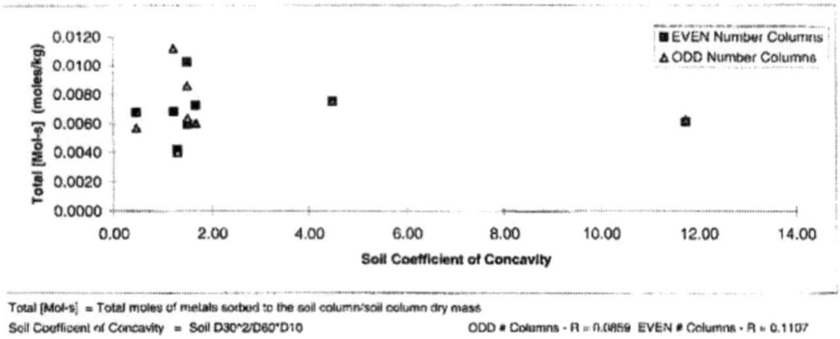

Figure 2. Total Sorbed Moles Concentration T[Mole -s] vs. Coefficient of Concavity Cc

<u>Coefficient of Uniformity</u> Data for T[Mol_S] and C_U are presented in Table 6. T[Mol_S] is plotted as a function of C_U in Figure 3. Regression analysis was not performed because the plot was very similar to that of C_C vs. T[Mol_S].

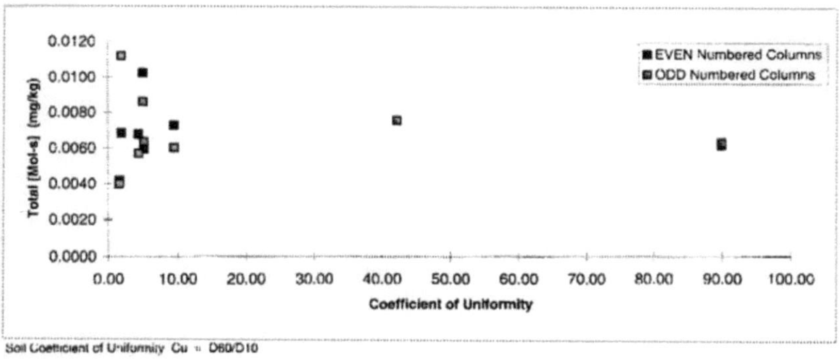

Figure 3. Total Sorbed Moles Concentration T[Mol -s] vs. Coefficient of Uniformity Cu

<u>Bulk Density</u> Data for T[Mol_S] and D_B are presented in Table 6. T[Mol_S] is plotted as a function of D_B in Figure 4. The regression coefficient R of 0.2288 indicates there is not a strong correlation between D_B and sorption.

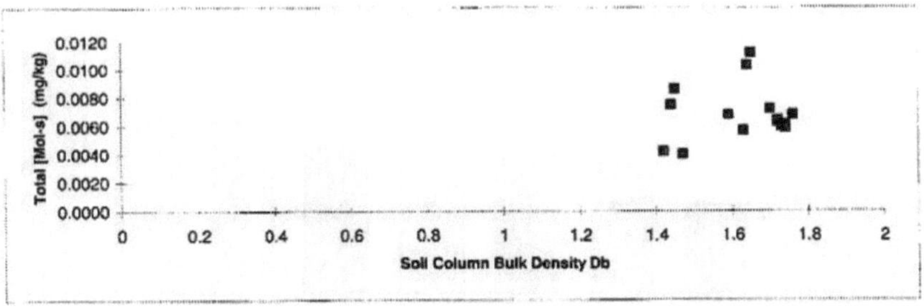

Figure 4. Total Sorbed Moles Concentration T[Mol -s] vs. Soil Conlumn bulk Density Db. R= 0.2288

The results of the investigations indicate that soil column mass fines and other general soil physical parameters are not good predictors of soil sorption. This probably is because the physical parameters do not distinguish between mass of silt and mass of clay minerals. As previously shown, the specific surface $\mathbf{S_S}$ increases by orders of magnitude as particle size decreases from silt to mineral clays.

3.5 Sorption and Soil Type

In Table 6 soil types for odd- and even-numbered columns are ranked in order of total sorbed moles concentrations $\mathbf{T[Mol_S]}$. Sorbed moles concentrations [Mol-s] for each metal, soil type and column are presented in Table 7. Histograms for sorption and soil type are presented in Figure 5a for odd numbered columns, and Figure 5b for even numbered columns. In both data sets, the Fine Sand, which did not contain fines exhibited the least sorption; 0.004 moles/kg soil. Sorption increases with the addition of non-plastic fines; however, the increase appears independent of the percentage (mass) of non-plastic fines. Sorption again increases with the addition of plastic fines. Soils 2, 7 and 8 that contain plastic fines generally exhibited the largest sorption, with maximum concentrations of about 0.011 moles/kg soil. This trend is best evidenced in the histogram for the odd-number columns. These results suggest that fines (silt- and clay-sized particles) and in particular plastic fines (clay minerals), are important sites for sorption of heavy metals. These results are not unexpected. In the absence of organic humus, mineral clay particles generally are the major sites for cation sorption.

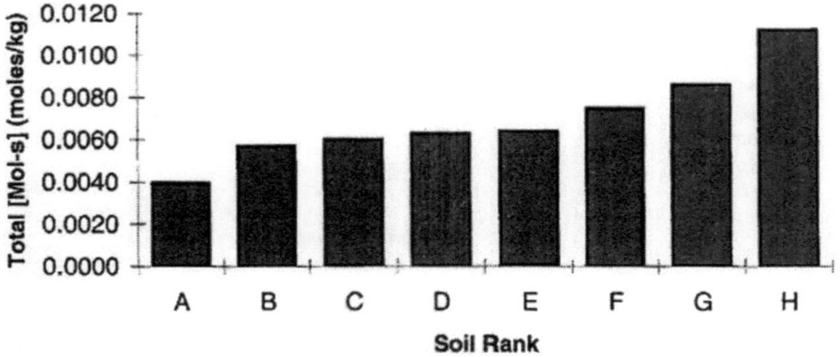

Figure 5a. Soil Rank in Order of Increasing Total Sorbed Moles Concentration T[Mol -s] Odd Numbered Columns

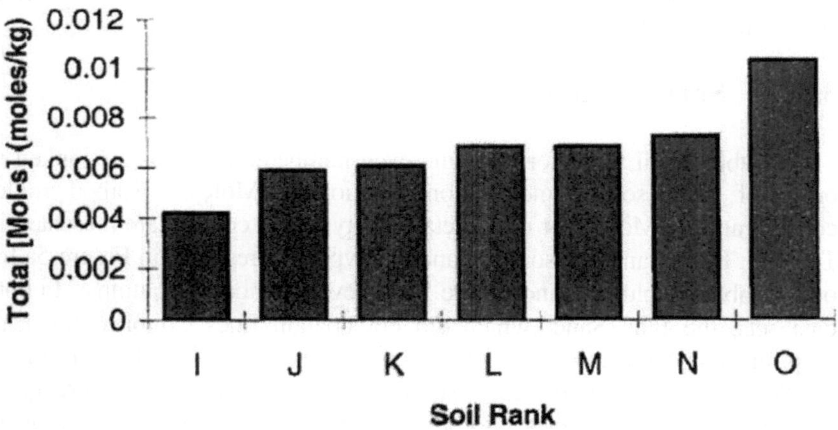

Figure 5b. Soil Rank in Order of Increasing Total Sorbed Moles Concentration T[Mol -s] Even Numbered Columns

Data for sorption of individual metals on each soil type was organized both by metal and soil type. In Table 6. data is organized by metal, presenting the sorbed moles concentrations [Mol_S] of a metal for each soil type and soil column. Figures 6a through 6e are histograms in which [Mol_S] for each soil type is presented for a specific metal. This data array highlights the affinity of a metal for a certain soil type. Figures 7a through 7e are histograms in which **[Mol$_S$]** for each metal is presented for a specific soil type. This data array highlights the sorptive capacity of a soil for a specific metal.

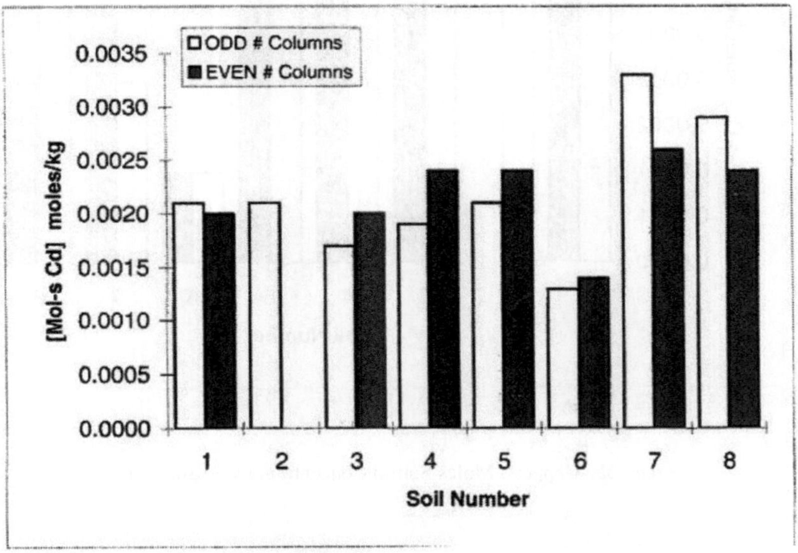

Legend for figures 6a -6e:

1 Outwash Sand + 10% NPF	5 Outwash Sand +21% NPF
2 Fuller Till	6 Fine Sand
3 Outwash Sand + 4% NPF	7 Outwash Sand + 5% PF
4 Outwash Sand +16% NPF	8 Outwash Sand + 5% NPF + 5% PF
NPF = non-plastic fines	PF= plastic fine

Figure 6a. Cadmium - Moles Sorbed Concentrations by Soil Type

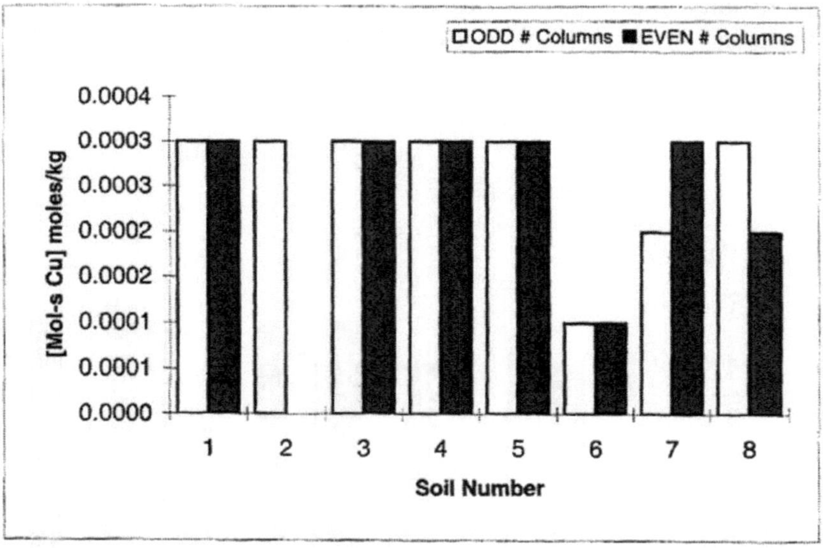

Figure6b. Copper - Moles Sorbed Concentrations by Soil Type

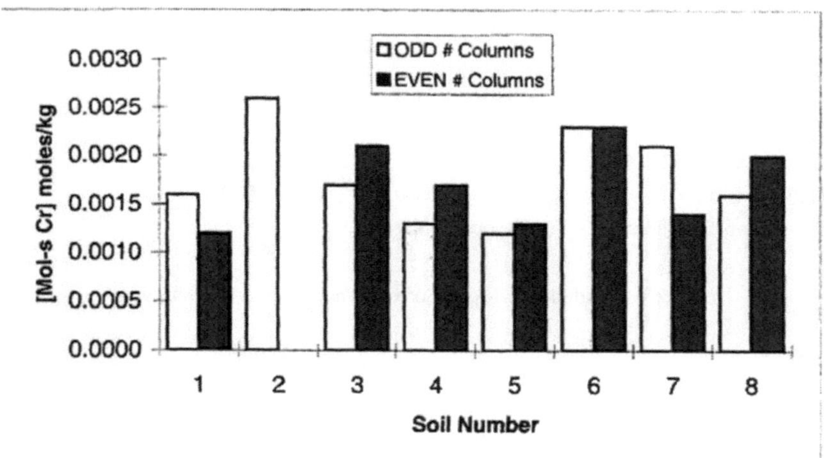

Figure 6c. Chromium - Moles Sorbed Concentrations by Soil Type

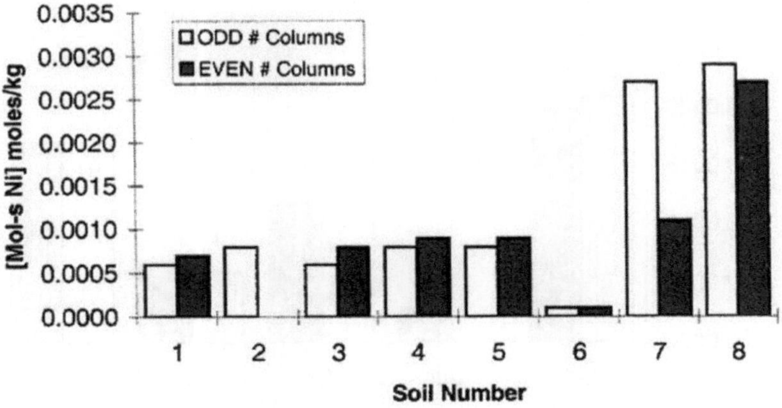

Figure 6d. Nickel - Moles Sorbed Concentrations by Soil Type

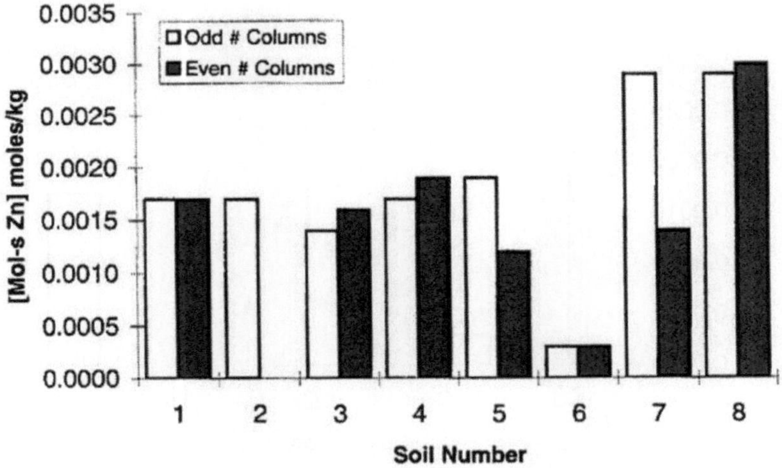

Figure 6e. Zinc - Moles Sorbed Concentrations by Soil Type

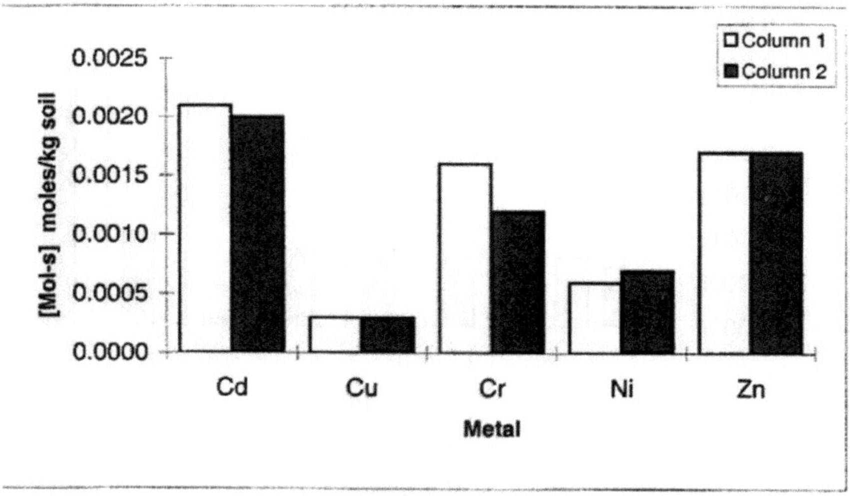

Figure 7a. Soil 1 - Outwash Sand + 10% Non-Plastic Fines

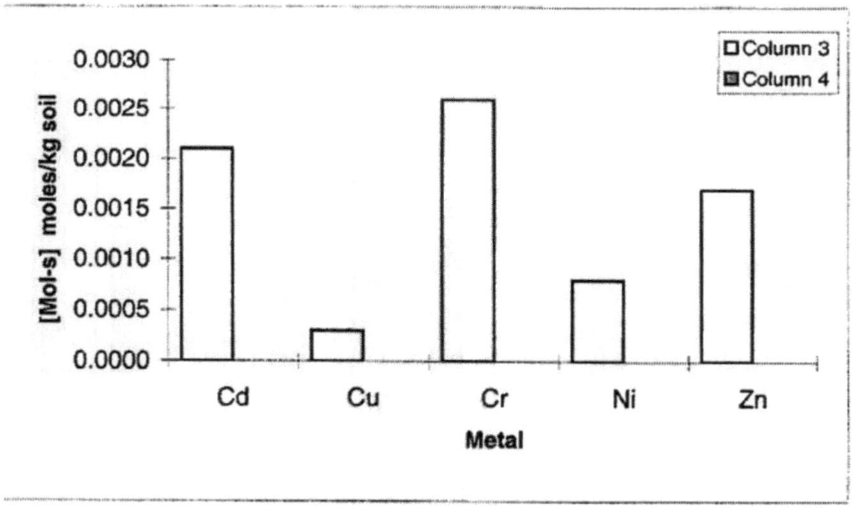

Figure 7b. Soil 2 - Fuller Till

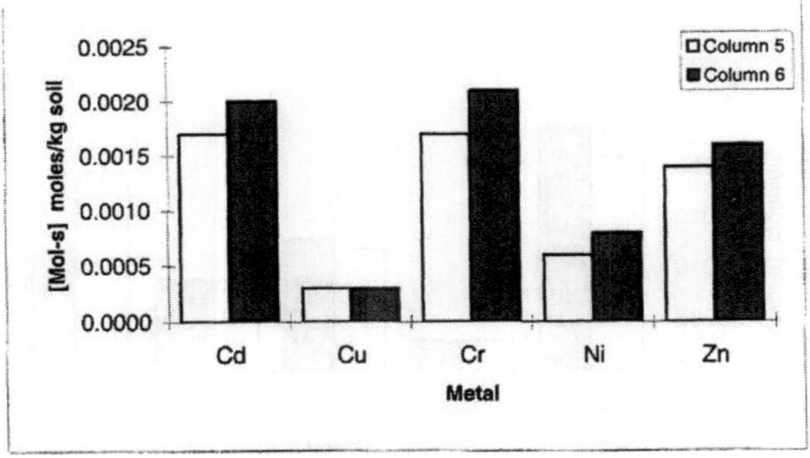

Figure 7c. Soil 3 - Outwash Sand + 4% Non-Plastic Fines

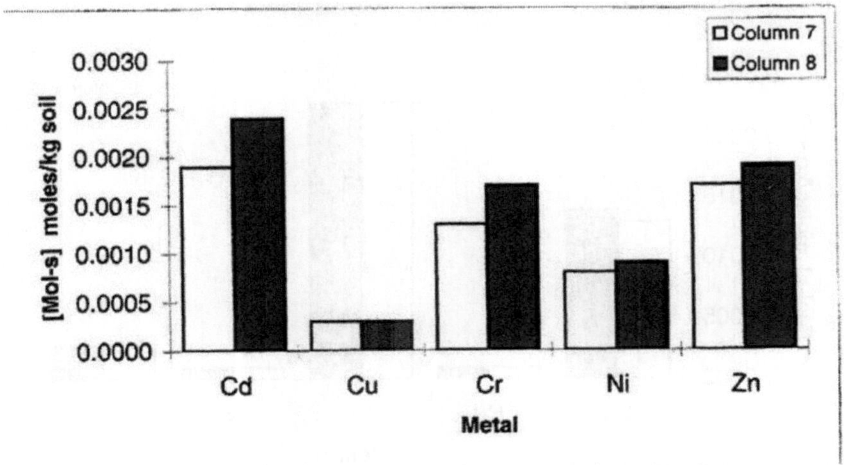

Figure 7d. Soil 4 - Outwash Sand + 16% Non-Plastic Fines

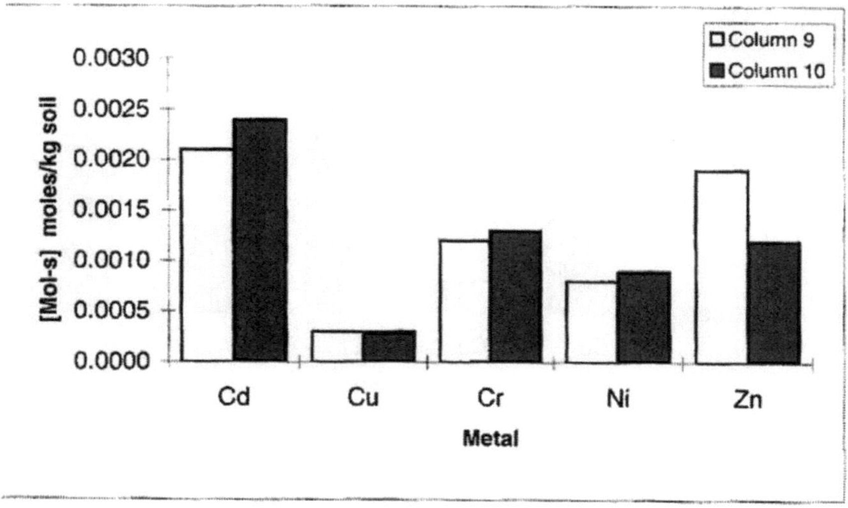

Figure 7e. Soil 5 - Outwash Sand + 21% Non-Plastic Fines

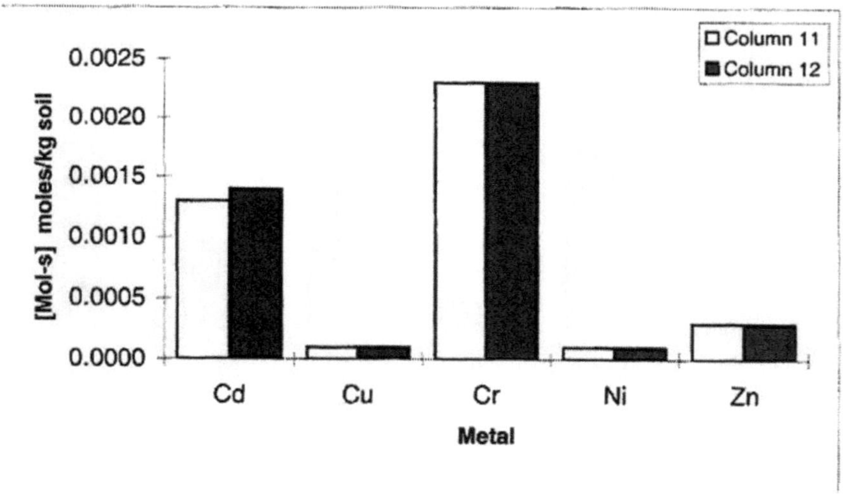

Figure 7f. Soil 6 - Fine Sand

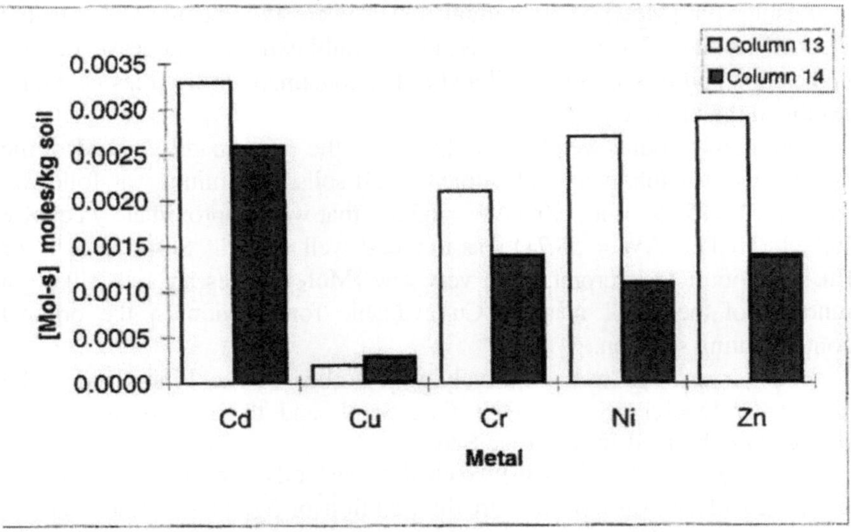

Figure 7g. Soil 7 - Outwash Sand + 5% Plastic Fines

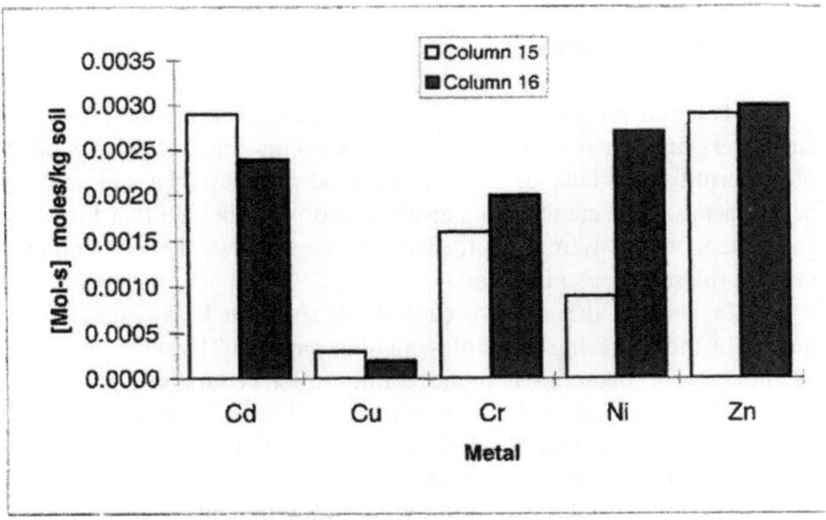

Figure 7h. Soil 8 - Outwash Sand + 5% Plastic Fines and 5% Non -Plastic Fines

Results for [Mol_S] of individual metals were in general similar to the T[Mol_S] results. Soil 6 the Fine Sand generally exhibited the least sorption of the eight soil types. Soils 2, 7 and 8 that contained plastic fines generally exhibited the highest sorptions.

Cadmium (atomic weight = 112.40) was the most readily sorbed of the metals, and was relatively well sorbed by all soils. Cadmium was followed by Cr (AW = 51.996) and Zn (AW = 65.37) that were approximately equally sorbed. Nickel (AW = 58.71) was the least well sorbed. Sorption data for Cu is difficult to interpret. The very low [Mol_S] values are most likely a function of the small mass of Cu available for sorption in the original contaminating solution.

Both Cr and Cd were relatively well sorbed to the Fine Sand. The highest Cr [Mol_S] occurred with Fine Sand, and the second highest Cd [Mol_S] was observed for the Fine Sand.

Nickel was relatively poorly sorbed to all soils except Soils 7 and 8, comprised of Outwash Sand with the addition of 5% plastic fines derived from the Cambridge Clay. In particular, Ni was not strongly sorbed to the Fuller Till which exhibited plasticity and contained approximately 5% clay that is native to the soil. Overcash (1979) notes that in soils Ni is sorbed on reactive surfaces such as hydrous oxides of Mn and Fe.

3.6 Conclusions

1. For mineral soils with very low organic content, the physical soil parameters mass or percentage of fines, coefficient of concavity, coefficient of uniformity, and bulk density are not good predictors for soil sorption of heavy metals. The cause of this result is probably the fact that the physical parameters which were investigated do not express the content of clay mineral (plastic) fines in the soil.

2. In general, the sorptive capacity of soils for heavy metals may be ranked in the following order of increasing capacity T[Mol_S]: Soils devoid of fines < Soils containing non-plastic fines < Soils containing plastic fines.

3. Total sorbed moles concentrations T[Mol_S] ranged from 0.004 moles/kg soil for the Fine Sand (Column 11) to 0.0112 moles/kg soil for the Outwash Sand + 5% plastic fines (Column 13).

4. The affinity of metals for sorption may be ranked in the following order of increasing affinity: Ni < Zn and Cr < Cd.

5. Cr and to a lesser extent Cd, were relatively well sorbed by the Fine Sand which did not contain fines.

6. The following ranges for sorption [Mol_S] of individual metals were observed (mols/kg soil):

Cd 0.0013 (Fine Sand) to 0.0033 (Outwash Sand + 5% PF)

Cr 0.0012 (Outwash Sand + 10% NPF, Outwash Sand + 21% NPF) to 0.0026 (Fuller Till)

Ni 0.0001 (Fine Sand) to 0.0029 (Outwash Sand + 5% NPF + 5% PF)

Zn 0.0003 (Fine Sand) to 0.0030 (Outwash Sand + 5% NPF + 5% PF)

4. DESORPTION AND LEACHING

4.1 Objectives and Overview

The objectives of the soil column desorption and leaching study were to investigate:

1. The overall removal of individual metals from soil contaminated by a high-concentration, mixed-metals liquid spill, when subjected to water leaching, under laboratory conditions which simulated those of an actual chemical spill in the field.

2. The effectiveness of water rinses to transport, desorb and leach sorbed metals from the soil, where desorption effectiveness was defined as the percent of the initial sorbed mass that was leached from the soil column by water.

The contamination process described in Section 2.0 was intended to duplicate the conditions of an actual environmental chemical spill, although in the field the soil column probably would not remain saturated for 4 days because the solution would percolate downward until field capacity was achieved. Columns were then drained which duplicated the flow of liquid after the spill from the upper soil column to a condition of field capacity. Under field conditions, liquid would drain to deeper soil until the gas-liquid-soil system reached hydraulic equilibrium. Field drainage would produce a contamination gradient downward through the soil column $d[Mol_S]/dz$. Drainage of the soil study columns produced (assumed) uniform contamination of the study column; d[Mols]/dz = 0. In effect, the study soil column duplicated the top section of the field soil column (uniform soil profile) exposed to the spill, where sorbed mass would be maximum.

The study soil columns were then subjected to a leaching process having 1 to n stages, where n equaled the following number of pore volume rinses: Columns 1 through 12, n = 20; Column 13 n = 14; Column 14, n = 16; Columns 15 and 16, n = 10. The tap water treatment was intended to duplicate the leaching of metals from the soil column by percolation of storm water runoff, which would naturally occur in the field if remediation of a spill was not completed within a short time period. Evaluation of the data provided baseline information on the ability of water to leach and transport sorbed heavy metals from the soil. However, the limitations of the experiment in approximating field leaching are noted. In the laboratory, application of the water rinse was relatively steady compared to the natural environment, and each water rinse was drained from the soil column to permit analysis and quantification of the leaching process. In the field, both the temporal and physical aspects of storm water application are random events. Due to the randomness of application, heterogeneity of soil and variability in hydraulic boundary conditions (e.g. depth to water table), actual field conditions could not be duplicated in the laboratory.

4.2 Overall Mass Removals

Overall mass removal data for each metal, as a function of the number of pore volume rinses, are presented in Table 8a through 8e, and plotted in Figures 8 through 15. Each table presents the percent of the initial mass retained iM_R at the conclusion of the contamination and two-day drainage period, mass removed by the first pore volume of water rinse M_{PV1} and by the total water treatment M_T, and the number of pore volumes of rinse employed for each treatment. Column 4 clogged during the leaching process and was not evaluated.

Table 8a. Overall Percent Removal - Cadmium

Column No.	Soil No.	iMr (mg)	% Rem PV 1	Add. % Rem	No. PV	% Rem Total	Mass Rem.
1	1	1072.4	33	10	20	43	460.36
3	2	908.5	34	12	20	46	416.30
5	3	844.5	37	12	20	49	416.25
7	4	1103.6	39	11	20	50	566.43
9	5	1170.2	38	11	20	49	581.08
11	6	732.1	50	3	20	53	383.13
13	7	1402.8	24	6	14	30	418.84
15	8	1029.9	20	6	10	26	272.26
2	1	1065.0	36	11	20	47	497.50
6	3	844.5	23	9	20	35	294.70
8	4	1124.5	27	8	20	35	398.82
10	5	1279.8	35	6	20	44	558.82
12	6	723.0	47	4	20	51	365.20
14	7	1256.9	27	10	16	37	463.78
16	8	1072.0	29	7	10	36	384.34

Table 8b. Overall Percent Removal - Copper

Column No.	Soil No.	iMr (mg)	% Rem PV 1	Add. % Rem	No. PV	% Rem Total	Mass Rem.
1	1	52.7	0	1	20	1	0.6
3	2	52.7	1	1	20	2	1.00
5	3	52.7	1	1	20	2	1.05
7	4	53.0	0	1	20	1	0.73
9	5	52.7	0	1	20	1	0.54
11	6	30.9	48	5	20	53	16.38
13	7	41.4	0	0	14	0	0.43
15	8	53.0	0	0	10	0	0.17
2	1	52.7	0	1	20	1	0.50
6	3	52.7	0	0	20	0	0.21
8	4	52.7	0	1	20	1	0.54
10	5	53.6	0	0	20	0	0.09
12	6	31.3	46	6	20	52	16.21
14	7	52.6	0	1	16	1	0.48
16	8	41.4	0	1	10	1	0.34

Table 8c. Overall Percent Removal - Chromium

Column No.	Soil No.	iMr (mg)	% Rem PV 1	Add. % Rem	No. PV	% Rem Total	Mass Rem.
1	1	505.7	50	10	20	60	302.09
3	2	554.7	39	8	20	47	260.59
5	3	476.4	46	9	20	55	260.59
7	4	495.5	60	10	20	70	346.09
9	5	483.5	61	10	20	71	346.42
11	6	594.0	49	1	20	50	293.07
13	7	592.6	47	12	14	59	348.63
15	8	386.5	46	14	10	60	229.00
2	1	432.2	56	10	20	66	285.00
6	3	476.4	37	24	20	61	291.66
8	4	479.0	46	10	20	56	266.36
10	5	508.7	60	10	20	70	359.59
12	6	582.9	48	1	20	49	283.27
14	7	549.2	58	13	16	71	390.17
16	8	588.9	50	9	10	59	348.50

Table 8d. Overall Percent Removal – Nickel

Column No.	Soil No.	iMr (mg)	% Rem PV 1	Add. % Rem	No. PV	% Rem Total	Mass Rem.
1	1	218.9	51	11	20	62	137.00
3	2	213.0	45	12	20	57	121.63
5	3	200.3	48	13	20	61	121.63
7	4	228.4	39	12	20	51	118.07
9	5	256.7	46	11	20	57	145.10
11	6	118.0	91	6	20	97	114.53

Column No.	Soil No.	iMr (mg)	% Rem PV 1	Add. % Rem	No. PV	% Rem Total	Mass Rem.
13	7	543.1	15	3	14	18	100.13
15	8	193.9	30	7	10	37	72.84
2	1	218.9	40	11	20	51	110.90
6	3	200.3	36	15	20	51	102.35
8	4	232.4	31	13	20	44	101.48
10	5	271.8	42	12	20	54	146.24
12	6	105.7	90	5	20	95	101.06
14	7	280.6	32	9	16	41	116.86
16	8	541.3	17	3	10	20	107.69

Table 8e. Overall Percent Removal - Zinc

Column No.	Soil No.	iMr (mg)	% Rem PV 1	Add. % Rem	No. PV	% Rem Total	Mass Rem.
1	1	479.7	28	9	20	37	179.40
3	2	403.9	30	7	20	37	147.48
5	3	380.2	32	7	20	39	147.48
7	4	495.4	31	8	20	39	192.09
9	5	522.7	29	8	20	37	194.84
11	6	168.9	73	6	20	79	132.50
13	7	653.5	16	4	14	20	132.66
15	8	533.7	12	3	10	15	86.31
2	1	479.7	27	10	20	37	180.54
6	3	380.2	22	13	20	35	133.38
8	4	490.8	24	9	20	33	158.69
10	5	361.3	31	7	20	38	138.55
12	6	149.2	70	6	20	76	114.66
14	7	358.0	20	6	16	26	91.88
16	8	682.9	17	3	10	20	135.40

Overall Percent Removal (5 Rem) is the percent of Initial Mass Retained (iMr) remov ed by tap water Rinses.
The % Rem is subdivided into the % Rem in Pore Volume 1 (PV 1) and the Additional % Rem by the rest of T1.

IMr = The Initial Mass Retained = the mass of a metal remaining in the soil column after the contamination an d initial drainage periods.
%Rem PV1 = % of iMr removed by rinse pore volume 1.
Add %Rem = additional % of iMr removed by rinses 2 through n.
No. PV = number of rinses of one pore volume each.
% Rem Total = total % of iMr removed by tap water rinses.
Mass Rem = total mass of a metal removed by tap water rinses after pore volume 1.

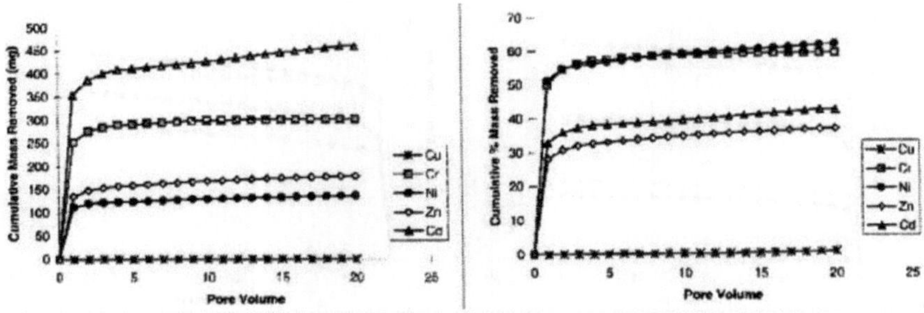

Figure 8a. Column 1 (Outwash Sand + 10% NFP) - Cumulative Mass Metal Removed and Cumulative Percent Mass Metal Removed

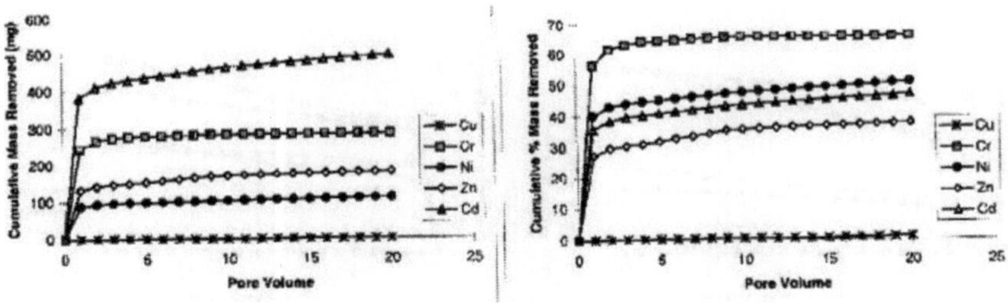

Figure 8b. Column 2 (Outwash Sand + 10% NFP) - Cumulative Mass Metal Removed and Cumulative Percent Mass Metal Removed

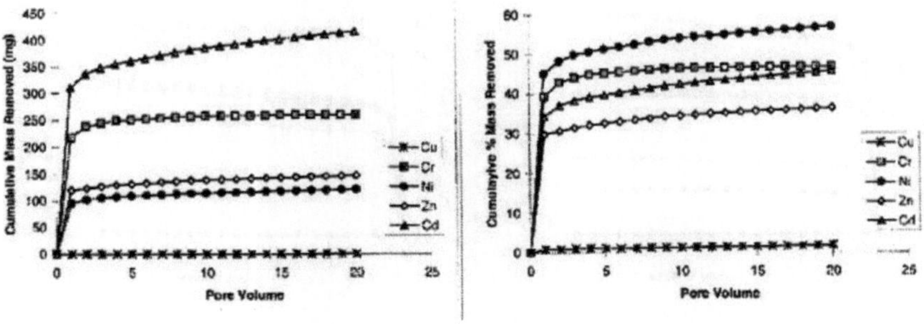

Figure 9. Column 3 (Fuller Till) - Cumulative Mass Metal Removed and Cumulative Percent Mass Metal Removed

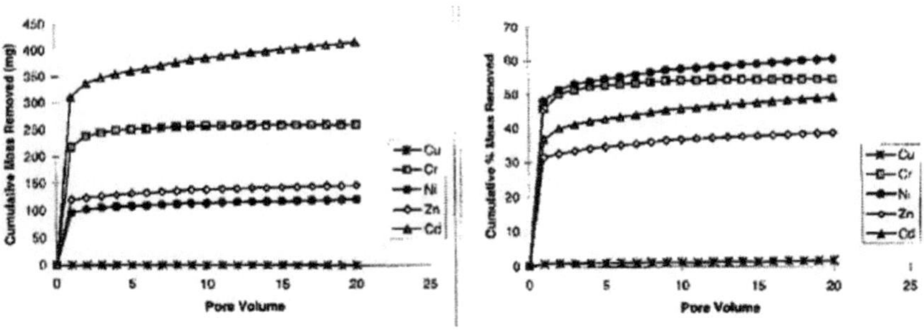

Figure 10a. Column 5 (Outwash Sand + 4% NFP) - Cumulative Mass Metal Removed and Cumulative Percent Mass Metal Removed

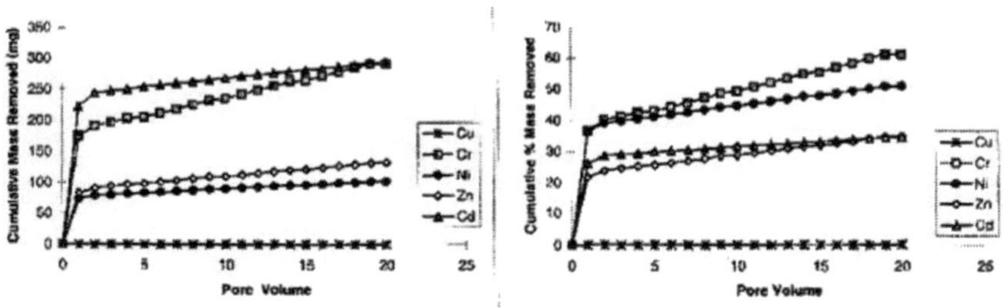

Figure 10b. Column 5 (Outwash Sand + 4% NFP) - Cumulative Mass Metal Removed and Cumulative Percent Mass Metal Removed

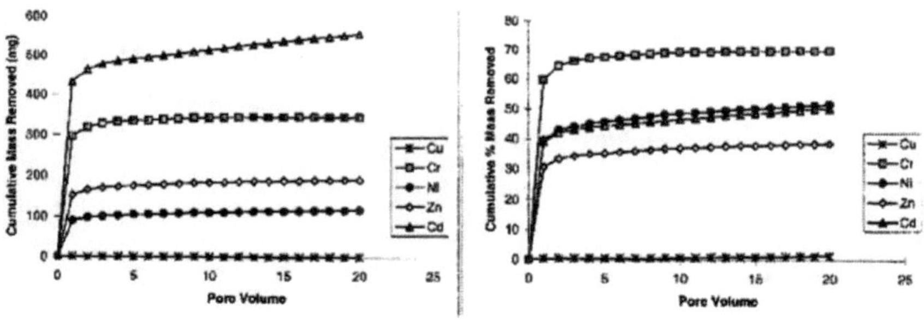

Figure 11a. Column 7 (Outwash Sand + 16% NFP) - Cumulative Mass Metal Removed and Cumulative Percent Mass Metal Removed

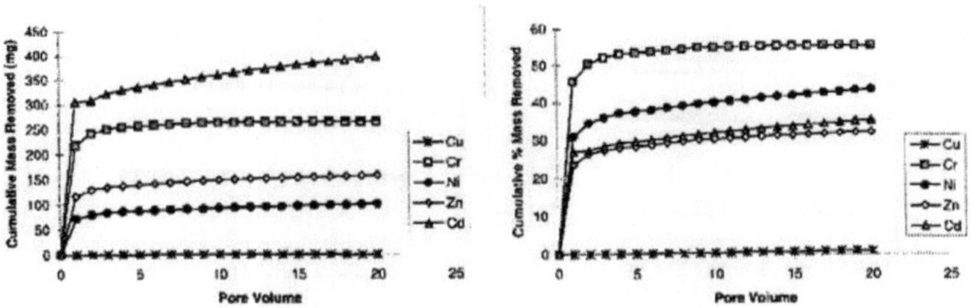

Figure 11b. Column 8 (Outwash Sand + 16% NFP) - Cumulative Mass Metal Removed and Cumulative Percent Mass Metal Removed

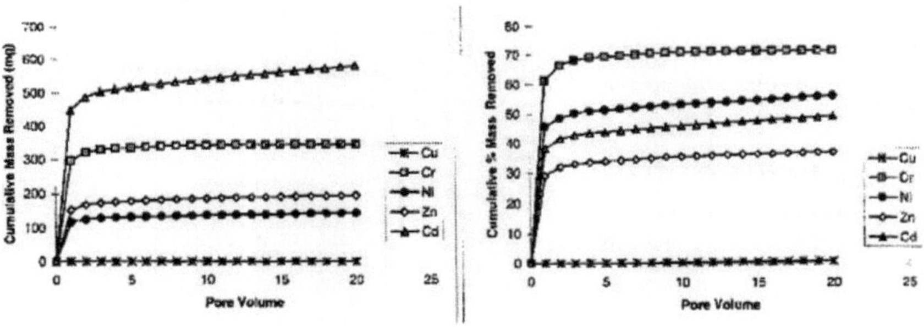

Figure 12a. Column 9 (Outwash Sand + 21% NFP) - Cumulative Mass Metal Removed and Cumulative Percent Mass Metal Removed

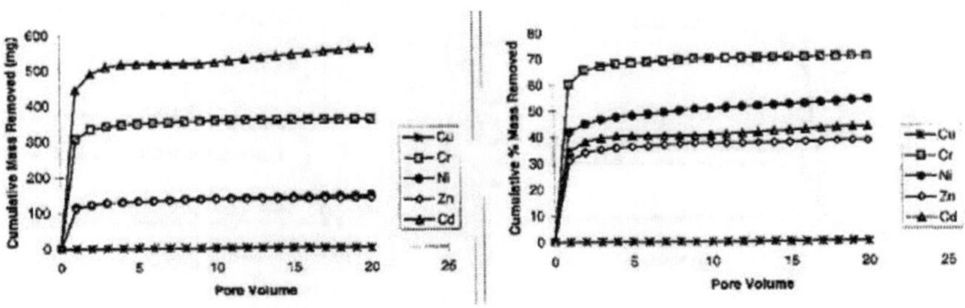

Figure 12b. Column 10 (Outwash Sand + 21% NFP) - Cumulative Mass Metal Removed and Cumulative Percent Mass Metal Removed

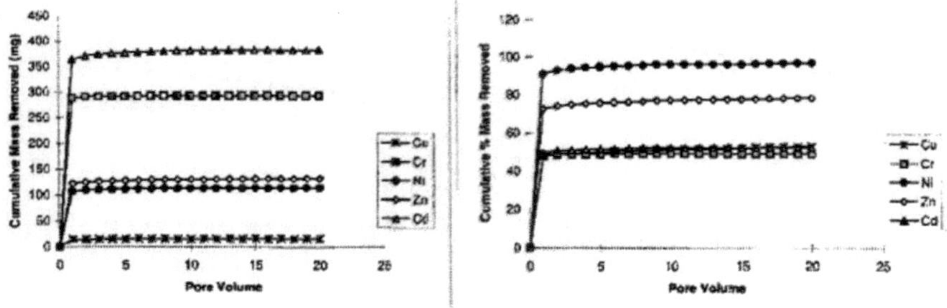

Figure 13a. Column 11 (Fine Sand) - Cumulative Mass Metal Removed and Cumulative Percent Mass Metal Removed

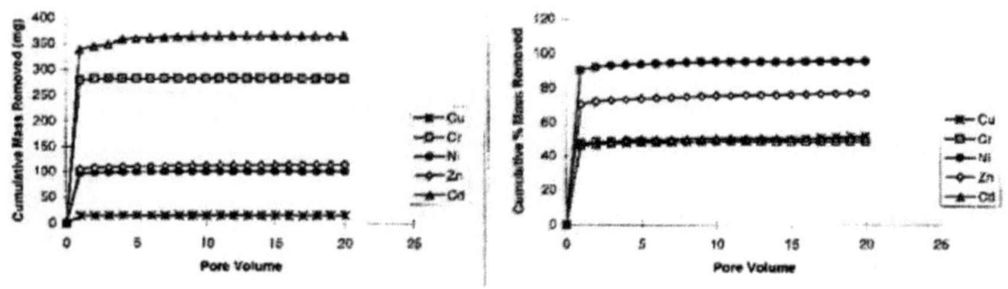

Figure 13b. Column 12 (Fine Sand) - Cumulative Mass Metal Removed and Cumulative Percent Mass Metal Removed

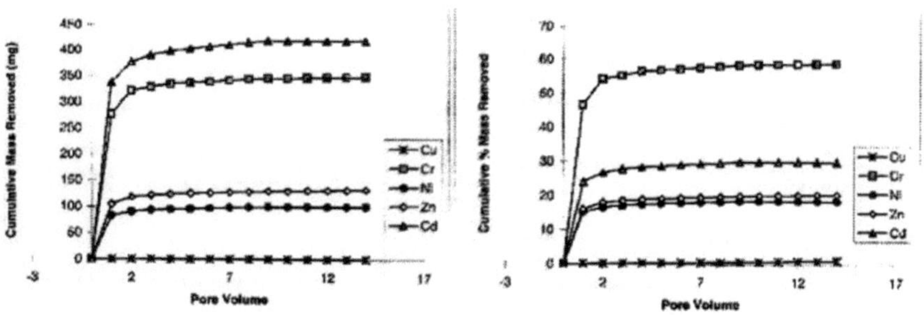

Figure 14a. Column 13 (Outwash Sand + 5% PF) - Cumulative Mass Metal Removed and Cumulative Percent Mass Metal Removed

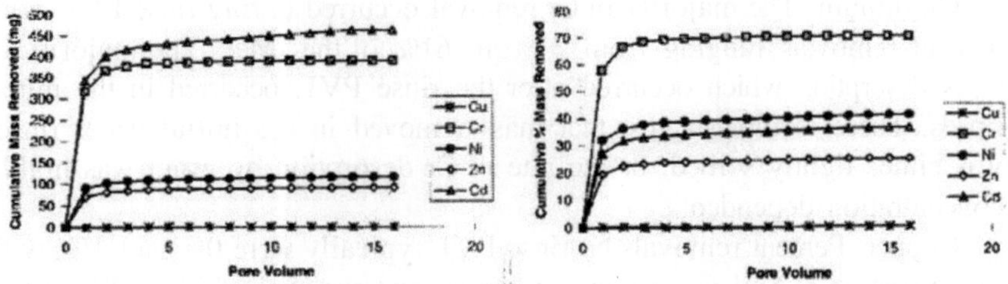

Figure 14b. Column 14 (Outwash Sand + 5% PF) - Cumulative Mass Metal Removed and Cumulative Percent Mass Metal Removed

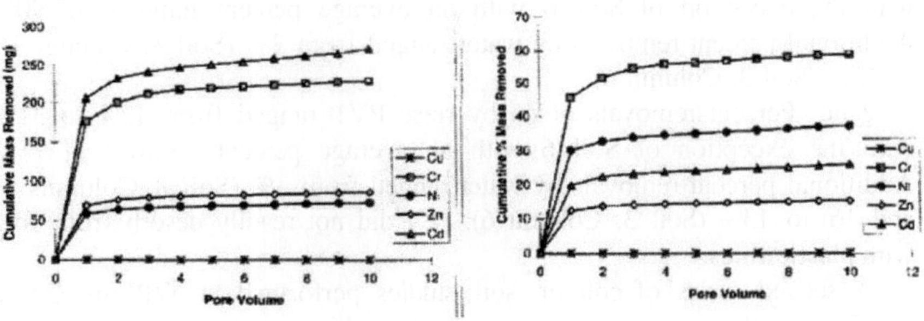

Figure 15a. Column 15 (Outwash Sand + 5% PF + 5% NPF) - Cumulative Mass Metal Removed and Cumulative Percent Mass Metal Removed

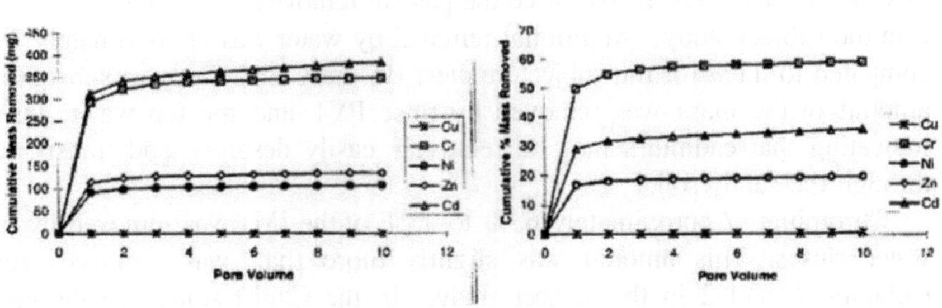

Figure 15b. Column 16 (Outwash Sand + 5% PF + 5% NPF) - Cumulative Mass Metal Removed and Cumulative Percent Mass Metal Removed

Cadmium The first pore volume rinse (PV1) removed 20% to 50% of the iM_R. Additional percent removals by Treatment 1 ranged from 3% to 12%.

Chromium The majority of Cr removal occurred during rinse PV1, with percent removals ranging from 37% to 61% of the **iM**$_R$. The majority of mass desorption which occurred after the rinse PV1, occurred in the initial rinses. These results suggest that mass removed in the initial water rinses was either lightly sorbed, or the rate of Cr desorption by water was highly concentration dependent.

Copper Percent removals by rinse PV1 typically were 0% and 1%, with the exception of Soil 6 with an average percent removal of 47%. Water was ineffective in desorbing Cu. This may have been a function of the low initial sorbed mass of Cu, due to the low Cu concentration in the contaminating solution.

Nickel Percent removals of Ni by rinse PV1 ranged from 15% to 48%, with the exception of Soil 6 with an average percent removal of 90%. Additional percent removals by water ranged from 3% (Soil 8, Column 16) to 15% (Soil 3, Column 6).

Zinc Percent removals of Zn by rinse PV1 ranged from 12% to 31%, with the exception of Soil 6 with an average percent removal of 71%. Additional percent removals by water ranged from 3% (Soil 8, Columns 15 and 16) to 13% (Soil 3, Column 6). Zn did not readily desorb from soils with plastic fines.

A second series of column soil studies performed at WPI by Caniff (1993) used the same procedures as in the subject study. However, in the Caniff studies soil columns were contaminated with only one metal, not a mixed-metal solution, and only one soil was used, Soil 1, the Outwash Sand + 10% NPF.

Cadmium Approximately 60% of the **iM**$_R$ was removed by tap water rinse PV1. This was almost twice the percent removed from Columns 1 and 2 in the subject study. Additional removal by water was approximately 3%, compared to 11% for the subject studies. In the Caniff study, a substantial amount of the mass was removed by rinse PV1 and the tap water rinses, indicating that cadmium may be relatively easily desorbed and transported through the sandy soil.

Chromium Approximately 63% to 73% of the **iM**$_R$ was removed by tap water rinses. This amount was slightly more than was removed from Columns 1 and 2 in the subject study. In the Caniff study, a substantial amount of the mass was removed by rinse PV1 and the tap water rinses, indicating chromium may be relatively easily desorbed and transported through the sandy soil.

Copper Results by Caniff were similar to the subject study. Cu was not removed by the tap water rinses.

Nickel Approximately 63% to 76% of the **iM**$_R$ was removed by tap water treatment, the same as for the subject study. The results indicate nickel may be relatively easily desorbed and transported through the sandy soil.

Zinc Approximately 61% to 63% of the $\mathbf{iM_R}$ was removed by the tap water treatment. This was almost twice the removal from Columns 1 and 2 in the subject study, indicating zinc may be relatively easily desorbed and transported through sandy soil.

4.3 Desorption Effectiveness

Desorption effectiveness is the percent of the residual mass sorbed $\mathbf{rM_S}$ for a metal which is removed by the water treatment, where $\mathbf{rM_S}$ equals the initial mass retained $\mathbf{iM_R}$ minus the mass in rinse pore volume one (PV1): $\mathbf{rM_S = iM_R - M_{PV1}}$. Mass removed in rinse PV1 was considered not to be sorbed, but to be dissolved-phase mass remaining in the pedicular water pore solution that did not drain during the drainage period. Thus, desorption effectiveness equals the total mass of the metal removed $\mathbf{TM_{Rem}}$ divided by residual mass sorbed $\mathbf{rM_S}$; **% Rem** = $\mathbf{TM_{Rem} / rM_S}$. Tables 9a through 9e present desorption effectiveness values.

Table 9a. Desorption Effectiveness - Cadmium

Column No.	Soil No.	Ms (mg)	% Rem	Mass Rem	No. PV
1	1	718.61	15	107.79	19
3	2	597.28	18	107.51	19
5	3	533.28	20	106.66	19
7	4	671.67	19	127.62	19
9	5	721.52	18	129.88	19
11	6	368.28	5	18.41	19
13	7	1064.5	8	85.16	13
15	8	823.76	8	65.90	9
2	1	684.57	17	116.38	19
6	3	622.16	12	74.66	19
8	4	819.39	11	90.13	19
10	5	835.59	14	116.98	19
12	6	384.65	7	26.93	19
14	7	914.16	13	118.84	15
16	8	760.15	10	76.02	9

Table 9b. Desorption Effectiveness - Chromium

Column No.	Soil No.	Ms (mg)	% Rem	Mass Rem	No. PV
1	1	235.50	20	47.10	19
3	2	336.68	13	43.77	19
5	3	258.38	16	41.34	19
7	4	199.02	25	49.76	19
9	5	187.18	27	50.54	19
11	6	305.25	1	3.05	19
13	7	316.06	23	72.69	13
15	8	210.29	25	52.57	9
2	1	189.25	22	41.64	19
6	3	301.44	39	117.56	19
8	4	260.15	18	46.83	19
10	5	201.46	26	52.39	19
12	6	303.44	1	3.03	19
14	7	231.33	31	71.71	15
16	8	295.99	19	56.24	9

Table 9c.. Desorption Effectiveness - Copper

Column No.	Soil No.	Ms (mg)	% Rem	Mass Rem	No. PV
1	1	52.69	1	0.53	19
3	2	52.24	1	0.52	19
5	3	52.24	1	0.52	19
7	4	52.77	1	0.53	19
9	5	52.64	1	0.53	19
11	6	16.06	10	1.61	19
13	7	41.30	1	0.41	13
15	8	52.99	0	0.00	9
2	1	52.67	1	0.53	19
6	3	52.64	0	0.00	19
8	4	52.63	1	0.53	19
10	5	53.52	0	0.00	19
12	6	16.91	11	1.86	19
14	7	52.39	1	0.52	15
16	8	41.30	1	0.41	9

Table 9d. Desorption Effectiveness - Nickel

Column No.	Soil No.	Ms (mg)	% Rem	Mass Rem	No. PV
1	1	106.34	23	24.46	19
3	2	116.81	22	25.70	19
5	3	104.11	24	24.99	19
7	4	138.31	20	27.66	19
9	5	138.56	19	26.33	19
11	6	10.22	66	6.75	19
13	7	460.32	4	18.41	13
15	8	135.42	11	14.90	9

Column No.	Soil No.	Ms (mg)	% Rem	Mass Rem	No. PV
2	1	131.02	18	23.58	19
6	3	127.22	23	29.26	19
8	4	160.38	18	28.87	19
10	5	157.35	20	31.47	19
12	6	10.40	54	5.62	19
14	7	190.23	14	26.63	15
16	8	449.56	4	17.98	9

Table 9e. Desorption Effectiveness - Zinc

Column No.	Soil No.	Ms (mg)	% Rem	Mass Rem	No. PV
1	1	344.18	13	44.74	19
3	2	283.49	10	28.35	19
5	3	259.79	10	25.98	19
7	4	342.94	12	41.15	19
9	5	369.58	11	40.65	19
11	6	46.09	21	9.68	19
13	7	548.23	5	27.41	13
15	8	484.70	4	19.39	9
2	1	348.73	14	48.82	19
6	3	297.04	17	50.50	19
8	4	374.61	11	41.21	19
10	5	248.77	10	24.88	19
12	6	44.20	22	9.72	19
14	7	287.75	8	23.02	15
16	8	568.59	4	22.74	9

Desorption Effectiveness is the percent of Mass Sorped (Ms) removed by the tap water rinses.
Desorption Effectiveness does not include mass removed in Pore Volume 1 (PV1)
Ms = Mass Sorbed = Initial Mass Retained - mass removed in Pore Volume 1 + iMr $- M_{PV1}$.
%Rem = % of Ms removed by tap water rinses.
Mass Rem = Mass Removed by the tap water rinses.

The following modifiers are used to describe desorption effectiveness:

- Ineffective, 0% to 3% removal
- Slightly effective, 3% to 10% removal
- Low effectiveness, 10% to 20% removal
- Low-moderately effective, 20% to 30% removal
- Moderately effective, 30% to 50% removal
- Highly effective, 50% to 80% removal
- Extremely effective, 80% to 100% removal

Cadmium Water had a slight to low effectiveness in desorbing Cd. Desorption ranged from 5% (Soil 6, Column 11) to 20% (Soil 3, Column 5), with an average of 8%. This data indicates that over time, water will leach some sorbed Cd from the soil column

Chromium With the exception of Soil 6 which had 1% desorption, water was low-moderate to moderately effective in desorbing Cr. Desorption (exclusive of Columns 11 and 12) ranged from 13% (Soil 2, Column 3) to 39% (Soil 3, Column 6), with an average of 15%. This data indicates that over time, water will leach sorbed Cr from most soils.

Copper With the exception of the Soil 6 which had an average of 10% desorption, water was not effective in desorbing Cu. Typical desorption was 1%. These data suggest that Cu is not readily desorbed by water and may not be particularly mobile in the soil column.

Nickel Desorption of Ni by water varied with soil type, ranging from 4% (Soil 7, Column 13) to 66% (Soil 6, Column 11), with an average desorption of 23%. Water was not effective in desorbing Ni from soils with plastic fines. Column 13 (Soil 7, Outwash Sand + 5% PF) and Column 16 (Soil 8, Outwash Sand + 5% PF and 5% NPF), which both had a high M_S (a function of the Ni concentration in the contaminating solution), both had only 4% desorption by water.

Zinc In general, water had a slight to low effectiveness in desorbing Zn from all soils. Desorption ranged from 4% (Soil 8, Column 15) to 22% (Soil 6, Column 12), with an average of 11%. Water was least effective in desorbing Zn from soils with plastic fines. Soils 7 and 8 which both had 5% PF had minimal desorption of Zn by water. Water was most effective in desorbing Zn from Soil 6.

4.4 Desorption Rate Expressions

In Figures 16 through 23, cumulative mass removed (y axis) is plotted as a function of the pore volume number. The graphs begin with PV2, PV3, or PV4 to eliminate mass initial removed as dissolved-phase in the contamination solution trapped in the pedicular water. The intent of the graphing was to evaluate long-term mass removal after the initial dissolved-phase and weekly sorbed mass were removed in the initial rinse flushes.

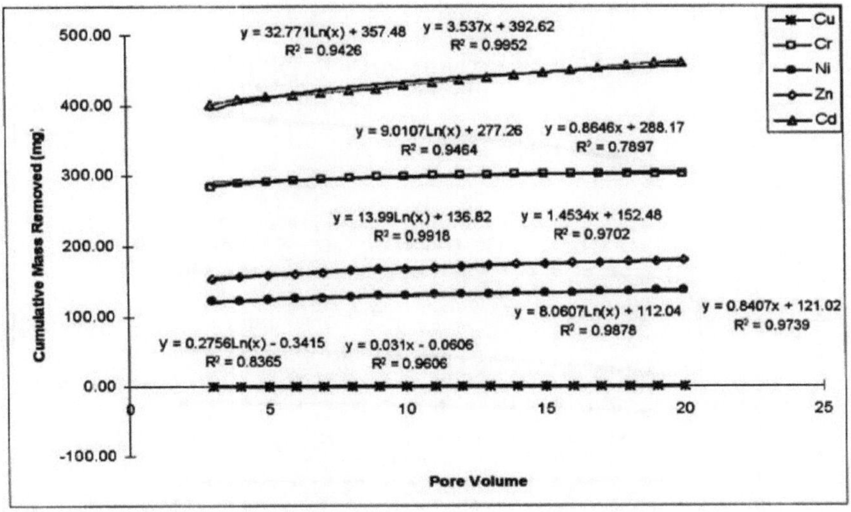

Figure16a. Column 1 (Outwash Sand + 10% NPF) - PV3 through PV20 - Cumulative Mass Metal Removed

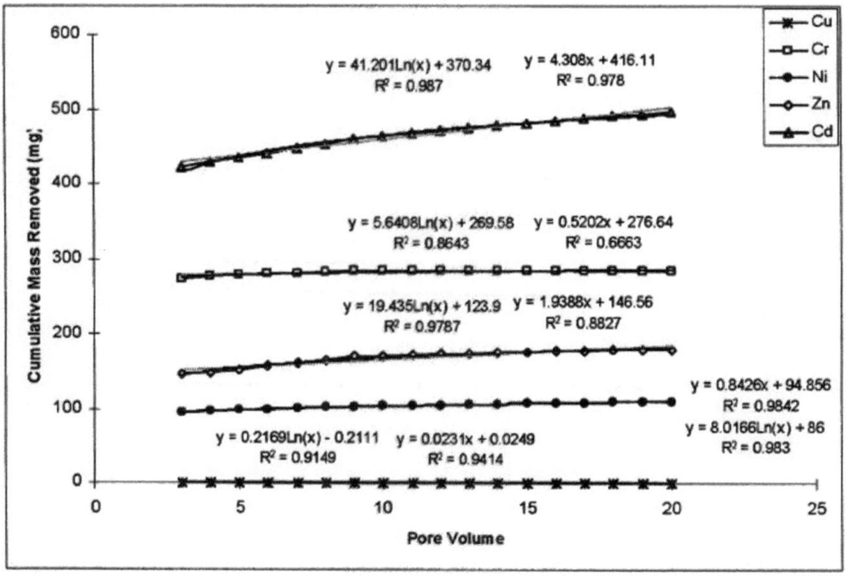

Figure 16b. Column 2 (Outwash Sand + 10% NPF) - PV3 through PV20 - Cumulative Mass Metal Removed

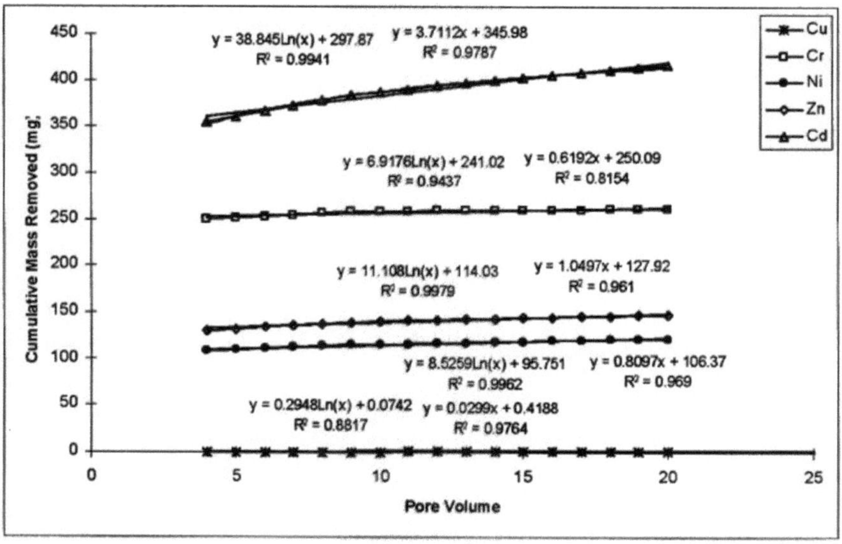

Figure 17. Column 3 (Soil 2 – Fuller Till) – PV4 through PV20 - Cumulative Mass Metal Removed

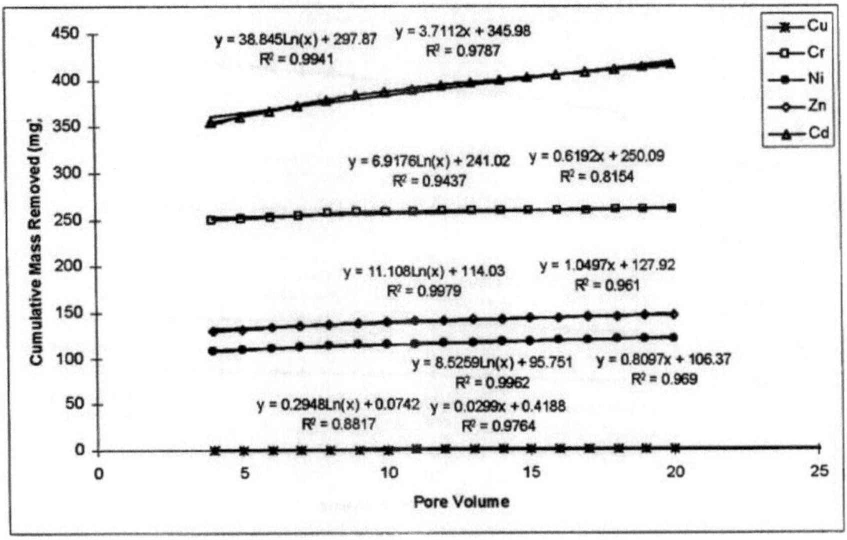

Figure 18a. Column 5 (Soil 3 - Outwash Sand + 4% NPF) - PV3 through PV20 - Cumulative Mass Metal Removed

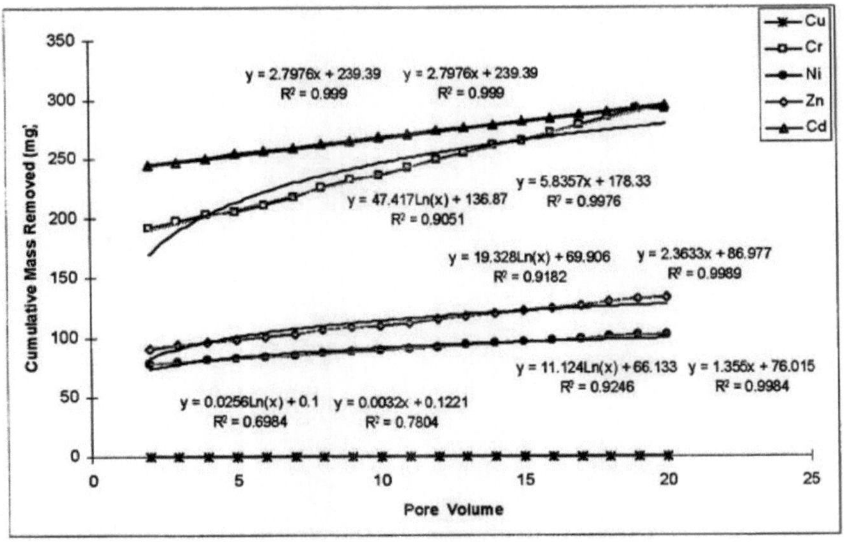

Figure 18b. Column 6 (Soil 3 - Outwash Sand + 4% NPF) - PV3 through PV20 - Cumulative Mass Metal Removed

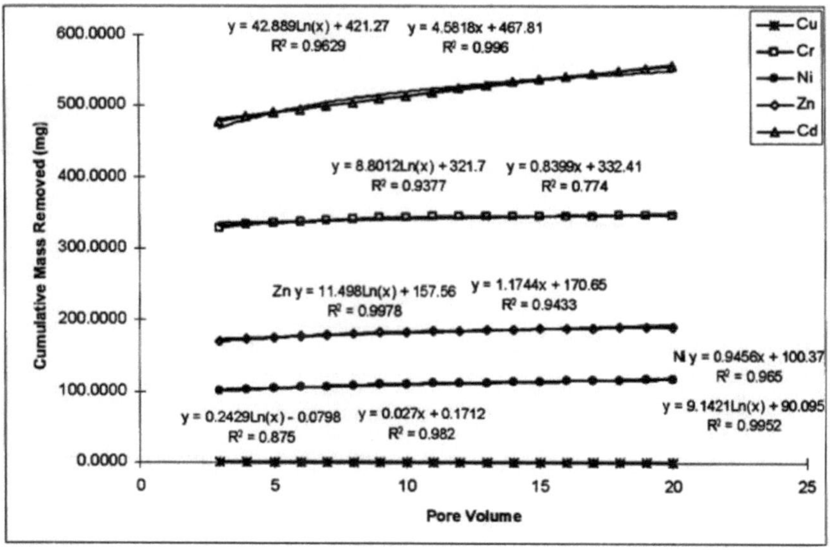

Figure 19a. Column 7 (Soil 4 - Outwash Sand + 16% NPF) - PV3 through PV20 - Cumulative Mass Metal Removed

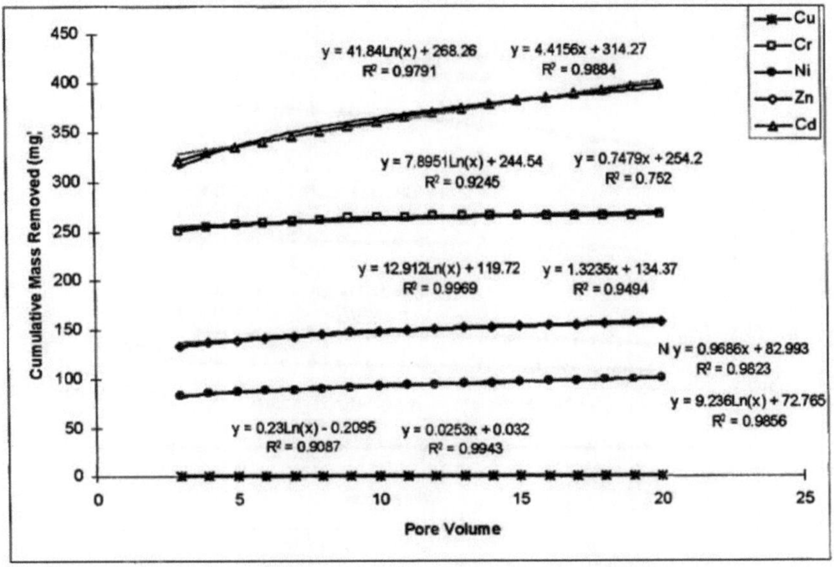

Figure 19b. Column 8 (Soil 4 - Outwash Sand + 16% NPF) - PV3 through PV20 - Cumulative Mass Metal Removed

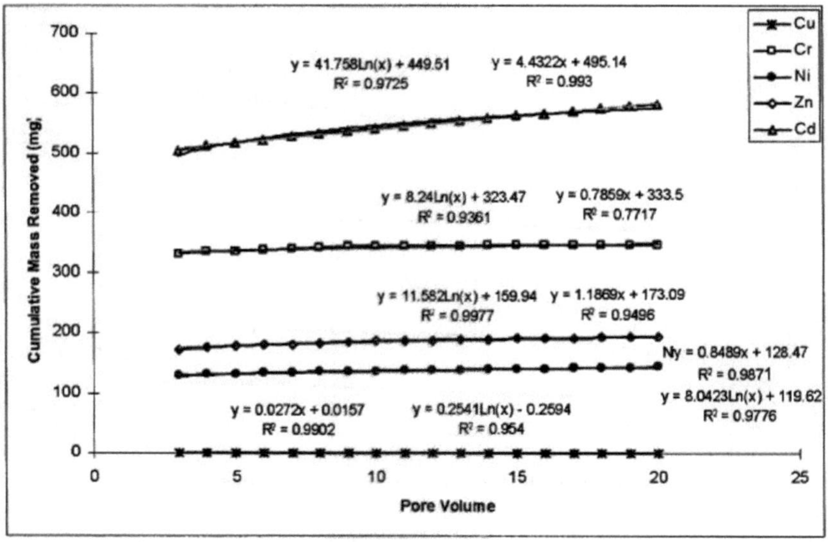

Figure 20a. Column 9 (Soil 5 - Outwash Sand + 21% NPF) - PV3 through PV20 - Cumulative Mass Metal Removed

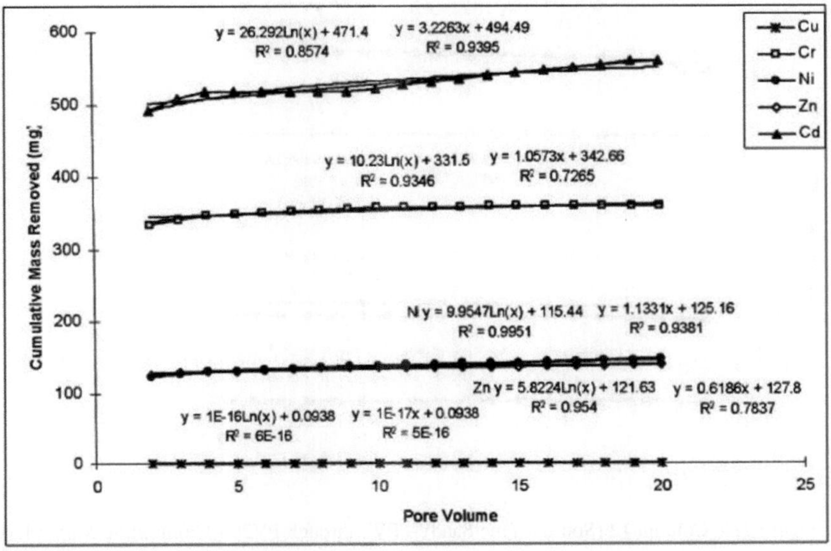

Figure 20b. Column 10 (Soil 3 - Outwash Sand + 21% NPF) – PV2 through PV20 - Cumulative Mass Metal Removed

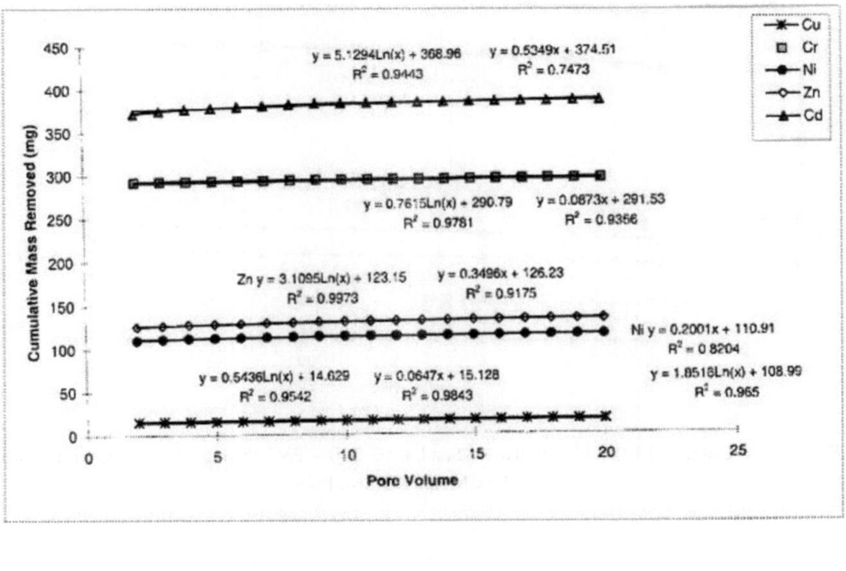

Figure 21a. Column 11 (Soil 6 – Fine Sand) – PV2 through PV20 - Cumulative Mass Metal Removed

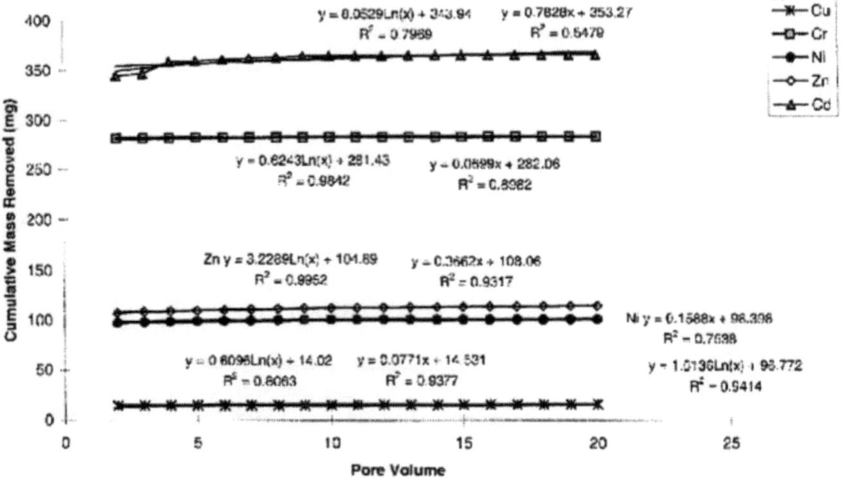

Figure 21b. Column 12 (Soil 6 – Fine Sand) – PV2 through PV20 - Cumulative Mass Metal Removed

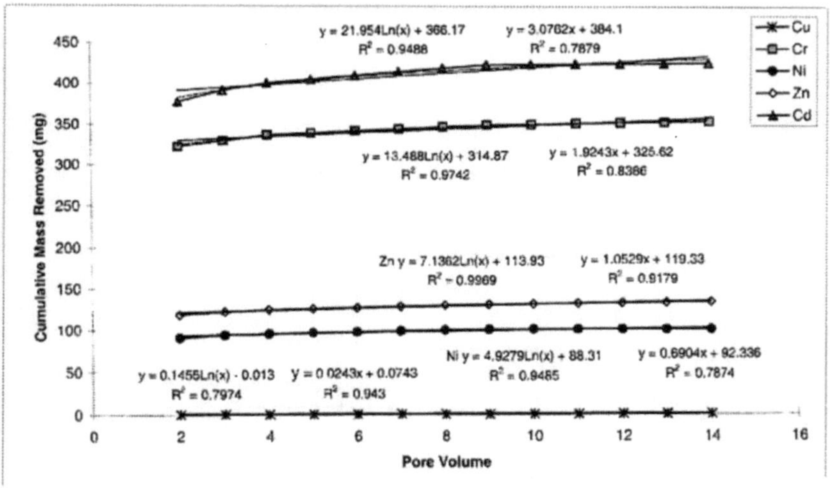

Figure 22a. Column 13 (Soil 7 - Outwash Sand + 5% PF) – PV2 through PV14 - Cumulative Mass Metal Removed

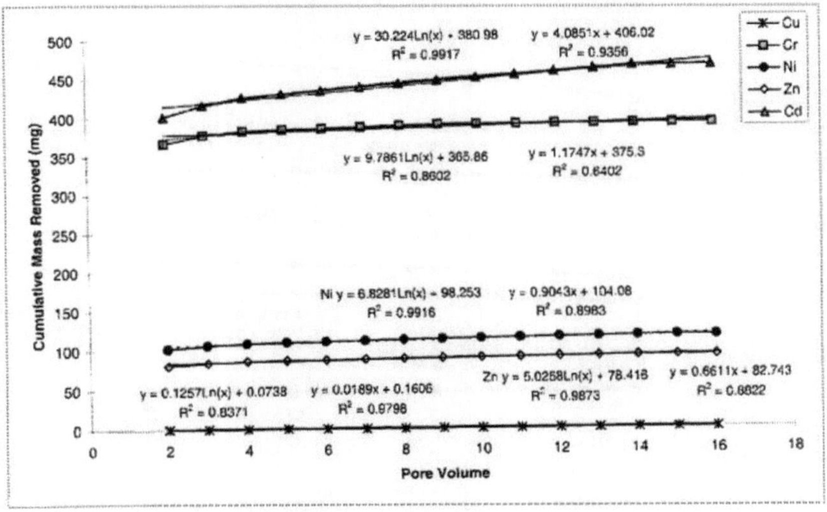

Figure 22b. Column 14 (Soil 7 - Outwash Sand + 5% PF) – PV2 through PV14 - Cumulative Mass Metal Removed

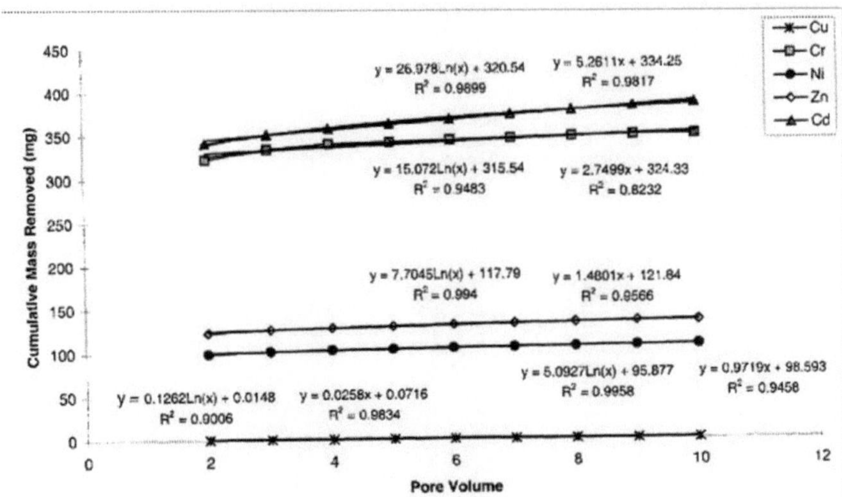

Figure 23a. Column 15 (Soil 8 - Outwash Sand + 5% PF + 5% NPF) – PV2 through PV10 - Cumulative Mass Metal Removed

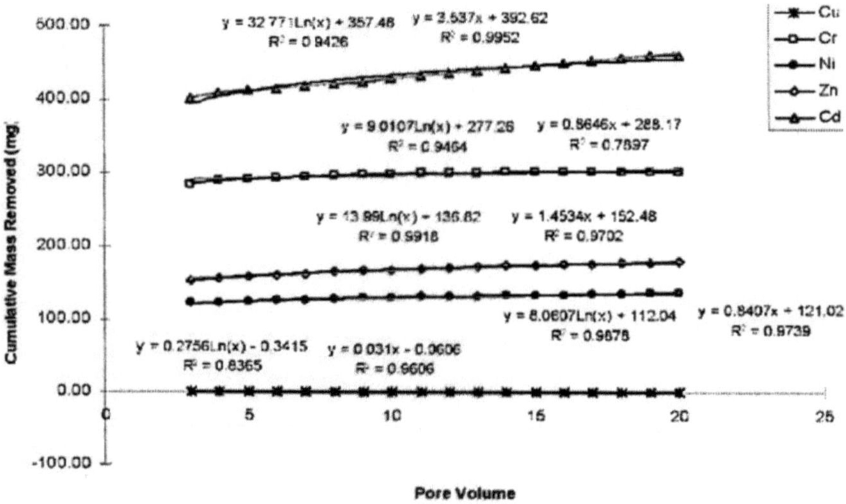

Figure 23b. Column 16 (Soil 8 - Outwash Sand + 5% PF + 5% NPF) – PV2 through PV10 - Cumulative Mass Metal Removed

Two curves were then fit to the data in an attempt to develop a predictive rate expression for each metal and soil type: (1) straight line regression function $y = mx + b$, and (2) natural logarithmic function $y = (C)Ln(x) + b$. A least squares value $\mathbf{r}^2$ was calculated for each equation to evaluate the quality of the data fit. Results are summarized in Table 10.

Table 10a. Summary of Curve Fitting Analysis – Cumulative Mass Removed – Cadmium, Copper, Chromium

				Cadmium				Copper				Chromium		
Col #	Soil #	PV	mX	R^2	(C)Ln(x)	R^2	mX	R^2	(C)Ln(x)	R^2	mX	R^2	(C)Ln(x)	R^2
1	1	3 - 20	3.540	0.9952	32.771	0.9426	0.031	0.9606	0.276	0.8365	0.865	0.7897	9.011	0.9464
2	1	3 - 20	4.308	0.9780	41.201	0.9870	0.023	0.9414	0.217	0.9149	0.520	0.6663	5.641	0.8643
3	2	4 - 20	3.711	0.9787	38.845	0.9941	0.030	0.9764	0.295	0.8817	0.619	0.8154	6.918	0.9437
5	3	3 - 20	3.711	0.9787	38.845	0.9941	0.03	0.9764	0.295	0.8817	0.619	0.8154	6.918	0.9437
6	3	2 - 20	2.798	0.999	2.798	0.999	0.003	0.7804	0.026	0.6984	5.836	0.9976	47.417	0.9051
7	4	3 – 20	4.582	0.9960	42.889	0.9629	0.027	0.9820	0.2429	0.8750	0.840	0.7740	8.801	0.9377
8	4	3 – 20	4.416	0.9884	41.84	0.9791	0.0253	0.9943	0.023	0.9087	0.7479	0.7520	7.895	0.9245
9	5	3 – 20	4.432	0.993	41.758	0.9725	0.254	0.9540	0.0272	0.9902	0.7859	0.7717	8.240	0.9361
10	5	2 - 20	3.226	0.9395	26.292	0.8574	-	-	-	-	1.057	0.7265	10.230	0.9346
11	6	2 - 20	0.535	0.7473	5.1294	0.9443	0.065	0.9843	0.544	0.9542	0.873	0.9356	0.762	0.9781
12	6	2 - 20	0.783	0.5479	8.053	0.7969	0.0771	0.9377	0.610	0.8063	0.070	0.8982	0.624	0.9842
13	7	2 - 14	3.076	0.7879	21.954	0.9488	0.024	0.9430	0.146	0.7974	1.924	0.8386	13.488	0.9742
14	7	2 - 16	4.085	0.9356	30.224	0.9917	0.0189	0.9798	0.1257	0.8371	1.175	0.6402	9.786	0.8602
15	8	2 - 10	4.572	0.9806	23.379	0.9834	0.018	0.9549	0.0917	0.9949	3.083	0.8703	16.608	0.9714
16	8	2 - 10	5.261	0.9817	26.978	0.9899	0.026	0.9834	0.126	0.9006	2.750	0.8232	15.072	0.9483

Table 10b. Summary of Curve Fitting Analysis – Cumulative Mass Removed – Nickel and Zinc

			Nickel				Zinc			
Col #	Soil #	PV	mX	R^2	(C)Ln(x)	R^2	mX	R^2	(C)Ln(x)	R^2
1	1	3 - 20	0.841	0.9739	8.061	0.9878	1.453	0.9606	13.99	0.9918
2	1	3 - 20	0.8426	0.9842	8.017	0.983	1.939	0.8827	19.435	0.9787
3	2	4 - 20	0.810	0.9690	8.526	0.9962	1.050	0.961	11.108	0.9979
5	3	3 - 20	0.81	0.9690	8.526	0.9962	1.050	0.961	11.108	0.9979
6	3	2 - 20	1.355	0.9984	11.124	0.9246	2.363	0.9989	19.328	0.9182
7	4	3 – 20	0.946	0.9650	9.142	0.9952	1.174	0.9433	11.498	0.9978
8	4	3 – 20	0.9686	0.9823	9.236	0.9856	1.324	0.9494	12.912	0.9969
9	5	3 – 20	0.849	0.9871	8.0426	0.9776	1.178	0.9496	11.582	0.9977
10	5	2 - 20	1.133	0.9381	9.955	0.9951	0.619	0.7837	5.822	0.9540
11	6	2 - 20	0.200	0.8204	1.852	0.9650	0.350	0.9175	3.120	0.9973
12	6	2 - 20	0.159	0.7538	1.514	0.9414	0.3662	0.9317	3.229	0.9952
13	7	2 - 14	0.690	0.7874	4.928	0.9485	1.053	0.9179	7.136	0.9969
14	7	2 - 16	0.904	0.8983	6.828	0.9916	0.661	0.8822	5.026	0.9873
15	8	2 - 10	1.066	0.9756	5.492	0.9925	1.176	0.9636	6.100	0.9948
16	8	2 - 10	0.972	0.9458	5.0927	0.9958	1.480	0.9566	7.705	0.9940

mX The slope value for the linear equation $y = mX + b$.

Ln(x) The coefficient C for the natural logarithmic equation $y = (C)Ln(x) + b$.

R^2 Least squares value

Soil	Columns		Soil	Columns		
1	1, 2	Outwash Sand + 10% NPF	5	9, 10	Outwash Sand + 21% NPF	PF = plastic fines
2	3, 4	Fuller Till	6	11, 12	Fine Sand	NPF = non plastic fines
3	5, 6	Outwash Sand + 4% NPF	7	13, 14	Outwash Sand + 5% PF	
4	7, 8	Outwash Sand + 16% NPF	8	15, 16	Outwash Sand + 5% PF + 5% NPF	

The equations are essentially kinetic rate expressions for zero and first order reactions. In the subject plots, the mass removed from soil is plotted rather than the mass (i.e. concentration) remaining in the soil. Given an initial concentration in soil, the equations theoretically could be used to estimate the pore volumes of water necessary to achieve some level of mass removal and resulting concentration in the soil. Local climate, hydrologic and physical soil data could then be used to convert pore volumes to time, providing a temporal expression.

It is noted the kinetic rate expressions are at the macro process or reactor level (e.g. soil column) not the actual micro intra-pore ion desorption level. The expressions are a function of the soil physical (e.g. hydraulic conductivity, porosity, and specific surface) and chemical (e.g. CEC) properties as well as the actual ionic activity at the sorption sites. In the soil column, kinetics may be largely transport controlled – i.e. the ability for ionic diffusion and/or advection transport – rather than sorbent/sorbate chemistry controlled.

The first order (logarithmic) expressions appear to have a very good fit for the nickel and zinc data, with r^2 values for both sets of equations being largely in the 0.98 to 0.99 range, and a moderate fit for the chromium data, with r^2 values in the 0.94 to 0.98 range. A combination of the first order and zero order (straight line) expressions appear to provide a good fit for the cadmium data, excepting the fine sand, with r^2 values generally in the 0.98 to 0.99 range. The zero order expression for copper has a moderate fit, with r^2 values generally greater the 0.94. Data fit for the other metals is scattered. None of the soils appeared to have a particular affinity for either of the rate expressions.

4.5 Conclusions

1. The first pore volume rinse of water was moderately to highly effective in removing dissolved-phase Cd, Cr, Ni, and Zn which remained in the undrained pore solution at the completion of the contamination period, or mass that was very weakly sorbed. This results indicates that rapid response is required for a high-concentration, liquid spill directly to soil. Infiltration and percolation of precipitation can rapidly transport dissolved-phase ions in pedicular water and weakly sorbed ions downward into the soil column and possibly to groundwater.

2. In general, application of water after rinse PV1 had a low effectiveness for desorbing Cd, Cr, Ni, and Zn. However, metal mass was more or less continually removed by water, which indicates that water will gradually leach sorbed metal mass from the soil column.

3. Water was not effective in desorbing Cu from any of the soils except Soil 6, Fine Sand. For this soil, substantial mass was removed by rinse PV1 and additional mass was desorbed by subsequent rinses.

4. Water was effective for desorbing Cr from all soils. However, the majority of mass was removed in the first few rinses, and probably was dissolved-phase mass or weakly sorbed mass.

5. Curve-fitting procedures produced very good reactor-level, first-order rate expressions for leaching of nickel and zinc, and a moderate first order expression for chromium. A combination of zero and first-order expressions provided a good fit for the cadmium data. The zero-order expression for copper produced a moderate fit. Data fit for the other metals is scattered. None of the soils appeared to have a particular affinity for either of the rate expressions.

REFERENCES

Brady, N.C., 1990. The Nature and Properties of Soils 10 th Addition. Macmillian Publishing Company, New York, NY.

Caniff, M.R., 1993. Heavy Metals Release From Contaminated Soil. M.S. Thesis, WPI, Worcester, MA.

Macaulay, M.J., 1994. Enhancement of Heavy Metals Removal From Contaminated Soils. M.S. Thesis, WPI, Worcester, MA.

O'Shaughnessy, J.C. and D'Andrea, R.A., 1994. Final Report On Enhancement of In Situ Soil Flushing for Heavy Metals Removal. Contract No. R 818627-01-0. Department of Civil Engineering, WPI, Worcester, MA.

O'Shaughnessy, J.C. and Stone, A.D., 1997. Leaching process for removal of heavy metals in soil. Proc. Ind. Waste Conf. 1997, 52 nd, 377-.

Overcash, M.R. and Pal, D., 1979. Designs of Land Treatment Systems for Industrial Wastes. Ann Arbor Science Publishers, Inc., Ann Arbor, MI.

Sims, R.C. and Wagner, K., 1983. In situ treatment technologies applicable to large quantities of hazardous waste contaminated soils. Proc. Nat. Conf. on Management of Uncontrolled Hazardous Waste Sites, Washington, D.C.

St. Germain, T.J., 1994 Sorption Phenomena in Heavy Metal Contaminated Soils. M.S. Thesis, WPI, Worcester, MA.

USEPA (U.S. Environmental Protection Agency). September 1987. The RPM Primer: An Introductory Guide to the Role and Responsibilities of the Superfund Remedial Project Manager. OSWER Directive 9344.1-02. Office of Emergency Remedial response, Washington D.C.

Venkataramanappa, S., 1994. Sorption and Desorption of Selected Heavy Metals in Soil. M.S. Thesis, WPI, Worcester, MA.

This page intentionally left blank

CHAPTER 16

SORPTION OF ZINC IN BENTONITE, ILLITE AND KAOLIN CLAY USING COLUMN LEACHING TESTS

Maria Antonia Tanchuling, Mohammod Ruhul Amin Khan and Osamu Kusakabe
Tokyo Institute of Technology, Dept of Civil Engineering, Tokyo, Japan

Abstract: Sorption is an important contaminant transport process with special significance in retarding the movement of contaminants. Determination of the capability of a clay liner to attenuate contaminants is necessary for the proper design of offshore waste containment sites. The sorption process in clay can be investigated using the column leaching test where a solution with a constant concentration is made to leach through a normally consolidated clay layer. In this study, two column tests were conducted using two different kinds of clay in each test. Bentonite and illite clay were used as the sorbent material and zinc nitrate solution was used as the contaminant made to leach through the clay layer. Results show that the amount of zinc retained in the clay decreases as the distance from the source of contamination increase s. Batch equilibrium tests were also conducted for three kinds of clay --- bentonite, illite and kaolin. For the range of concentrations used, linear adsorption isotherms were formed. Comparing the three kinds of clay, bentonite shows t he highest attenuating capacity while kaolin clay has the least capacity to retain zinc. Comparing the results of the batch tests with the column leaching tests, it can be shown that batch equilibrium tests overestimate the attenuation capacity of the soil.

Keywords: sorption, partitioning coeffiecients, column leaching tests, batch tests

1. INTRODUCTION

Sorption is an important process affecting the fate and transport of contaminants in the subsurface. Sorption includes the processes of 1) adsorption, where a solute clings to a solid surface, 2) ion exchange, where cations may be attracted to the region close to a negatively charged clay mineral surface, 3) chemisorption which occurs when the solute is incorporated on a sediment, soil or rock surface by a chemical reaction, and 4) absorption where the solute diffuses into the interior surfaces of porous aquifer materials (Fetter, 1999). Because of sorption, the advance of the contaminant plume is retarded compared to the solute front where advection is the only transport process occurring.

One of the important parameters to quantify sorption is the retardation factor which is the ratio of the average linear groundwater velocity to that of the average velocity of the solute front where the concentration is one-half of the original (Fetter, 1999). One way of determining its value is through the use of column leaching tests where a solution with a constant input concentration is made to leach through a normally consolidated clay layer. The clay is normally consolidated to simulate the stress conditions found at the bottom of landfills where the effective stresses prior to the landfilling operations have never been higher than the stresses after the landfilling has started. This study used the column leaching test to study parameters related to sorption, such as retardation factors, and also to be able to quantify the amount of contaminant retained in the layers.

2. MATERIALS AND METHODS

Clay materials used were bentonite, illite and kaolin. The physical properties of the clay used are shown in Table 1 and were based on data reported in Reddi and Inyang (2000). All types of clay were supplied by Hayashi Kasei Corporation. Bentonite (montmorillonite) is characterized by a high surface area, 80 percent of which can be attributed to internal surfaces. The montmorillonite clays have a high activity and high liquid limit. Water enters easily between the layers, resulting in swelling. Illite is bonded together by potassium ions. The potassium ions, fixed between the layers, prevent swelling of the mineral and provide little interlayer surface area for cation exchange. Kaolinite has the least amount of substitution and the lowest surface area and cation-exchange capacity. The combination of the high specific surface area and the significant surface charge make clay important in the development of contaminant attenuating processes or contaminant accumulation mechanism in soil.

Table 1. Properties of Clay Used

	Bentonite (Montmorillonite)	Illite	Kaolinite
Specific surface (m^2/g)	700 – 840	65 – 100	10 – 20
Specific gravity	2.35 – 2.7	2.6 – 3	2.6 – 2.68
CEC (mEq/100g)	80 – 150	10 – 40	3 – 15
Liquid limit (%)	100 – 900	60 – 120	30 – 100
Plastic limit (%)	50 – 100	35 – 60	25 – 40

Zinc used was from the compound zinc nitrate hexahydrate supplied by Kaso Chemical Co. Ltd. Tokyo (Dog Brand Pure Chermicals). The properties of the compound used are shown in Table 2. Zinc was chosen as the contaminant to be used since it is commonly found in municipal landfill leachate as reported in several publications. Moreover, it is also relatively less toxic than other heavy metals, and thus pose less risks in handling and disposal.

Table 2. Properties of Zinc

Compound used	Zinc nitrate hexahydrate
Formula	$Zn(NO_3)_2 \cdot 6 H_2O$
Molecular Weight	297.49 g
Density	2.065 mg/L
Solubility	0.5 part water
pH (100 g/L, 25 °C)	3.5 – 5.4

The column test system is illustrated in Figure 1. The column is made of translucent acrylic resin material, has an inner diameter of 150 mm and a height of 500 mm. It is attached to a base plate and sealed with a top lid with a hole for the piston of the air cylinder. An air cylinder is placed on top to supply the vertical loading to apply the pressure directly on the clay layer. The vertical pressure applied was 30 kPa which approximates the stress condition at a depth of 7.5 meters of soft clay deposit in natural sea bed assuming the unit weight of the saturated soil is 14 kN/m^3.

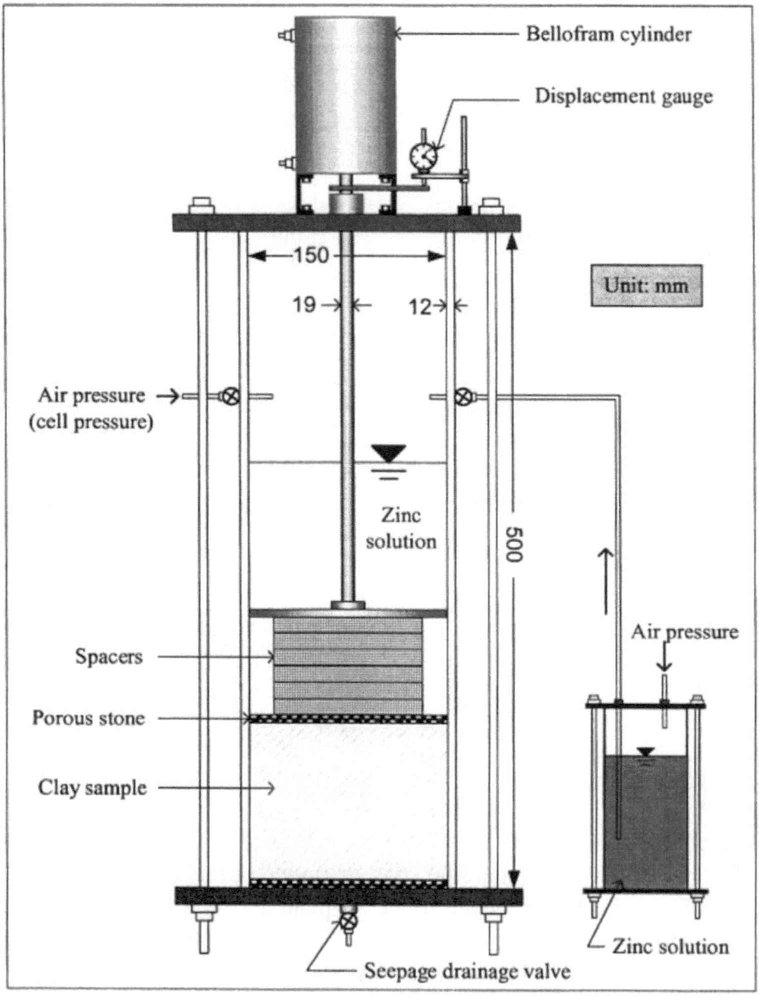

Figure 1. Column set up

Hydraulic loading is applied by introducing air pressure through an opening at one point on the column wall 150 mm from the top. To make vertical load distribution and seepage uniform across the clay layer, a porous metal plate is placed on top of the clay layer. The porous plate has 300 holes of 2 mm diameter and has a thickness of 5 mm. At the neck of the piston, a plastic plate is inserted. The tip of the settlement gauge is placed on this plate to measure settlement. The settlement gauge is used to establish that the layer has been fully consolidated (Ozawa, Takemura and Kusakabe, 2000; Tanchuling et al., 2000).

The clay was mixed with de-ionized water. The clay slurry was then mixed and de-aired for 30 minutes using a vacuum pump. The column walls were lined with a silicone-seal bond to prevent water from flowing along the side walls. A round filter paper was placed on top of the bottom plate. Using bare hands, the clay slurry was then placed inside the column. Another round filter paper was placed on top of the clay layer, on top of which was laid the upper porous plate. To consolidate the clay, force is applied through a loading plate which is attached to a piston which is being pushed by air pressure through a bellofram cylinder. The ambient temperature in the testing room was kept constant at 20°C.

Vertical and hydraulic loading were applied independently. The clay was subjected to a vertical loading of 30 kPa applied in increments of 10 kPa at a time. For the last increment, a displacement-time graph ($\sqrt{t}$ method) was prepared. To determine whether full consolidation has been achieved, a line was drawn on the linear part of the curve. Another line starting from the same origin of the linear part of the curve but with a slope 115 times less than the first line was drawn. The intersection of this line with the curve signifies the point at which 90% of the full consolidation was achieved. Using this point, the settlement corresponding to 100% theoretical consolidation was defined. When full consolidation with 30 kPa (σ_P') was achieved, hydraulic loading was applied in increments of 10kPa up to 40kPa (σ_{cell}) until full consolidation from hydraulic loading of 40 kPa was attained. As in vertical loading, full consolidation from hydraulic loading was defined in the same manner. The clay remained fully saturated at all times during the consolidation. The effective stress distribution at the end of the consolidation is shown in Figure 2. Z denotes the final height of the clay layer which is 35 mm for the bentonite and 65 mm for the illite.

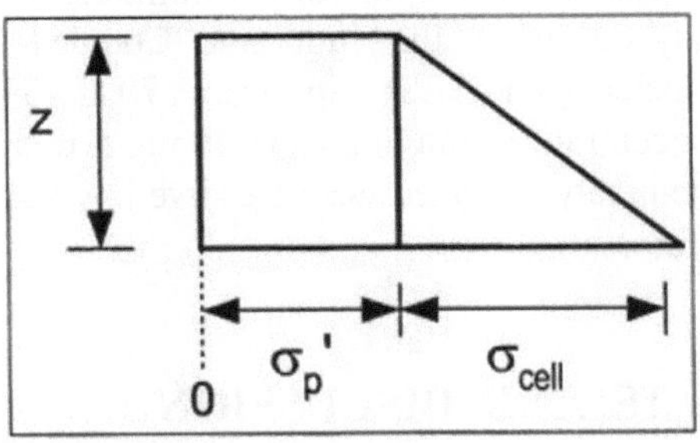

Figure 2. Effective stress distribution after consolidation

After full consolidation has been achieved, the remaining water on top of the porous plate was drained by using a suction pipe. A solution was prepared to make an input zinc concentration of 475 mg/L. The input contaminant was placed inside the column through an opening on the upper part of the column. Leaching was commenced. Effluent concentrations were collected regularly, approximately every 24 hours. The effluent was analyzed for zinc concentration. After a pre-determined number of pore volumes, the test was stopped.

The top lid was unfastened and taken off the column set-up. Moisture on top pf the layer was dried off manually as quickly as practicable to prevent swelling of clay. By using a tube, a sample that cut through the entire layer of the clay was collected. With a spatula, the sample was partitioned along its depth, weighed and placed inside the oven for 24 hours. After drying the sample in the oven, the weight of the samples were taken, and the moisture content was computed for each partition of the sample.

The amount of solids retained in the layers was determined by extraction of the zinc content through acid digestion. After drying, the samples were ground into fine powder. In a beaker, 0.5 g of the powder was mixed with 2 mL of nitric acid and 4 mL of hydrochloric acid and placed in a hot plate at a temperature of 120 °C for 2 hours. A watchglass covered the beaker to keep the liquid from evaporating. The supernatant was diluted, then filtered using a microfiltration system. Blank samples were also prepared to determine background concentrations and were used to adjust the concentrations that were determined by the ICP-AES.

Batch tests were done by mixing 5 g of clay with 100 mL of solution with a known initial concentration, then agitated for 24 hours using a magnetic stirrer. The solids were separated from the liquid using a laboratory centrifuge with a speed of 5000 rpm for 10 minutes. The supernatant was collected and analyzed using the Inductive Coupled Plasma – Atomic Emission Spectrometer (ICP-AES) Spectrum 7000. This was done for different initial concentrations ranging from 40 mg/L to 5000 mg/L. Samples that were not immediately analyzed were preserved by adding nitric acid and kept in the refrigerator.

3. RESULTS AND DISCUSSION

3.1 Column tests of bentonoite and illite

Two column tests were conducted, one for bentonite and another for illite. The test conditions and the flow characteristics of the column leaching tests are shown in Table 3. The graph showing the void ratio **e** – **log σ'** curves are shown in Figure 3.

Table 3. **Test conditions and flow characteristics of column tests**

Material used	Bentonite	Illite
Final measured height of clay layer (mm)	35	65
Diameter of sample (mm)	150	150
Moisture content (%)	285	67.33
Constant head difference across layer (m)	4.08	4.08
Volumetric flow rate (mL/hr)	0.236	2.93
Starting Influent concentration (mg/L)	1085	475
Consolidation Time (days)	17.58	4.79
Leaching time (days)	23.9	16.91
Total pore volume	0.19	1.62
Porosity, n	0.86	0.64
Interstitial velocity, v int (m/s)	3.70×10^{-9}	7.19×10^{-8}
Permeability (m/s)	3.69×10^{-11}	1.14×10^{-10}
Hydraulic gradient, i	116.62	62.79

The total leachate flow is expressed in terms of pore volume to be able to compare results of different cases. It is computed as $U = \frac{tv_{int}}{L}$ where U = pore volume, t = time, v_{int} = **interstitial** velocity, and L- length of the clay layer. The hydraulic permeability was computed using Darcy's law, k = v/i where k = hydraulic permeability and i = hydraulic gradient.

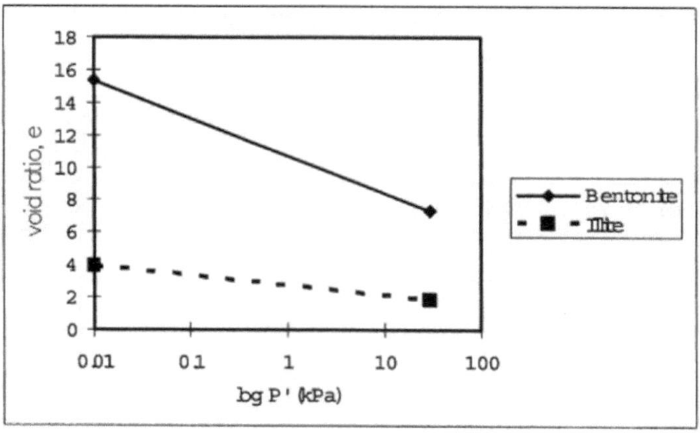

Figure 3. Void ratio e – log P curves

The amount of solute retained in the layers after the bentonite leaching tests was determined using acid digestion. Results are shown in Table 4 and are graphically illustrated in Figure 4. Three core samples were taken from the clay, with each sample divided into 4 slices each. The amount of solute retained decreased along the depth of the layer; i.e. the closer the slice is to the top, the more solute were retained. As the contaminant moves through the layer, some of it sorbs into the surfaces and is retained, thus reducing the concentration of contaminant moving on the lower parts of the layer. For a one-dimensional flow downward in the column with a homogenous soil media, it is expected that the three samples on the same layer would contain the same amount of zinc retained. On the topmost layer, more solute was retained at the sides than at the center. This can be attributed to the influent-sampling method where the column was tilted for a few minutes to be able to collect influent which was done by letting the solution pass through a hole on the side of the column, which was kept closed while testing was going on. This may also be the reason why the coefficient of variation at the topmost layer is highest at 24 percent. The lower layers do not show a smaller variation, and thus indicate uniform flow throughout the layer.

Table 4. Amount of solute retained in bentonite

Distance from Top (mm)	Solute retained (mg/g) Sample A (Center)	Sample B (Side)	Sample C (Side)	Mean	cv
10	0.1300	0.2130	0.1778	0.1736	23.9972
20	0.0727	0.0804	0.0667	0.0733	9.3768
30	0.0766	0.0686	0.0545	0.0665	16.8536
35	0.0744	0.0577	0.0607	0.0643	13.8532

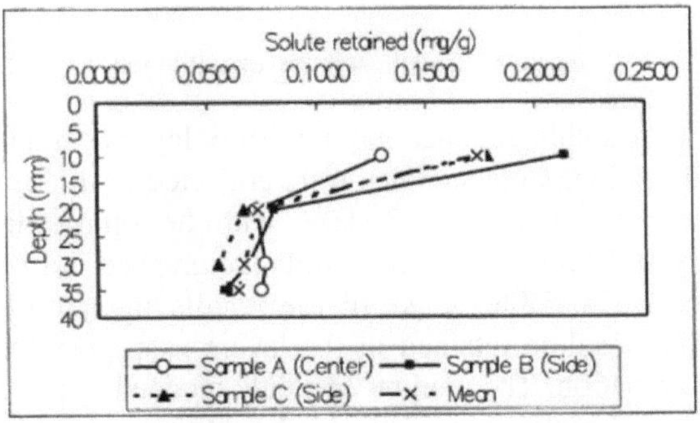

Figure 4. Amount of solute retained in bentonite

The concentrations of the effluent in the illite leaching tests are shown in Figure 5. (Due to a poor effluent sampling system of the bentonite test, the data was deemed erroneous and will not be included here.) Using an initial concentration C_o equal to 475 mg/L, relative concentrations were computed as C_t/Co (where Ct is concentration at any time t) shown at the y-axis and graphed against pore volume at the x-axis. The trend shows a very small relative concentration and slow increase in the concentrations at pore volume less than 1, but showed a rapid increase with a linear trend after that. The effluent concentration reached half of the initial concentration after leaching for about 1.5 pore volumes.

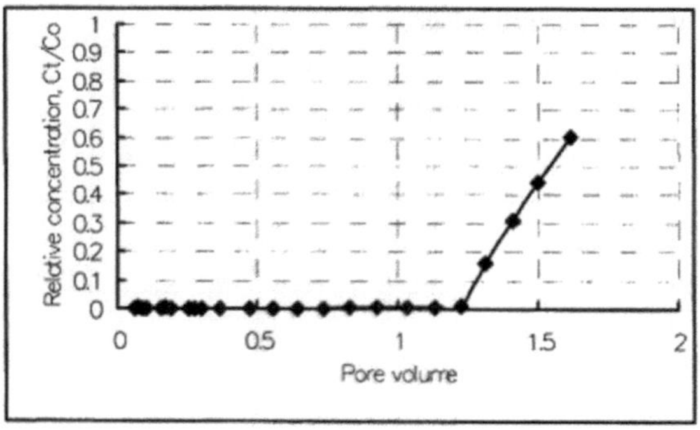

Figure 5. Breakthrough curve of illite test

The amount of solids retained in the illite layer was also determined through acid digestion. Core samples were collected at the center and at the sides. The layer was divided into 7 slices, with the topmost layer measuring 5 mm and the rest 10 mm each. The results are graphed in Figure 6 with the data points shown in Table 5. As in the results in bentonite, there is a decreasing trend of solute retained in the layers as it goes farther from the source of contamination. The amount of solute retained at the center was less than those collected at the sides. This can be attributed to side wall leakage, which causes a higher velocity at the sides thus allowing more solution to flow at the sides, and more contact time with the soil. The disparity of the values as illustrated by the coefficient of variation is highest also at the topmost layer (17.35 percent) although is not remarkably higher than the rest. It is interesting to note that in both column tests, the highest variation is found at the topmost layer, and the lowest is in the middle (second layer from the top for bentonite, and third from the top for illite).

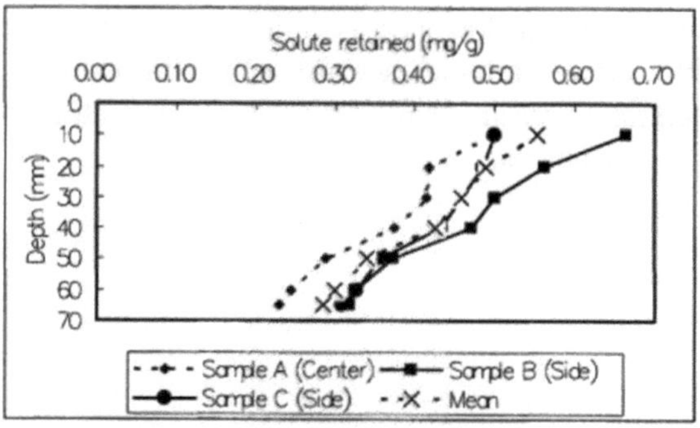

Figure 6. Amount of solute retained in illite

Table 5. **Amount of solute retained in illite**

	Solute retained (mg/g)				
Distance from top (mm)	Sample A (center)	Sample B (side)	Sample C (side)	Mean	cv
10	0.4951	0.6632	0.4991	0.5525	17.3594
20	0.4181	0.5607	0.4852	0.4880	14.6179
30	0.4139	0.4994	0.4581	0.4571	9.3500
40	0.3741	0.4699	0.4313	0.4251	11.3377
50	0.2888	0.3723	0.3551	0.3387	13.0114
60	0.2448	0.3241	0.3261	0.2984	15.5361
65	0.2300	0.3166	0.3087	0.2851	16.7970

3.2 Batch Tests of Bentonite, Illite and Kaolin

Batch tests were conducted for three types of clay: bentonite, illite and kaolin. The amount of solids sorbed was graphed against the equilibrium concentration. For the concentrations of the solution that were prepared, the data fit the linear isotherm model, as illustrated in Figure 7. It should be noted that linear isotherms are applicable only for the range of data that were experimentally derived and should not be extrapolated beyond the range of concentrations that were tested. The linear relationship was defined using the least squares fit method and is expressed as $\mathbf{C^* = K_dC}$ where C* is amount of solids sorbed, C is the equilibrium concentration, and $\mathbf{K_d}$ is a constant termed as the partitioning coefficient. Bentonite has the highest partitioning coefficient (0.0109 L/g), followed by illite (0.0073 L/g) and lastly by kaolin (.0005 L/g). The partitioning coefficient is a measure of the affinity of the contaminant to the clay surface, and is affected by the soil's cation exchange capacity (CEC) and surface area. Referring to Table 1, the physical properties of the clay show that bentonite has the highest CEC and surface area, followed by illite and lastly by kaolinite.

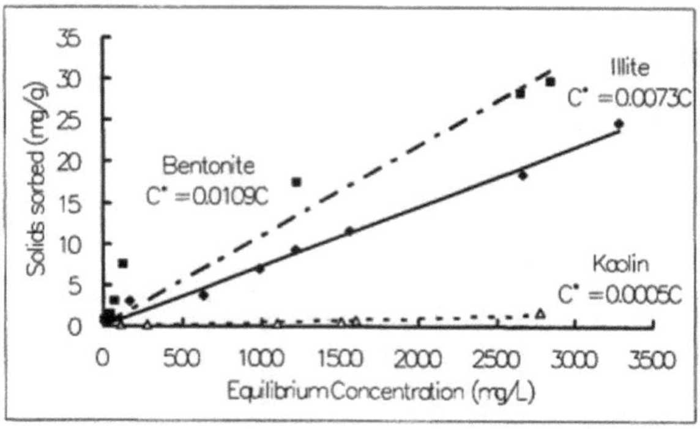

Figure 7. Linear isotherms of bentonite, illite and kaolin

3.3 Comparison of column tests with batch tests

To compare the partitioning coefficients derived from the column tests with those derived from the batch tests, the illite column test will be used. The retardation factor is computed as the ratio of the velocity of the flow of the water to the velocity of the contaminant, i. e. $R = v_{int}/v_c$ where v_c is the velocity of the contaminant at half of the input concentration, v_c for the illite case is 4.75×10^{-8} m/s, thus R is equal to 1.51. The retardation factor R is related to the partitioning coefficient K_d, by the equation $R = 1 + \frac{\rho_b}{\theta} K_d$ (Freeze and Cherry, 1979) where ρ_b is bulk density and θ is porosity of the soil. To compute for the dry bulk density, the equation $\rho_b = (1 - \theta)\rho_s$ where ρ_s is the specific gravity of the soil, was used. ρ_s was assumed to be at the typical value of 2.65 g/cm^3 (Coduto, 1999). The ρ_b value was computed to be 0.954 g/cm^3. Using θ equal to 0.64 (obtained experimentally, the average of the porosities in the layers taken after the column test), then K_d has been computed to be 0.3625 mL/g. From the batch tests, K_d has been obtained as 7.3 mL/g. Comparing the two values and assuming that there is no sidewall leakage, the batch tests have been overestimated by as much as 2000 percent, which is a gross error if used as a design parameter in the design of liners since a higher partitioning coefficient assumes a higher retardation of contaminant thus overestimating the attenuating capacity of the soil. The difference in the values obtained can be explained by the fact that in batch equilibrium tests, the soil particles are completely dispersed making all sides of the soil particles exposed to the contaminant and available for sorption. In column tests, the soils under a certain level of effective stress are packed and thus not all sites are available for sorption (Yong, Mohamed and Warkentin, 1992).

It should be noted however, that in this case, the column test may have had sidewall leakage thus also underestimating the partitioning coefficient. With this limitation notwithstanding, this study is valuable in illustrating the difference in values obtained from the two kinds of tests.

4. CONCLUSIONS

This study has demonstrated that the column leaching test can be used for the determination of the amount of inorganic contaminant such as zinc that can be sorbed in the soil. Moreover, the derivation of the partitioning coefficient of zinc using a column test was also illustrated. By comparing the partitioning coefficient derived from column leaching tests and those from batch tests, it was shown that batch tests overestimate the sorption capacity of the soil by as much as 2000 percent in the case of illite. From among the clays that were used in batch equilibrium tests, it was shown that bentonite has the best sportive capacity, followed by illite, and lastly by kaolin.

REFERENCES

Coduto, D.P. 1999. Geotechnical Engineering Principles and Practices. Prentice -Hall, N.J.; 759 pp.

Fetter, C.W. 1999. Contaminant Hydrogeology. Prentice -Hall, Englewood Cliffs, N.J.; 500 p.

Freeze, R.A. & Cherry, J. A. 1979. Groundwater . Prentice-Hall, Englewood Cliffs, N.J.; 604 p.

Ozawa, K., Takemura, J. & Kusakabe, O. 2000. An experiment al study of effects of void ration on contaminant transport in clay. In: Physical Modeling and Testing in Environmental Geotechnics, pp. 93 – 100.

Reddi, L. N. & Inyang, H.I. 2000. Geoenvironmental Engineering: Principles and Applications. Marcel Dekker, N.Y. 494 p.

Tanchuling, M., Saquing, J., Ozawa, K., Kusakabe, O. & Takemura, J. 2000. Determination of hydrodynamic dispersion coefficient of chloride in kaolin clay for a range of low interstitial velocities. Tokyo Institute of Technology Technical Report 61, 49 -66.

Yong, R.N., Mohamed A.M.O. & Warkentin, B.P. 1992. Principles of Contaminant Transport in Soils. Elsevier Science, Amsterdam; 327 p.

This page intentionally left blank

CHAPTER 17

PASSIVE DIFFUSION SAMPLING FOR METALS

John Tunks, John Hicks Parsons, Rafael Vazquez, Don Vroblesky
Air Force Center for Environmental Excellence (AFCEE), US Geological Survey (USGS)

Abstract: Groundwater sample collection using passive diffusion samplers (PDSs) represents a relatively new technology that employs passive sampling methods for monitoring selected dissolved constituents in groundwater . To date, the most common application of diffusion sampling has been for long -term monitoring of volatile organic compounds. Results of a field-scale PDS demonstration for inorganics at Grissom Air Reserve Base, Indiana are presented. The primary objective of this PDS demonstration was to assess the effectiveness of the PDS method by comparing groundwater analytical results for metals obtained using the current (conventional) sampling method with results obtained using the PDS method. The comparison between the conventional and diffusion sampling results allows for an assessment of the appropriateness of implementing diffusion sampling for metals at each sampled well. Details include a general description of the work performed, a summary of how PDS and conventional results compared, and a comparative cost analysis of the two sampling approaches.

Key words: No-purge, Diffusion, Passive, Sampling, Long-term monitoring, Inorganic, Metals

1. INTRODUCTION

This report presents the results of a passive diffusion sampling (PDS) demonstration consisting of groundwater monitoring of metals at Grissom Air Reserve Base (ARB), Indiana. The demonstration was performed by Parsons for the Air Force Center for Environmental Excellence, Science and Engineering Division (AFCEE/ERS). The objectives of the PDS demonstration were to:

Develop vertical profiles of metals concentrations across the screened intervals of the sampled monitoring wells;

Assess the effectiveness of the PDS method by statistically comparing groundwater analytical results for metals obtained using the current (conventional) sampling method (i.e., micropurge/sample) during regularly scheduled long-term monitoring (LTM) events with results obtained using the PDS method; and

Compare the costs of passive diffusion (PD) and conventional sampling.

Diffusion sampling is a relatively new technology designed to utilize passive sampling techniques that eliminate the need for well purging. Specifically, a capsule is filled with purified water, sealed with a diffusive-membrane, suspended in a well-installation device, lowered to a specified depth below the water level in a monitoring well, and left undisturbed for a period of time (typically no less than two weeks). Over time, the metals in the groundwater diffuse across the membrane, and the metals content of the water inside the sampler reaches equilibrium with groundwater in the well casing. The sampler is subsequently removed from the well, and the water in the diffusion sampler is transferred to a sample container and submitted for laboratory analysis of the target metals. Potential benefits of diffusion sampling include reduced sampling costs and reduced generation of investigation-derived waste.

PDS relies on the natural flow of groundwater through a well screen, and therefore the results obtained from this method will not always be comparable to results obtained using conventional sampling methods which induce groundwater flow into a well by creating a hydraulic gradient through well purging. A variety of factors can cause mixing in an unpumped well, such as vertical flow, convection, and diffusion. In a mixed well, the PDS will represent the mixture or the zone of inflow. In addition, contaminant stratification sometimes is observed in unpumped wells (Vroblesky and Peters, 2000). In this case, the PDS may represent the depth interval at which the sampler is placed.

A similar technology (passive diffusion bag sampling [PDBS]) has been developed for monitoring volatile organic compounds exclusively. That technology has been evaluated through various studies (Vroblesky and Hyde, 1997; Parsons, 1999; Tunks *et al.,* 2000; Church, 2000; Hare, 2000; McClellan AFB, 2000; Vroblesky *et al.,* 2000; Vroblesky and Peters, 2000; Vroblesky and Petkewich, 2000; Tunks *et al.,* 2003), and a guidance document for their use has been developed (Vroblesky, 2001). The investigation described in this report is one of the first field-scale evaluations of the PDS technology for metals sampling. However, laboratory-scale evaluations of the technology have been performed (Vroblesky *et al.,* 2002; Vroblesky and Sheible, 2003).

2. SCOPE AND DESCRIPTION OF FIELD ACTIVITIES

Grissom ARB is located approximately 80 miles north of Indianapolis, Indiana. During this demonstration, a total of 60 PDSs were deployed in 19 monitoring wells at the Base for analysis of metals. PDSs were deployed from October 9 through 11, 2002. Prior to deployment, dissolved oxygen (DO) concentrations in each well were measured and recorded. If the DO concentration was less than 1 milligram per liter (mg/L), then anaerobic water was used to fill the PDSs. Conversely, aerobic water was used in PDSs for monitoring wells that indicated a DO concentration of 1 mg/L or greater. Each metals PDS was comprised of five 30-milliliter (mL) polyethylene bottles filled with either aerobic or anaerobic deionized water (150 mL total volume). Anaerobic water was generated by bubbling nitrogen through deionized water in a clean 5-gallon plastic bucket until DO concentrations were less than 1 mg/L. After filling the bottles with water, a 125-micron nylon mesh screen was placed over the bottle top and the mesh was secured in-place with an open-type cap. The bottles were then placed upside-down into a low-density polyethylene (LDPE) mesh sleeve, maintaining a spacing of approximately 1 inch between the top of one bottle and the bottom of the adjacent bottle. The loaded LDPE mesh sleeve was subsequently secured with zip-ties to a braided polypropylene rope which was weighted at the bottom using a stainless-steel weight sealed inside polyethylene tubing. Each loaded LDPE mesh sleeve was approximately 1.5 feet long.

In order to have enough sample volume upon PDS retrieval to permit both total and dissolved metals analyses, two PDSs were deployed end-to-end such that a total of 3 feet of saturated well screen were monitored. Water from the two PDSs deployed in each 3-foot interval was composited, resulting in a sample for total and dissolved metals analysis that was reflective of a 3-foot-long portion of the saturated screen interval. For instances where less than 3 feet of saturated screen was available, only one PDS was deployed. The resulting sample was analyzed for either total or dissolved metals, and the sampling interval was only 1.5 feet long. Vertical distribution of metals in the wells was evaluated by deploying a separate pair of PDSs for each 3-foot-long section of the saturated screen interval.

The assembled PDS string was lowered into the well until the bottom weight was just resting on the well bottom. After PDS deployment, the well was capped and locked (if a lock was present at the well).

Sample retrieval occurred on November 4 and 5, 2002. As described above, samples were collected by compositing the contents of all bottles from two contiguous samplers (a total of 10 30-mL bottles representing a 3-

foot depth interval) into a clean container. Samples designated for dissolved metals analyses were collected in sample containers after being filtered in the field using a peristaltic pump, silicon tubing, and an in-line 0.45-micron filter. Samples designated for total metals analyses were transferred directly to sample containers without filtration.

Studies have shown that nylon-screen samplers (similar to those used at Grissom ARB for metals sampling) equilibrate in 20.5 hours or less (Vroblesky and Scheible, 2003). Therefore, the total deployment period of 24 to 26 days is presumed to have been adequate to allow local groundwater conditions to re-equilibrate following PDS deployment, and to allow adequate time for the diffusion process to occur.

Samples were analyzed for metals using US Environmental Protection Agency (USEPA) Methods 6010, 7041, 7470 and 7841. The laboratory and analysis methods used for PD samples were identical to those used for the corresponding conventional samples to minimize analytical variability.

PDSs were retrieved as close in time to the conventional sampling of the same wells as feasible. However, the incumbent environmental sampling contractor performed the conventional sampling from November 12 to 20, 2002, and therefore lag times ranging from 7 to 16 days between the two sampling events occurred.

3. RESULTS AND ANALYSIS

Figure 3-1 is a plot of the maximum PDS result (i.e., if multiple PDSs were deployed in a well) versus the corresponding conventional sample result for the same well for all metals. Instances where either result (e.g., PD or conventional sample) was qualified as non-detect were assigned a value equal to the analytical method detection limit (MDL).

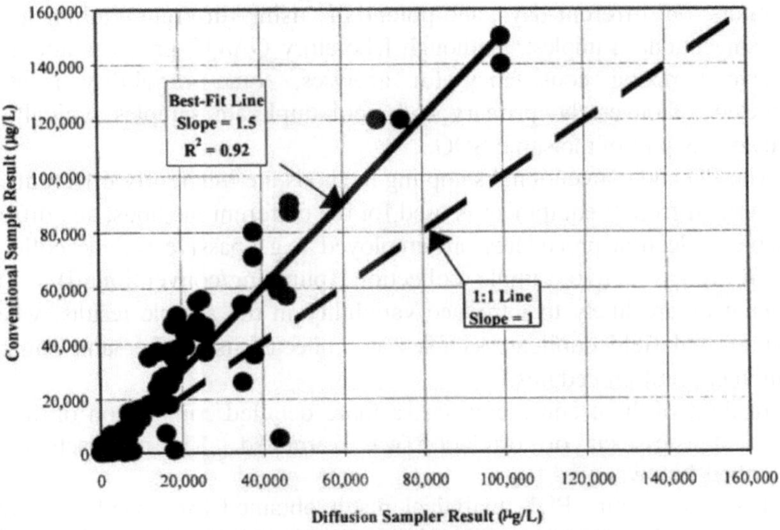

Figure 3-1.

Figure 3-1 is a presentation format typically used when comparing two paired data sets that are expected to be similar (e.g., primary and duplicate samples), since it is a visual depiction of the correlation between the two data sets. However, the variance between the results of PD and conventional samples would be expected to be greater than the variance between a primary and field duplicate sample. This is due in part to the following considerations:

PD and conventional sampling were performed by different contractors. Although the same standard operating procedures for tasks common to both sampling approaches (e.g., sample handling and preservation) were followed by both sampling teams, some variability is to be expected simply because different individuals were involved. Conversely, primary and field duplicate samples are collected by the same sampling personnel.

Logistically, it was not possible to perform PD sampling immediately prior to conventional sampling. A time lag of 7 to 16 days occurred between PD and conventional sampling. Although it is unlikely that groundwater chemistry will change significantly over the course of a few hours, some variability can occur over the course of several days or weeks. In contrast, primary and field duplicate samples are collected at the same time.

PD and conventional samples were not included in the same laboratory sample delivery group (SDG). Therefore, even though the same laboratory was used, the PD samples may have been analyzed by different instrument operators, on different days, and potentially using different equipment than the conventional samples. Although laboratory QA/QC practices attempt to achieve temporal consistency for analyses, some variability is to be expected. Conversely, primary and field duplicate samples typically are analyzed as part of the same SDG.

The PD and conventional sampling methods are inherently different. For example, different equipment is used for the different methods, and different sample collection procedures are employed (e.g., passive sample collection [PDS] versus active sample collection [pumping/conventional]). These differences are likely to introduce variability in the sample results, whereas primary and field duplicate samples are collected using the same sampling equipment and procedures.

In light of these considerations, a more detailed comparison of the data sets and assessment of correlation was warranted. The approach used is described below.

If the maximum PDS analytical result obtained for a well was greater than or equal to the conventional sampling result for that well, the PDS method was inferred to be appropriate for use in that well (Vroblesky, 2001). Nonetheless, comparison to the conventional sampling results was performed in all instances. Specifically, analytical results for all samples collected using the diffusion samplers were compared to results from the conventional sampling using the following relative-percent-difference (RPD) equation:

$$RPD = 100*[abs(D-C)]/[(D+C)/2]$$

Where:
abs = absolute value,
D = diffusion sampler result, and
C = conventional sample result.

Based on standard conventions for comparing primary and field duplicate samples (USEPA, 1996, 1999, and 2000; Radian International, 1997 and 1998; EarthTech, 1998; McClellan AFB, 2000; Goad, 2001), and the inherent differences that were expected between the two data sets due to the differences in sample collection procedures, an RPD of 50 was selected as the criterion for this PDS demonstration. This RPD criterion, in conjunction with three other criteria developed specifically for the PD/conventional sample comparison, was used to determine the suitability of PDS for monitoring a specific metal in a specific well. All four correlation criteria are described below.

PDS ≥ Conventional Criterion: If at least one PDS result for a given metal in a given well was equal to or greater than the conventional sampling result, PD sampling was deemed appropriate for monitoring that metal in that well.

RPD Criterion: If either the PD or the conventional sample result was greater than three times the laboratory reporting limit (RL), and the PDS result was less than the conventional result, then an RPD of 50 was used as the acceptance criterion.

RL Criterion: If both the PD and conventional sample results were less than or equal to three times the laboratory RL, a value of ± the RL was used as the range of acceptance between the two values. If the RLs for the conventional and PD samples were different, the lowest RL was used to determine the acceptance range.

Low-Magnitude Concentrations Criterion: In cases where the PD and conventional sample results were low-magnitude and differed by less than 5 micrograms per liter (μg/L), an acceptable correlation between sample results was achieved.

Each metal concentration from each PDS in each well evaluated was compared to the conventional sample concentration of the same metal in the same well using the four correlation criteria. If any one of the correlation criteria were met, then PDS was considered to be an acceptable alternative sampling method for that metal in that well.

A correlation ratio was then calculated by dividing the number of comparisons in which the correlation criteria were met by the total number of comparisons made. For example, if total aluminum was detected in samples from 19 wells, and at least one of the correlation criteria was met for 18 of these samples, a correlation ratio of 95 percent was realized.

Total and/or dissolved concentrations of 20 different metals (17 as dissolved phase, 18 as total phase) were detected in either a PDS or conventional sample in at least one of the 19 wells evaluated. Concentrations of nine of the 17 dissolved metals (aluminum, cadmium, chromium, copper, lead, mercury, potassium, selenium, and zinc) met the correlation criteria in all of the wells sampled. Concentrations of four additional dissolved metals (arsenic, cobalt, nickel, and sodium) met the correlation criteria in at least 80 percent of the wells sampled. Concentrations of the remaining four dissolved metals (barium, iron, magnesium, and manganese) met the correlation criteria in fewer than 80 percent of the wells sampled.

Concentrations of 10 of the 18 total metals (antimony, arsenic, beryllium, cadmium, chromium, copper, lead, potassium, vanadium, and zinc) met the correlation criteria in all of the wells sampled. Concentrations of four additional total metals (aluminum, cobalt, nickel, and sodium) met the correlation criteria in at least 80 percent of the wells sampled. Concentrations of the remaining four total metals (barium, iron, magnesium, and manganese) met the correlation criteria in fewer than 80 percent of the wells sampled.

Generally, similar correlation ratios for both total and dissolved metal concentrations were realized, suggesting that correlation was not dependent on the phase (e.g., total or dissolved) of the metal.

Figure 3-2 is a graphical presentation of the correlation ratio for each detected metal. Generally, most of the metals (27 of the 35 detected [including total and dissolved phases]) had correlation ratios of at least 80 percent.

Well-specific correlation criteria ranged from 57 to 100 percent. The correlation criteria were met for at least 80 percent of the metals detected in 11 of the 19 wells evaluated. Overall, correlation criteria were met in 354 of 428 comparison instances (83 percent).

Further review of data was performed to identify specific variables that may have led to reduced correlation ratios including review of compound, well, and technology-specific variables. A brief description of 11 reasons that were evaluated to explain instances where correlation criteria were not met, in addition to the summary application of these reasons to the data set, is provided in Table 3-1.

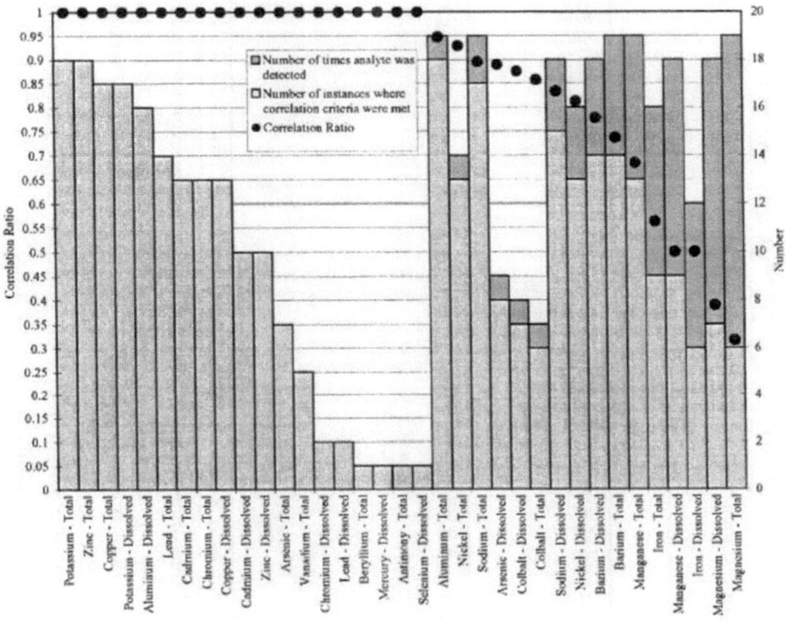
Number of times analyte was detected
Number of instances where correlation criteria were met
Correlation Ratio
Correlation Ratio
Number
1
0.95
0.9
0.85
0.8
0.75
0.7
0.65
0.6
0.55
0.5
0.45
0.4
0.35
0.3
0.25
0.2
0.15
0.1
0.05
0
20
18
16
14
12
10
8
6
4
2
0
Potassium - Total
Zinc - Total
Copper - Total
Potassium - Dissolved
Aluminum - Dissolved
Lead - Total
Cadmium - Total
Chromium - Total
Copper - Dissolved
Cadmium - Dissolved
Zinc - Dissolved
Arsenic - Total
Vanadium - Total
Chromium - Dissolved
Lead - Dissolved
Beryllium - Total
Mercury - Dissolved
Antimony - Total
Selenium - Dissolved
Aluminum - Total
Nickel - Total
Sodium - Total
Arsenic - Dissolved
Colbalt - Dissolved
Colbalt - Total
Sodium - Dissolved
Nickel - Dissolved
Barium - Dissolved
Barium - Total
Manganese - Total
Iron - Total
Manganese - Dissolved
Iron - Dissolved
Magnesium - Dissolved
Magnesium - Total

Figure 3-2.

TABLE 3-1

POTENTIAL REASONS FOR LOW CORRELATION BETWEEN PD AND CONVENTIONAL SAMPLES

Number	Reason for Low Correlation Between PD and Conventional Samples	Summary Evaluation of the Impact of this Reason
1	Excessive time lag between passive diffusion and conventional sampling events.	This reason was considered and it was determined that it did not adversely influence correlation ratios.
2	Laboratory- or field-induced contamination that may be present at varying levels in different samples, but are not representative of site-related contamination.	This reason was considered and it was determined that it did not adversely influence correlation ratios.
3	Inherent differences between passive diffusion and conventional sampling methods (e.g., the fact that some conventional samples are drawn from a much larger volume of the aquifer than passive	Conventional sampling at Grissom ARB was performed using a low-flow purge technique. Harte (2002) notes that analyte concentrations obtained using the micropurge technique varied with the pumping rate, with the higher concentrations associated with the higher pumping rates. PD sampling relies on the natural (i.e., passive) flow of groundwater through a well, whereas the conventional method actively draws groundwater into a well from the surrounding aquifer through pumping. Therefore, the potential exists for conventional sample results to be representative of water quality from a portion of the aquifer not monitored by the PDS. Although, it is difficult to quantify the effects of the different sampling methods on the

Number	Reason for Low Correlation Between PD and Conventional Samples	Summary Evaluation of the Impact of this Reason
	diffusion samplers).	resulting data sets, the two data sets should be expected to demonstrate some variability due to the sampling method differences.
4	Significant contaminant stratification in which dissolved contaminants are localized in discrete layers that may not be sampled by passive diffusion samplers. Also, non-uniform distribution of dissolved contaminants in the aquifer due to variations in solubility and sorptive properties.	The vertical distribution of contamination could only be adequately evaluated in one of the six wells having a correlation ratio less than 70 percent since the other five wells had very short saturated screen intervals. In that vertically profiled well, all detected metal concentrations in the deeper PDS, except total potassium and total lead, were equal to or higher than in the shallower PDS. This suggests that metal concentrations in the water-bearing zone screened by this well are stratified, with higher concentrations near the bottom of the well. This stratification may be the result of lithologic variability across the screened portion of the aquifer. Alternately, considering the relatively high concentrations of metals in this well, this may be a density-driven stratification in which relatively high-density (i.e., greater than water) material such as metals tends to sink. This could affect correlation between PD and conventional samples if the conventional sample is drawn from a deeper portion of the aquifer with higher metals concentrations than the water that flows across the PDS.
5	Vertical flow in the well resulting from the presence of vertical hydraulic gradients in the screened portion of the aquifer.	This reason was considered and it was determined that it did not adversely influence correlation ratios.
6	Well-specific conditions that would limit groundwater flow through the well, such as where the	The field staff noted that there was standing water (presumably surface-water runoff) in the vaults of three flush-mounted wells during sampler retrieval. If surface-water runoff had drained into the well during the deployment period, it may have diluted both the water inside the well and the PD samples. Conversely, the conventional sampling method collected

Number	Reason for Low Correlation Between PD and Conventional Samples	Summary Evaluation of the Impact of this Reason
	the well, such as where the filter-pack permeability is lower than that of the aquifer, causing groundwater flow lines to diverge around the well, or the presence of low-permeability aquifer materials opposite the screen interval.	groundwater directly from the aquifer via pumping, and therefore would have been less affected by dilution of the water in the well casing.
7	Laboratory-induced variability resulting from the passive diffusion and conventional samples being analyzed in different sample delivery groups, and therefore being potentially analyzed on different days, by different operators, on different equipment.	To minimize laboratory-induced variability, each of the conventional and PD samples were analyzed by the same laboratory using the same procedures. However, although difficult to quantify, there is a potential that laboratory-induced variability may have affected the comparability of the two data sets. Conventional and PD samples collected at Grissom ARB were analyzed in different SDGs. It is reasonable to assume that samples analyzed in different SDGs will be exposed to more laboratory-induced variability than samples analyzed as part of the same SDG.
8	Inappropriate (too short or too long) deployment	Although the PDSs were left in-place for 24 to 26 days, the relatively high concentrations of some metals in the wells (e.g., see Reason 4), coupled with the potential effects of the bottle

Number	Reason for Low Correlation Between PD and Conventional Samples	Summary Evaluation of the Impact of this Reason
	period.	orientation described in Reason 11, may have necessitated a longer deployment period in some cases. Metals with correlation ratios less than 70 percent tended to be present in the groundwater at higher concentrations than those metals which had a correlation ratio greater than 70 percent (Figure 3.3). Furthermore, the best-fit trend lines for the data series indicate that, for metals with correlation ratios less than 70 percent, the conventional sample result was typically approximately 1.5 times greater than the PDS result (as evidenced by the slope of 1.5 for the best-fit line). Conversely, for those metals having a correlation ratio greater than 70 percent, the PD and conventional sample results were more similar (as evidenced by the slope of the best-fit line [0.96]). These observations suggest that the PDS was not as effective at accurately monitoring relatively high concentrations of metals. A longer deployment period may have enabled more complete diffusion to occur and may have resulted in higher correlation ratios.
9	Deployment of the sampler above or below the screened interval of the well where natural groundwater flow through the well does not occur.	The screened intervals of three wells were not known. Although it was assumed that the bottom 10 feet of these wells were screened, it is possible that some of the samplers deployed in these wells were not located within the screened interval, and would therefore not reflect the quality of groundwater flowing through the well.
10	Chemical reactions between the natural groundwater and the water contained inside the diffusion sampler. For example, if the pH or DO	The solubility of metals in groundwater is greatly influenced by the geochemical conditions of the groundwater. For example, small changes in DO concentrations, pH, and/or oxidation-reduction potential (ORP) can cause metals to either dissolve into solution or precipitate as a solid. Although the pH of the groundwater was not measured during this demonstration, concentrations of DO were measured. Observed trends suggest that better correlation was achieved in more highly aerobic conditions. Although efforts were made to fill the PDSs with

Number	Reason for Low Correlation Between PD and Conventional Samples	Summary Evaluation of the Impact of this Reason
	example, if the pH or DO of the groundwater is significantly different than the pH or DO of the water inside the diffusion sampler, precipitation of some dissolved metals could occur as they migrate into the diffusion sampler.	water that had a similar DO concentration to that of the groundwater, it is possible that the de-oxygenation process may not have been adequate, and residual DO in the PDSs may have enabled chemical reactions to occur at the PDS-groundwater interface during the equilibration period. Geochemical differences (e.g., differences in DO concentrations) between the groundwater and the water inside the PDSs may have contributed to instances where correlation criteria were not met. However, no visible evidence of precipitation on or inside the PDSs was noted.
11	PDSs are still undergoing development. Refinements to their construction (e.g., the size of the opening in the nylon screen and/or the ratio of the bottle opening to the bottle volume) and use (e.g., orientation of the deployed bottle [upside-down, right-side-up, sideways]) may still be necessary.	In all wells, the PDSs were deployed such that the bottle openings were facing downward. Typically, it is preferred to orient the bottle openings sideways, but this was not possible since the diameter of the wells was only 2 inches, and the bottles would not have fit into the well in this orientation. Orienting the bottle opening in the upward direction may allow mixing of the bottle contents with the groundwater in the well during sampler retrieval, and is generally not the preferred approach (Vroblesky, 2003). Based on the density differences between the groundwater (higher density due to presence of metals and other dissolved constituents) and the deionized water in the bottles (lower density), the process of diffusion may have been hindered by density differences (i.e., the upward motive force applied by the process of diffusion would have to overcome the downward motive force of gravity) (Webster, 1998). This process is not expected to affect bottles where the opening is oriented upwards or sideways, but could affect bottles where the opening is oriented downward. Additionally, in wells where lower concentrations of metals were reported, the density differences between PDS water and groundwater may not have been as significant. In these wells, the effects of

Number	Reason for Low Correlation Between PD and Conventional Samples	Summary Evaluation of the Impact of this Reason
		bottle orientation may not have been as significant as in those wells where the density differences were greater (i.e., wells with higher metals concentrations). This observation is supported by the data presented in Figure 3.3 in which wells with higher concentrations of metals tended to meet correlation criteria less frequently than wells in which the metals concentrations were lower. Additionally, the PDS technology is still in the development stage and final design criteria have not been determined. As such, the size of the openings in the nylon mesh and the physical dimensions of the bottle (i.e., the ratio of the area of the bottle mouth to the volume of the bottle) may not be the final recommended design for PDS. However, testing of different nylon mesh screens and bottle sizes was not incorporated into this demonstration, and therefore, no conclusions can be made regarding the impact of these design criteria on the ultimate performance of the technology.

4. COST ANALYSIS

A comparative cost analysis was performed for PD and conventional sampling. The four-step cost-analysis procedure consisted of an estimate of PDS implementation costs, a comparison of long-term PDS to long-term conventional sampling costs, an analysis of the costs that would be avoided if PDS were implemented, and a calculation of return on investment (ROI).

4.1 Estimate of Implementation Costs

The first step of the cost analysis involved estimating the per-well PDS implementation cost. This step reflected the per-well cost of the PDS demonstration, assuming it had been performed by the current Base contractor (with normalized labor rates). Therefore, the PDS implementation included costs to develop a work plan, procure field equipment, perform field sampling activities, and prepare the PDS results report. Additional implementation costs include the development of well-specific recommendations for long-term PDS deployment, interaction with regulatory agencies, and revision of sampling-related guidance documents such as sampling and analysis plans, records of decision, and quality-assurance plans. The magnitude of these future costs was not known, so estimates were used instead.

The implementation cost calculated for this demonstration was $2,543 per well.

4.2 PD and Conventional Sampling Cost Comparison

The next step was to estimate and compare long-term cost differences between PD and conventional sampling methods. A capital cost depreciation period of 10 years (i.e., the assumed duration of the LTM program) was used for dedicated conventional sampling equipment. Because some dedicated equipment (e.g., submersible pumps, generators, meters) also have limited life expectancies, replacement rates were considered in the analysis where appropriate. Ultimately, a cost per well per sampling event for each sampling method was calculated.

4.2.1 Long-Term PD Sampling Costs

Two of the primary assumptions made in calculating the long-term costs for PD sampling were:

Only one PDS per well would be used during each event. This is in contrast to the use of multiple PDSs per well for vertical profiling, which is usually recommended for the initial implementation evaluation only.

Each time a PDS is retrieved from a well, it would be replaced with a new PDS such that only one mobilization would be required per sampling event. A PDS would therefore remain in the well for the entire period between sampling events.

The long-term PD sampling cost per well per event calculated during this demonstration was $117.

4.2.2 Long-Term Conventional Sampling Costs

Most of the cost information for conventional sampling was provided by the Base contractor, and any information gaps were filled using professional judgment. The long-term conventional sampling cost per well per event calculated during this demonstration was $345.

4.3 Sampling Cost Avoidance

This cost analysis assumed that the diffusion sampling technology could be used for LTM of metals in 37 wells at Grissom ARB on a semi-annual sampling frequency. Based on the per-well, per-event cost estimates provided in Sections 4.2.1 and 4.2.2, the cost difference between using diffusion sampling and conventional sampling for LTM of metals is $228. Assuming semi-annual sampling of 37 monitoring wells, an annual sampling cost avoidance of approximately $16,900 could be realized.

4.4 Return on investment

This step compared the cost to implement a PDS demonstration (i.e., the up-front investment required to implement long-term PD sampling) with the estimated cost avoidance over the duration of the LTM program.

Considering an implementation cost of $2,543 per well (Section 4.1), and a total of 37 wells to evaluate, a total implementation cost of approximately $94,100 is estimated. Also, considering the potential annual cost savings of $16,900 calculated in Section 4.3, and a 10-year semi-annual LTM program, a total LTM program cost avoidance of $169,000 could be realized by switching from conventional to passive diffusion sampling for metals.

The ROI is calculated by dividing the potential cost savings by the PDS implementation cost and is 180 percent.

5. CONCLUSIONS AND RECOMMENDATIONS

The relatively high degree of correlation summarized in Section 3 indicates that the PDS method is reasonably robust and capable of accurately monitoring concentrations of certain metals in groundwater in most instances. However, concentrations of three metals (iron, magnesium, and manganese) detected during this demonstration were consistently higher in conventional samples than in PD samples. These three metals, particularly iron and magnesium, also were present at much higher concentrations in the groundwater than the other metals detected, suggesting that the PDS method was not as effective at monitoring either those specific metals or concentrations of that magnitude. Accordingly, more evaluation is warranted for these "outlier" situations prior to full-scale implementation of the technology.

Available information (summarized in Table 3-1) suggests that the most likely causes of non-correlation include:

Inherent differences between the two sampling methods (i.e., PDS and low-flow purge/sample);

Stratification or non-uniform vertical distribution of metals concentrations in the aquifer;

Possible dilution of the PD sample as a result of surface-water runoff entering the well during the deployment period (one well only);

Insufficient deployment time due to the effects of high metals concentrations in some wells and upside-down bottle orientation;

PDS(s) deployed outside the screen interval, causing analytical results to be unrepresentative of the quality of groundwater flowing through the well (one well only);

Laboratory-induced variability due to PD and conventional samples being analyzed in different SDGs;

Geochemical differences between the groundwater and the water used in the PDS that may influence metals solubility or drive geochemical reactions (e.g., DO concentration differences); and

Technology development issues that may still need refinement. For example, bottle orientation (particularly as it applies to situations where high concentrations of metals present in the groundwater [i.e., high density] may hinder diffusion in an upward direction) and construction materials (e.g., nylon screen mesh opening size and bottle opening area to bottle volume ratio) may be refined as the technology continues to develop.

Correlation ratios for 27 of the 35 metals detected (including total and dissolved phases) were between 80 and 100 percent. However, even where correlation criteria were not met for a given metal or well, PDS may still be a viable alternative to conventional sampling depending on the degree to which concentrations of that particular metal must be quantified to achieve LTM objectives. In these instances, it may be desirable to perform additional evaluations to determine whether the instances of reduced correlation were a one-time occurrence, whether correlation of PD to conventional samples is even reasonable to expect or appropriate in that instance, or whether the wells or analytes are poorly suited to the PDS method. Additional evaluations could include one or more of the following activities:

Comparison of PDS data to historical analytical results to determine whether the PDS results are anomalously low;

Review of available stratigraphic and hydrogeologic data to assess whether these wells are screened in low-permeability zones that could restrict groundwater flow through the well;

Performance of a second PD sampling event in these wells that implements certain technology modifications (e.g., bottle type/size, bottle spacing, bottle orientation) and comparison of the results to conventional sampling results;

Use of a borehole flow meter to assess the degree to which vertical flow in the well may be impacting PDS results; or

Vertical profiling of contaminant concentrations in selected wells using an alternative method (e.g., micropurge) and comparing the results to PDS results.

Excluding the costs of additional field testing of PDSs to clarify outlier (i.e., low-correlation) situations, the PDS method can provide significant long-term cost savings compared to conventional sampling methods. The cost to evaluate the implementability of PDS per well during this demonstration was approximately $2,500. Furthermore, the estimated cost difference between LTM using the PDS method as opposed to the conventional method was $228 per well per sampling event. Therefore, even if PDS is implemented on a limited basis, significant long-term cost savings could be realized. The ROI will depend on the number of wells sampled, the frequency of sampling, and the duration of the LTM program. However, using the assumptions detailed in Section 4, the ROI calculated in this demonstration was 180 percent.

REFERENCES

Church, P.E., 2000. *Evaluation of a Diffusion Sampling Method for Determining Concentrations of Volatile Organic Compounds in Ground Water, Hanscom Air Force Base, Bedford, Massachusetts:* U.S. Geological Survey Water-Resources Investigations Report 00-4242, 20 p.

Earth Tech, 1998. Draft *Quality Assurance Project Plan (QAPP) For Operable Units (OUs) 5, 6, 8, And 10 At Edwards Air Force Base, California* . Version 2.0. November.

Goad, Linda, 2001. Telephone conversation between John Tunks (Parsons) and Linda Goad (Montgomery Watson) regarding QAPP requirements for March ARB. December.

Hare, P.W., 2000. *Passive Diffusion Bag Samplers for Monitoring Chlorinated Solvents in Ground Water*: The Second International Conference on Remediation of Chlorinated and Recalcitrant Compounds, Battelle, May 22 -25, 2000, Monterey, California, v.2, no 1, p. 377-386.

Harte, Philip T. 2002. *Comparison of Temporal Trends in VOCs as Measured with PDB Samplers and Low-Flow Sampling Methods.* Ground Water Monitoring and Remediation . Volume 22, Number 2. Spring, 2002.

McClellan AFB, 2000. Final *Passive Diffusion Membrane Samplers Technology Application Analysis Report.* National Environmental Technology Test Sites (NETTS). August.

Parsons Engineering Science, Inc. (Parsons), 1999. *Final Technical Report for the Evaluation of Groundwater Diffusion Samplers.* December.

Radian International (Radian), 1997. Basewide *Remedial Investigation /Feasibility Study (RI/FS) Quality Assurance Project Plan (QAPP), Revision 3* . Report prepared for McClellan AFB, California. Final, April.

Radian, 1998. *Defense Distribution Depot San Joaquin (DDJC) – Sharpe/Tracy Sites, California. Comprehensive Field Work Plan - Volume 2 of 2, Quality Assurance Project Plan. Version 2.0.* Report prepared for U.S. Army Corps of Engineers, Huntsville, Alabama. Final, October.

Tunks, J.P., Guest, P.R., Faile, M., and Santillan, J., 2000. Diffusion Sampler Evaluation of Chlorinated VOCs in Groundwater. In: Proceedings of the Second International Conference on Remediation of Chlorinated and Recalcitrant Compounds. May.

Tunks, J.P., Hicks, J.R., Santillan, J., and Vazquez, R., 2003. A Passive Diffusion Bag Sampler Demonstration at Multiple US Department of Defense Installations. In: Proceedings of the Annual Conference on Contaminated Soils at the University of Massachusetts at Amherst. October.

USEPA, 1996. *Laboratory Data Validation Functional Guidelines for Evaluating Organics Analysis.* EPA Order No. EPA540R94082.

USEPA, 1999. *SOP No. HW-24, Standard Operating Procedure for the Validation of Organic Data Acquired Using SW-846 Method 8260B (Rev 2, Dec 1996).* Region II. Revision 1, June.

USEPA, 2000. Method 526. *Determination of Selected Semivolatile Organic Compounds in Drinking Water by Solid Phase Extraction and Capillary Column Gas Chromotography/Mass Spectrometry (GC/MS).* Revision 1.0. National Exposure Research Laboratory, Office of Research and Development. Cincinnati, Ohio. June.

Vroblesky, D. A. 2003. Personal communication between Don Vroblesky (USGS) and John Tunks (Parsons). April.

Vroblesky, D. A. 2002. *Passive Diffusion Bag Sampler Protocol, Training, and Regulatory Considerations.* Presented at the AFCEE Cleanup Technology Transfer Workshop. San Antonio, Texas. March 4-7.

Vroblesky, D. A. 2001. *User's Guide for Polyethylene-Based Passive Diffusion Bag Samplers to Obtain Volatile Organic Compound Concentrations in Wells.* US Geological Survey Water-Resources Investigations Report 01 -4060. Columbia, South Carolina.

Vroblesky, D.A., and Hyde, W.T., 1997. *Diffusion samplers as an Inexpensive Approach to Monitoring VOCs in Ground Water.* Ground Water Monitoring and Remediation, v. 17, no. 3, p. 177-184.

Vroblesky, D.A., Borchers, J.W., Campbell, T.R., and Kinsey, Willey, 2000. *Investigation of Polyethylene Passive Diffusion Samplers for Sampling Volatile Organic Compounds in Ground Water at Davis Global Communication, Sacramento, California* . August 1998 to February 1999: U.S. Geological Survey Open-File Report 00-307, 13 p.

Vroblesky, D.A., and Peters, B.C., 2000. *Diffusion Sampler Testing at Naval Air Station North Island, San Diego County, California.* November 1999 to January 2000: U.S. Geological Survey Water-Resources Investigations Report 00-4182, 34 p.

Vroblesky, D.A., and Petkewich, M.D., 2000. *Field Testing of Passive Diffusion Bag Samplers for Volatile Organic Compound Concentrations in Ground Water, Naval Industrial Reserve Ordnance Plant, Fridley, Minnesota.* November 1999 and May 2000: U.S. Geological Survey Water-Resources Investigations Report 00 -4246, 10 p.

Vroblesky, D.A., Petkewich, M.D., and Campbell, T.R., 2002. *Field tests of diffusion samplers for inorganic constituents in wells and at a ground-water-discharge zone: U.S. Geological Survey Water Resources Investigation Report 02 -4031,* 24 p.

Vroblesky, D.A., and Scheible, Walt, 2003. *Laboratory equilibration tests of nylon -screen samplers: Spring Meeting of the Inter state Technology Remediation Cooperative,* March 25, 2003, Charlotte, NC.

Webster, Ian T. 1998. *Theoretical and Experimental Analysis of Peeper Equilibration Dynamics.* Environmental Science & Technology, Volume 32.

This page intentionally left blank

CHAPTER 18

BACKGROUND STUDY OF ARSENIC CONCENTRATIONS AT A RHODE ISLAND SITE

Stephen G. Zemba[1], Edmund A.C. Crouch[1], Shailesh R. Sahay[1], Eirlyss Vanderhoff[2], and Ambrose Donovan[2]
[1]Cambridge Environmental Inc., Cambridge, MA and [2] McPhail Associates, Inc., Cambridge, MA

Abstract: Arsenic concentrations in soils at a property in Newport, Rhode Island, were found to exceed both statewide background levels and remediation criteria. Mean arsenic concentrations of 12.0 mg/kg and 11.7 mg/kg in site and background soils, respectively, are both about four times greater than a background level of 2.7 mg/kg published by the Rhode Island Department of Environmental Management (DEM). A risk-based concentration of 0.4 mg/kg is necessary to meet DEM's incremental cancer-risk target of 1×10–6. Given the similarity of arsenic concentrations in site and background soils and a lack of evidence of any anthropogenic releases of arsenic, a background study was undertaken to gauge whether concentrations of arsenic at the site are likely due to background conditions. The study was based on measured arsenic concentrations in 20 soil samples from the site and 21 soil samples determined by site investigators to represent background conditions. The mean arsenic concentrations in both site and background soils were found to be statistically similar, but the distributions of on- and off-site data differed, possibly due to land -development patterns. Although the results of the study do not indicate that site arsenic levels differed from background, the regulatory decision required installation of engineered barriers at the site to minimize contact with the soils.

Key words: arsenic, background, soil, risk management, Rhode Island

1. INTRODUCTION

The Rhode Island Department of Environmental Management (DEM) has developed a detailed "Guidance for Arsenic in Soil" (DEM, 2001) to aid in the assessment of arsenic in soil at potential disposal sites. The "Guidance for Arsenic in Soil" recognizes the possibility that some sites may have arsenic levels in excess of site remediation criteria due to natural background conditions. A plan was developed to conduct a background study for a site in Newport, Rhode Island, in accordance with DEM's arsenic policy to determine whether concentrations of arsenic at the site are due to natural background conditions (Zemba, 2002).

To support the background study, 20 soil samples from the site and 21 soil samples from off-site locations were collected and analyzed for arsenic. The site is part of a parcel of land that was partially developed as residences in the 1940s and has not seen major development since. Most of the off-site samples were obtained from an adjacent area of the same parcel that was not developed residentially. It appears from examination that the native soils at the parcel were moved during the residential development and mixed to varying degrees, so that any spatial variation in arsenic concentration (for example, a regular variation with depth or trend across the parcel) may have been altered. Moreover, it is likely that the soil in the currently developed area (including the site) and the soil in the undeveloped portion would have been moved and mixed to differing degrees, so that there is no expectation that the spatial distribution of arsenic concentrations is similar on the site and in the background area.

In such circumstances, the spatial distribution of arsenic concentrations resulting at the site and background locations may differ, since any mixing of soil may have been different at different locations. The only constant through such development would be the average concentration of arsenic — while mixing of soil may change the spatial distribution of arsenic, it cannot change the average concentration. The effect of such mixing on the distribution of concentrations observed in sets of soil samples would be similar. Thorough mixing of soil would be expected to reduce the variation in concentrations between samples, but incomplete mixing (as expected here) may have unpredictable effects on the distribution. Consequently, the background study focuses on evaluation of the average concentration of arsenic, and primarily tests for any difference in average concentration between the site and background.

2. STATISTICAL METHODS

This analysis evaluates whether average concentrations of arsenic in soil at the Rhode Island site are comparable to background levels of arsenic in local soils. Our analysis is similar to those suggested in *Guidance for Characterizing Background Chemicals in Soil at Superfund Sites* (EPA, 2001) with implementation details from *Guidance for Data Quality Assessment, Practical Methods for Data Analysis* (EPA, 1998), except where noted.

Twenty arsenic measurements were collected on the site. Twenty-one were collected from background areas adjacent to the site that, based on indications of historical information, have never been developed.

The original plan for the background study (Zemba, 2002) called for comparison of samples collected on site and in background locations at all depths. Subsequent discussions with regulatory authorities (RIDEM, 2002; McPhail Associates, 2002) requested that certain statistics and comparisons be provided for surface and subsurface samples separately. Consequently, these comparisons are developed,[2] although the site history of earth movement indicates that comparisons contrasting concentrations at different depths may have little meaning.

To examine the null hypothesis that the mean concentration of arsenic is the same at the site and in the background, and separately to characterize the distributions of surface, subsurface, and combined surface and subsurface data from the site and background, we use the following tests and procedures.

Parametric tests or procedures:

Calculation of means, standard deviations, and kurtosis for each distribution.

Shapiro-Wilk (Royston, 1982, 1993, 1995) and Studentized range tests to examine the normality or lognormality of all the data distributions.

Outlier tests for each distribution.

- A parametric F-test to test the equivalence of the variances of subsurface and surface data and of combined background and combined site data.
- Parametric t-test to test the equivalence of the means of surface and subsurface data and combined site and combined background data.

[2] No specific definition of "surface" and "subsurface" was requested, so surface and subsurface data sets were distinguished by depth — "surface" data were defined to be measured between 0 and 2 feet below ground surface (bgs) or in depth ranges from less than 2 feet to 2 feet bgs inclusive, while "subsurface" data were al 1 measured at 2 feet or more bgs.

- Non-parametric tests or procedures:
- Construction of normal probability plots to qualitatively compare the combined site and combined background distributions of concentrations (and logarithms of concentrations), and evaluate their normality. The plotting positions on these normal probability plots follow the recommendations of Cunnane (1978) — that is, for x_i, the ith of n order statistics, the plotted location is

$$\left(x_i, \Phi^{-1}\left(\frac{i-3/8}{n+1/4}\right)\right)$$

  where Φ^{-1} is the inverse of the standard normal distribution. The ordinate is here chosen to approximate the expected value of the ith (of n) normal order statistic.
- The Wilcoxon Rank Sum test is performed to test the equivalence of distributions for site and background data.

The distribution of soil concentrations of metals in samples at individual sites is often found to be approximately normal or lognormal. However, there is no *a priori* reason to expect normality, lognormality, or any other particular distributional shape in a sample set from any particular site. The tests for normality or lognormality are primarily performed to evaluate whether other statistical tests may be substantially affected by deviations from the assumptions underlying their construction.

3. RESULTS

3.1 Summary Statistics

Summary statistics for the measurements are shown in Table 1 and for log-transformed data in Table 2. The mean concentrations for the site and background (all data) differ very little, with the site marginally higher than the background. Examination of normal probability plots for the concentration and the logarithm of the concentration (Figures 1 and 2) indicate that the distributions are closer to lognormal than normal, with the background concentration more dispersed than the site concentrations. That the distributions are closer to lognormal than normal is confirmed by Shapiro-Wilk tests, although the Studentized range tests cannot usefully distinguish these particular distributions (see Table 3).

Table 1. Summary statistics for arsenic concentrations.

	Number	Mean (mg/kg)	Standard Deviation (mg/kg)	Kurtosis	Minimum (mg/kg)	Maximum (mg/kg
			Background			
Surface	9	10.2	2.7	-1.06	5.8	13.0
Subsurface	12	12.8	8.5	0.98	5.8	32.0
All data	21	11.7	6.7	3.76	5.8	32.0
			Site			
Surface	8	12.2	3.1	1.26	7.7	18.0
Subsurface	12	11.8	3.5	3.68	7.6	21.0
All data	20	12.0	3.3	2.09	7.6	21.0

Table 2. Summary statistics for arsenic concentrations (log transformed data).

	Number	Mean (mg/kg)	Standard Deviation (mg/kg)	Kurtosis	Minimum (mg/kg)	Maximum (mg/kg
			Background			
Surface	9	2.28	0.30	-0.61	1.76	2.56
Subsurface	12	2.39	0.56	-0.24	1.76	3.47
All data	21	2.34	0.46	0.85	1.76	3.47
			Site			
Surface	8	2.48	0.25	1.21	2.04	2.89
Subsurface	12	2.44	0.27	1.17	2.03	3.04
All data	20	2.45	0.26	0.49	2.03	3.04

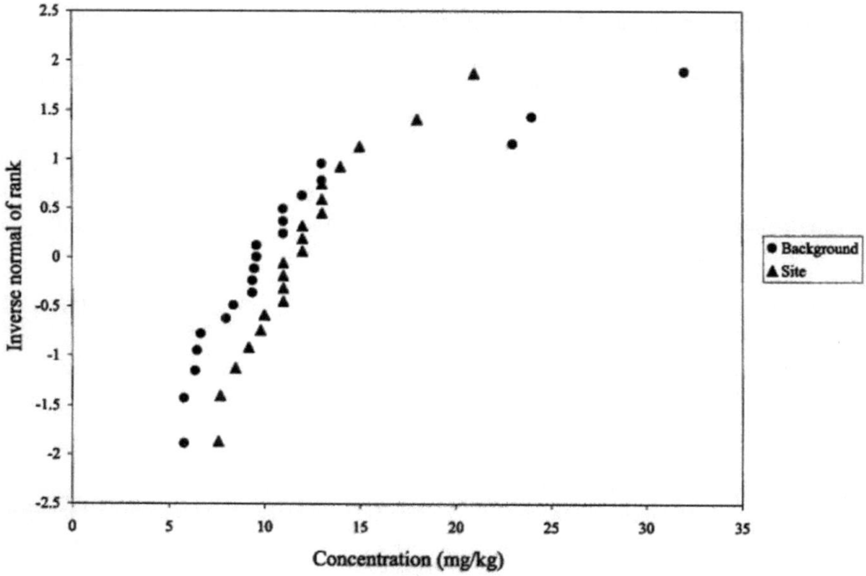

Figure 1. Normal probability plot for the site and background concentrations

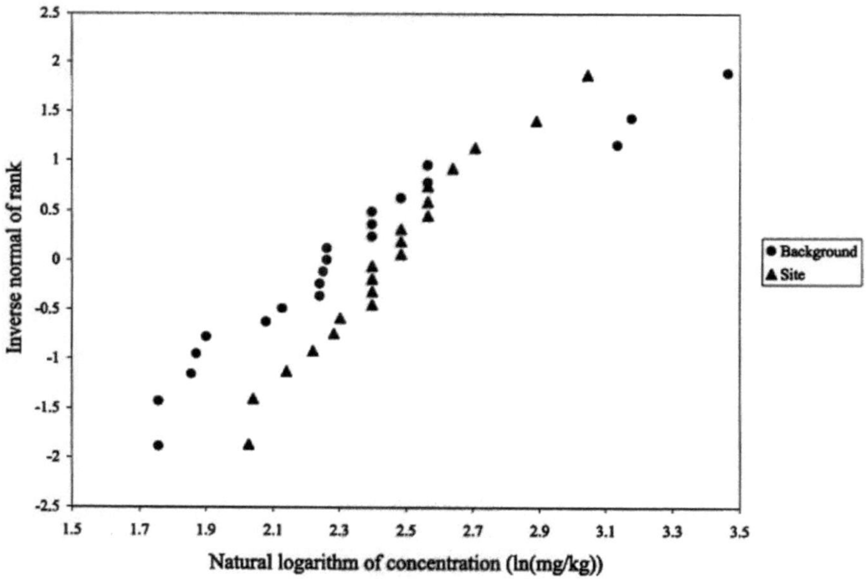

Figure 2. Normal probability plot for logarithms of concentrations of site and background

A simple evaluation for outliers was performed on the logarithmic scale (since the distributions were closer to lognormal than normal) by finding those that would be shown as outliers to a lognormal distribution on a box-and-whisker plot (EPA, 1998). The test for outliers indicated that the values shown in Table 4 might be outliers. However, we have no *a priori* reason for rejecting any of these values as invalid; there appears to be nothing wrong with the samples, and the hypothesis of lognormality for these data is entirely *ad hoc.* All the listed data were therefore included in the comparisons that were performed. The results observed are not considered abnormal, given the site history of earth movement.

Table 3: Tests for normality and lognormality

	Shapiro-Wilk p-value	Studentized range p-value[a, b]	Shapiro-Wilk p-value	Studentized range p-value
	Normal		Lognormal	
Background				
Surface	0.13	>0.05	0.067	>0.05
Subsurface	0.0019	>0.05	0.022	>0.05
All data	0.00010	>0.05	0.032	>0.05
Site				
Surface	0.31	>0.05	0.45	>0.05
Subsurface	0.053	>0.05	0.62	>0.05
All data	0.054	>0.05	0.66	>0.05

[a] Two-sided

[b] The critical range for the Studentized range test was obtained from a re -calculation and extension of Tab le A-2 of EPA (1998), using 10,000,000 samples for each value of n in a Monte Carlo simulation.

Table 4: Potential outliers to lognormal distributions

Location	Potential outlier values (mg/kg)
Background surface	none
Background subsurface	none
Background combined	23, 24, 32
Site surface	7.7
Site subsurface	21
Site combined	21

3.2 Comparison of Means

The primary comparison to be performed is between the mean of the concentrations observed on the site and the mean background concentration. This comparison was performed using one-sided t-tests on the untransformed data. It was qualitatively judged that the distribution for the sample mean obtained from distributions like those observed would be sufficiently close to normal for the sample sizes involved that the t-test would not be compromised. The untransformed scale was chosen in order to ensure comparison of the correct statistics — the sample arithmetic means.[3]

Results of multiple t-tests are shown in Table 5. A two-sided F-test has been conducted first to test for equality of standard deviations, with p-values lower than 0.05 indicating the use of an unequal-variance t-test approximation.

Table 5: Comparison between data distributions

	F-test p-value	t-test p-value
Surface versus subsurface		
Background	0.0029	0.33 [a]
Site	0.74	0.81 [a]
Site versus background		
Surface	0.70	0.081 [b]
Subsurface	0.0068	NA [c]
All data	0.0029	0.42 [b]

SE: Estimated standard error of the mean
[a] Two-sided (the test is whether the surface concentration differs from the subsurface)
[b] One-sided (the test is for whether the site has higher concentration than background)
[c] Site subsurface concentration is lower than background , so a one-sided t-test for the site being higher than background is inappropriate.

None of the p-values suggest any difference between site and background (or between surface and subsurface). The smallest t-test p-value obtained (0.081, one-sided, for site versus background surface concentration) is likely to be among the more reliable, since both surface concentration distributions were consistent with normal (Table 3).

The distributions of concentrations in site soil and background are clearly distinct and different, as apparent in Figures 1 and 2. In addition, the F-tests shown in Table 5 indicate differences in the distribution shapes. Nevertheless, a further comparison between site

[3] Transformation, for example by taking the logarithm, may give distributions that are more normal. But a standard t-test then would compare the transformed statistics, not the arithmetic means. For example, a logarithmic transformation would result in comparisons of medians, not means.

and background was performed using a Wilcoxon Rank Sum test,[4] for which comparison between site and background indicated a p-value of 0.054.[5] Thus although the distributions are different, there is insufficient evidence to indicate that the site concentrations are systematically higher than background concentrations.

4. CONCLUSIONS

The site-specific background study indicates that the average arsenic concentration detected at the Rhode Island site is consistent with and hence appears to be due to natural background conditions. The distribution of concentrations found on the site and in a nearby background area differ, but the difference (a narrower distribution on site versus in the background area) is consistent with the history of earth movement on the site and in the background area. We thus conclude that the arsenic concentrations do not indicate a release of hazardous materials at the site.

5. RISK MANAGEMENT IMPLICATIONS

Compared to other areas of New England, Rhode Island soils are generally low in arsenic. Even low arsenic levels, however, present potentially significant risks to human health. As stated in the abstract, a concentration of 0.4 mg/kg of arsenic in soil corresponds to a one in a million incremental cancer risk (based on standard screening-level risk assessment assumptions for a residential use scenario). By allowing background levels of arsenic of up to 2.7 mg/kg, the Rhode Island Site Remediation program permits a risk level of more than six times their target goal of one in a million. Average arsenic concentrations of the order of 12 mg/kg in soil (such as those found at the site examined in this study) translate to a screening-level risk of about thirty in a million, or roughly thirty times Rhode Island's target-risk criterion.

This background study was developed in the context of Rhode Island's Site Remediation program. Conceptually, by demonstrating the presence of arsenic in soils is due to natural conditions, soils with elevated levels of

[4] The procedures given in EPA (1998) for the Wilcoxon Rank Sum test are clearly incorrect. Correct procedures may be found in, for example, NBS (1963).

[5] The assumptions required for a Wilcoxon Rank Sum test — namely that the distributions differ only by location, but otherwise have the same shape — are almost certainly violated, as indicated by their difference in variance.

natural arsenic can be exempted from remediation requirements. Given the magnitude of potential health risks, however, substantial evidence is needed to demonstrate that higher-than-average arsenic concentrations in soil constitute background.

Even if background is demonstrated, it can in practice be a difficult decision to leave elevated arsenic levels unmitigated. Ultimately, the Rhode Island Department of Environmental Management determined that the arsenic at this site constituted a release subject to remediation requirements (RIDEM, 2003).

It is not clear, however, that in the end the issue of background is a controlling factor. Even if exempted from site-remediation requirements, conditions on site development would likely have been imposed by other regulatory authorities (*e.g.*, the Rhode Island Department of Health) to protect human health. Similar risk-management measures would likely have been required for the site irrespective of a formal determination on the source of the arsenic in soil. In fact, choosing to regulate the arsenic as a release probably facilitated site development, because it provided clear requirements within the context of an established regulatory program.

REFERENCES

Cunnane, C. 1978. Unbiased plotting positions —a review. J. Hydrology 37:205–222.

DEM. 2001. Rhode Island Department of Environmental Management. "Appendix I: guidance for arsenic in soil." Arsenic policy. Office of Waste Management. July 13, 2001.

EPA (United States Environmental Protection Agency). 1998. Guidance for data quality assessment, practical methods for data analysis. Office of Research and Development, Washington, D.C. 20460. EPA/600/R-96/084. January 1998.

EPA. 2001. Guidance for characterizing background chemicals in soil at superfund sites. Office of Emergency and Remedial Response. Washington, DC 20460. EPA 540-R-01-003. OSWER 9285.7-41. External Review Draft. June 2001.

McPhail Associates. 2002. Letter from Vanderhoff, E.H., and Donovan, A.J., McPhail Associates to S. Martino, Rhode Island Department of Environmental Management. October 9, 2002.

NBS. 1963. National Bureau of Standards. Handbook 91. Experimental Statistics. United States Department of Commerce, August 1, 1963.

RIDEM. 2002. Letter from S.R. Martino, Rhode Island Department of Environmental Management (RIDEM), to D.W. Marvelle, The Housing Authority of the City of Newport, Rhode Island. September 27, 2002.

RIDEM. 2003. Letter from S.R. Martino and K. Owens, Rhode Island Department of Environmental Management (RIDEM) to Patrick Lee, Trinity Financial. September 27, 2002.

Royston, J.P. 1982. Algorithm AS 181: The W test for normality. Appl. Statist. 31:176 –180.

Royston, P. 1993. A toolkit for testing for non-normality in complete and censored samples. The Statistician 42:37–43.

Royston, P. 1995. Remark AS R94. A remark on algorithm AS 181: The W -test for normality. Appl. Statist. 44:547–551.

Zemba, S. G. 2002. Background Study Proposal: Phase 1 Revitalization of Tonomy Hill, Newport, Rhode Island. Cambridge Environmental Inc. Cambridge, MA. July 17, 2002.

PART IV: MTBE

CHAPTER 19

HIPOX ADVANCED OXIDATION OF TBA AND MTBE IN GROUNDWATER

Reid H. Bowman, Ph.D
Applied Process Technology, Inc., Pleasant Hill, CA

Abstract: This paper presents results from several pilot studies and full -scale remediation sites in which an Advanced Oxidation Process (ozone/hydrogen peroxide) was used to remove TBA and MTBE from contaminated groundwater. The concentration of TBA in the groundwater at the various sites ranged from 29 μg/L to 4,800 μg/L. The MTBE concentrations ranged from 42 to 110,000 μg/L. The sites include abandoned gas stations, active gas stations and gasoline terminals located in California, Nevada and New Jersey. At the full-scale remediation sites, the flow of contaminated water treated ranged from 5 to 60 gallons per minute (gpm). While discharge requirements vary from site to site, we have demonstrated that the TBA and MTBE concentrations can be reduced to less than each site's respective detection limits. Pilot work at each site allowed for the development of an empirical model to determine the concentrations of ozone and hydrogen peroxide demanded to meet the TBA and MTBE discharge requirements. Comparisons between the pilot study results and the full -scale remediation site results demonstrate the scalability of this process.

Key words: t-butyl alcohol (TBA); methyl t-butyl ether (MTBE); Advanced Oxidation; HiPOx; Groundwater; Applied Process Technology

1. INTRODUCTION

The widespread use of methyl *t*-butyl ether (MTBE) and to a lesser degree *t*-butyl alcohol (TBA) as oxygenates in gasoline has resulted in groundwater contamination by these two compounds. Leaking underground-storage tanks are the major source of groundwater contamination by these compounds. However, MTBE and TBA are also released into the environment through other types of sources (Melin, 2000). Both MTBE and TBA are poorly biodegraded, very soluble in water, and very mobile in subsurface environments (Squillance et al., 1997).

Several recently published reports study the use of Advance Oxidation Processes (AOP) based on ozone and hydrogen peroxide to remediate MTBE (Karpel Vel Leitner et al., 1994; Liang et al., 1999; Acero et al., 2001). AOPs generate hydroxyl radicals as their primary oxidant. In all of these recently published studies, influent waters contained relatively low concentrations of MTBE and did not contain other contaminants. Typically, groundwater contaminated with MTBE at or near a gas station will also contain fuel components including benzene, toluene, ethylbenzene and xylene (BTEX) and usually some TBA, which forms as a natural degradation product of MTBE. The purpose of our work was to examine the effectiveness of ozone and hydrogen peroxide in remediating groundwater at gas station sites, where all of the contaminants associated with a leaking underground storage tank were present.

2. EXPERIMENTAL CONDITIONS

The treatment system is a continuous, in-line process that operates at feed water pressure. Ozone and hydrogen peroxide are injected at mg/L levels into the contaminated water to be treated. The ozone is produced at 40 to 50 psig on-site from a solid-state ozone generator. Pure oxygen is supplied to the ozone generator to yield high concentrations of ozone (6-10 weight percent in oxygen). The injection of ozone takes place at various points along the treatment system's flow path in order to increase the effectiveness of the process and minimize by-product formation. This process is trademarked as HiPOx™ and is described in U.S. Patent No. 5,851,407 issued December 1998. Figure 1 shows the schematic of the overall treatment system.

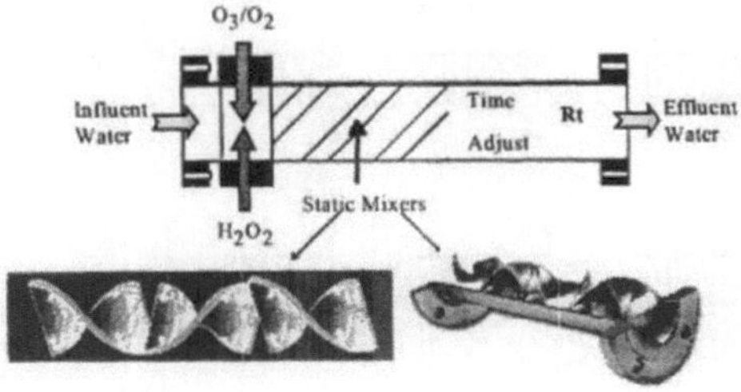

Figure 1. Schematic of HIPOx advanced oxidation system

The system's continuous reactor is comprised of individual reactors. The individual reactors consist of an injection section where ozone and hydrogen peroxide are injected, a mixing section and a reaction section. A sampling point (SP) is located at the end of each reactor section so that the degree of contaminant destruction can be monitored as a function of applied oxidant (ozone and hydrogen peroxide). The residence time in each individual reactor ranges from 3 to 10 seconds. This is diagramed in Figures 2 and 3.

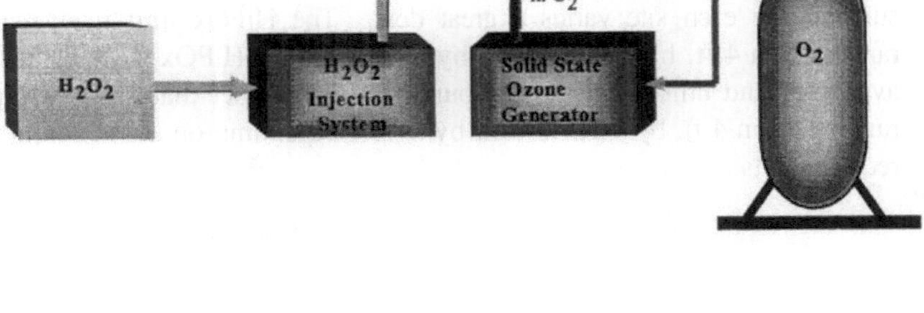

Figure 2. Schematic diagram of individual reactor

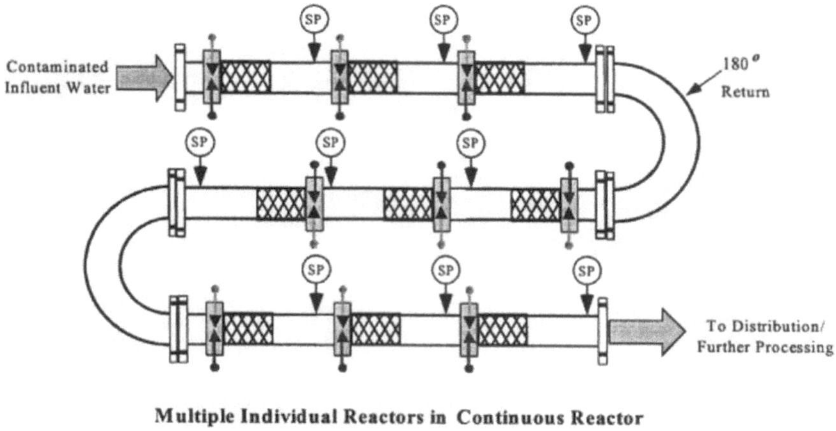

Figure 3. Schematic diagram of multiple individual reactors

The initial work was conducted at flow rates of 3 to 5 gpm of the contaminated water using a manually controlled mobile HiPOx Pilot Unit. The pilot unit was configured with 18 individual reactors. Each reactor was equipped with an ozone injector and a 1/2-inch static mixer and had a residence time of 5 to 30 seconds. The hydrogen peroxide was injected into the contaminated water upstream of the first reactor. The flow of hydrogen peroxide was controlled with a metering pump, and was monitored and recorded. The flow of oxygen was monitored and recorded by an oxygen mass flow indicator. The concentration of ozone injected was measured using an on-line ozone analyzer. The pilot unit was provided water directly from the contaminated wells.

For typical service station applications, the size of the enclosure surrounding each site varies a great deal. The HiPOx unit itself usually ranges from 4 ft. by 8 ft. to 8 ft. by 20 ft. In-situ HiPOx systems are also available, and utilize an above-ground control cabinet that has a footprint ranging from 4 ft. by 4 ft. to 8 ft. by 10 ft. depending on a site's treatment requirements.

3. RESULTS AND DISCUSSION

3.1 Northern Nevada Site

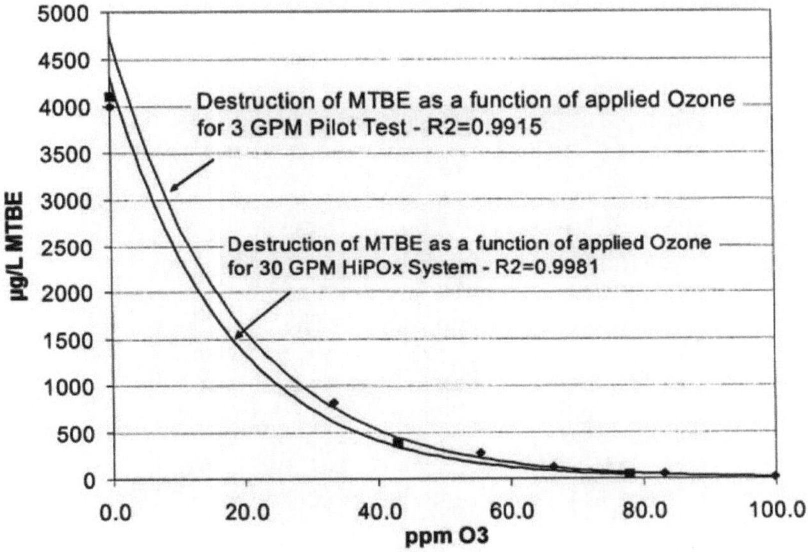

Figure 4. Comparison of MTBE destruction as a function of applied ozone at two different flow rates

At a former petroleum terminal in Northern Nevada, a bio-reactor system had been installed to degrade MTBE in the contaminated water from 30 extraction wells. The MTBE concentrations were too high in some of the wells to be treated by the existing bio-reactor system. The requirement at the Northern Nevada site was to reduce the MTBE from 4,000 µg/L to less than 100 µg/L. A pilot test was conducted using a HiPOx mobile pilot unit. The results of this trial are shown graphically in Figure 4. In Figure 4, MTBE destruction is plotted versus applied ozone. This data was obtained by removing samples at known points in the continuous reactor. During the pilot test, the MTBE concentration was reduced to less than 2 µg/L (the detection limit at the time of the pilot test). As a result of the pilot testing, a 40-gpm HiPOx unit was built and is currently operating as wellhead pretreatment to the existing water treatment system at the Northern Nevada site. The 40-gpm HiPOx unit is computer controlled, allowing for unattended operation 24 hours a day, 7 days a week. A photograph of the installation at the Northern Nevada site is shown in Figure 5. Figure 4 also shows a comparison between the ozone efficacy achieved by the pilot unit and that achieved by the full-scale system. The degree of MTBE destruction versus applied ozone is essentially identical for the 3- and 40-gallon-per-minute units.

Figure 5. Photograph of the Northern Nevada HiPOx installation.

The capability of the HiPOx system to handle changes in the influent concentration of MTBE is shown in Figure 6. The concentration of MTBE entering the HiPOx unit varies at this location because there are 30 extraction wells that feed the system but no buffer tank. The wells feeding the system are not always the same wells, and each well has a different MTBE concentration. The influent MTBE concentration varied from 1,700 to 7,000 µg/L (Left Y-Axis Dark Line Figure 6). However, the effluent MTBE concentration remains relatively constant (Right Y-Axis, Light Line, Figure 6).

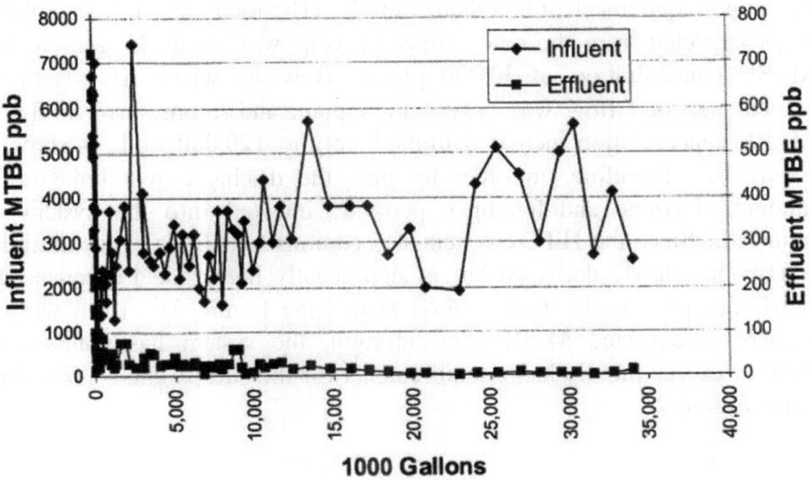

Figure 6. Effect of MTBE influent concentrations on MTBE effluent concentrations

3.2 Northern California Gas Station

At an operating gas station in Northern California, a pilot test was conducted at 2.5 gpm. The concentration of MTBE in the groundwater ranged from 30,000 to 53,000 µg/L. The influent concentration of TBA ranged from 3,200 to 9,100 µg/L. The concentration of *t*-Amyl Methyl Ether (TAME), another oxygenate found at this site, was 3,000 to 6,200 µg/L. In addition to these oxygenates, BTEX concentrations ranged from approximately 4,500 to 6,500 µg/L, and TPHg was present at approximately 30,000 µg/L. The results of this pilot test are shown in Table 1.

***Table 1.* HiPOx Pilot Results from Northern California Gas Station**

	MTBE	TBA	TAME	BTEX	TPHG
Influent Concentrations (µg/L)	33,000	3,200	3,000	6,700	31,000
Effluent Concentrations (µg/L)	<5	<100	<10	<5	<500

Based on the success of the pilot test, a site plan was implemented using a HiPOx AOP system with the addition of several extraction wells. The installed HiPOx unit is computer controlled and operates 24 hours a day, 7 days a week. This location had a total petroleum hydrocarbons (TPH) discharge requirement of less than 2 mg/L. The total flow of contaminated water expected from the new extraction wells was 10 to 15 gpm with an MTBE concentration of 30,000 μg/L. However when the system was started, the total flow was only 4 to 5 gpm, and, more surprisingly, the MTBE concentration increased immediately to 120,000 μg/L as shown in Figure 7. Therefore, in order to meet the discharge requirements, the amount of ozone and hydrogen peroxide injected into the system was adjusted. Since the HiPOx system was commissioned, the concentration of MTBE has slowly decreased but is significantly higher than experienced in the initial pilot trials (Left Y-Axis Dark Line Figure 7). Even with the higher-than-expected MTBE concentration, the system has continued to meet or exceed the discharge requirements for the site (Right Y-Axis, Light Line, Figure 7).

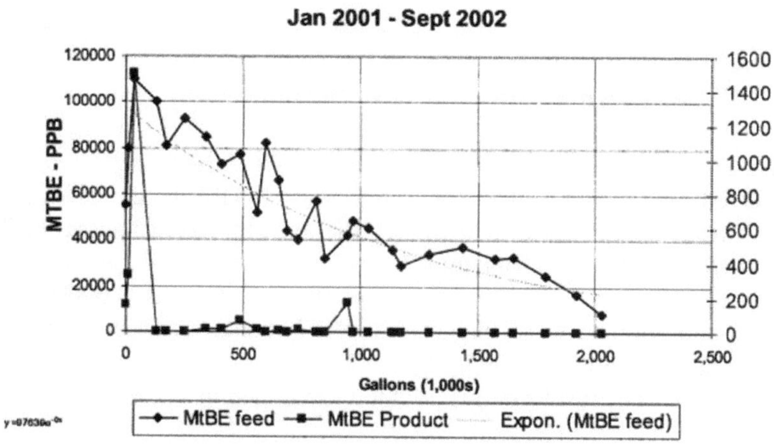

Figure 7. Influent and effluent MTBE concentrations as a function of cumulative gallons treated

Although there was no specific requirement for TBA destruction, in the course of removing the MTBE, the HiPOx unit reduced TBA from an average influent concentration of approximately 4,800 μg/L to an average concentration in the effluent of 325 µg/L. A photograph of the installation at the Northern California gas station is provided in Figure 8.

Figure 8. Photograph of the HiPOx insatallation at the Northern Califonrnia gas station

3.3 New Jersey Terminal

At a gas station in New Jersey, a remediation system had been installed to treat the groundwater, which had been contaminated with MTBE. After the MTBE remediation system had been in operation, a significant concentration of TBA was found to be present. The concentration of TBA was as high as 13,000 µg/L. The existing MTBE remediation system had little to no effect on reducing the concentration of TBA in the contaminated groundwater. A HiPOx pilot test was conducted, and the results demonstrated the ability of ozone and hydrogen peroxide to oxidize TBA to less than 3.3 µg/L. The oxidation by-product of TBA is acetone (Karpel Vel Leitner et al., 1994; Liang et al., 1999; Acero et al., 2001), which is slowly oxidized to carbon dioxide and water. As shown in Figure 9, the oxidation of acetone requires the majority of the applied ozone. After the addition of approximately 284 mg/L of ozone, the TBA concentration was < 3.3 µg/L but an additional 428 mg/L of ozone was required to reduce the acetone to 150 µg/L. Therefore, a more effective method to remove the acetone has been planned. This involves the conjunctive use of a bioreactor, which has been shown to be a very effective technology for acetone destruction, on the effluent of the HiPOx system.

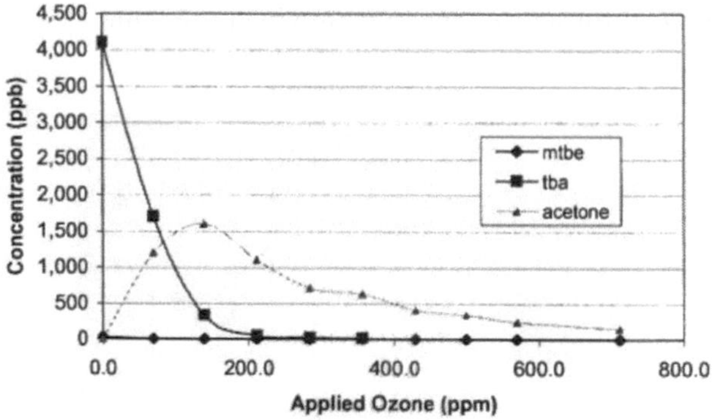

Figure 9. TBA destruction as a function of applied ozone

3.4 Former Military Base Gas Station

A HiPOx pilot test was conducted at a former gas station site on a closed military base in Southern California. The HiPOx unit reduced MTBE in the groundwater from approximately 35,400 μg/L to 1 μg/L. At the time of the pilot study, the concentrations of TBA, TBF and most of the BTEX components were not detectable in the influent groundwater. The results of the pilot test are reported in Table 2.

Table 2. HiPOx Pilot Results from former military base gas stations

	MTBE	TBA	TBF	BENZENE	TOLUENE	ETHEL-BENZENE	XYLENE
Influent (μg/L)	35,400	<20	N.D.	<5	<5	<5	0.8
Effluent (μg/L)	1	<20	N.D.	<5	0.9	<5	0.5

After the pilot tests were completed, a full-scale 30-gpm unit was installed on the site. The installation was then expanded to operate at flow rates of 50 to 70-gpm and has been operating at 60-gpm since February 2002. Contaminant concentrations in the groundwater at this site have fluctuated significantly since the initial pilot test was conducted. Influent MTBE concentrations have decreased to approximately 17,000 μg/L, and influent TBA has increased to approximately 200 μg/L. TBF was not detected in the water, and BTEX components were present at levels less than 1 ppb. Acetone was also present in the influent at a concentration of 110 μg/L, and additional acetone was formed as a destruction by-product of MTBE and TBA. Similar to the previously described application at the New Jersey terminal, the destruction of acetone requires a significant amount of ozone, and it was determined that a bioreactor would provide a more effective method of removing the additional acetone. Therefore, a bioreactor was installed on the effluent of the HiPOx system at this site. Results from treatment by the HiPOx system and the bioreactor are shown in Table 3.

Table 3. **Results from HiPox System and Bioreactor at former military base gas station**

	MTBE	TBA	ACETONE
Influent (μg/L)	17,000	200	110
HiPOx Effluent (μg/L)	<2	<4	680
Bioreactor Effluent (μg/L)	<2	<4	8.2*

***Trip blank acetone: 5.7 ppb.**

A photograph of the installation is provided in Figure 10. Total treatment costs for this installation are $25.13 per 1,000 gallons of water treated. Of this, $20.50 is attributed to system lease costs, and $4.63 is attributed to operating costs.

Figure 10. Photograph of the HiPOx installation at the former military base service station

3.5 Central Valley Gas Station

Since March 2002, a 10-gpm commercial HiPOx unit has been removing MTBE and TBA from groundwater at a gas station site in the Central Valley of California.

Table 4. HiPOx System and Bioreactor Results from Central Valley gas station

May 10, 2002	MTBE	TBA	ACETONE
Influent (μg/L)	3,400	220	<200
HiPOx Effluent (μg/L)	<0.5	<5	110
Bioreactor Effluent (μg/L)	<0.5	<5	<10

The most recent analytical data is provided in Table 4. During the first six weeks after the system was commissioned, influent concentrations of MTBE gradually decreased from approximately 19,000 μg/L to 3,700 μg/L. In each case, the HiPOx system effectively destroyed this range of MTBE contamination to non-detect levels, which ranged from less than 5 μg/L to less than 0.5 μg/L. Influent TBA concentrations were initially difficult to ascertain because high concentrations of MTBE and other contaminants in the influent interfered with the detection of TBA, making detection limits abnormally high. As treatment progressed and other influent contaminants decreased, the detection limit for influent TBA decreased as well. During the first week of treatment, influent TBA could not be detected at levels above the detection limit of 4,000 μg/L. During the second week of treatment, it could not be detected at levels above the detection limit of 1,000 μg/L. By the third week of treatment, as other influent contaminants decreased, TBA could be detected in the influent at levels of approximately 550 μg/L. The HiPOx system destroyed the TBA to below the effluent detection limit of 5 μg/L. None of the samples taken from the effluent of the HiPOx system since it was commissioned contains TBA at detectable limits. Acetone is formed during destruction of MTBE and TBA and has been detected at levels of up to 160 μg/L in the effluent of the HiPOx system. Therefore, a bioreactor removes the excess acetone from the HiPOx system effluent. A photograph of the installation is provided in Figure 11.

Figure 11. Photograph of the HiPox installation at the Central Valley gas station

3.6 Southern California Terminal

In November 2002, a 2.5-gpm HiPOx pilot test was conducted at an operating terminal in Southern California for the purpose of evaluating the system's removal of MTBE and TBA. The influent MTBE and TBA concentrations at this site were 30 to 60 µg/l. The HiPOx pilot unit reduced MTBE and TBA concentrations to below the detection limits of <0.5 µg/l and <5 µg/l, respectively. The pilot unit also reduced TPHg from 2,100 µg/l to <50 µg/l, Di-isopropyl ether (DIPE) from 11 µg/l to <2 µg/l, and BTEX from 50 µg/l to <0.5 µg/l (all below their respective detection limits). Based on successful pilot test results, a site plan was developed in which a 10-gpm HiPOx system was installed as a protective barrier at the toe of the contaminant plume. This was done to protect several drinking-water production wells in operation a half mile from the toe of the plume. Results from the 10-gpm barrier system are shown in Table 5. A photograph of the barrier system is shown in Figure 12.

Table 5. **HiPOx Barrier System Results from Southern California terminal**

	MTBE	TBA	DIPE	BTEX
Influent (µg/L)	42	29	6.6	39.7
Effluent (µg/L)	<1	<3	<2	<0.5

Figure 12. Photograph of the 10-gpm HiPOx installation as the Southern California Terminal

The site plan also calls for installation of a 200-gpm HiPOx system to be positioned at the source of the contaminant plume. The 200-gpm HiPOx system is scheduled to commence operation toward the end of 2003.

4. CONCLUSIONS

Advanced oxidation, using ozone and hydrogen peroxide as the oxidant, is very effective for removing MTBE and TBA as well as the other constituents of gasoline to non-detect levels in groundwater contaminated with gasoline. This has been demonstrated at several sites. The AOP in the form of HiPOx, is effective for removing MTBE or TBA independent of the influent concentration. The HiPOx technology is directly scaleable from pilot trials to full-scale systems. The full-scale systems are compact, portable, quiet, computer controlled, and operate unattended 24 hours a day, 7 days a week.

REFERENCES

Acero, J.L., Haderlein, S.B., Schmidt, T.C., Suter, M.J.F., and Von Gunten, U. 2001. *Environ. Sci. Technol.* 35, 21, 4252.

Karpel Vel Leitner, N., Papailhou, A.L., Croue, J.P., Peyrot, J., and Dore, M. 1994. *Ozone Science & Engineering* 16, 41.

Liang, S., Palencia, L.S., Yates, R.S., Davis, M.K., Bruno, J.M., and Wolfe, R.L. 1999. *Journal AWWA* 91, June, 104.

Melin, G. (Ed.). 2000. *Treatment Technologies for Removal of Methyl Tertiary Butyl Ether (MTBE) from Drinking Water: Air Stripping, Advanced Oxidation Processes, Granular Activated Carbon, Synthetic Resin Sorbents*, 2nd ed. Prepared for the California MTBE Research Partnership by the National Water Research Institute, Fountain Valley, CA.

Squillance, P.J., Pankow, J.F., Korte, N.E., and Zogorski, J.S. 1997. *Environmental Toxicology & Chemistry* 16, 1836.

This page intentionally left blank

CHAPTER 20

ENHANCED NATURAL ATTENUATION OF MTBE & BENZENE AT A LOW PERMEABILITY SITE USING ISOC TECHNOLOGY

Walter S. Mulica[1], Nick Mathis[2], James F. Begley[3]
[1]Global Technologies, Inc., Fort Collins, CO, [2]O&G Environmental Consulting, LLC, [3]MT Environmental Restoration, Plymouth, MA

Abstract: The operation and performance of an iSOC ™ (*in situ* Submerged Oxygen Curtain) groundwater bioremediation system were evaluated over a period of twenty -one months at a low-permeability Colorado gasoline release site. The iSOC™ system was initially evaluated with a pilot study and then selected to replace a failing air -sparging system. The iSOC™ system was designed to enhance natural attenuation and control a hot spot of benzene and MTBE contamination in groundwater. The results demonstrate that the system has been successful in achieving significant reductions in benzene and MTBE concentrations and has substantially reduced the size of the contaminant hot spot.

Key Terms: *in situ* bioremediation, Isoc, groundwater remediation, case study, benzene, MTBE

1. INTRODUCTION

The performance of an iSOC™ (*in situ* Submerged Oxygen Curtain) groundwater bioremediation system was evaluated over a period of twenty-one months at a low-permeability gasoline release site in Colarado. The results demonstrate that the system has been successful in achieving significant reductions in benzene and MTBE concentrations and has substantially reduced the size of the contaminant hot spot.

Attempts to remediate the plume of dissolved benzene and MTBE in the late 1990s had proven difficult and operation of an air-sparging system at the site was suspected to be causing increased contaminant migration. An alternative approach was necessary, and the site consultant, O&G Environmental, and the State of Colorado decided to evaluate the iSOC™ oxygen delivery system to stimulate *in situ* bioremediation. A pilot system was installed in January 2002 and based on a six-month trial, iSOC™ was selected for full scale hot spot treatment at the site.

2. CASE STUDY SITE

Gasoline was released from a leaking underground storage tank (UST) at a Colorado industrial site resulting in soil and groundwater contamination. The UST was removed along with contaminated soil from around and below the tank; however, the full extent of contaminated soil could not be removed due to the depth of contamination. Geologic conditions at the site include 40 feet of silty clay (permeability between 10^{-4} and 10^{-6} cm/sec) over fractured bedrock. The water table was encountered about 20 feet below the land surface. Initial attempts at remediation included the construction and operation of an air-sparging and soil-vapor extraction system. Monitoring during treatment system operations indicated that air sparging was causing increased contaminant migration, and air sparging operations were discontinued. Generally, air-sparging treatment is not recommended where the permeability is below 10^{-4} cm/sec (Norris et al. 1993). The increased contaminant migration may have been caused by excessive physical displacement of groundwater by injected air. The soil-vapor-extraction system was operated until off-gas concentrations declined and was shut down prior to the start of the iSOC bioremediation pilot study.

As of December 2001, prior to the initiation of the iSOC™ pilot study, the area of groundwater contaminated with benzene, toluene, ethylbenzene and xylenes (BTEX) and MTBE was well characterized with a set of twelve monitoring wells. The benzene plume extended approximately two hundred feet to the northwest of the former UST location and was roughly 100 feet wide. The maximum concentrations of benzene (5200 ppb) and total BTEX (5589 ppb) were located at MW-8 (Figure 1). The maximum concentration of MTBE in December of 2001 was also measured in groundwater at MW-8 (195,000 ppb; Figure 2). The MTBE plume was considerably wider than the benzene plume, with a diameter of just over 200 feet. The difference in the shapes of the benzene and MTBE plumes was most likely due to MTBE's higher solubility in water resulting in more lateral spread.

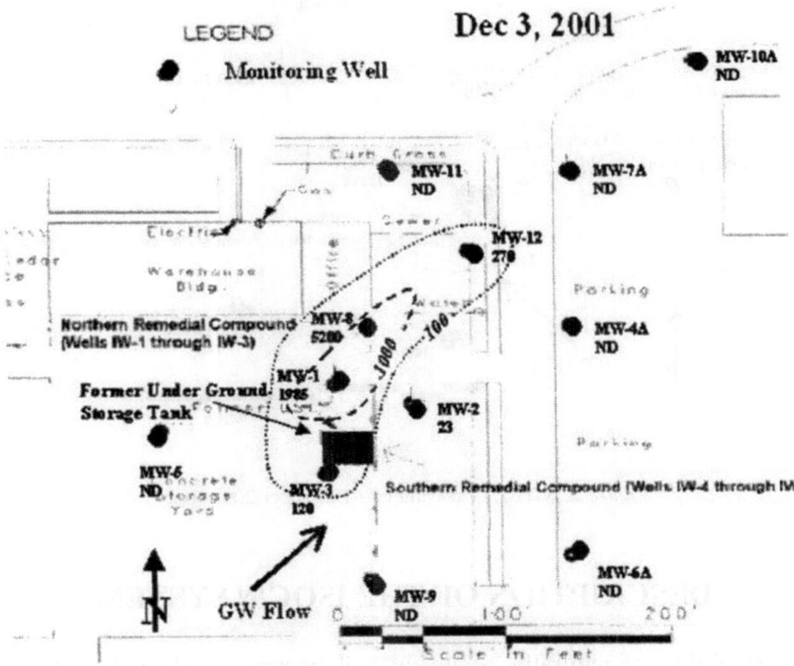

Figure 1. Benzene Concentrations December 2001 (ppb)

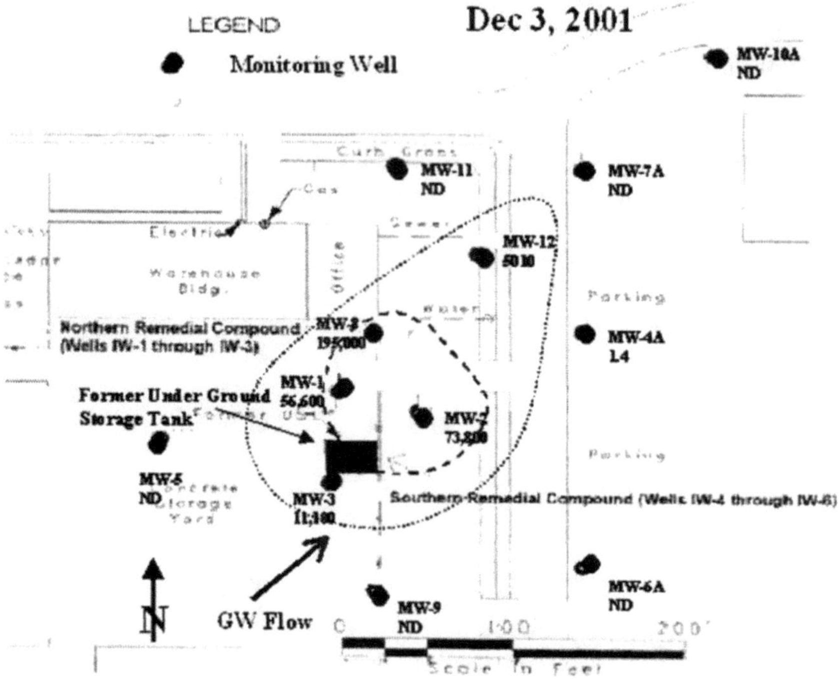

Figure 2. MTBE Concentrations December 2001 (ppb)

3. DESCRIPTION OF THE ISOC™ SYSTEM

The ability of naturally occurring soil bacteria to degrade petroleum hydrocarbons in groundwater has been studied since the 1970s and bioremediation systems have been implemented at numerous contaminated sites since the 1980s. Aerobic degradation of MTBE as a result of oxygen addition has been demonstrated more recently (Nyer et al. 2002). The iSOC™ system was introduced in 2000 as a tool for delivery of pure oxygen to groundwater without sparging. The iSOC™ system includes a mass transfer device (iSOC™ unit) that is 1.62 inches in diameter and 15 inches in length and is placed in monitoring or treatment wells to transfer oxygen from the gas phase directly into groundwater. Treatment wells are generally designed to fully screen the thickness of the plume.

The iSOC™ unit is constructed of stainless steel and contains micropourous tubing that is continuously filled with pure oxygen during operation and provides a high surface area for mass transfer. The iSOC™

units are connected with 0.25 inch polyurethane tubing to a control panel and a two-stage oxygen regulator on a pressurized oxygen cylinder. Gas delivery is managed with the iSOC™ control panel that includes a gas flowmeter and pressure gauges. The gauges are used to monitor gas inlet pressure and the pressure on the iSOC™ units due to the head of water in the treatment well. No electric power is required to operate the system and there is no practical depth limitation on the system beyond limitations on the installation of treatment wells (Figure 3).

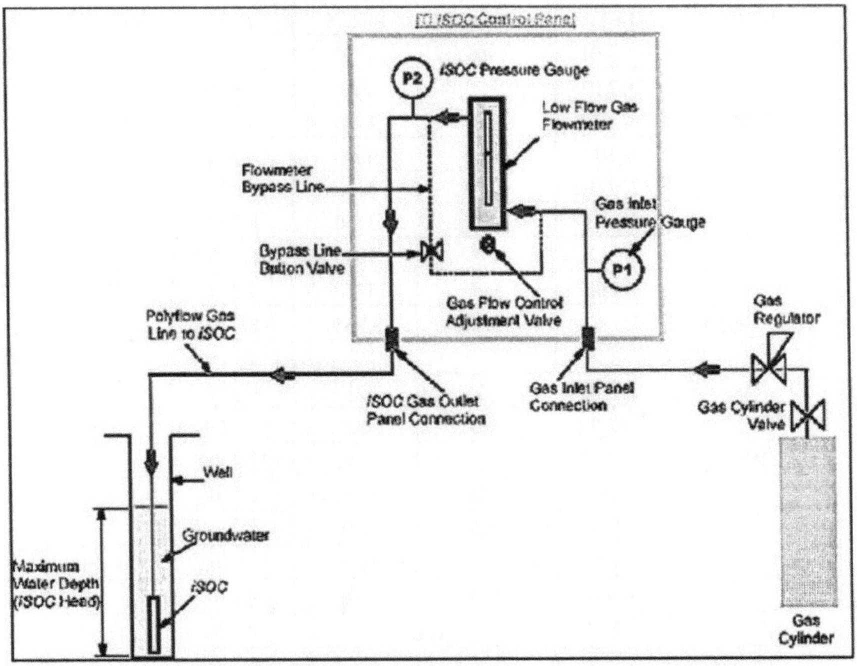

Figure 3. iSOC Flow Diagram

4. ISOC™ PILOT TEST

The bioremediation pilot test was funded by the State of Colorado and included one iSOC™ unit installation in a newly installed well (IW-1) located 15 feet upgradient of MW-8. The oxygen supply cylinder and iSOC™ control panel were housed in a small shed placed near the treatment well. Oxygen was delivered to the IW-1 iSOC™ unit at the standard rate of 15 cm^3/min. A dissolved oxygen (DO) concentration between 45 and 60 ppm was developed and maintained in the treatment well. Groundwater test results indicated a 97% reduction in the benzene concentration at MW-8 from December 2001, prior to the start of the pilot test, to June of 2002. MTBE concentrations were reduced approximately 30% during the same period (Table 1).

Table 1: Groundwater Monitoring Data (ppb)

WELL #	DATE	BENZENE	TOTAL BTEX	MTBE	Water Table (ft)
	12/3/01	1985	4525	56,600	72.5
	3/12/02	2940	620	52,400	72.0
MW-1	6/13/02	2420	4900	94,600	72.0
	9/5/02	1950	4690	59,000	72.5
	12/16/02	1360	3110	39,300	72.5
	3/3/03	1620	3756	8,000	71.5
	6/9/03	538	1986	8,910	74.5
	9/9/03	884	3008	9,760	74.2
% Change		**-55.5%**	**-33.5%**	**-82.8%**	
	12/3/01	23	23	73,800	
	3/12/02	ND	BDL	121,300	
MW-2	6/13/02	ND	BDL	103,000	
	9/5/02	ND	BDL	328,000	
	12/16/02	ND	BDL	128,000	
	3/3/03	ND	BDL	155,000	
	6/9/03	ND	BDL	92,400	
	9/9/03	8	8	97,800	72.9
% Change				**32.5%**	
	12/3/01	120	120	11,180	73.3
	3/12/02	26.3	26.3	10,910	73.0
MW-3	6/13/02	40.7	40.7	7,540	73.0
	9/5/02	21	21	5,840	73.1
	12/16/02	10.9	109	3,420	73.0
	3/3/03	8.3	8.3	3,780	72.1
	6/9/03	4.8	4.8	3,520	75.5
	9/9/03	3	3	3,260	75.2
% Change		**-97.5%**	**-97.5%**	**-70.8%**	

	12/3/01	ND	ND	ND	
	3/12/02	ND	ND	ND	
MW-5	6/13/02	ND	ND	ND	
	9/5/02	ND	ND	ND	
	12/16/02	ND	ND	ND	
	3/3/03	ND	ND	ND	
	6/9/03	ND	ND	ND	
	9/9/03	ND	ND	ND	
% Change					
	12/3/01	ND	ND	ND	
	3/12/02	ND	ND	ND	
MW-7A	6/13/02	ND	ND	ND	
	9/5/02	ND	ND	ND	
	12/16/02	ND	ND	ND	
	3/3/03	ND	ND	ND	
	6/9/03	ND	ND	ND	
	9/9/03	3	3	33	
% Change					
	12/3/01	ND	ND	ND	
	3/12/02	ND	ND	ND	
MW-10A	6/13/02	ND	ND	ND	
	9/5/02	ND	ND	ND	
	12/16/02	ND	ND	ND	
	3/3/03	ND	ND	ND	
	6/9/03	ND	ND	ND	
	9/9/03	ND	ND	ND	
% Change					
	12/3/01	5200	5589	195,000	71.5
	3/12/02	350	350	96,500	71.0
MW-8	6/13/02	160	189.3	138,000	71.0
	9/5/02	330	330	489,000	70.5
	12/16/02	20	20	118,000	71.5
	3/3/03	1	1	135,000	70.5
	6/9/03	7.9	15.4	8,980	73.5
	9/9/03	11	11	79,600	73.4
% Change		**-99.8%**	**-99.8%**	**-59.2%**	

	12/3/01	270	278.6	5,010	
	3/12/02	195	195	4,605	
MW-12	6/13/02	111	111	3,650	
	9/5/02	64.5	64.5	3,380	
	12/16/02	230	230	3,070	
	3/3/03	439	442.3	3,780	
	6/9/03	30.3	30.3	374	
	9/9/03	144	144	1,000	
% Change		**-46.7%**	**-48.3%**	**-80.0%**	
	6/13/02	ND	BDL	1	
	9/5/02	ND	BDL	1	
MW-4A	12/16/02	ND	BDL	ND	
	3/3/03	ND	BDL	ND	
	9/9/03	ND	ND	ND	73.6
% Change				**-100%**	
	12/3/01	ND	ND	ND	
	3/12/02	ND	ND	ND	
MW-6A	6/13/02	ND	ND	ND	
	9/5/02	ND	ND	ND	
	12/16/02	ND	ND	ND	
	3/3/03	ND	ND	ND	
	6/9/03	ND	ND	ND	
	9/9/03	ND	ND	ND	
% Change					
	12/3/01	ND	ND	ND	
	3/12/02	ND	ND	ND	
MW-9	6/13/02	ND	ND	ND	
	9/5/02	ND	ND	ND	
	12/16/02	ND	ND	ND	
	3/3/03	ND	ND	ND	
	6/9/03	ND	ND	ND	
	9/9/03	ND	ND	ND	
% Change					

	6/13/02	ND	ND	ND	
	9/5/02	ND	ND	ND	
MW-11	12/16/02	ND	ND	ND	
	3/3/03	ND	ND	ND	
	9/9/03	ND	ND	ND	
% Change					

Based on these results and the fact that the iSOC™ system would operate in the low permeability setting without sparging and associated groundwater displacement, a decision was made to install five additional iSOC™ treatment wells for hot spot treatment. The additional iSOC™ units were installed in August 2002. The full scale system included the pilot test injection well and was comprised of two sets of three iSOC™ wells each that were operated independently with separate 3-Unit Control Panels and oxygen-supply storage locations. Each set of three wells utilized approximately one 250 cubic-foot oxygen cylinder per month. The layout of the treatment wells was based on the highest concentration benzene and MTBE areas, and was designed to enhance natural attenuation by reducing the hot spot concentrations within the treatment area (Figure 4).

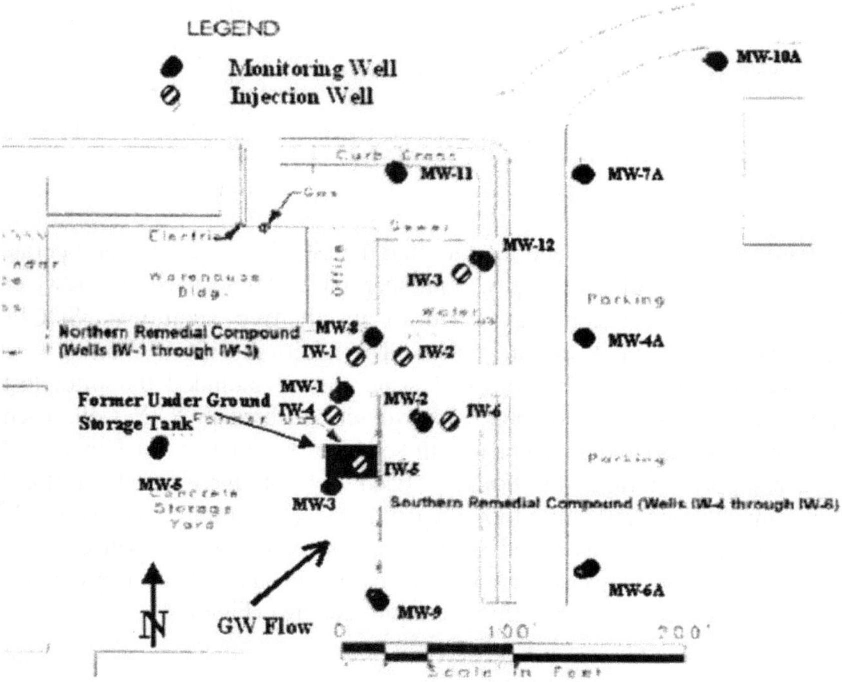

Figure 4. Case Study Treatment Well Layout

5. FULL SCALE TREATMENT RESULTS

The longitudinal axis of the benzene hot spot area, defined by four monitoring wells where the groundwater concentrations were greater than 100 ppb, was reduced approximately 25 % during the iSOC™ pilot test operated from January 2002 to August 2002. Following the startup of the full-scale treatment system in August 2002, the hot spot area was reduced to one well (MW-1) near the former UST location by June 2003. The most recent groundwater-monitoring well test results from September 2003 show that concentrations remain greater than 100 ppb at MW-1 and MW-12 (Figure 5).

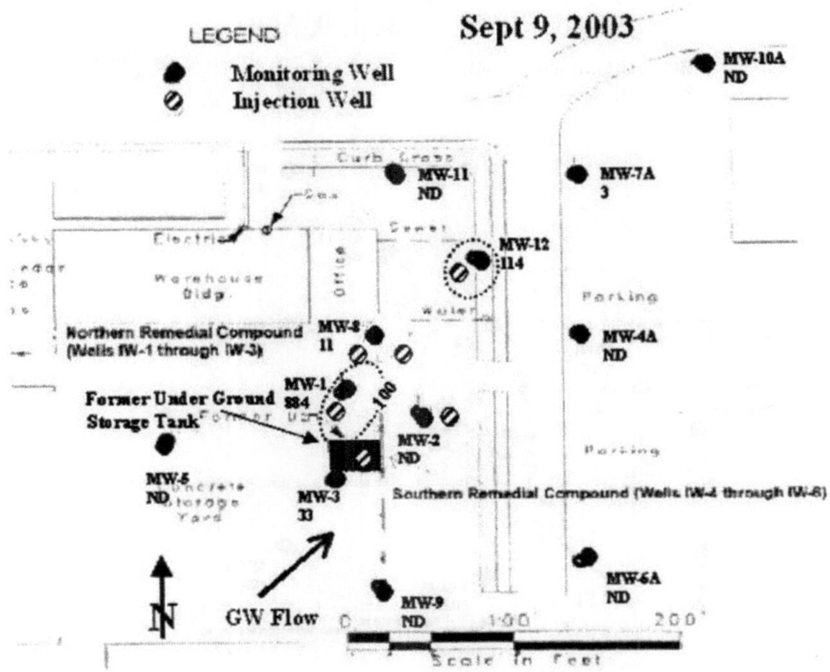

Figure 5. Benzene Concentrations September 2003 (ppb)

The MTBE hot spot area was characterized by groundwater concentrations greater than 50,000 ppb at three monitoring-well locations (MW-1, MW-2, and MW-8) in December 2001. The hot spot area was not reduced in size during the pilot test. During full-scale treatment, groundwater test data indicated that the MTBE hot spot was reduced down to the area around MW-2 by June 2003. In the most recent round of groundwater analyses the MTBE hot spot included MW-2 and MW-8 (Figure 6).

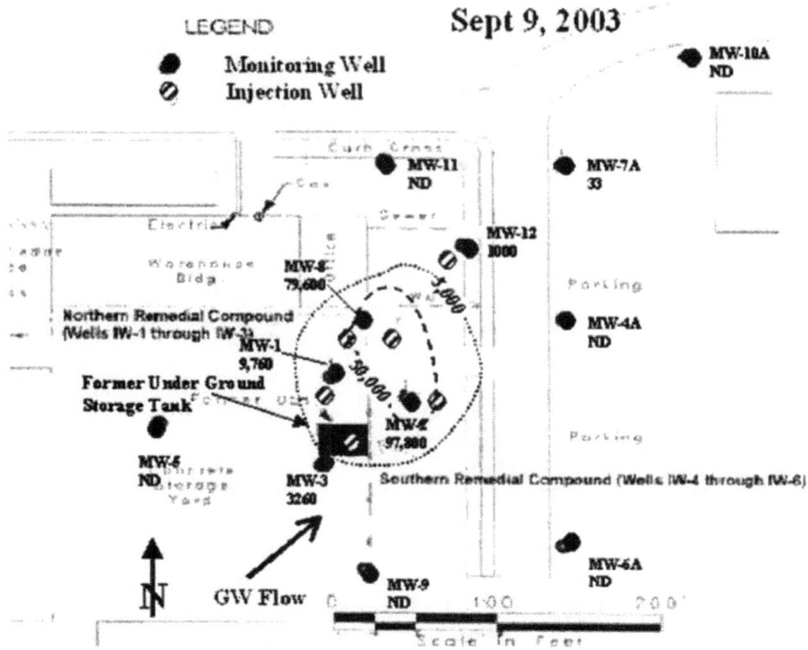

Figure 6. Site MTBE Concentrations September 2003 (ppb)

6. CONCLUSIONS

Benzene concentrations were rapidly reduced while MTBE degradation showed more lag time before significant concentration reductions were detected. The observed fluctuations in benzene and MTBE concentrations in June and September 2003 may be related to rebound of gasoline constituents from an untreated residual source area in the unsaturated zone, or a possible smear zone of contaminants at the water table. Rebound can result from changes in seasonal recharge conditions and longer-term variations in the water table elevation. Significant rainfall events occurring in the spring of 2003 produced a four-foot rise in the water table elevation at the site between March and September 2003. Overall, the site data show that the treatment is reducing the mass of contaminants in groundwater and preventing the migration of significant concentrations to downgradient areas.

It is interesting to note that during the operation of the iSOC™ treatment wells, while the concentrations of dissolved oxygen in the treatment wells has been typically between 40 and 50 ppm, the dissolved oxygen concentration measured in the core of the hot spot area has remained below 1 ppm. However, the downgradient extent of the low-dissolved oxygen area has been reduced based on the observed increases in DO from less than 1 to greater than 4 ppm at MW-7A. Oxygen delivered to the treatment zone is fully utilized by bacteria during contaminant degradation and by geochemical sources of oxygen demand in the plume. The iSOC™ system efficiently delivers oxygen directly into groundwater; continued oxygen delivery is expected to maintain control of the hot spot area until natural attenuation processes eliminate the residual source.

REFERENCES

Norris, Robert D. et al. Handbook of Bioremediation, 1993, CRC Press, Inc., Boca Raton, Florida

Nyer, Evan K., Davis, J. Scott and King, Isabel. 2002, 'Oxygen Revisited', Groundwater Monitoring & Remediation, Volume 22 No3/Summer 2002, pp.44 -49.

This page intentionally left blank

PART V: PHYTOREMEDIATION

CHAPTER 21

PHYTOREMEDIATION OF ZINC AND LEAD CONTAMINATED SOILS USING MIRABILIS JALAPA

Alessandra Carucci, Alessia Cao, Giuseppe Fois, Aldo Muntoni
University of Cagliari, Cagliari, Italy

Abstract: This paper describes a laboratory scale experiment that aims to verify the possibility of using the vegetable species *Mirabilis jalapa* for the decontamination of soils with different zinc and lead content. The use of the phytoextraction or phytostabilization strategy was evaluated.

The experiment was divided into two parts. During the first part *Mirabilis jalapa* was planted in soils artificially contaminated with different Zn (500 ppm, 1000 ppm, 2000 ppm) and Pb concentrations (400 ppm, 600 ppm, 800 ppm). Every experiment was made using two replicates. In the second part Montevecchio soils were diluted with vegetative soil and compost with a dilution factor 4 and 8 (600 -1700 ppm Zn; 5700-19000 ppm Pb). Each part of the experiment lasted one year.

Mirabilis jalapa extraction capacity was demonstrated to be lower than liter ature values, and in particular seemed to be higher during the first growing season in comparison with the second one.

The preferential accumulation of metals in the roots and the reduction of bioavailable metal fraction seem to demonstrate that *Mirabilis jalapa* can be used for phytostabilization processes.

Key words: compost, heavy metals, mining activity, *Mirabilisjalapa,* phytoremediation

1. INTRODUCTION

The presence of high levels of heavy metal contamination in soil constitutes a considerable problem for animal and human health as a consequence of both their toxicity and persistence in the environment. After chemical processes, these elements can pass into the soil solution where plant uptake or leaching to groundwater can contaminate the food chain. At present there is a critical need for the development of cost-effective remediation technologies that reduce such risks.

In situ remediation techniques are generally less expensive and disruptive to the natural landscape, hydrology and the ecosystem than conventional excavation, treatment and disposal methods. Among *in situ* techniques phytoremediation can be considered a good choice to remediate large areas with a low metal contamination depth (Knox et al., 2001). The applicability of phytoremediation depends on the possibility of finding a fast growing plant that tolerates heavy metals and is able to extract and accumulate contaminants in the above-ground part.

Sardinia is characterised by the presence of abandoned mining activities that determined the diffusion of the contaminants in huge areas where traditional remediation techniques appear inapplicable. Selection of plants to apply phytoremediation in these areas must take into account the use of native plants. Previous studies have demonstrated that native plants can be considered good hyperaccumulators but are characterised by a low biomass production (Cao et al., 2002).

The contaminated area object of this work is located near Montevecchio, one of the areas with the richest galena and blende ores in Italy. The weathering on the mining products combined with wind and water erosion and transport created a wide area of contamination 20 km downstream from the Picalinna flotation tailing dam (Dessì et al., 1999). Soils used for the experiment were taken from this area. Heavy metals concentrations in this area are very high (2000 – 11000 ppm of zinc and 5000 – 16000 ppm of lead).

This study aims to verify the possibility of using a fast growing plant, *Mirabilis jalapa,* to decontaminate soils with high lead and zinc contamination levels and to test its hyperaccumulating performances. A parallel study was made to understand how metal removal efficiency could be influenced by different lead and zinc concentrations in the soil.

2. MATERIALS AND METHODS

2.1 Plant selection

The researches on phytoremediation focused initially on the study of the hyperaccumulating plants: plants which have the ability to concentrate high amounts of heavy metals in their plant tissues, and that are usually native to the areas characterized by a high concentration of metals in the soils. Moreover native plants of contaminated areas often accumulate only a specific element, and are as a rule slow growing, low biomass producing plants with few known agronomic characteristics (Cunningham et al.,1995). This constrains their practical use for phytoremediation, since the total metal extraction is the product of plant biomass and heavy metals tissue concentration.

Previous studies on native plants of Picalinna tailing dam (Cao et al., 2002) demonstrated the ability of *Helichrysum italicum* as an hyperaccumulator but showed that this plant, as the majority of plants that survive in the abandoned mines, was characterized by a low biomass production.

Blaylock et al. (1997) used *Brassica juncea* for phytoremediation studies on Pb. *Brassica juncea* demonstrated to extract 1.5% lead in shoots from soil containing 600 mg/kg Pb amended with EDTA. *Mirabilis jalapa* is a fast-growing plant that demonstrated the ability to accumulate lead from contaminated soils extracting about 1500 mg/kg Pb from soils containing 3000 mg/kg Pb (Kambhampati and Williams, 2001).

Mirabilis jalapa was chosen for this study both for its adaptability to the Sardinian climate and its capacity to extract lead from contaminated soils.

2.2 The experiments

The experiment was divided into two parts: during the first part zinc and lead hyperaccumulation capacity was evaluated using soils with different lead and zinc concentrations; in the second part contaminated soils diluted with compost and with vegetative soil were used to verify how soil characteristics can influence phytoremediation performances.

2.2.1 Vegetative soils with different zinc and lead concentrations

Vegetative soils were contaminated by the addition of solutions of either $Zn(NO_3)_2 \cdot 6\ H_2O$, for zinc contamination, or $PbCl_2$ for lead, then equilibrated for one month. Two-litre pots with zinc concentration of 500,

1000 and 2000 ppm and with lead concentrations of 200, 400 and 600 ppm were prepared. The use of soil with different heavy metals concentrations enables the researcher to determine the response of plants to heavy metals contamination and the different plant-soil relationships (Baker, 1981). Every experiment was performed using two replicates. The vases were watered every other day using tap water. Zinc and lead concentration in contaminated soils were evaluated before and after the experiment using both the *aqua regia* and the sequential extraction method (Barbafieri et al., 1996). Root and shoot metal concentrations were evaluated after the experiment. The experiment lasted one year.

2.2.2 Dilution of the contaminated soil with vegetative soils and compost

The soil used in this part of the experiment was taken from an area called Corti Baccas, 20 km downstream from Picalinna flotation tailing dam. The results of tests on bioavailability and toxicity of heavy metals in these soils show both the presence of high concentrations of the toxic elements (2000 - 3000 ppm of zinc and 8000 – 9000 ppm of lead) and a remarkable fraction of easily exchangeable metals causing a high risk to the ecosystem (Dessi et al., 1999).

The contaminated soils were mixed with two types of substrata: vegetative soil and compost in order to reduce soil toxicity with dilution factors 4 and 8.

Concentration of zinc and lead in soil before and after the experiment was evaluated using both the *aqua regia* extraction method (for total concentration) and the sequential extraction method (Barbafieri et al., 1996). Root and shoot metal concentrations were evaluated after the experiment. The experiment lasted one year.

2.3 Soil and plant analyses

A partial chemical and physical characterization was made to verify how the characteristics of soil can influence metal bioavailability to plants, plant growth and phytoremediation performances. Soil was physically characterised through the determination of cation exchange capacity and pH while chemical characteristics investigated during the work were carbon and nitrogen content. Cation exchange capacity is an important factor regulating metal bioavailability while pH dramatically affects the cation exchange capacity of soil by limiting the available exchange sites at low pH. The content of nitrogen and carbon can influence plant growth and can limit heavy metal toxicity effect. Organic carbon analysis was made through the

Springer-Klee method while cation exchange capacity was carried out using $BaCl_2$ method (G.U. n° 248, 21/10/1999).

Both total heavy metal content (G.U. n° 248, 21/10/1999) and bioavailable fraction (Barbafieri et al., 1996) of metals in soil were evaluated.

After the experiment, plants were harvested, washed with tap water and dried at 105 °C. The dried material was digested with aqua regia.

Both zinc and lead concentrations were evaluated in all the samples.

3. RESULTS AND DISCUSSION

3.1 Plant growth

Mirabilis jalapa seeds were planted in the contaminated soils in September 2002. A germination test was made on the original soil from *Corti Baccas* and on the diluted soils. The seeds succeeded in spreading in all the types of soil but were able to grow only in the ones with a dilution factor 8.

The seeds sown in the vegetative soils with different lead and zinc concentrations were all able to spread and grow even though the plants that grew in the soil contaminated with lead with a theoretical concentration of 400 ppm and 600 ppm were not able to grow again after the first growing season. The experiment finished in August 2003.

It was decided to study the metal content in the plants during two growing seasons. In fact, since *Mirabilis jalapa* is an annual plant, it passes through a period of quiescence during winter and grows again in spring. While in the first period the plant grew directly from the seeds, in the second the plant spread from the quiescent bulb.

During the two growing seasons plants were characterised by different growing modes (Figure 1). During the first period the growth was rapid and leaves were large and healthy. The stalk was thin but strong. During spring the plant grew more slowly. The stalk was generally larger than the previous period but leaves were very small and characterised by a light green colour. The difference in plant growth during the two periods can be a reaction by the plant to soil toxicity as a consequence of the adaptation to the environment. It is in fact known that plants that have adapted to contaminated areas are characterized by poor growth, the presence of the symptoms of chlorosis and perhaps by a slow variation in the DNA.

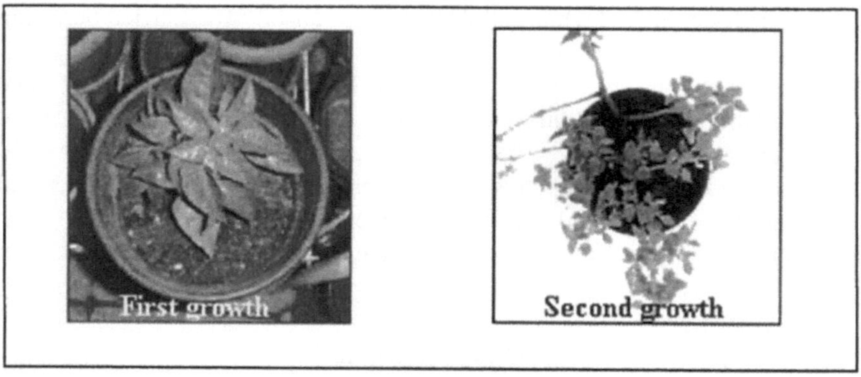

Figure 1. Growth modes in the two growing periods

3.2 Bioavailability of metals in soil

An analysis of total metal content and bioavailability was made before and after the experiment. Figure 2 shows that the amount of heavy metals bioavailable to plants in all the soils used in the experiment generally decreases during the experiment, probably as a consequence of both the effect of the use of a soil amendment and of plant extraction. It is in fact demonstrated that metal bioavailability can be decreased by the use of soil amendments. The effect of bioavailable metal fraction decrease can determine, as a consequence, a stabilization of the toxic metals in soil and prevent their dispersion in the environment.

In the reactor with a mixed substratum (vegetative soil 1:8) a remarkable zinc and lead removal efficiency was obtained both on the total and bioavailable metal concentration (Figure 2). The need for further investigations on the reactor with *Corti Baccas* soil mixed with compost (compost 1:8) did not allow the eradication of the plants and the measurement of the metal content in the plants and soil after the experiment.

The variation in zinc bioavailable concentration in the artificially contaminated reactors was substantially lower. The difference in metal-removal efficiency can be due to the different soil metal concentrations. It is in fact known that plant uptake increases with soil concentration.

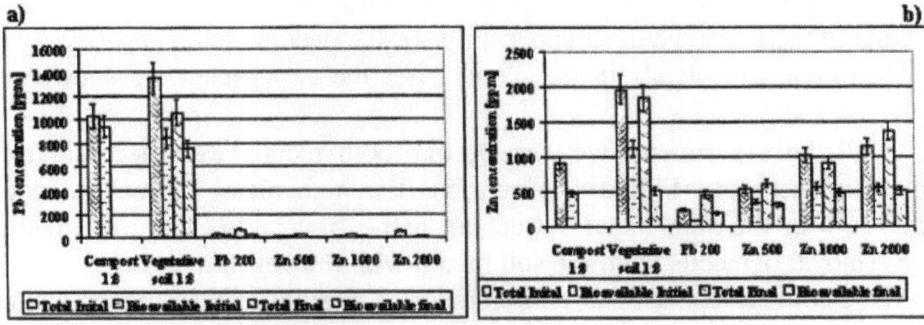

Figure 2. Bioavailable and total metal content before and after the experiment; a) lead; b) zinc

3.3 Distribution of Zn and Pb in the plant

A comparison between bioavailable metal content before and after the experiment and metal accumulated in the shoots in the two periods of growth was made (Figure 3). From Figure 3 we can easily argue that the concentration of both metals in the plant is higher in the first period than in the second. This can mean that the plants can have acquired, in the second period, the capacity to exclude the metal and to accumulate it in the roots, not translocating it to the upper part of the plant. The quantity of zinc and lead accumulated in the upper part of the plant is indeed very low in the second period. The lead content in particular is one order of magnitude lower than that encountered in literature (Kambhampati and Williams, 2001).

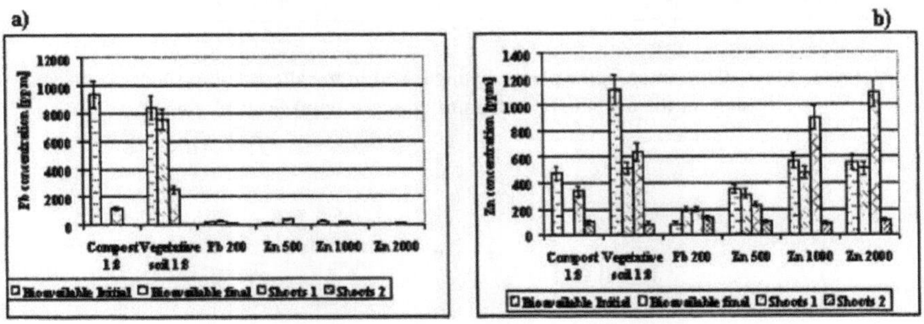

Figure 3. Bioavailable content in soil before and after the experiment compared with metal content in shoots after the first and second growth period; a) lead; b) zinc. Shoots 1 = concentration of metal after the first period of growth; Shoots 2 = concentration of metal after the second period of growth.

On the other hand, metal content in the plant after the first period was relatively high even though it was much lower than metal content in the soil.

This can be a consequence of the high concentration of metal in the selected soil and of the existence of an upper limit for the accumulation capacity.

Distribution of metals in the plants after the experiment are shown in Figures 4 and 5.

In all the experiments we can see that both metals were accumulated mostly in the roots (Figure 4, Figure 5). The high metal content in roots together with the high metal extraction in the first period can explain the high removal of metals from the soil mixed with the vegetative soil.

We can observe that lead concentration was significantly higher in the reactors artificially contaminated with lead (Figure 4a), and was similar in the reactors mixed with vegetative soil and compost (Figura 5a), showing that the choice of soil mixture could not be considered significant.

The study of soils with different zinc content allowed to determine the correlation between the metal content in the soil and in the plant. In Figure 6 it is clear that metal content in soil has no evident correlation with metal content in shoots and stalks of the examined plants while it has a noticeable correlation with root metal concentration.

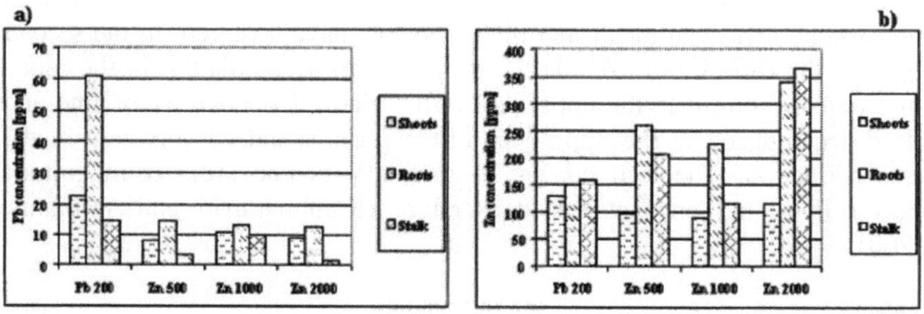

Figure 4. Metal distribution in the plants distinguished in metal content in shoots, roots and stalk in the artificially contaminated reactors; a) lead; b) zinc.

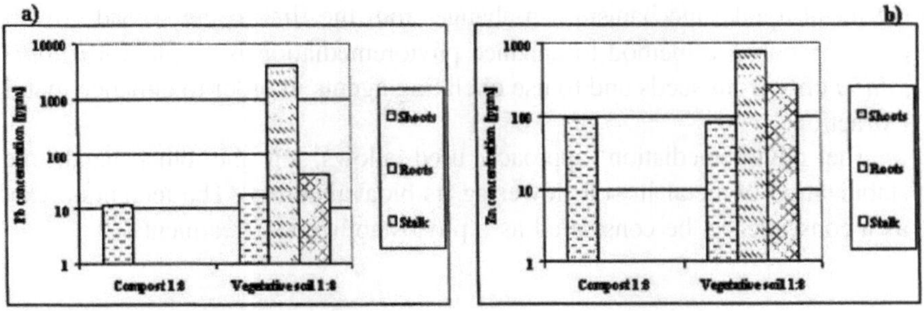

Figure 5. Metal distribution in the plants distinguished in metal content in shoots, roots and stalk in the reactors with diluted soil; a) lead; b) zinc.

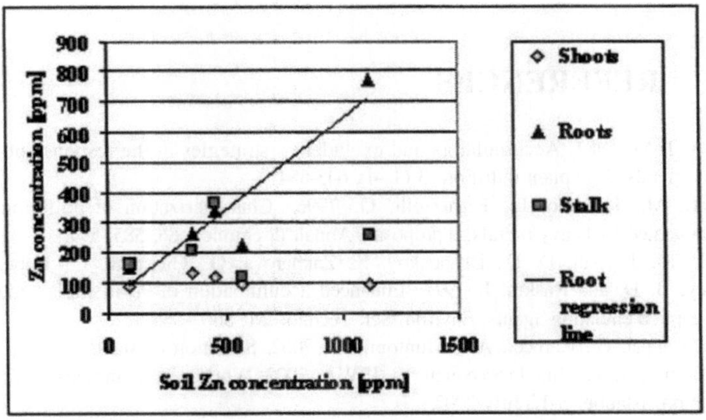

Figure 6. Relation between metal content in the plant and in soil .

4. CONCLUSIONS

Mirabilis jalapa demonstrated a low capacity of accumulating both zinc and lead from soil in the second period of growth while during the first period metal accumulation can be considered significant. The decrease in zinc and lead bioavailable and total concentration can be explained both by the high metals extraction rate by roots, and by the high metal accumulation in shoots during the first period.

The difference in the amount of metal uptake by the plant can be a consequence of a possible variation in plant uptake mechanism that is determined by the high metal concentration in soil. The plant in fact acted in the second period as a metal excluder and accumulated the metals in roots. If the metal uptake mechanism can change from the first to the second growth period, a possible method to enhance phytoremediation is to plant *Mirabilis jalapa* only from seeds and to use chelating agents in order to enhance metal extraction.

The phytoremediation approach used allows, on the other hand, the stabilization of metal in soil, lowering its bioavailability. The technique can, as a consequence, be considered as a phytostabilization treatment.

ACKNOWLEDGMENTS

The authors wish to thank Carlo Desole, who assisted in performing experiments and Martina Piredda for laboratory analyses.

REFERENCES

Baker, A. J. M. 1981. Accumulators and excluders – strategies in the response of plants to heavy metals. J. of plant nutrition. 3 (1 -4), 643-654.

Barbafieri, M., Lubrano, L., Petruzzelli, G. 1996. Characterization of pollution in sites contaminated by heavy metals: a proposal. Annali di chimica, 86, 585 -594.

Blaylock, M. J., Salt, D. E., Dushenkov, S., Zakharova, O., Gussman, C., Kapulnik, Y., Ensley, B. D. and Raskin, I. 1997. Enhanced accumulation of Pb in Indian mustard by soil-applied chelating agents. Environ. Sci. Technol. 31, 860 –865.

Cao, A., Cappai, G., Carucci, A. , Muntoni, A.. 2002. Selection of plants for zinc and lead phytoremediation. In: Proceedings of ISWA 2002 World Environmental Congress & Exibition, Istanbul 8-12 July 2002.

Cunningham S C, Berti W R and Huang J W 1995 Remediation of contaminated soils and sludges by green plants. In: Bioremediation of Inorganics, pp 33–54. (E Hinchee, J L Means and D Burris Eds). Columbus -Richland, Batelle Press.

Dessì, R., Boi, M., Persod, P., Ullu, F. 1999. Environmental characterisation and rehabilitation proposals of Montevecchio mine district. In: Proc. of REWAS 99, San Sebastian, Vol. III, 2541 -2550.

Gazzetta Ufficiale della Repubblica Italiana, Supplemento ordinario n° 248, 21/10/1999. Metodi ufficiali di analisi chimica del suolo.

Kambhampati, M.S., Williams, L. 2001. Phytoremediation of lead -contaminated soils using *Mirabilis jalapa.* In: Phytoremediation, wetlands, and sediments, pp. 145 -150 (Leeson, A.,Foote, E. A., Banks, M. K., and Magar, V. S., Eds.). Battelle Press.

Knox, A.S., Seaman, J.C., Mench, M.J., Vangronsveld, J. 2001. Remediation of Metal- and Radionuclides- Contaminated Soils by In Situ Stabilization Techniques. In: Environmental restoration of metals contaminated soils, pp.21 -60. (I.K.. Iskandar Eds.). CRC Press LLC.

CHAPTER 22

LIGNINOLYTIC ENZYMES AND PHYTOTOXIC EFFECTS FOLLOWING GROWTH OF STRAINS OF PLEUROTUS IN MSWI FLY ASH CONTAMINATED WITH PCCD/FS

M. F. Dos Reis- Gorny and L.R. Durrant
Universidade Estadual de Campinas – UNICAMP, Campinas, SP, Brazil

Abstract: Three Pleurotus strains were grown in a medium containing MSWI fly ash having high concentrations of PCDD/Fs. The supernatants were collected every seven days up to 42 days, and used for the determination of lignin - and manganese-peroxidases (LIP and MnP), and laccase (LAC), and also for toxicity levels. Enzyme activities were determined by spectrophotometric assays. The toxicity levels were determined using a bioassay employing Lettuce (Latuca sativa) seeds. When 2.5 g were used, low levels of LIP, LAC e M nP were detected, regardless of the strain. When Pleurotus sp. CCB 068 was grown in medium containing 3.5 g of fly ash, high LAC activities (+/ - 20 U/L) was detected up to the 7th day. Lower activities were observed for P. sajor caju 020 (+/- 15 U/L) e P. tailandia (+/- 10 U/L). P. sp. CCB 068 was the only strain showing a good decrease in the phytotoxic effect after 42 days of treatment (35.90% effect). Our results indicate that P. sp. CCB 068 produced metabolites which were less toxic to the lettuce seeds than the other two strains, and that the higher levels of LAC and MnP produced may play a role in the degradation of PCDD/Fs.
Keywords: PCDD/Fs, fungi, ligninolytic enzymes, toxicity.

1. INTRODUCTION

Municipal solid waste incineration (MSWI) fly ash has recently attracted significant attention because of the large quantities produced and its high content of low boiling point heavy metals and organic micropollutants such as polychlorinated biphenyls (PCBs), polychlorinated dibenzo-p-dioxins (PCDDs) and dibenzofurans (PCDFs), polycyciclic aromatic hydrocarbons (PAHs), etc. (Hester & Harrison, 1994). These micropollutants are generated as a result of incomplete combustion in the combustion and post-combustion processes. Since the disposal of MSWI fly ash with these highly toxic organic contaminants can lead to significant environmental problems, such as contamination of sub-soils in land filling, it is necessary to remove the organic contaminants from MSWI fly ash before its final disposal (Huang et al., 2003).

The major goal of environmental biotechnology is to establish biological processes that use the naturally existing catabolic potential for the elimination and detoxification of many toxic compounds. Microbiological techniques are useful tools for the decontamination of toxic materials, which not only save energy but also reduce the need for additional chemicals. The degradation of PCDD/Fs has been examined using white rot fungi (Bumpus, 1989; Hammel et al., 1986; Szklaarz et al., 1989; VanOvermiere et al., 2000). The ability of these microorganisms to mineralize complex aromatic polymeric substances, such as lignin, is well known (Köller et al., 2000). The use of white rot fungi in bioremediation is particularly of interest, due to their extracellular enzymes (Eggen & Majchercyk, 1998), and their non-specific radical mechanisms to attack toxic compounds such as DDT (Bumpus et al., 1985), PCB (Takada et al., 1996), chlorinated phenols (Huang et al., 2003), chlorinated dibenzodioxins (PCDD/Fs) (Van Overmiere et al., 2000), etc. The initial reactions of degradation by white-rot fungi are thought to be catalyzed by extracellular ligninolytic enzymes, i.e., oxidases (laccases) and peroxidases (Pointing, 2001). Purified extracellular enzymes are able to carry out in-vitro transformation of compounds such as dibenzodioxins and polycyclic aromatic hydrocarbons (Camarero et al., 1996; Joshi & Gold, 1994; Van Overmiere et al., 2000; Szklaarz et al., 1989).

The aim of this work was to determine the production of ligninolytic enzymes by three Pleurotus strains following their growth in a medium containing PCDD/Fs and also to evaluate the toxicity produced during 42 days of treatment with these microorganisms.

2. MATERIALS AND METHODS

2.1 Fly ash

The fly ash used was obtained from a municipal waste incinerator (MWI) plant in São Paulo city -SP, Brazil. The same batch of fly ash was used in all experiments; it was treated according to Takada et al. (1996).

2.2 Fungal strains and inocula preparation

The *Pleurotus* strains used here belong to the Culture Collection of the Systematic and Physiology Laboratory- UNICAMP, Campinas -SP, Brazil. The strains were maintained on 2% Difco potato dextrose agar (PDA) slants. Petri dishes containing PDA were inoculated with a mycelium portion of each strain and incubated at 30° C for 7 days, after which the agar was cut and either 1 cm^2 or 5 x 1 cm^2 mycelial pieces were removed with a thin spatula and used as the inocula for the experiments in solid medium and liquid medium, respectively.

2.3 Determination of enzymes production in liquid medium

The fungal strains were inoculated (5 x 1cm^2) in liquid medium (50 mL/ 250 mL Erlenmyer flasks) containing per liter: 0.5g $(NH_4)HPO_4$; 0.8g KH_2PO_4; 0.3g KH_2PO_4; 0.3g $MgSO_4$. 7 H_2O; 0.004g $ZnSO_4$. $6H_2O$; 0.005g $MnSO_4$; 0.0055g CaCl. $2H_2O$; 0.005 g $FeSO_4$. $7H_2O$; 0.002g $CoSO_4$; 0.2 g yeast extract and 1.0 mL thiamine (2,0 mg/mL) and incubated for at 30°C. Either 2.5 g or 3.5 g of fly ash containing PCDD/Fs were added on the 3rd.day of cultivation (t_0) and incubated for 7(t_7), 14(t_{14}), 21(t_{21}), 28(t_{28}), 35 (t_{35}) and 42(t_{42}) days following the addition of the fly ash. The final pH of the culture media was 5.0. After the addition of the fly ash the flasks were sealed with a Teflon cap and were flushed with oxygen, and subsequently at each 7-day interval. Controls flasks, without the fungi, were also run. All experiments were in triplicate.

2.4 Enzymes Assays

Lignin peroxidase (LiP), Manganese peroxidase (MnP) and Laccase (Lac) activities were assayed spectrophotometrically in the extracelullar fluid of the culture's supernatants. LiP was determined by measuring veratryl alcohol oxidation according to Tien & Kirk (Tien & Kirk, 1984). MnP determinations were based on the oxidation of phenol red, according to Kuwahara et al. (Kuwahara et al., 1984). Lac was determined via the H_2O_2 - independent oxidation of syringaldazine (ethanol solution) to its quinone form (Sanglard et al., 1986). The supernatants were obtained following centrifugation of the content of whole flasks at 32.000 x g for 15 minutes at 4°C. All enzyme activities were expressed in units per liter (U/L), where one unit corresponds to 1 μmol of substrate oxidized per minute.

2.5 Toxicity bioassays

Petri dishes containing filter paper disc were previously sterilized at 121° C for 30 minutes. Fifty seeds of Lettuce *(Latuca sativa)* were placed in each Petri dish and treated with 2 mL of the culture supernatants. Controls treated with 2 mL of distilled water were also run. All experiments were in triplicates. All Petri dishes were incubated at room temperature in darkness. Results were calculated as % of phytotoxic effect: [(root length in control-root length in treatment)/root length in control]* 100.

3. RESULTS AND DISCUSSION

Tables 1, 2 and 3 show enzymes production for *P.* sp CCB 068 (BCCB068), *P sajor caju* 020 (PSC020) and *P. tailandia* (PT), respectively. All the fungal strains used here exhibited some ligninolytic activity in 2.5 g and 3.5 g of fly ash. When 2.5 g were added, low levels of LiP, LAC and MnP were detected, regardless of the strain (Tables 1, 2, and 3). The highest LiP activity was detected on the 7th day of growth of *P.* sp CCB 068. Low levels of LAC activity were detected in all days assayed. MnP was only detected after the 21th day of growth (Table 1). When 3.5 g fly ash was used, high levels of LAC activity were detected at the beginning of the incubation period (first week). MnP was produced during all days assayed and the highest activity was detected after 28 days of growth (Table 1). The highest LiP, LAC and MnP activities were observed on the 42nd day of growth of *P. sajor caju* 020 in 2.5 g fly ash (Table 2). *P. sajor caju* 020 produced LiP activity when grown in 3.5 g when assayed. Higher MnP activities were detected on the 28th day. When *P. tailandia* was grown in 2.5 g of fly ash, LiP and LAC were only detected on the 14th day. The highest MnP activity was detected on the 21st day (12.5 U/L), (Table 3).

Table 1. **Enzyme production by P. sp CCB68 grown in liquid medium containing 2.5 g and 3.5 g fly ash.**

Days	LiP		Laccase		MnP	
	2.5 g	3.5 g	2.5 g	3.5 g	2.5 g	3.5 g
7	0.98	0.11	0.90	17.35	0.00	10.73
14	0.05	0.34	1.48	6.50	0.00	7.03
21	0.00	0.00	1.46	2.25	4.50	5.40
28	0.00	0.00	1.46	1.70	0.01	14.89
35	0.00	0.00	0.76	0.00	0.00	0.96
42	0.00	0.00	2.30	2.19	3.80	5.75

Table 2. . Enzyme production by *P. sajor caju* 020 grown in liquid medium containing 2.5 g and 3.5 g fly ash.

Days	LiP		Laccase		MnP	
	2.5 g	3.5 g	2.5 g	3.5 g	2.5 g	3.5 g
7	0.05	0.40	0.00	0.00	0.00	5.83
14	0.05	2.64	0.02	2.45	0.00	4.45
21	0.00	0.02	0.00	0.00	2.00	3.99
28	0.00	0.05	0.00	0.14	3.00	14.90
35	0.00	0.00	0.00	0.00	0.00	2.03
42	4.5	0.18	0.48	0.00	8.20	1.61

Table 3. . Enzyme production by *P. tailandia* grown in liquid medium containing 2.5 g and 3.5 g fly ash.

Days	LiP		Laccase		MnP	
	2.5 g	3.5 g	2.5 g	3.5 g	2.5 g	3.5 g
7	0.00	0.88	0.00	0.00	0.00	10.70
14	0.10	0.04	0.01	0.00	0.00	3.02
21	0.00	0.00	0.00	0.00	12.5	7.89
28	0.00	0.03	0.00	0.03	0.00	10.24
35	0.00	1.83	0.00	0.10	0.00	3.20
42	0.00	0.17	0.25	0.00	3.00	0.42

The phytotoxic effects produced in Lettuce after 42 hours are shown in Figure 1. *P.* sp CCB 068 produced the lowest phytotoxic effect in almost all days in which it was evaluated, with the exception ofthe 14th and 28th day. *P. sp* CCB 068 was the only strain able to produce a decrease in the phytotoxic effect following 42 days of growth. *P. tailandia* produced lower phytotoxic effects only after 7 and 21 days of growth. P *sajor caju* 020 produced a supernatant with phytotoxic effects, regardless of the time of growth.

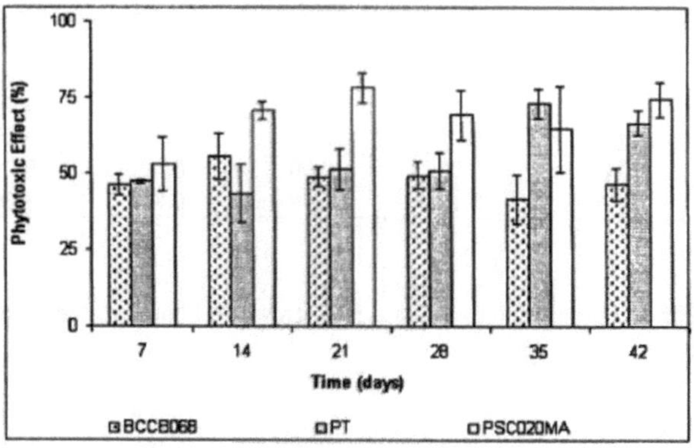

Figure 1. Phytotoxic effect of cultures supernatants on Lettuce seeds (Latuca sativa). Pleurotus sp. CCB 068 = BCCB 068; Pleurotus tailandia = PT and Pleurotus sajor caju 020 (PSC 020 MA).

White-rot fungi are capable of degrading a variety of recalcitrant aromatic pollutants, such as polyaromatic hydrocarbons (PAHs), polychlorinated phenols, and polychlorinated dibenzo-p-dioxins. The degradation of these recalcitrant pollutants by white-rot fungi is well correlated with their ligninolytic activities. Valli et al. (1992) demonstrated that the 2,7-diCDD degrading mechanisms included a LiP – catalyzed initial oxidation by *P. chrysosporium..* In the metabolic reation catalysed by LIP, cleavage products were anticipated as metabolites (Van Overmiere et al., 2000). Takada et al. (1996) examined the corresponding metabolites were formed for 2,3,7,8-TCDD and OCDD. The cleavage products (4,5-dichloro-1,2-benzoquinone and / or 4,5-dichlorocatechol for 2,3,7,8-TCDD and tetrachloro-1,2-benzoquinone and /or tetrachlorocatecol for OCDD) were observed in the degradation of 2,3,7,8-TCDD and OCDD by *P. sordida* , although the fungus secreted no LiP (Szklaarz et al., 1989).

Our results showed MnP to be the predominant enzyme produced by the three fungal strains grown in the 3.5 g fly ash. MnP was produced in all days assayed. *Pleurotus sp.* CCB 068 produced the highest MnP and laccase activities and was able to decrease the phytotoxic effect when checked against lettuce seeds, indicating that this strain may have produced different metabolites in the culture medium less toxic than the other two strains.

ACKNOWLEDGEMENTS

The authors wish to express their thanks to Fundação de Amparo à Pesquisa do Estado de São Paulo (FAPESP) and CNPq for financial support.

REFERENCES

Assunção, J V de; Pesquero, R.C. Dioxinas e furanos : origens e risco. Rev. Saúde Pública. v. 33(5), p. 523-530,1999.

Barr, D. P.; Aust, S. D. Mechanisms white-rot fungi use the degrade pollutants. Environmental Science & Technology. v. 28(2), p. 78-87, 1994.

Behnischi, P. A.; Hosoe, K.; Sakai, S-I. Combinatorial bio/ chemical analysis od dioxin and dioxin-like compounds in waste recycling, feed/ food, humans/ wildilife and the environment. Environment International. v.27.p.495-519. 2001.

Bumpus, J. A. Biodegradation of polycyclic aromatic hydrocarbons by *Phanerochaete chrysosporium.* Appl. Environ. Microbiol. v. 55, p. 154-158, 1989.

Bumpus, J. A; Tien, M.; Whiight, D.; Aust, S. D. Oxidation of persistent environmental pollutants by a white-rot fungus. Science, v. 228, p. 1434-1436, 1985.

Camarero, S.; Böckle, B.; Martinez, M.J.; Martinez, A. Z. Manganese -mediated lignin degradation by *Pleurotus pulmonaris.* Appl. Environ. Microbiol. v.62. p. 1070-1072., 1996.

Chang, M-B; Chung, Y. T. Dioxins contents in fly ashes of MSW Incinerator in Taiwan. Chemosphere. v. 36(9), p. 1959-1968, 1998.

Clemente, A. R.; Anazawa, T. A.; Durrant, L. R. Biodegradation of polycyclic aromatic hydrocarbons by soil fungi. Brasilian Journal of Microbiology. v. 32, p. 255-261, 2001.

Domingo, J. L.; Granero, S.; Schuhmscher, M. Congener profiles of PCDD/Fs in soil and vegetation samples collected near to a municioal waste incinerator. Chemosphere. v. 43, p. 517-524, 2001.

Eggen, T. Application of fungal substrate from commercial mushroom production – *Pleurotus ostreatus*- for bioremediation of creosoto contaminated soil. International Biodeterioration & Biodegradation. v. 44,p. 117-126, 1999.

Eggen, T.; Majchercyk, A.. Removal of polycyclic aromatic hidrocarbons (PAH) in contaminated soil by white-rot fungus *Pleurotus ostreatus.* International Biodeterioration & Biodegradation. v. 41, p. 111 -117, 1998.

Fries, G. F. A review of the significance of animal food products as potencial pathways of human exposures to dioxins. J. Anim. Sci. v.73, p. 1639-1650, 1995.

Hammel, K. E.; Kalyanaraman, B.; Kirk, T. K.. Oxidation of polycyclic aromatic hydrocarbons and dibenzo -p-dioxins by *Phanerochaete chrysosporium* ligninase. J. Bio. Chem. v.261.p. 16948-16952, 1986

Hatakka, A.. Lignin-modifying enzymes from selected white-rot fungi: production and role in lignin degradation. FEMS Microbiology Reviews. v. 13, p. 125-135, 1994.

Hester, R. E.; Harrison, R. M. 1994. Issues in Environmental Science and Technology, vol. 2. Waste incineration and the environment. Edujee, G. H, Bath Press.

Höfer, C. & Scholosser, D.. Novel enzymatici oxidation of Mn^{2+} to Mn^{3+} catalysed by a fungal laccase. FEBS Letters, v. 451, p. 186-190, 1999.

Hong, H. B.; Chang, Y. S.; Choi, D. S.; Park, Y. H.. Degradation of Dibenzofuran by *Pseudomonas putida* PH 01. Was. Res. v. 34 (8), p. 24024-2407, 2000.

Hoogenboom, R.; Portier, L; Onstenk, C; Polman, T.; Hamers, A., Traag, W. Determinatio n of dioxins and planar PCBs in fish using the CG/MS and CALUX bioassay. Introduction of the screening approach for control purposes. Organohalogen Compound . v.45, p. 236-239., 2000.

Huang, Y.; Takada, M.; Takeda, N. Chlorobenzenes removal from Municipal solid waste incineration fly ash by surfactant-assisted column flotation. Chemosphere. v. 52, p. 735-743, 2003.

Ingol, C. T.; Hudson, H. J. (1993). Growth and Nutrition. In: The Biology of Fungi, 6 ed. Champman& Hall, London, p. 7 -24.

Joshi, D. K.; Gold, M. H. Oxidation of dibenzo -p-dioxin by lignin peroxidase from the basidiomycete *Phanerochaete chrysosporium* . Biochemestry. v.33, p. 10969-10976, 1994.

Kao, C. M.; Chen, S. C.; Liu; J. K.; Wu, M. J. Evaluation of TCDD biodegradability under different redox conditions. Chemosphere. v. 44, p. 1447-1454, 2001.

Kasuga, K.; Nojiri, H.; Yamane, H.; Omori, T.. Genes of enzymes involved in the biodegradation of Carbazole, Dibenzofuran, Fluorene and Dibenzo -p-dioxin by bacteria. Wat. Sci. Tech. v. 36 (10), p. 9-16, 1997.

Köller, G.; Möder, M.; Czihal, K. Peroxidative degradation of selected PCB: a mechanistic study. Chemosphere.v.41, p. 1827-1834, 2000.

Kubátová, A.; Erbanová, P.; Eichlerová, I.; Homolka, L.; Nerud, F.; Sasek, V. PCB congenere selective biodegradation by the white-rot fungus *Pleurotus ostreatus* in contaminated soil. Chemosphere. v.43, p. 207-215, 2001.

Kuwahara, M.; Glenn, J. K; Morgan, M. A; Gold, M.H Separation and characterization of two extracelullar H_20_2 dependent oxidases from lygninolytic cultures of *Phanerochaete chrysosporium.* FEBS Lett. v. 169. p. 247-250, 1984.

Li; W.; Wu, W. Z.; Barbara, R.B., Schramm, K. W.; Kettrup, A. A new enzyme imunoassay for PCDD/F TEQ scrrening in environmental samples: comparision to micro -EROD assay and to chemical analysis. Chemosphere. v. 38(14), p. 499-506, 1999.

Mätzing, H.; Baumann, W.; Becker, B.; Jay, K.; Paur, H -R; Seifert, H. Adsorption of PCDD/F on MWI fly ash. Chemosphere. v. 42, p. 803-809, 2001.

Szklaarz, G. D.; Antibus, R. K, Sinsabaugaugh, R.L; Linkins, A. E. Production of phenoloxidases and peroxidases by wood -rotting fungi. Mycolo.v.81, p. 234-240, 1989.

Takada, S.; Nakamura, M.; Matsueda, T.; Kondo, R.; Sakai, K. Degradation of polychlorinated dibenzo -p-dioxins and polychlorinated dibenzofurans by the white -rot fungus *Phanerochaete sordida* YK-624. Appl Environm. Microbiol. v.62, p. 4323-4328, 1996.

Thomas, D.R.; Carswell, K.. S.; Georgiou, G. Mineralization of biphenyle and PCBs by white-rot fungus Phanerochaete chrysosporium. Biotech. Bioeng. v. 40: p. 1395 -1402, 1992.(36)TIEN, M. and KIRK, T. K. (1988). " Lignin peroxidase of *Phanerochaete chrysosporium".* Methods Enzymol. 161: 238-249. & KIRK 1984

Van Overmiere, I.; Goeyenes, L.; Beernaert, H.; Srebrnik, S.; De Poorter, G. Bayeens, W.; Clark, G.; Chu, A., Chu, D; Morris, R.; Brown, D. A comparative study of GC -HRMS and CALUX-TEQ determinations in food samples by the Belgian Federal Ministries of Public Health and Agriculture. Organohalogen Compond. v.45p. 196-199, 2000

Valli, K.; Wariishi, H.; Gold, M.H Degradation of 2,7-diclorodibenzo-p-dioxin by the lignin-degrading basidiomycete *Phanerochaete chrysosporium.* J. Bacteriol. v. 174, p. 2131-2137, 1992

Wittichi, R-M. Degradation of dioxin -like compounds by microorganisms. Appl. Microbiol. Biotechnol. v. 49, p. 489-499, 1998.

This page intentionally left blank

CHAPTER 23

ACCUMULATION OF HEAVY METALS BY JAPANESE WEEDS AND THEIR SEASONAL MOVEMENT

Ryuji Takeda; Noriyoshi; Sadayoshi Matsumoto; Sadao Komemushi; Akiyoshi Sawabe
Kinki University

Abstract: Phytoremediation is a technique for the removal of contaminants in the environment using plants, which is currently is being researched worldwide. We focus attention on certain Japanese weeds, which have a large biomass and high environmental adaptability as a hyperaccumulator. Specifically, we investigated the seasonal metals movement of roots, stems and leaves on *Artemisia princeps,* a common Japanese weeds, and examined methods of heavy metal removal from the soil. Plants and surface soils were collected at watersides such as a reservoir and adjustment pond around our University at Nara Prefecture, and around the Kizu river of Seika town at Kyoto Prefecture in Japan. The soils and plants were ashed using H_2NO_3, HCl, and H_2O_2. Mn, Ni, Cr, Fe, Cu, Zn and Li in the ashes were measured by AAS. Hyperaccumulators have not been found yet at our investigation sites. However, *Aster leiophyllus, Artemisia princeps* and *Stenactis annuus* accumulated 2 or 3 times as many metals (Cr, Cu, Mn) as other plants in the same s ites. As to the seasonal metals movement of roots, stems and leaves on *Artemisia princeps,* plants concentration and accumulation ratio of Cu that compared the concentration between soil and plants are high in autumn. As a result, we think that Cu plays special role different from other metals. Accordingly, this movement is important in examination of the treatment stage after having taken in heavy metal.

Key words: phytoremediation, Artemisia princeps, seasonal metals movement, Cu

1. INTRODUCTION

Pollution by heavy metals from industries, the storage of polluted wastes, and agricultural fertilizer have all contributed to pose a serious threat to human health. These pollutants may pass into the soil solution where plant uptake or leaching to groundwater can contaminate the food chain. Thus, most pollution is artificial; however, natural sources also cause pollution. The natural pollutions are deposits and mother rocks. In a volcanic country such as Japan, arsenic content in the soil is high (Fausto et al., 1981; Marumo, 2003). Heavy metals are naturally present in soils and elevated levels may be a consequence of human activity. Soils are contaminated due to direct deposition of pollutants from industries, waste disposal sites or dumps and agriculture, as well as by deposition via wastewater treatment and air pollution. And although chemical and physical treatments can remove these heavy metals, the cost is enormously high.

Phytoremediation is a technique that removes contaminants in the environment using plants, and is currently being researched worldwide. (Scott et al., 1996). Phytoremediation can be used for metal removal. This method has been successfully used for different toxic trace elements, for instance Cd and Zn or Ni. Water, evaporating from plant leaves, serves as a pump to absorb nutrients and other soil substances into plant roots. This process, termed evapotranspiration, is responsible for moving contamination into the plant shoots as well. Because contamination is translocated from roots to the shoots, which are then harvested, contamination is removed while leaving the original soil undisturbed. Some plants that are used in phytoextraction strategies are termed "hyperaccumulators", which are plants that achieve a shoot-to-root metal-concentration ratio greater than one. Non-accumulating plants typically have a shoot to root ratio considerably less than one. Ideally, hyperaccumulators should thrive in toxic environments, require little maintenance and produce a high biomass, although few plants perfectly meet these requirements. (McGraph et al., 2002; Salido et al., 2003)

Recently, the cruciferous plant was identified as a metal hyperaccumulator. The cruciferous plant has the capacity to accumulate heavy metals more than 100-fold above plants normally used. More than 50 species of cruciferous plants currently are reported. (Baker et al., 1994) However, these species grow only in the ultrabasic rock zone, and their biomass is very small. (Brooks et al., 1977; Reeves et al.; 1988; Mizuno et al., 2001) making it is difficult to apply these species to a polluted area. EDTA is being investigated as soil additive to increase the absorbance of heavy metals, (Blaylock et al., 1997; Vassil et al., 1998; Matsuo et al., 2001); however, there are fears of secondary pollution caused by using EDTA and of rising costs.

In the 20th century, the environment around the city has worsened due to increased urbanization and industrialization (Shiraishi et al., 2002). In Tokyo, Japan, high concentrations of Cu, Zn and Pb have been detected in main-road dusts and in the leaves of roadside trees. However, one special plant grows robustly by the side of the main roads - weeds. Therefore, we focused attention on the impact of several common Japanese weeds' growth on polluted environments such those beside a main road or factory. The advantages of applying weeds to phytoremediation are the large biomass, the adaptability of the environment and the early growing season. We investigated the concentration of heavy metals (Mn, Ni, Cr, Fe, Cu, Zn and Li) in the weeds and in the soil located on the main roads, reservoirs, ponds and riverside around Nara city, Japan. Moreover, *Artemisia princeps* is grown in the 80% of the collection place. We investigated the seasonal movement of heavy metals concentration in the same collection sites.

1.1 Materials and Methods

1.1.1 Plant and Surface Soil Sampling

Plants and surface soils were collected at the side of a factory, main roads and watersides such as reservoir and adjustment pond around the Faculty of Agriculture, Kinki University at Nara Prefecture, and around the Kizu river of Seika town at Kyoto Prefecture in Japan (Fig. 1). The sampling term was the growth period of the weeds. *Artemisia princes* Pampan were collected from August 2002 to July 2003.

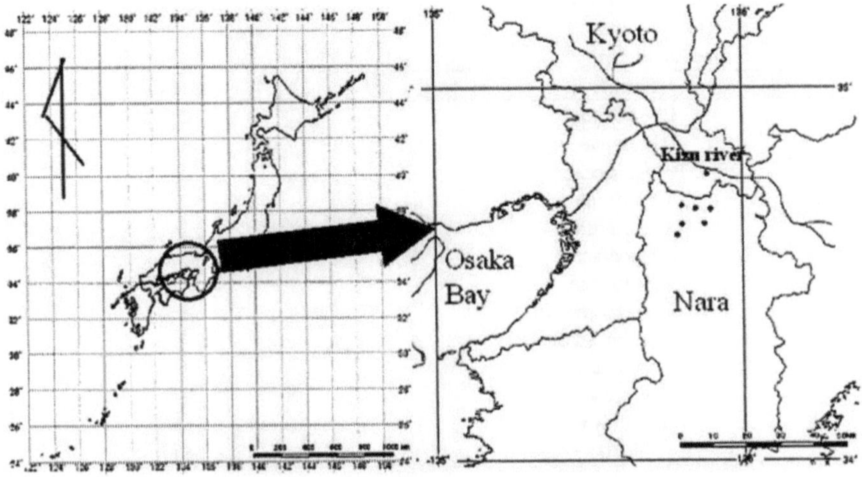

Figure 1. Location of Sampling Places

Plants and surface soils were collected. The plants were washed with distilled water, and then the roots, stems and leaves were separated, and dried in an oven at 80°C for 24hrs. After drying, they were blended using a blender with titanium blades. The soils were air-dried and sieved using a 2.0mm nylon mesh sieve.

1.1.2 Heavy Metal Analysis

Samples (~500mg) were digested by heating (200°C, 120mins) in 15ml of mixed acids consisting of nitric acid, hydrochloric acid and hydrogen peroxide (5:5:5). After cooling, the digest was diluted to 50 ml with distilled water, filtered through Advantec paper No.6 and analyzed for heavy metal concentration by Atomic Absorbance Spectroscopy (AAS) (Z-6100, Hitachi). All the measurements by AAS were used with the air-acetylene frame.

1.2 Results

1.2.1 Sampling Analysis

40 kinds of plants were collected between September 2002 and August 2003. Among these, 25 kinds of weeds were identified. At the collection sites, there were 11 kinds of Compositae, the first number of species; 3 kinds of Gramineae, the second number of species; 2 kinds of Cruciferae, Leguminosae, etc (Table 1).

Table 1. Collecting Plants

Family	Number
Compositae	11
Graminae	3
Polygonaceae	3
Cruciferae	2
Leguminosae	2
Chenopodiaceae	1
Cyperraceae	1
Equisetaceae	1
Scrophulariaceae	1
Labiatae	1
Liliaceae	1
Unknown	13
Total	**40**

1.2.2 Heavy Metals in Soils

Table 2 shows the amount of heavy metals in the soils from the collection sites. Although low levels of pollutants were found at the collection sites, high amounts of Cr was found at the main-road side site. But the concentration of Mn and Cr in this area was rather lower than that in the other area measured by Asami et al., 1988. However, other heavy metals were almost correct.

Table 2. Concentration of Heavy Metals in Soils

	Concentration (mg/kg DW)						
Area	Cr	Cu	Zn	Ni	Mn	Fe	Li
Adjustment pond (Kinki Univ.)	9.0	13.1	77.2	65.5	350	31870	12.0
Ground (Kinki Univ.)	11.5	10.2	44.1	63.5	238	2798	9.0
Parking (Kinki Univ.)	24.1	43	94.3	39.0	336	2645	6.2
The side of factory (Nara City)	ND	17.1	73	23.5	242	14745	7.0
The side of main road (Nara City)	41.2	40.2	163	14.1	285	3248	11.5
Oike pond (Nara City)	ND	19.5	70	4.5	275	1078	7.4
Ikaruga pond (Ikaruga, Nara)	ND	24	159.5	1.5	211	3673	11.9
River side of Kizu (Seika, Kyoto)	ND	27.5	174.5	33.0	210	2163	5.5

ND: Not Detected

1.2.3 Heavy Metals in Plants

Fig. 2 summarizes the amount of heavy metals in plants from the same collection sites. Metal hyperaccumulators were not yet found in the samples from the investigation sites. However, *Aster leiophyllus, Artemisia princeps* Pampan and *Stenactis annuus* Cass had accumulated 2 or 3 times more metals (Cr, Cu, Mn) than other plants sampled at the same collection sites. Table 3 showed the large amount of absorption of each heavy metal on the absorbance and the accumulation ratio.

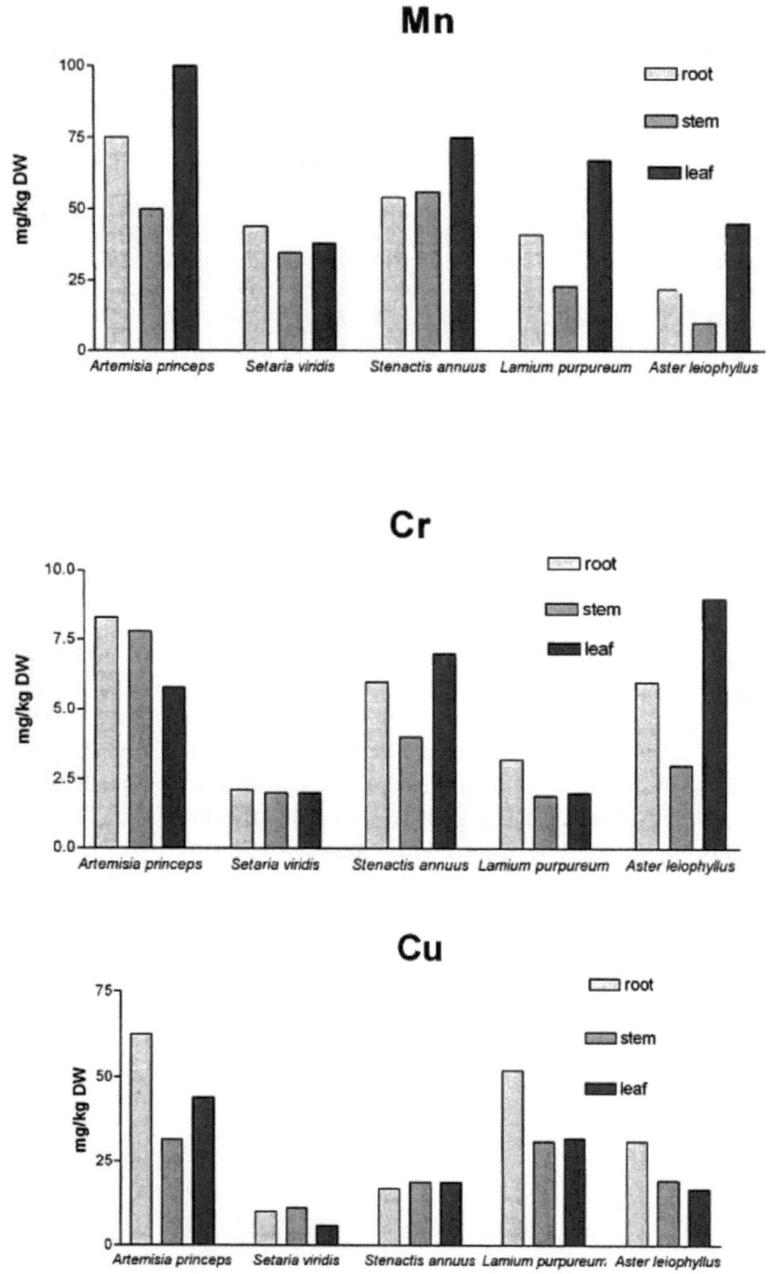

Figure 2. Concentration of Heavy Metals in Plants

Table 3. High Concentration of Heavy Metals in Plants

Elements	Family	Species	Parts	Concentration	Ratio
Chromium					
	Gramineae	*Alopecurus aequalis*	Root	142	5.92
	Compositae	*Taraxacum officinale*	Leaf	80	1.86
	Gramineae	*Alopecurus aequalis*	Flower	56	2.33
Copper					
	Compositae	*Artemisia princeps*	Root	382	2.49
	Compositae	*Taraxacum officinale*	Leaf	167	4.35
	Gramineae	*Alopecurus aequalis*	Root	163	4.07
Zinc					
	Compositae	*Bidens pilosa*	Root	279	1.75
	Gramineae	*Taraxacum officinale*	Root	248	4.77
	Chenopodiaceae	*Chenopodium album*	Leaf	225	4.83
Manganese					
	Chenopodiaceae	*Chenopodium album*	Leaf	513	2.05
	Polygonaceae	*Rumex acetosa*	Leaf	487	0.23
	Compositae	*Artemisia princeps*	Leaf	474	0.22
Nickel					
	Compositae	*Bidens pilosa*	Root	185	9.67
	Compositae	*Aster leiophyllus*	Leaf	172	7.3
	Compositae	*Artemisia princeps*	Root	115	7.67
Lithium					
	Compositae	*Artemisia princeps*	Root	5.0	0.10
	Gramineae	*Poa annua*	Leaf	4.2	0.70
	Compositae	*Bidens pilosa*	Root	4.0	0.57
Iron					
	Compositae	*Aster leiophyllus*	Root	4081	3.32
	Polygonaceae	*Rumex acetosa*	Root	3063	1.11
	Compositae	*Conyza bonariensis*	Stem	2360	3.35

1.2.4 Seasonal Movement of *Artemisia princeps*

Figure 3 shows the seasonal movement of each heavy metal in *Artemisia princeps* Pampan. The accumulation ratio of Cu indicates a rise in the autumn. The accumulation ratio of Mg indicates a rise in the summer. The accumulations of Fe, Ni and Zn was found to have no seasonal movement. The amount of Li in *Artemisia princeps* is smaller than detection limit.

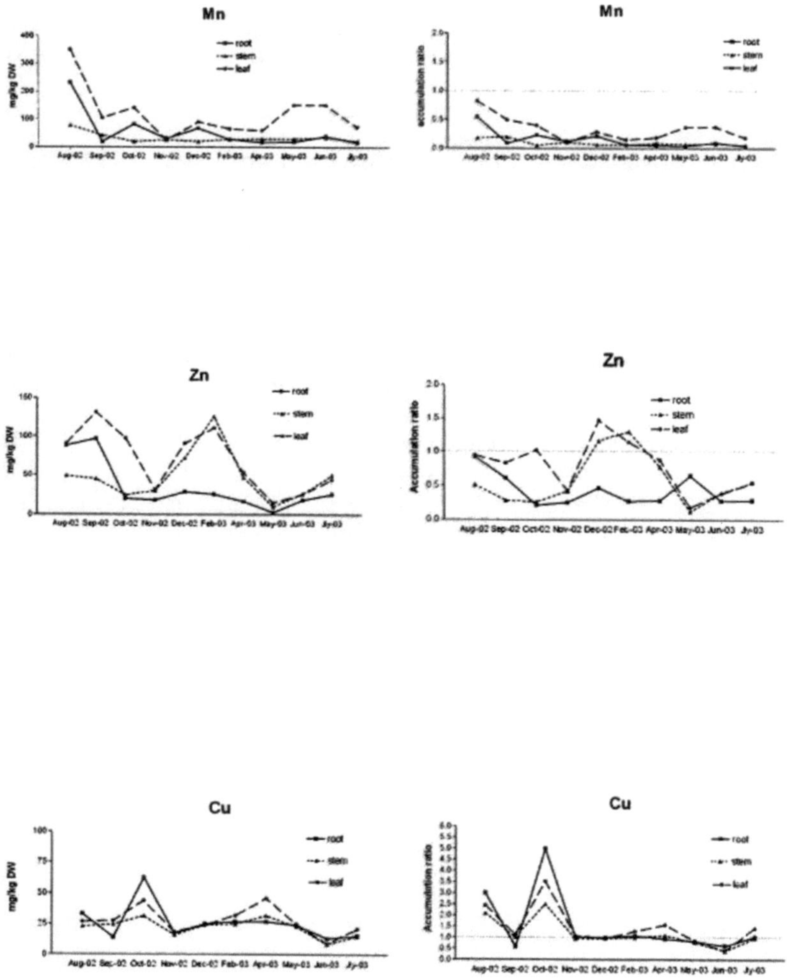

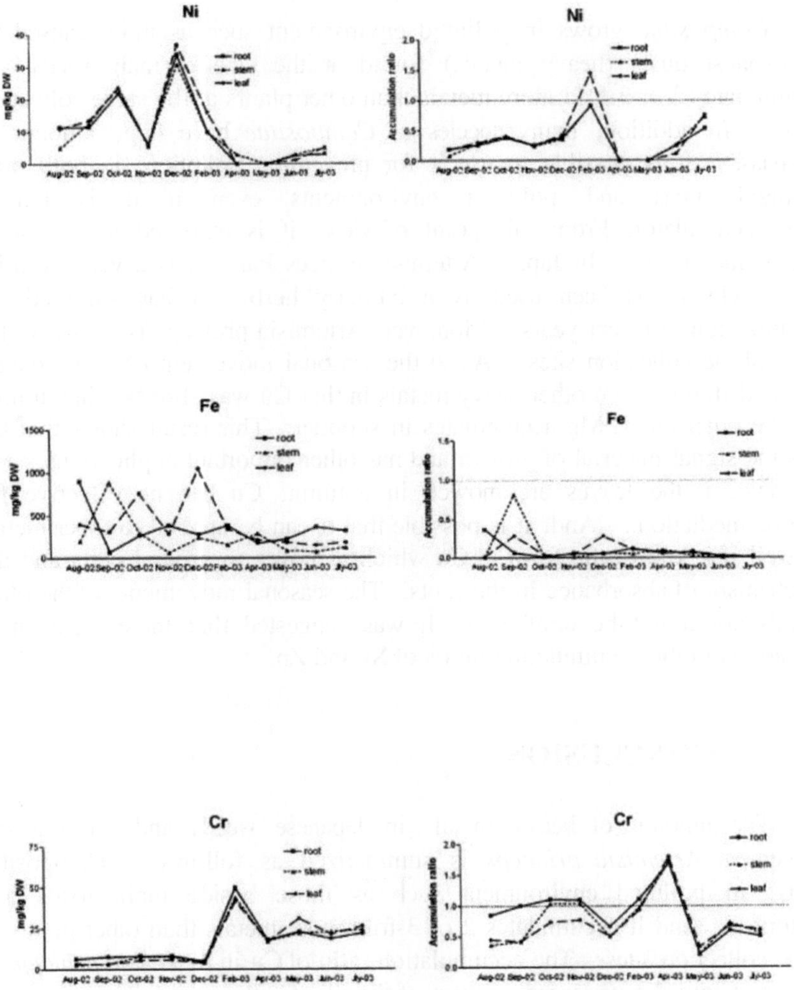

Figure 3. Concentration and Accumulation Ration of Seasonal Metals Movement: Roots, Stems and Leaves on *Artemisia Princeps*

1.3 Discussions

Compositae grows in polluted environment such as those caused by car-exhaust dusts (heavy metals) found at the side of main roads. It accumulates 2 or 3-fold more metals than other plants at the same collection sites. In addition, many species of *Compositae* have large amount of biomass. It is possible to use it for preventing pollution, in both non-polluted areas and polluted environments even if it is not a hyperaccumulator. From this point of view, it is expected to be useful phytoremediation. In Japan, Artemisia princes Pampan is a very familiar plant, which has been used as a culinary herb. It has survived the urbanization of recent years. Moreover, Artemisia princeps is grown in the 80% of the collection sites. As to the seasonal movement of heavy metal, Cu is different from other heavy metals in that Cu was absorbed in autumn. On the other hand, Mg accumulates in summer. This result shows that Cu was the signal material of growth and has other important applications. For example, if the leaves are mowed in autumn, Cu can be effective for phytoremediation.. And, it is possible that it can be applied to other metals as well if it knows the form of Cu which it moves to ground parts and the mechanism of absorbance in the roots. The seasonal movements of the other metals could not be confirmed. It was suggested that there were small variations in the accumulation ratios of Ni and Zn.

2. CONCLUSION

The amount of heavy metals in Japanese weeds and the seasonal movement *Artemisia princeps* is summarized as follows. Compositae grows in polluted environment such as those beside main roads and factories. And it accumulates 2 or 3-fold more metals than other plants in same collection sites. The accumulation ratio of Cu in *Artemisia princeps* is absorbed into the underground part of the plant - the root - in the autumn, and moves to the ground part - the shoot - in the spring. As a result, it is thought that Cu has special role different from that of other heavy metals. Accordingly, this movement is important in the examination of the treatment stage after having taken in heavy metal, and is useful in the efficient employment of phytoremediation.

REFERENCES

Arthur L. Salido, Kelly L. Hasty, Jae-Min Lim and David J. Butcher. 2003. Phytoremediation of arsenic and lead in contaminated soil using Chinese brake ferns (Pteris vittata) and Indian mustard (Brassica juncea), *Int. J. Phytoremediation.* 5(2), 89-103

Asami T., Kubota M. and Minamisawa K. 1988. Natural abundance of cadmium, antimon y, bismuth and some other heavy metals in Japanese soils, *Japanese Journal of Soil Science and Plant Nutrition.* 59, 197-199.

Blaylock, M. J., Salt, D. E., Dushenkov, S., Zahkharova, O., Gussman, C., Kapulnik, Y., Ensley, B. D. and Raskin, I. 1997. Enhanced accumulation of Pb in Indian mustard by soil-applied chelating agents. *Environ. Sci. Tech.,* 31 860-865

Baker A. J. M., McGrath S. P., Sidori C. M. D. 1994. The possibility of in situ heavy metal decontamination of polluted soils using crops of metal -accumulating plants. *Resources Conservation and Recycling.* 11, 41-49

Brooks, R. R., Lee, J., Reeves, R. D. and Jaffre, T. 1977. Determination of nickeliferous rocks by analysis of herbarium specimens of indicato r plants. *J. Geochem. Explor.* 7, 49-77

Fausto, S.D. and Aly, O.M. 1981. Chemistry of Natural Waters, Ann Arbor Science Publishers. Inc. Ann Arbor, Michigan.

Matsuo K. and Egashira K. 2002. Lead removal by Leaf Mustard toward Phytoremediation. *Japanese Journal of Soil Science and Plant Nutrition.* 73, 769-771

Marumo K., 2003. Role of Clays in Controlling Geochemical Cycling of Toxic Heavy Metals in the Lithosphere and Hydrosphere. *Journal of the Clay Science Society of Japan.* 42, 120-128.

McGrath, S. P., Zhao, F. J., and Lombi, E. 2002. Phytoremediation of metals. *Adv. Agron.* 75, 1-56.

Mizuno N., Horie K., Mizuno T. and Nosaka S. 2001. Chemical Composition and Ni Compound Crystals in Ni-Hyperaccumulator "Thlaspi japonicum". *Japanese Journal of Soil Science and Plant Nutrition* . 74, 529-534

Reeves, R. D. 1988. Nickel and zinc accumulation by species of Thalaspi L., Cochlearia L., and other genera of the Brassicaceae. *Taxon,* 37, 309-318

Salt D. E., Smith R. D. and Raskin I. 1998. Phytoremediation. *Annu. Rev. Plant Physiol. Plant Mol. Biol.* 49, 643-668

Scott D. Cunningham, Todd A. Anderson, A. Paul Schwab, and F. C. Hsu. 1996. Phytoremediation of Soils Contaminated with Organic Pollutants. *Advances in Agronomy.* 56, 55-114

Shiraishi S., Watanabe I. and Kuno K.. 2002. Heavy metal accumulation in the street dust, roadside soil and roadside tree leaves near by main streets in Tokyo, Japan. *J. Environ. Chem.* 12, 829-837.

Vassil, A. D., Kapulaik, Y., Raskin, I. and Salt. 1998. The role of EDTA in lead transport and accumulation by Indian mustard. *Plant Physiology.* 117, 447-453

This page intentionally left blank

PART VI: RADIATION

CHAPTER 24

CORRELATION TEST BETWEEN INDOOR RADON AND SURFICIAL GAMMA RADIATION IN NORTHERN VIRGINIA

George Saiway[1], George Mushrush[2] and Douglas Mose[2]
[1]George Mason University, Chemistry Department, Fairfax, VA, [2]Center for Basic and Applied Science, Culpeper, VA

Abstract: In northern Virginia, significant indoor radon differences occur between nearby communities. Studies reveal these differences to be correlated with soil chemistry and permeability. Homes constructed in the soil that developed over the Peters Creek Schist almost all exceed the USEPA 's MCL of 2 pCi/L for homeowners, and over 50% exceed the USEPA's MCL of 4 pCi/L for home buyers. In this study group, home construction seems not important except that homes with electrical heating systems averaged almost 2 pCi/L higher that homes with gas an oil furnaces. It was unanticipted that total -gamma radiation in the county soil was not homogeneous, and comparisons between indoor radon and soil radon show a positive correlation, suggesting that soil aeroradioactivity measurements can delineate areas with a high potential for indoor radon.

Key words: precipitation, radon monitors, radon, aeroradioactivity

1. INTRODUCTION

The natural radioactive decay series beginning with Uranium-238 is the major source of natural radiation exposure in the environment. The most geologically significant uranium daughter is Radium-226, which precipitates in near-surface oxidizing conditions. Radium's daughter, radon, is a radioactive noble gas and is widely dispersed in the environment. Because of health implications, the early work on indoor radon has given rise to a broad range of research characterizing Radon-222 progeny and their occurence and control in inhabited structures (Nero, 1988). With the tendency for the gas to concentrate in buildings where air exchange is limited, radon is becoming identified as a major form of indoor air pollution.

National interest in the relationship between radon and lung cancer developed because an estimated 8 to 25 percent of all current lung cancer deaths are thought to be related to past inhalation of airborne radon (Puskin et al., 1988). The concern has intensified since the discovery that inhaled radon and its progeny pass from the lungs into the blood and body tissues (Pohl et al., 1967; Lykken et al., 1989; Henshaw et al., 1990), and may initiate many types of soft tissue cancers.

Soil and rock are the sources of most radon to which people are exposed. The importance of soil as a source of indoor radon combined with the increasing evidence of unacceptably high radon concentrations in a significant fraction of houses has been discussed in several Virginia and Maryland studies (Mose and Mushrush, 1988, Mose et al., 1988; Mose etal., 1989; Mushrush et al., 1988a, 1988b). These studies are designed to predict, on a geological basis, where high indoor radon levels might be found. Homes with high levels of indoor radon have also been discovered in many parts of the Appalachain Mountain system (Watson et al., 1988; Hess et al., 1985; Froelich et al., 1988; Washington, 1988; Smith et al., 1989; Perritt et al, 1990; George et al., 1983; Rose et al., 1988). These and other studies in other areas have shown that geology, seasonal weather changes and home construction can each contribute to elevated indoor radon levels (Cohen et al., 1988; Buchli et al., 1989; Steck, 1990).

In the following study regional levels of indoor radon are compared to precipitation, geological unit, home construction factors and soil radioactivity, as measured from an airplane (aeroradioactivity survey). In this study, about 70% of the homes were in Fairfax County, Virginia located on the southwest and northwest margin of Washington, DC. About 30% were in the surrounding counties in Virginia and Maryland. The following discussion will discuss primarily indoor radon measurements within Fairfax County.

Fairfax County is composed of rock units of the Coastal Plain, the Culpeper Basin and the Piedmont Province (Froelich, 1985a; Froelich, 1985b; Froelich and Zenone, 1985b; Leavy et al., 1983; Obermeier and Langer, 1986). The Coastal Plain Province is present along the eastern edge of the study area. It consists of poorly cemented clastic sedimentary strata, mostly layers of clay and sand, that were deposited during the opening of the modern Atlantic Ocean. These deposits were formed between about 130 million years ago and the present. The western margin of the study area is part of the Culpeper Basin, a fault-bounded valley containing terrestrial clastic rocks (siltstone, sandstone, conglomerate) along with extrusive and intrusive igneous rocks (basalt and diabase) that were deposited during the Mesozoic Era, about 190-150 million years ago. The Piedmont Province extends from Maine to Georgia, and rock units of this province underlie most of the central part of the Virginia study area. These rock units are composed of metamorphic and igneous rocks that were formed when the Appalachian Mountains were created during the Paleozoic Era, about 600-300 million years ago.

2. METHODOLOGIES

Measurement of airborne radioactivity in homes is normally reported in terms of radon concentration. This is somewhat confusing to homeowners, who learn that most inhaled radon is almost immediately exhaled, and that it is the polonium daughters of radon that mainly contribute to the lung cancer risk. The observed indoor concentration of Rn-222 and its progeny depends on three factors: the entry or production rate from radium in the source material, the ventilation rate, and the rate of Rn-222 decay. Because of radium's long half-life and lack of chemical reactivity, Rn-222 itself acts much like a stable pollutant whose concentration can be determined by a comparison of the entry and ventilation rates. The decay product concentration is somewhat more complex, but as a practical matter, the decay product concentration is indicated approximately by the overall radon concentration. By convention radon concentrations are reported, because in a home the radon-to-radon progeny ratio is relatively constant. This is unlike a mine, where rock releases radon and not radon progeny, and the radon-to-radon progeny ratio is very dependent on the rates of ventilation and rock removal.

The indoor radon monitors used in this study are from the Tech/Ops Landauer Corporation in Illinois. They are called "alpha-track monitors" and are of a type long used for geological investigations (Alter and Oswald, 1987; Hess et al., 1985). With an adequate soil shield, they have been used for hydrocarbon exploration and earthquake prediction, and in the search for uranium and gold. The indoor radon monitors do not require the "soil shield," which is a permeable membrane required on soil monitors to keep out Rn-219 and Rn-220. These radon isotopes fortunately have very short half-lives and are not found in indoor air. The indoor radon monitor does have a dust filter, through which the radon can pass. A fraction of the radon produces alpha particles that penetrate the small square of plastic film inside the monitors, and produce alpha-tracks that are enhanced by chemical etching.

The nuclear tracks recorded on the small square of plastic film inside these monitors are not affected by normal variation in home humidity and temperature, and the dislocation sites (more commonly called alpha-tracks) are permanently recorded on the film. The humidity and temperature insensitivity and the permanent record keeping are probably the major advantages of the alpha-track monitors. The inexpensive alpha-track monitors with their small fragment of film require at least one month for enough tracks to accumulate in a typical home to generate a useful measurement. Estimates of analytical uncertainty for the alpha-track monitors are related to the measurement interval, so intervals of three months to a full year are often utilized.

We have seen no evidence that suggests that these measurements are analytically biased towards too-high or too-low results. This leads us to believe that individual homeowners may have obtained some measurements that were significantly higher or lower than their actual concentrations, and for that reason we advise homeowners to measure indoor radon over several seasons. The random nature of the possible inaccuracies leads us to believe that the compilations obtained for large groups of homes such as are presented later in this paper are reasonably correct.

Homeowners in Fairfax County were informed of the testing program through announcements by area newspapers, radio and television stations. All homeowners who expressed interest were sent a package of literature about radon and about the testing program. All the homeowners who subsequently requested to join the study were accepted, probably because those with potentially low indoor radon (occupants who live more than one story above the ground) were self-screened from the study. Most joined because of newspaper coverage. The preliminary questionnaire showed that only about 10% of the test participants had obtained a radon measurement before joining the study.

Participation in the test series required that the homeowner participate in the entire four-season testing period. The homeowner provided an exact location on a county map which, when compared to a geological map, made it possible to identify the geologic rock unit underlying the home. A questionnaire completed by the homeowner at the start of the test series serves to quantify home construction factors (type of basement, age of home, etc.). Another questionnaire is completed at the end of each seasonal measurement interval to describe home use (number of heating days, number of days with opened windows, location of monitor, etc.).

Homeowners were allowed to place their radon monitors at any location in their home, but were advised that a basement is probably the location of the greatest radon concentration. About 90% of the homes had basements, and of these homes, about 90% of the homeowners placed the monitors in a basement location.

In this study of indoor radon, each homeowner was sent his or her first alpha-track radon monitor near the start of a season to facilitate a study of seasonal variations (winter was November, December and January, spring was February, March and April, summer was May, June and July, and fall was August, September and October). At the end of each seasonal interval, the homeowner was sent a quarterly report concerning the radon "picture" in Virginia and Maryland during the previous quarter, and a request to send in the exposed monitor. Homeowners who had not finished their series of four monitors were also sent their next monitor.

3. RESULTS

3.1 Precipitation

In northern Virginia and southern Maryland, the local weather is monitored by about 300 volunteer weather stations. The weather reports are compiled by the National Oceanic and Atmospheric Administration, and distributed as monthly summaries. According to the weather summaries for the summer intervals, the two subsequent study summers had essentially the same maximum temperatures and average atmospheric pressures, but the second summer had about 20% more precipitation (Table 1). Although there is some disagreement about the effect of rainfall, we think that when the land surface is capped with intergranular water, soil radon cannot move vertically and escape directly from the soil to the atmosphere. When radon accumulates in the soil around a home, below this near-surface layer that is saturated with water, one would expect that indoor radon would increase.

Table 1. Compliation of Summer Weather Conditions

Season of Each Year	Total Amount of Precipitation	Average Maximum Temperature	Average Maximum Pressure
First Summer	9.9"	84.4 F	30.3 mm
Following Summer	12.0"	83.9 F	30.3 mm

Note: This chart was generated from monthly weather summaries titled METROPOLITAN WASHINGTON CLIMATE REVIEW that cover Fairfax County, Montgomer y County and adjacent areas. These summaries are available at no charge from: Cooperative Program Branch W/OSO141X4, U.S.Department of Commerce, National Oceanic and Atmospheric

Administration, National Weather Service, Silver Spring, MD 20910. The compila tions in this table are averages, obtained using measurements from the Washington -National Airport, Baltimore-Washington Airport, and Washington -Dulles Airports.

An overview of the seasonal indoor radon variations can be seen by compiling basement radon measurements in Fairfax County (Table 2). Although it is commonly thought that indoor radon is always at its greatest concentration during the winter (Mose and Mushrush, 1988), this is clearly not always the case. The data show that the second (more precipitation) summer had a higher regional radon average than the first (less precipitation) summer, and the second summer also had higher regional radon average than the intervening winter, which had relatively little precipitation.

Table 2. Seasonal Indoor Radon Measurements in Fairfax County

Basement Level Indoor Radon Measurements:

Season and Year	Average Radon (pCi/l)	Median Radon (pCi/l)	Percent over 4 pCi/l	Percent over 10 pCi/l	Number of Homes
Winter No. 1	5.1	3.9	49	9	286
Spring No. 1	4.1	2.9	33	5	487
Summer No. 1	3.0	2.4	23	2	735
Fall No. 1	3.8	3.0	34	3	772
Winter No. 2	4.0	2.8	33	5	525
Spring No. 2	3.9	3.0	33	5	334
Summer No. 2	4.2	3.5	41	5	126
Fall No. 2	6.2	4.2	53	9	108

First Floor Indoor Radon Measurements:

Season and Year	Average Radon (pCi/l)	Median Radon (pCi/l)	Percent over 4 pCi/l	Percent over 10 pCi/l	Number of Homes
Winter No. 1	3.4	2.4	23	5	39
Spring No. 1	2.6	1.6	16	4	76
Summer No. 1	2.1	1.6	11	0	125
Fall No. 1	2.7	2.2	21	0	115
Winter No. 2	2.9	2.1	23	2	100
Spring No. 2	3.1	2.0	24	2	59
Summer No. 2	3.0	1.8	32	0	22
Fall No. 2	4.5	3.9	50	5	20

3.2 Geologic Unit

The extent to which each geological unit contributed to this summer-to-summer increase in indoor radon provides a better estimate of the regional increase in radon. From a geologist's point of view, the study area is very useful. There is a large diversity of rock units which includes poorly consolidated sediments, sedimentary rocks, igneous rocks and metamorphic rocks. However, although the average indoor radon varies considerably from unit-to-unit, the summer-to-summer radon increase caused by precipitation is quite uniform (Table 3).

Table 3. Summer Indoor Radon Compilations for Geological Units in Farifax County, Virginia

Indoor Radon During the First Summer

Geological Rock Unit	Number of Homes	Average Radon (pCi/L)	Median Radon (pCi/L)	Percent over 4 pCi/L
Coastal Plain Province				
Sedimentary Strata	92	1.9	1.7	7
Culpeper Basin				
Diabase	22	1.9	1.7	0
Sandstone/Conglomerate	40	2.6	1.9	13
Siltstone/Shale	8	5.4	2.7	38
Piedmont Province				
Meta-Mafic Rock	12	2.2	1.4	25
Occoquan Granite	54	2.6	2.0	13
Falls Church Tonolite	21	2.2	2.3	10
Sykesville Formation	107	2.9	2.4	20
Indian Run Formation	36	2.6	1.8	14
Annandale Group	52	3.2	2.6	31
Popes Head Formation	86	3.5	2.8	24
Peters Creek Schist	193	3.8	3.3	36

Indoor Radon During the Following Summer

Geological Rock Unit	Number of Homes	Average Radon (pCi/L)	Median Radon (pCi/L)	Percent over 4 pCi/L
Coastal Plain Province				
Sedimentary Strata	7	2.4	2.0	14
Culpeper Basin				
Diabase	4	4.8	2.2	25
Sandstone/Conglomerate	3	5.5	6.5	67
Siltstone/Shale	4	3.5	2.6	25
Piedmont Province				
Meta-Mafic Rock	1	2.9	2.9	0
Occoquan Granite	6	4.0	1.7	33
Falls Church Tonolite	3	2.1	2.5	0
Sykesville Formation	9	3.0	2.1	22
Indian Run Formation	6	4.5	4.0	67
Annandale Group	11	4.7	4.4	64
Popes Head Formation	17	3.7	3.5	41
Peters Creek Schist	42	5.0	3.9	45

3.3 Home Construction

Other comparisons can be made using home-construction factors. Most of the area homes have basements with walls composed of concrete blocks or poured concrete. Earlier studies have shown that basements with concrete-block walls tend to have higher indoor radon concentrations during the winter (Mose et al., 1988; Mushrush and Mose, 1988). This is presumably because the blocks are joined by mortar which tends to crack, allowing soil-gas enriched in radon to enter. During the study summers, this difference was noted (Table 4). Apparently the effect of basement-wall construction persists during the summer, even though the summer is the interval when windows are most frequently left open to admit cooling (and low-radon) outside air.

Table 4. Comparison of basement wall construction wit h summer indoor radon.

Season	Type of Basement Wall	Average Radon (pCi/l)	Median Radon (pCi/l)	Percent over 4 pCi/l	Number of Homes
First Summer	Concrete Block	3.0	2.5	24	469
	Poured Concrete	3.0	2.3	19	237
Following Summer	Concrete Block	3.8	3.4	41	81
	Poured Concrete	4.6	3.5	40	40

Another comparison involves the type of home heating system. Most homes have either a gas or oil furnace or use electrical heat, normally in the form of a heat pump. Earlier studies had shown that the electrical-heating-system homes tend to have higher indoor radon concentrations (Mose et al., 1988; Mushrush and Mose, 1988). This is presumably because homes with a furnace develop a partial vacuum within the home when the furnace operates, and this partial vacuum draws into the home more radon-poor outside air than radon-enriched soil air. However, although the homes are not heated during the summer, the measurements taken during the study summers show that homes with electrical heating systems have higher indoor radon, particularly during the second summer (Table 5).

Table 5. Comparison of home heating system with summer indoor radon .

Season	Heating System	Average Radon (pCi/l)	Median Radon (pCi/l)	Percent over 4 pCi/l	Number of Homes
First Summer	Oil or Gas	2.7	2.3	18	381
	Electrical	3.7	2.8	31	215
Following Summer	Oil or Gas	3.1	2.5	32	59
	Electrical	5.0	3.9	48	61

In the study area, the electrical h eating system is usually a heat pump. Homes with a heat pump normally use the device for whole -home cooling in the summer. A small percentage of the homes with both a heat pump (used mainly for temperatures above 40 degrees Fahrenheit) also had an oil or g as combustion system. These homes were listed under "Oil and Gas Heating System" since the oil or gas heater is more frequently used during the heating season.

Since homes are usually not heated during the summer, the higher summer indoor radon readings in the group with electrical heating systems probably occurs because heat pumps are commonly used for whole-home cooling during the summer. When a house is closed to preserve cooled air, indoor radon derived from the underlying soil increases.

3.4 Soil Radon (Aeroradioactivity)

The summer-to-summer variation is particularly important for scientists who are currently trying to identify radon problem areas and to develop what have been called radon-potential maps. It now appears that if indoor radon measurements are used to develop maps for adjacent areas, the work should be conducted by using measurements from a single season and not by using measurements from different seasons. For example, the U.S. Geological Survey (Otton et al., 1988) has recently published a radon potential map for Fairfax County. As shown in Table 6, the correlation between the U.S.G.S. Radon Potential Number (first column on Table 6) and the indoor radon for each set of homes is rather good. However, this might not have been the case if the radon measurements had been conducted in different seasons, or during the same season in different years in different parts of the county. As shown in Table 6, the radon relationship to each U.S.G.S. Radon Potential Number is much higher for the second summer when compared to the first summer in our study.

Table 6. Comparison between summer basement indoor radon measurements and the U.S. Geological Survey radon potential numbers used in Otten et al.(1988).

Indoor radon measurements from the first summer

USGS Radon Potential	Number of Homes	Average Radon (pCi/L)	Median Radon (pCi/L)	Percent over 4 pCi/L
1	123	2.0	1.7	9
2	98	2.8	2.5	17
3	188	3.3	2.6	26
4	129	3.2	2.8	23
5	131	4.2	3.4	39

Indoor radon measurements from the following summer

USGS Radon Potential	Number of Homes	Average Radon (pCi/L)	Median Radon (pCi/L)	Percent over 4 pCi/L
1	9	2.4	2.3	0
2	8	3.0	2.0	25
3	26	3.8	3.4	42
4	21	5.2	3.6	43
5	25	4.7	4.0	50

The problem of seasonal changes in indoor radon has some implications for the federal and state radon surveys currently being conducted, particularly if the season-to-season change noted in the present study is typical. One implication is that if two areas (e.g., state- or county-size areas) which in fact have the same radon potential are tested for indoor radon during different seasons, they will probably be seen to have different indoor radon potential. This now appears to be true, even if the same season was used but in different years. Furthermore, if soil radon varies with weather in the same way that indoor radon does, it appears likely that radon potential maps based on soil radon characteristics such as the U.S. Geological Survey map (Otton et al., 1988) would apply most directly to intervals when the weather simulates the conditions that prevailed when the soil characteristics were compiled. Alternatively, it would be necessary to develop several radon-potential maps, corresponding to times of cool and wet weather, warm and dry weather, and so on. To create national or regional radon-potential maps based on soil characteristics that relate particular radon potential numbers to particular median radon levels or to a particular percentages of homes over 4 pCi/l, it would be necessary for the entire map area be examined under similar weather conditions. To whatever extent the weather conditions varied throughout the area of a radon-potential map during the time of soil characterization, the map will contain inherent inaccuracies that over- or underestimate the relative radon potentials.

We have found that a better radon potential map can be derived by a survey of surface radioactivity, measured from an airplane (Daniels, 1980; Mose et al., 1988). An aeroradioactivity map is available for Fairfax County.

As shown in Table 7, an increase in aeroradioactivity compares well with an increase in indoor radon.

Table 7. Comparison between total gamma aeroradioactivity and summer basement indoor radon in Fairfax County, Virginia.

Indoor radon measurements from the first summer

Aeroradio-Radioactivity	Number of Homes	Average Radon (pCi/L)	Median Radon (pCi/L)	Percent over 4 pCi/L
100-200 cps	51	2.7	2.1	18
200-300 cps	142	3.0	2.1	23
300-400 cps	262	3.1	2.6	24
400-500 cps	94	4.0	3.1	35
500-600 cps	8	4.8	4.5	63

Indoor radon measurements from the following summer

Aeroradio-Radioactivity	Number of Homes	Average Radon (pCi/L)	Median Radon (pCi/L)	Percent over 4 pCi/L
100-200 cps	5	2.2	2.2	0
200-300 cps	16	3.0	2.3	13
300-400 cps	45	4.3	3.7	47
400-500 cps	14	5.5	4.7	57
500-600 cps	1	-	-	-

4. CONCLUSIONS

The indoor radon concentrations in several hundred homes in northern Virginia show a strong dependence on weather conditions. A comparison between the indoor radon signature of the study area during adjacent summers shows generally higher levels of radon during the second summer, based on comparisons using geological units, basement wall construction and home heating system. The higher levels of indoor radon apparently occurred because the second summer had more rainfall, which prevented a significant amount of soil radon from escaping directly to the atmosphere.

Prior to the present study, only two reports examined season-to-season variations. In Maine, it has been shown that there can be a measurable difference by comparing indoor radon in about 60 homes during the 1980-81 winter with about 60 homes during the 1981-82 winter (Hess et al., 1985). Our study of indoor radon in northern Virginia reaches a similar conclusion, but is based on several hundred homes, under well-documented weather changes. It now appears that the concentration of indoor radon is related to the amount of rainwater (or snow), which "caps" the ground and prevents soil radon from escaping out of the soil surrounding the home.

One implication of seasonal variations involves the estimation of annual radiation dose related to radon. It now appears that during some intervals, when rainfall is unusually low, indoor radon is reduced. Intervals with more precipitation would be characterized by a higher indoor radon signature. However, as shown in our study, one might find that dry winter conditions and wet summer conditions could result in a similar radiation dose. To realistically estimate variations in the annual radiation dose for a large area like a county or state, it would be necessary to create a series of maps which present the radon situation for several combinations of temperature and precipitation.

The discovery of significant summer-to-summer variations, and the observation that sometimes the winter can closely simulate a summer in terms of indoor radon, indicate that a single season measurements may not be very useful. This is apparent, even if the season of measurement is the winter. This reinforces the idea that "closed-home" conditions which are thought to simulate winter conditions can sometimes incorrectly estimate the annual dose, and that single short-term measurements under this condition should be interpreted with caution.

Perhaps the most important conclusion is that it is very difficult to predict lifetime health risks due to exposure to radon based on a measurement of a single week, a single season, or even an entire year. It is becoming increasingly obvious that seasonal and year-to-year variations in weather play an important role in determining the exposure to indoor radon. The best approach is probably the regular placement of alpha-track indoor radon monitors, used with one year exposure intervals. In this fashion, the home occupant accumulates a permanent record of radon exposure, and can detect and rectify unanticipated increases in indoor radon.

Relatively few characterizations of soil over a large area can actually be made during a single season. Aeroradioactivity compares well with an increase in indoor radon. The only characterization of which the authors are aware is aeroradioactivity, in that several hundred square kilometers can be measured in a week, and several thousand square kilometers can be measured in a single season. We therefore recommend that should regional radon potential maps be constructed by state and federal agencies, strong consideration should be given to first obtaining regional spectral aeroradioactivity data using 5- to 10-kilometer wide, flight-line spacings (or less). A map of this type provides a mechanism to compare widely separated areas. This would be followed by data acquisition using 1/2 to 1 kilometer flight-line spacing, to accurately determine relative radon potential on a local scale. In this fashion, regional radon potential-maps can be constructed that do not contain a seasonal bias caused by varying weather conditions.

REFERENCES

Buchli R. and Burkart W. 1989. Influence of subsoil geology and construction technique in indoor air Rn-222 levels in 80 houses of the central Swiss Alps. *Health Physics* 56,423-429.

Cohen, B.L., and Gromicko, N. 1988. Variation of radon levels in U.S. homes with various factors. *Journal of the Air Pollution Control Association* 38, 129-134.

Daniels, D.L. 1980. Geophysical-geological analysis of Fairfax County, Virginia: *U.S. Geological Survey Reports* 80-1165, I-64.

Froelich, A.J. 1985a. Folio of geologic and hydrologic maps for land use planning in the Coastal Plain of Fairfax County and Vicinity, Virginia. *U.S. Geological Survey Miscellaneous Investigations Series* Map No. I-1423.

Froelich, A.J. 1985b. Map and geotechnical properties of surface materials of the Culpeper basin and vicinity, Virginia and Maryland. *U.S. Geol. Survey Miscellaneous Investigation Series* Map No. I-1313-E.

Froelich, R.K., and Pearson C.A. 1988. Potential radon hazard in Rhode Island estimated from geophysical and geological surveys. *Northeastern Environmental Science* 7, 30-34.

Froelich, A.J., and Zenone, C. 1985. The relation of water quality to ge ology and land use changes in Fairfax County and vicinity, Virginia. *U.S. Geological Survey Miscellaneous Investigation Series* Map No. I-1561.

George, A.C., and Eng, J. 1983. Indoor radon measurements in New Jersey, New York and Pennsylvania. *Health Physics* 45, 374-400.

Henshaw, D.L., Eatough J.P., and Richardson R.B. 1990. Radon as a causative factor in induction of myeloid leukaemia and other cancers. *The Lancet* 335, 1008-1012.

Hess, C.T., Fleischer, R.L., and Turner, L.G. 1985. Field and laboratory tests of etched track detectors for Rn-222: Summer vs. winter variations and tightness effects in Maine houses. *Health Physics* 49, 65-79.

Leavy, B.D., Froelich, A.J., and Abram, E.C. 1983. Bedrock map Leavy B.D., Froelich, A.J. and Abram, E.C., Bedrock map and geotechnical properties of rocks of the Culpeper Basin and vicinity, Virginia and Maryland. *U.S. Geological Survey Miscellaneous Investigation Series,* Map No. I-1313-I.

Lykken, G.I., and Ong, W.S. 1989 Evidence of exhalation of stored environmental Rn determined by whole -body counting. *Health Physics* 57, 161 -162.

Nero, A.V. Jr. 1988. Radon and its decay products in indoor air: An overview. In: *Radon and Its Decay Products in Indoor Air,* pp. 1 -56. (Nazaroff , W.W. and Nero, A.V. Jr, Eds.). New York: Wiley-Interscience.

Mose, D.G., and Mushrush, G.W. 1988. Regional levels of indoor radon in Virginia and Maryland. *Environmental Geology and Water Sciences* 12, 197-201.

Mose, D.G., Mushrush, G.W., and Kline S.W. 1988. The interaction of geology, weather and home construction on radon in northern Virginia and southern Maryland. *Northeastern Environmental Science* 7, 15-29.

Mose D.G., Mushrush, G.W., and Kline, S.W. 1989. Winter-to-winter variations in indoor radon.: *Journal of Toxicology and Human Health* 28, 129-150.

Mushrush, G.W., and Mose D.G. 1988a. Regional variation of indoor radon over three seasons. *Environmental Toxicology and Chemistry* 7, 879-887.

Mushrush, G.W., and Mose D.G. 1988. The effect of home construction on indoor radon in Virginia and Maryland. *Environment International* 14, 395-402.

Obermeier, S.F. and Langer, W.H. 1986. Relationships between geology and engineering characteristics of soils and weathered rocks of Fairfax County and vicinity, Virginia. *U.S. Geological Survey Professional Papers* No. 1344, 1-30.

Otton, J.K., Schumann, R.R., Owen, D.E., Thurman, N., and Rice, C.A. 1988. Map showing radon potential of rocks and soils in Fairfax County, Virginia. *U.S. Geological Survey Miscellaneous Investigation Series* , Map No. MF-2047.

Perritt, R., Hartwell, T., Sheldon, L., Cox, B., Clayton, C., Jones, S., and Smith, M. 1990. Radon-222 levels in New York State homes. *Health Physics* 8, 147-155.

Pohl, E., and Pohl-Ruling, J. 1967. The radiation dose received by inhalation of air containing Rn222, Rn220, Pb212 (ThB) and their decay products. *Annals da. Academia de Ciencias* 39, 393-404.

Puskin, J.S., and Yang Y. 1988. A retrospective look at Rn-induced lung cancer mortality from the viewpoint of a relative risk model. *Health Physics* 54, 635-643.

Rose, A.W., Washington, J.W., and Greeman, D.J. 1988. Variability of radon with depth and season in a central Pennsylvania soil developed on limestone. *Northeastern Environmental Science* 7, 35-39.

Smith, D.L., and Hansen, J.K. 1989. Distribution of potentially elevated radon levels in Florida based on surficial geology. *Southeastern Geology* 30, 49-58.

Steck, D 1990. A comparison of EPA screening measurements and annual Rn -222 concentrations in statewide surveys. *Health Physics* 58, 523-530.

Washington, P.A. 1988. Proper scaling of radon surveys, a perspective based on natural radon sources in southern Vermont. *Northeastern Environmental Science* 7, 40-44.

Watson, J.E. Jr., Adams, W.C., Xie, Y., Fong, S.W., and Newman H.J. 1988. Survey of Rn-222 in North Carolina Homes. *Health Physics* 55, 71-75.

CHAPTER 25

REDUCTION OF RADON IN MUNICIPAL WELLS IN VIRGINIA AND MARYLAND

Fiorella Simoni[1], George Mushrush[2] and Douglas Mose[2]
[1]*George Mason University, Chemistry Department, Fairfax, VA and* [2]*Center for Basic and Applied Science, Culpeper, VA*

Abstract: Approximately 10% of the homes in northern Virginia use well water. Our measurements show that some exceed 4000 pCi/L, that the average is about 2000 pCi/L, and that a few homes have waterborne radon as low as 300 pCi/L, the USEPA's recommended Maximum Concentration Level. In this area, granitic rocks yield well water with about 5000 pCi/L, felsic metamorphic rocks yield well water with about 2000 pCi/L, and quartz -rich sedimentary rocks average about 1000 pCi/L. Tests of remediation technology showed that passing water through large tanks of activated charcoal removes about 90% of the waterborne radon, but the charcoal becomes less efficient over a few months. Aeration experiments using a blade and experiments using a splash box in large community storage tanks removed about 60-70% of the waterborne radon, but using both removed about 90%, and the combination does not become less effective through time.

1. INTRODUCTION

Indoor radon is a mixture of radon that comes through basement soil-facing walls and the basement floor, and radon that escapes from the water in homes with private water wells. The only difference is that the radon in well water originates from somewhat deeper geological material beneath the home. Differences in weather, home use and home design strongly influence the concentration of indoor radon derived from the soil, but these factors probably do not significantly influence the quantity of radon in the home water supply, nor the rate at which dissolved radon escapes the home water supply.

The alpha dose from drinking naturally radioactive water is primarily due to the presence of uranium decay chain nuclides, including Ra-226,

Rn-222, Po-218 and Po-214. Of these radionuclides, radon most easily escapes from the surfaces of cracks in rocks and from the surfaces of sediment grains to enter ground water, because radon is a noble gas. For this reason, its concentration is usually many times greater then the other alpha emitters in drinking water.

Radon is a noble gas, chemically inert and highly soluble in groundwater. Alpha decay recoil and diffusion are thought to be the mechanism by which radon moves from grain and fracture surfaces into groundwater. In much of the United States, municipal water supplies derived from well water are low in radon. Yet such is not the case in some areas, such as the communities in the rolling hills of the Appalachian Mountain System.

The biological significance of radon is related to the fact that radon and its radioactive daughter products comprise the major portion of the natural internal radiation dose to humans. For the general population, excess amounts of these radionuclides can be inhaled when in buildings built over soils with high radon emanation properties, or ingested by consuming well or spring water.

The number of deaths from cancer in the U.S. has doubled since 1950, from about 200,000/year to 400,000/year, and has increased an estimated 10-fold since 1900 (Moolenaar, 1988). This observation is largely responsible for the upsurge in public fear and environmental regulation. However, most of the increase in cancer mortality is due to the growth and aging of the U.S. population. The only major cause of cancer mortality that has increased since 1950 is lung cancer, which is believed to be largely due to smoking. Cancer mortality, adjusted for increasing population and age, and exclusive of respiratory cancer, has decreased by about 10% since 1950, and now stands at about 10/year per 10,000 population. The mortality is about equally divided between breast, digestive and gland cancer (Moolenaar, 1988).

The search for the cause of cancer has promoted the identification of natural and industrial-age carcinogens. This approach has proven successful when high carcinogen levels are present. For low carcinogen levels, the currently used no-threshold model of chemical carcinogenesis is often applied, particularly for radiation. In this model, mortality at high exposure levels is extrapolated to estimate mortality at lower levels. Unfortunately, this has resulted in public health policies implemented in the absence of scientific certainty.

Few of the radionuclides occurring in natural water are sufficiently studied to facilitate the adoption of regulatory guides. The interim Maximum Contaminant Level (MCL) for the sum of Ra-226 and Ra-228 has been set by the U.S. Environmental Protection Agency (USEPA) at 5 pCi/L(Cothern, 1987a). USEPA drinking water standards have not yet been set for uranium or radon. In the summer of 1991, the USEPA proposed a MCL of 300 pCi/L for radon in municipal water supplies. This would be consistent with the standards applied to other contaminants (i.e., the added risk at the MCL must add only one chance in a million if possible, or one chance in 10,000 at worst, of dying from a contaminant-related disease).

Several studies, starting with reports of Kahlos and Asikainen (1980), have suggested that radon dissolved in the home water supply can substantially contribute to the indoor radon of the home. Several subsequent studies have estimated that a water supply with a radon concentration of 10,000 pCi/L would add about 1 pCi/L to the indoor radon due to the outgassing of the domestic water supply (Cothern, 1987; Prichard, 1987). Three booklets published by the USEPA (USEPA 1987a, 1987b, 1987c) state that radon-enriched domestic water must have about 10,000 pCi/L of radon to contribute 1 pCi/L to indoor radon (a 10,000:1 ratio). For suppliers of municipal water systems, and by inference for the purchasers of a home and current homeowners, the USEPA recommends that waterborne radon be less than 300 pCi/l. The following paper, which reports on the waterborne radon concentration in many homes and municipal water wells in northern Virginia and southcentral Maryland, indicates that on average, the water wells are supplying potable water usually greater than the USEPA's MCL.

It should be noted that several studies have suggested that radon in the bloodstream is a more potent carcinogen than generally realized. Although the ingestion of radon-enriched water adds radionuclides to the blood, the possibility that cancer can occur due to drinking naturally radioactive water has not been studied nearly as extensively as might be expected. Pohl and Pohl-Ruling (1967) and Lykken and Ong (1989) showed that inhaled radon passes through the lungs and enters the bloodstream, where it is transmitted throughout the body. Henshaw and others (1990) were able to correlate cancers other than lung cancer with inhaled radon in several states and cities. Hess and others (1983, 1985) showed a correlation between radon in potable water and the incidence of cancer of many types, as did Mose and others (1990). Gosink and others (1990) showed that radon consumed by drinking well water has a long biological residence time (e.g., a sedentary or sleeping person would require about 12 hours to eliminate the radon consumed in a glass of well water typical of our study area).

2. REGIONAL GEOLOGY

Most of the geological units in the Appalachian Mountain System are found over vast areas and were formed by well-understood geological processes. The oldest unit in the study area is the quartzite, which was deposited as beach sand on the Blue Ridge Province approximately 600 million years ago. The deposition of this sand marks the opening of a narrow seaway, a new ocean between the Americas to the east and Europe and Africa to the west, where one had not existed before. The seaway widened, forming a pre-Atlantic ocean basin (proto-Atlantic) which continued to open over the subsequent 100 million years. During this time interval, an increasingly deep and wide proto-Atlantic formed, and many additional layers of sedimentary and volcanic strata were deposited in this ancient pre-Atlantic ocean basin. From about 500-400 million years ago, this ancient ocean began to close, and between approximately 400 and 300 million years ago, these ancient sedimentary and interlayered volcanic strata were deeply buried and severely compressed.

At depths of about 5 miles, these strata were recrystallized into the Blue Ridge quartzite (originally beach sand) and the Piedmont schist and phyllite (originally clay-rich and volcanic strata). In some areas of particularly high temperatures, the metamorphosed sedimentary and volcanic rocks were partially melted, and when this material cooled it formed the Piedmont granite of the Appalachian mountain system.

Following a long interval of uplift and erosion between about 300 and 200 million years ago, these deformed rocks were uplifted and eroded. About 200 million years ago, when the join between the America and the now-adjacent continents of Europe and Africa was worn sufficiently thin, the Americas were pulled away from Europe and Africa. At first, narrow basins developed from cracks, formed along the length of the Appalachians, and grew in length and depth between about 200 to 150 million years ago. One of these narrow basins, locally known as the Culpeper Basin, accumulated layers of red terrestrial sandstone. At about 150 million years ago, an alignment of cracks formed and the modern Atlantic Ocean opened to form a narrow seaway (the Atlantic is still opening today, at about 1 inch/year). Shortly after the Americas broke away from Europe and Africa, and the Atlantic Ocean formed, the eastern edge of America was covered by Atlantic sea water. Between about 150 million years ago and today, the eastern edge of America has slowly risen, causing the Atlantic shoreline to move eastward. Sand and clay strata deposited during this interval formed the Coastal Plain Province.

3. PRIVATE WATER WELLS: FAIRFAX AND MONTGOMERY COUNTIES

As part of an ongoing study of indoor and soil radon in Fairfax County and Montgomery homes, measurements were also made of well-water radon in these homes. The measurements were made in homes using reservoir water and in homes using well water. When this study was almost completed, waterborne radon was also measured are in municipal wells in Prince William County, immediately south of Fairfax County. These measurements are discussed in the next section.

At the present time, over 1200 Fairfax County and Montgomery County homes have been examined in our indoor radon study. The indoor radon measurements are year-long averages, obtained using three-month exposure intervals, and using a series of four alpha-track indoor radon monitors (Tech/Ops Landauer Corporation, Type SF).

Participants in the indoor radon project were offered a test kit with which to measure drinking-water radon. About 700 homeowners (more than half of the participants in the indoor radon project) joined the radon-in-water study. The homeowners who participated filled out a questionnaire about the size of their homes and their home water supply (e.g., from where is the water obtained, if from a well how deep and how far away is the well, is the water treated, and if so how, etc.). The participant homeowners were each provided with a syringe, a capped vial with 5 mL of toluene-based liquid scintillation fluid, and directions about how to collect 10 mL of drinking water from a commonly used water tap.

During the indoor radon study, and prior to the start of the county-wide investigations of waterborne radon, one of the Fairfax County homes with a private water well was monitored to examine variations in drinking-water radon. At that point, it was anticipated that the variations in waterborne radon would be below a factor of 2 (Prichard and Gesell, 1981). However, since indoor radon varies seasonally in this home, this particular home was studied to see if its drinking water radon varied in a fashion similar to that of its indoor radon. Over the four seasons of measurement, indoor radon in the basement averaged about 9 pCi/L and the first-floor average was about 4 pCi/L. Over the year, using monthly measurements, the drinking water radon only varied from about 3200 to 3700 pCi/L (average was about 3500 pCi/L).

To examine the contribution by waterborne radon outgassing from the home water supply to the airborne radon in this home, a continuous indoor radon monitor (Femto-Tech Indoor Radon Monitor) was placed in a first-floor, 50-square-foot bathroom with a shower. The shower was turned on and left on for several hours, during which time the bathroom door was left closed and the indoor radon concentration was measured each hour (Table 1). This is similar to experiments reported by McGregor and Gourgon (1980), Hess and others (1982), and Kearfott (1989).

Table 1. Compilation of airborne radon increase due to outgassing of a shower in a 50 -square-foot bathroom.

Indoor Radon Test	Hour-Long Test Intervals for Indoor Radon before Water Shower was Turned on (pCi/L)			Hour-Long Averages for Indoor Radon During the Time that Water Shower was Turned on (Indoor Radon in pCi/L)				
	3rd Hour	2nd Hour	1st Hour	Hour 1	Hour 2	Hour3	Hour 4	Hour 5
Test 1	15	6	8	71	190	Data Not Taken		
Test 2	27	5	4	24	134	226	293	Data Not Taken
Test 3	35	4	5	29	134	238	317	350

As shown in Table 1, there was a significant increase in the radon content of the bathroom. The degree to which this shower would increase the overall radon concentration of the home can be estimated. For example, if the air handling system circulated the radon-enriched air (at about 350 pCi/L) from the 50 square foot bathroom throughout the 2500 square-foot home, the increase to the whole home would have been about 1/50 of the bathroom concentration, which would be about 7 pCi/L. If the shower had been used for 1/20 of 5 hours (a shower duration of about 15 minutes), one shower per day would add only about 0.35 pCi/L to the indoor radon of the home, which is 1/10,000 of the average radon-in-water concentration in this home.

The 10,000:1 ratio of waterborne radon to airborne indoor radon in this home is obviously fortuitous. If each day there were more than one shower, plus a cycle of a dishwasher and a cycle of a clothes washer, the indoor radon concentration would be increased by more than 1/10,000 of the radon-in-water level. Alternatively, the increase could be less with different parameters. To learn more about the relationship between waterborne radon and indoor radon, waterborne radon measurements obtained from many other Fairfax and Montgomery County homes were compared to their indoor radon measurements. Of these approximately 700 homes, about 500 used potable water originating in a local reservoir, and because radon quickly leaves surface water, waterborne radon was absent in these homes. Of the remaining approximately 200 homes, where well water is used, the indoor radon concentrations ranged from about 1 to 40 pCi/L and the waterborne radon concentrations ranged from about 100 pCi/L to about 8,000 pCi/L. It was noted that smaller homes seemed to have higher indoor radon concentrations. To determine the importance of home size, homes in Fairfax and Montgomery County with water-wells radon were divided into groups determined by the radon concentration of the home water supply.

In the homes that used well water, it became apparent that as the waterborne radon concentration increased, the importance of the size of the home increased. In short, the smaller homes generally had higher indoor radon, but the effect was greater if the waterborne radon was greater. This effect was apparent, even though the wells in this set of homes were all less than 10,000 pCi/L. Waterborne radon concentrations of less than 10,000 pCi/L do exert a measurable effect on indoor radon, if the home is small. Conversely, when home volumes reach 100,000 cubic feet (a large home), the indoor radon concentration is relatively low, even if the waterborne radon concentration is relatively high.

From a health department point of view, the fact that waterborne radon is most likely to present a problem for people with water wells and small homes is of concern. People who have small homes tend to be less financially able to make changes in their water supply.

4. MUNICIPAL WATER WELLS: PRINCE WILLIAM COUNTY

The Prince William County Well Water Radon Study began as a review and a retest of water wells studied by the U.S. Geological Survey (USGS). The USGS investigation revealed that many of the Prince William County Service Authority (PWCSA) wells carried waterborne radon in excess of 300 pCi/l, the USEPA's MCL. Since little was known about many other wells in the county, and little was known about possible variations in waterborne radon concentrations over time, a two-year investigation was conducted using many of the PWCSA municipal water wells. After two years, it was possible to quantify the magnitude of variation through time, and the variations between geological units in which the wells were placed. It was also possible to conduct radon-removal experiments on large storage tanks served by relatively large municipal wells.

During the first year of study, waterborne radon measurements every two months were obtained from the wells measured earlier by the USGS, and from a home serviced by one of these wells. To concentrate on the more radioactive wells, the low-radon wells (defined as having less than 1500 pCi/L) were not during the second year of measurements. Also, since the first year of study gave a rough estimate of the well-to-home decrease in waterborne radon (about 30%; discussed below), the second year of study did not include measurements of waterborne radon at homes. Instead, the second-year study concentrated on obtaining frequent (1/month) waterborne radon measurements from the higher-radon wells (higher than 1500 pCi/L), and on radon-removal experiments.

During the first year of study in Prince William County, testing of wells was completed in Fairfax and Prince William Counties. These data can be characterized as measurements from low-productivity wells, because each well typically serves about 100 gallons/day to only one home (Table 2). Conversely, the compilation of radon measurements from PWCSA municipal wells can be characterized as measurements from high-productivity wells, because each serves thousands of gallons/day to tens or hundreds of homes (Tables 3 and 4).

Table 2. Average waterborne radon in geological units in Fairfax and Montgomery Counties

Geological Unit	Number of Wells	Waterborne Radon (pCi/l) Average	Median	Minimum	Maximum
Quartzite	4	1017	860	720	1630
Schist/Phyllite	111	2541	2379	<100	7402
Sandstone	20	1787	1540	129	4699
Granite	21	3292	2912	805	6946

Table 3. . Summary of average waterborne radon measurements and relatively typical weekly productivity of municipal wells in Prince William County, for which both measurements were available. The radon measurement is given with a +/- calculation which represents the range of measurements, compared to the average measurement.

Well Number	Radon (pCi/L)	Productivity (gallon/week)	Geology
Center of County: Algonquin Hills			
WO6 (38 tests)	2320 + / – 32%	98,600	Schist
WO7 (16 tests)	3800 + / – 19%	61,000	Granite
WO8 (16 tests)	2630 + / – 14%	411,900	Granite
WO10(15 tests)	2180 + / – 46%	59,000	Granite
WO14(16 tests)	3280 + / – 16%	174,000	Granite
WO15(16 tests)	1320 + / – 43%	56,900	Schist
WO2 (14 tests)	1710 + / – 21%	17,800	Schist
WO4 (6 tests)	240 + / – 19%	53,400	Schist
WO13(16 tests)	1460 + / – 32%	157,900	Schist
East Side of County			
WOF1(12 tests)	2300 + / – 35%	not avail.	Granite
WOF2(11 tests)	2760 + / – 12%	not avail.	Granite
WOF3(10 tests)	3460 + / – 12%	not avail.	Granite
WOF4(10 tests)	2240 + / – 29%	not avail.	Granite
Northcentral Part of County: Manassas			
WG1 (16 tests)	2610 + / – 25%	99,600	Sandstone
WG2 (14 tests)	1560 + / – 9%	1,030,800	Sandstone
WG4 (6 tests)	1460 + / – 17%	193,600	Sandstone

Well Number	Radon (pCi/L)	Productivity (gallon/week)	Geology
WG5 (16 tests)	1600 + / – 14%	413,700	Sandstone
WG6 (16 tests)	2580 + / – 8%	450,500	Sandstone
WG7 (16 tests)	1950 + / – 21%	1,734,600	Sandstone
WG8 (2 tests)	1080	not avail.	Sandstone
WG9 (16 tests)	1970 + / – 21%	265,100	Sandstone
WG10(15 tests)	2160 + / – 26%	1,593,800	Sandstone
WG11(16 tests)	2060 + / – 17%	1,901,300	Sandstone
West Side of County: Bull Run Mountain Area			
WB1 (11 tests)	1870 + / – 59%	not avail.	Quartzite
WB2 (2 tests)	1330	not avail.	Quartzite
WB5 (2 tests)	810	not avail.	Quartzite
WB7 (6 tests)	1280 + / – 36%	292,700	Quartzite
WB8 (6 tests)	760 + / – 13%	8,100	Quartzite
WB9 (15 tests)	1670 + / – 19%	42,300	Quartzite
WB10(6 tests)	830 + / – 21%	53,700	Quartzite
Center of County: Haymarket Area			
WH1 (16 tests)	4190 + / – 33%	47,400	Sandstone
WH2 (14 tests)	1490 + / – 21%	97,000	Sandstone
Center of County: Gainesville Area			
WG14(3 tests)	1200	not avail.	
WG16(16 tests)	1670 + / – 25%	48,700	Sandstone
WG17(6 tests)	910 + / – 25%	35,100	Sandstone
Center of County: Yorkshire Area			
WY6 (6 tests)	1000 + / – 18	1,658,200	Sandstone
WY9 (6 tests)	1390 + / – 20	645,900	Sandstone
WY11(5 tests)	740 + / – 14	1,097,400	Sandstone

Table 4. Average waterborne radon in geological units in Prince William County.

Geological Unit	Number of Wells	Waterborne Radon - Average	Waterborne Radon - Range Around Average (+ / – %)
Quartzite	7	1220	35
Schist	5	1420	32
Sandstone	18	1760	24
Granite	8	2790	36

5. WATERBORNE RADON AND GEOLOGY

In Fairfax and Montgomery Counties (Table 2), and in Prince William County (Table 4), the comparison between geological-unit and waterborne radon shows that rocks characteristically more radioactive produce more radioactive well water. For example, granite is the most radioactive of all the common rock types, and the wells in Piedmont granite produce the most radon-enriched water in all three counties.

Elevated radon in groundwater from the schist and phyllite probably occurs because as the uranium-enriched granite melted and moved to higher crustal levels, its radioactive component could escape as vapors which moved into and enriched the solid rock (i.e. into the schist and phyllite) surrounding the still-molten granite. This movement of uranium from granite to host rock continued even as the granite cooled and became solid. This is because, to a great extent, the uranium atoms in the solidifying granite are located on grain boundaries. When very hot groundwater escapes from the still-hot granitic rock, uranium is relatively easily removed from its grain boundary locations, and carried into the surrounding host rock.

The grain surface effect is important today. In the Piedmont rocks like granite, schist and phyllite, much of the uranium is still on grain boundaries. As the uranium decays, radon forming on grain boundaries can easily accumulate in groundwater that passes through the rock. In this fashion, water wells in granite, schist and phyllite tend to contain more waterborne radon than wells in the other rocks of Prince William County.

Sandstone of the Culpeper Basin appears on average to have an intermediate level of waterborne radon. This is probably because the sand and clay in this sandstone were deposited by rivers. Such sediment tends to retain some but not all of its original uranium and radium.

Quartzite has the lowest waterborne radon, probably because although the quartzite was formed from sand, it formed in an environment different from the sandstones of the Culpeper Basin. The sand that formed the quartzite accumulated along an ancient ocean, where wave action quickly dissolved away most of the uranium in the sediment. Sand depleted in uranium tends to produce low levels of waterborne radon.

6. WELL-TO-HOME DECREASE IN WATERBORNE RADON

In Prince William County, a relatively short interval (less than 1 day) occurs between the time that well water is pumped into storage tanks and the tank water is pumped to the surrounding community. Put another way, the half-life for radon-222 is 3.8 days, so little loss of radioactivity will occur between the time that water is taken out of the wells and when the water arrives at the distribution sites. This problem has been noted by others (Brutsaert and others, 1981; Smith and others, 1961). To estimate the magnitude of the decrease, a typical home-to-well comparison was made using well WO-6, which has the most waterborne radon measurements. Based on 5 comparisons over two months, the well-to-home loss of radioactivity in this particular case is about 3O% (Table 5).

Table 5. **Same-day measurements from municipal we ll WO6 and one of the homes served by this well.**

Day of Year	Well Measurement (pCi/L)	Home Measurement (pCi/L)
35	2450	1750
42	3510	2140
45	2010	940
88	2610	1990
124	2640	1830
186	2430	1780
187	2150	1780
222	3260	2680
235	2630	1830
325	2230	1320
350	2130	1730
	Average = 2550 pCi/L	Average = 1800 pCi/L

7. WELL SITE RADON REMOVAL BY ACTIVATED CHARCOAL

Several methods of radon removal can be successful. Of these methods, activated charcoal and aeration have been used successfully, both at well sites and at homes. Investigations were conducted to compare the radon-removal effectiveness of these two methods, using some of the PWCSA municipal wells.

Activated charcoal is often used to remove volatile organic compounds (VOC) from groundwater. In fact, for many years activated charcoal has been used to remove VOC present in several PWCSA wells (Nelms and Richardson, 1990). One of the wells in the radon study, well WG-7, is a VOC-contaminated well; however, its water is rendered VOC-free by passing through two large tanks containing activated charcoal.

Well WG-7 was selected because of the already functioning charcoal-removal system. Since it is known that charcoal retains radionuclides, tests using a field radiation meter were conducted on the outside of the two charcoal-containing tanks. The measurements showed that while the charcoal does retain radionuclides, the intensity of gamma-radiation outside the tanks is not a significant health risk even assuming long-term exposure. That is, a person whose office is inside the building containing the tanks would not be exposed to a significant gamma-dose assuming 40 hours/week in the office. Other forms of radiation, such as beta and alpha radiation, cannot escape the metal tanks as long as the charcoal remains within.

At well WG-7, the activated charcoal in each of the tanks is changed in alternating years. That is, once a year the charcoal in one of the tanks is replaced, and the flow of well water is directed so as to first pass through the tank containing the older charcoal, after which it passes through the newer recently-changed charcoal. After one year, the tank of older charcoal is filled with new charcoal (the old charcoal goes to a landfill), and the water pathway is again changed so as to have the water pass through the one-year-old charcoal, then through the tank ofjust-replaced charcoal. Well WG-7 delivers about 1,7000,000 gallons of VOC-free potable water to Prince William County residents, and although the well water that comes into the WG-7 storage tank exceeds 2000 pCi/L, after the water passes through the two tanks, the waterborne radon is less than 800 pCi/L (Table 6).

Table 6. Measurements of groundwater radon out of well WG-7, plus radon measurements of water after it passes through a tank of activated charcoal that has been in use at least 1 year (T-1), and radon measurements of water after it passes through a second -in-line tank containing charcoal that has been in use for less than a year.

Date Situation	Well Head (pCi/L)	Out of T-1 (pCi/L)	Out of T-2 (pCi/L)
3/28	1800	1880	730
6/7	1290	1640	800
7/25	2330	2040	750
11/21	2230	N/A	N/A
The charcoal is replaced in the "older" tank.			
2/10	2190	830	90
2/25	N/A	1250	310
3/03	1680	1280	350
3/23	1760	1250	430
4/22	1990	1080	460
5/27	2150	1720	570
6/24	2160	1650	560
8/11	2090	1540	620
9/28	1650	1440	690
11/4	1950	1260	580

Date Situation	Well Head (pCi/L)	Out of T-1 (pCi/L)	Out of T-2 (pCi/L)
12/2	2160	1780	740
1/27	2110	1690	N/A
3/04	1830	1820	810

8. WELL SITE RADON REMOVAL BY AERATION

In some cases, activated charcoal is not viewed as a viable method, because the installation and maintenance of the holding tanks is expensive. Another problem is the responsibility of disposing of radiation-enhanced charcoal.

Other than activated charcoal, the only other popular method of radon removal in municipal water supplies (and homes) is aeration. The aeration method does carry an installation cost. However, the majority of the waterborne radioactivity escapes into the atmosphere, and this air contaminant is not viewed as significant by any regulatory agency.

To evaluate aeration as a method of radon reduction for PWCSA wells, aeration experiments were done on the Occoquan Forest System (WOF 1-4) and on the Algonquin Hills System (well WO-6).

8.1 Occoquan Forest System (Wells WOF 1-4)

Waterborne radon levels in wells WOF 1-4, which were drilled into granite, are among the highest in the county. Waterborne radon measurements at the wells typically are about 2500 pCi/L, though monthly measurements close to 4000 pCi/L have been recorded during the study interval. These four wells are closely placed, and well water from the four wells is combined in a single storage tank. However, to keep the tank completely filled most of the time, the water sent to homes in the surrounding community is a mix of water that flows out of the storage tank and water pumped directly from the wells. When the aeration experiment described below was in progress, the "aerated water" measurements were made using this mixture of aerated water from inside the tank and non-aerated water coming directly from the wells. The "mixed" post-aeration-tank measurements were useful because they describe the water sent to the area homes.

Two aeration experiments were conducted using an inverted T-shaped pipe which delivered compressed air to the water inside the storage tank for wells WOF 1-4. The T-shaped pipe was constructed by the PWCSA technical staff, and was designed to induce a rolling boil to the water in the tank, starting near the bottom of the tank.

The first experiment began during a time when the water going into the storage tank carried a waterborne radon level of about 2500 pCi/L. Over a two-week interval, during which the aeration unit was continuously operated, the waterborne radon coming out of the storage tank was reduced by about 65%. The second experiment was a duplication of the protocol and the results of the first experiment (Table 7).

Table 7. Measurements of groundwater radon in water entering the storage tank for wells WOF 1-4, plus radon measurements of the water leaving the storage tank while aeration is and is not in progress.

Date	Water Entering Tank (pCi/L)	Water Leaving Tank (pCi/L)	% Decrease
4/21	Aeration Process Begins		
4/22	2460	1250	49
4/24	N/A	850	
4/27	N/A	1240	
4/29	N/A	1080	
5/6	N/A	1180	
5/6	Aeration Process Ends		
5/27	2690	2280	15
6/4	3210	2470	23
6/29	Aeration Process Begins		
7/1	2600	770	70
7/6	2620	1150	66
7/8	2210	940	58
7/10	2680	1000	63
7/13	2180	860	61
7/13	Aeration Process Ends		
8/11	2960	2560	14

8.2 Algonquin Hills System (Well WO 6)

The waterborne radon in well WO-6 is about 2320 pCi/L. An aeration unit similar to that described above for WOF 1-4 (though with a lower air-flow capacity) was used in the storage tank for well WO-6. The first experiment ran for only 4 days, and removed about 65% of the waterborne radon. The second aeration experiment with well WO-6 used a larger air compressor and a larger inverted T-shaped air bubble dispersal system inside the storage tank. The experiment ran for 7 days, during which about 65% of the radon was again removed. The third aeration experiment using WO-6 ran for 36 days. In this experiment, the storage tank was again aerated using the inverted T-shaped air bubble-dispersal system as described above. However, the tank was also fitted with a splash box, designed by the PWCSA technical staff, and an exhaust fan was fitted on a port at the top of the storage tank. With all three of the components (splash box, fan and aeration by the air bubble dispersal system), over 90% of the waterborne radon was removed (Table 8).

Table 8 shows that the use of a splash box reduces radon by about 60%. Aeration reduces the radon by 60-70%, and does the use of a splash box along with an exhaust fan. Simultaneous use aeration, a splash box and an exhaust fan reduces the radon by over 90%. Although these are impressive reductions, it should be noted that merely letting the well sit in a storage tank for a day or less causes the waterborne radon to decrease by about 10% (see Tables 7 and 8).

Table 8. Measurements of radon in groundwater leaving the storage tank for WO -6, while aeration of the water in the tank is and is not in progress. Measurements were taken using water coming out of the well, and of water coming out of the storage tank.

Date	Water From Well (pCi/L)	Water From Storage Tank (pCi/L)	% Decrease
2/5	2540	N/A	
2/12	3510	N/A	
2/15	2010	N/A	
3/28	2610	N/A	
4/2	2640	2160	18
6/6	2430	N/A	
6/7	2150	1870	13
6/24	2460	2260	9
6/24	Aeration of Storage Tank Begins		
6/26	N/A	1060	
6/28	N/A	750	
6/29	Aeration of Storage Tank Ends		
7/10	3260	2950	10
7/25	2630	2340	9
11/19	2230	N/A	
12/16	2130	1810	15
12/16	Aeration of Storage Tank Begins		
12/17	2330	N/A	
12/19	1780	710	60
12/23	1590	500	69
12/26	Aeration of Storage Tank Ends		
3/23	2190	N/A	
4/22	2390	N/A	
5/27	2290	1970	14
6/24	2360	N/A	
8/11	2210	2110	5
9/23	2360	2190	7
9/23	Aeration, Splash Box & Exhaust Fan Begins		
9/24	2340	150	94

Date	Water From Well (pCi/L)	Water From Storage Tank (pCi/L)	% Decrease
9/25	2360	160	93
9/28	2450	180	93
9/30	2410	220	91
10/2	2140	150	93
10/2	Aeration, Ends, Splash Box & Exhaust Fan Continues		
10/5	2240	750	67
10/6	2390	620	74
10/7	2260	620	73
10/9	2370	730	69
10/13	2470	710	71
10/14	2380	780	67
10/14	Aeration, Splash Box & Exhaust Fan Continues		
10/20	2440	950	61
10/22	2190	850	61
10/26	2330	940	60
10/28	2390	940	61
11/1	Aeration, Splash Box & Exhaust Fan All End		
1/29	2190	N/A	

9. CONCLUSIONS

A study of waterborne radon in northern Virginia and southcentral Maryland showed that on average, the waterborne radon in homes supplied by wells exceeded the USEPA's MCL by a factor of 10. The study also showed that the extent to which the waterborne-radon concentration affected the concentration of indoor radon was determined by the volume of the home.

Measurements over a two-year study reveal that wells located in Piedmont granite usually contain the highest waterborne radon, and average about 3000 pCi/L. Wells in Piedmont schist and phyllite average about 2000 pCi/L, wells in Culpeper Basin sandstone average about 1500 pCi/l, and wells in Blue Ridge quartzite average about 1000 pCi/L.

About 10% of the waterborne radon in the well water is lost during the short time that the water sits in storage tanks, prior to being sent to surrounding homes. Additional time passes between when the water comes out of the ground and when it arrives in a surrounding home, and water at the home contains about 30% less radon than when it comes out of the ground.

Removal of radon at the well site was examined, using tanks of activated charcoal and using aeration system constructed in the water storage tanks. Tanks of activated charcoal removed over 90% of the waterborne radon, but the efficiency of removal decreased monthly. After one year, the removal declined to about 50%. Aeration using an air-bubble dispersal system, supplemented by a splash box and a storage tank exhaust fan, also reached 90% reduction in waterborne radon and the percentage removal did not decrease over time. Use of only an air-bubble dispersal system or the splash box reached about 70% removal. The exhaust fan on a storage tank is estimated to improve radon removal by about 10%.

Unfortunately for the general population, the perception of what represents an unreasonable risk depends on many factors, only one of which is generally agreed standards. At this stage of public awareness, airborne radon clearly garners more attention than waterborne radon. We suggest, pending the gathering of additional data of the type compiled in this report, that the risks of radon in both air and water be equally emphasized by public health officials.

REFERENCES

Brutsaert W F, Norton S W, Hess C T and Williams J S 1981 Geologic and hydrologic factors controlling radon-222 in ground water in Maine: Groundwater v. 19, p. 407-417.

Cothern C R 1987a Development of regulations for radionuclides in drinking water: Radon, Radium and Other Radioactivity in Ground Water ed. B. Graves (Lewis Publishers, Chelsea MI). p. 1-11.

Cothern C R 1987b Estimating the health risks of radon in drinking water: Journal of the American Water Works Association v. 79, p. 153 -158.

Gosink T A , Baskaran M and Holleman D F 1990 Radon in the human body from drinking water: Health Physics, v. 59, p. 919-924.

Henshaw D L, Eatough J P and Richardson R B 1990 Radon as a causative factor in induction of myleoid leukemia and other cancers: The Lancet, v. 335, p. 1008-1012.

Hess, C.T., Weiffenbach, C.V., and Norton, S.A., (1982). Variations of airborne and waterborne Rn-222 in houses in Maine, Environ. Int., 8, 59-66.

Hess C T, Weiffenbach C V and Norton S A 1982 Variations of airborne and waterborne Rn-222 in houses in Maine: Environment International, v. 8, p. 59-66.

Hess C T, Weiffenbach C V and Norton S A 1983 Environmental radon and cancer correlations in Maine: Health Physics, v. 45, p. 339-348.

Hess C T, Michael J, Horton T R, Prichard H D and Coniglio W A 1985 The occurrence of radioactivity in public water supplies in the United States: Health Physics, v. 48, p. 553-586.

Kahlos, H., and Asikainen, M., (1980). Internal radiation doses from radioactivity of drinking water in Finland, Health Phys., 39, 108 -111.

Kearfott, K. J., (1989). Preliminary experiences with ^{222}**Rn** gas in Arizona homes, Health Phys., 56,.169-179.

Lykken G I and Ong W S 1989 Evidence of exhalation of stored environmental Rn determined by whole-body counting: Health Physics, v. 57, p. 161-162.

McGregor, R.G., and Gourgon, L.A., (1980). Radon and radon daughters in homes utilizing deep wellwater supplies, Halifax County, Nova Scotia, J . Environ. Sci. and Health, 15, 25-35.

Moolenaar R J 1988 AIHC, an industrial contribution to public policy: Toxic Substances Journal, v. 8, p. 13-20.

Mose D G, Mushrush G W and Chrosniak C 1990 Radioactive hazard of potable water in Virginia and Maryland: Bulletin of Environmental Contamination and Toxicology, v. 44, p. 508-513.

Nelms D L and D L Richardson 1990 Geohydrology and the occurrence of volatile organic compounds in ground water, Culpeper Basin of Prince William County, Virginia: U.S. Geological Survey Water-Resources Investigations Report 90-4032, 94 p.

Pohl E and J. Pohl-Ruling 1967 The radiation dose received by inhalation of air containing Rn222, Rn220, Pb212 (ThB) and their decay products: Annals da. Academia Brasileira de Ciencias v. 39, p. 393-404.

Prichard, H.M., and Gesell, T.F., (1981). An estimate of population exposure due to radon in public water supplies in the area of Houston, Texas, Health Phys., 41, 599-606.

Prichard H M 1987 The transfer of radon from domestic water to indoor air: Journal of the American Water Works Association, v. 79, p. 159 -161.

Smith B M, Grune W N, Higgins F B Jr and Terrill J G Jr 1961 Natural radioactivity and ground water supplies in Maine and New Hampshire: Journal of the American Water Works Association, v. 53, p. 75 -88.

USEPA, 1987a, United States Environmental Protection Agency, Radon reduction methods, a homeowner's guide (second edition),U.S. Environmental Protection Agency Publication OPA-87-010, 21pp.

USEPA, 1987b, Removal of radon from household water:U.S. Environmental Protection Agency Publication OPA -87-011, 10 pp.

USEPA, 1987c, Radon reference manual: U.S. Environmental Protection Agency Publication 520/1-87-20, 140 pp.

PART VII: REGULATORY/LEGAL

CHAPTER 26

AN OVERVIEW OF STATE PROGRAMS TO INVESTIGATE AND REMEDIATE CONTAMINATED DRY CLEANER SITES

Steve Goins[1] and Dale Trippler[2]

Enviornmental Program Manager, Tennesee Department of Enviornment and Conservation, Nashville, TN[1]; Majors and Remediation Division, Minnesota Pollution Control Agency, St. Paul, MN[2]

Abstract: It is estimated that between 75 and 90 percent of dry cleaner sites are contaminated due to improper use and disposal of solvents. Since 1994, twelve states have passed legislation creating programs to investigate, remediate, and prevent contamination at these facilities. These programs are similar to insurance pools in that dry cleaners pay fees that go into a fund used to remediate contaminated sites.

In June 1998, the US Environmental Protection Agency, Technology Innovation Office brought the states together in the belief that technical issues and problems concerning solvent contamination at dry cleaner sites are not unique to one state and that the free exchange of information would benefit all. The result of that effort is the State Coalition for the Remediation of Dry Cleaners (SCRD), a forum for sharing technical and programmatic information regarding the remediation of dry cleaner sites. Member states include Alabama, Florida, Illinois, Kansas, Minnesota, Missouri, North Carolina, Oregon, South Carolina, Tennessee, Texas, and Wisconsin. In addition, participation in SCRD as "Represented States" is open to states without dry cleaner -specific programs; however, they are actively remediating dry cleaner sites under other authorities. Maryland, New York, and California currently participate as "Represented States".

Even though the programs are fairly new and most have very limited budgets, they have been effective in performing necessary remedial actions in a timely manner. State dry cleaning programs have performed at least 630 assessments, 197 response actions, and closed 83 dry cleaning sites. These numbers are increasing rapidly as each state program develops.

Key words: Dry Cleaners, Dry Cleaning, Solvent Contamination, Remediation, SCRD

1. MANUSCRIPT: COMPARISON OF STATE DRY CLEANER PROGRAMS

Currently there are twelve states that have dedicated dry cleaner remediation programs. Although each state program is unique, most consist of five primary components. The five components are funding mechanisms, eligibility, prioritization, implementation, and benefits.

1.1 Funding Mechanisms

There are a number of different ways states pay for their dry cleaner remediation programs. There is no one structure that has been found to be the best. Each has positive and negative aspects to be considered. Some states assess annual fees based on full-time employees at active facilities, some assess a flat annual operating or registration fee, and others base their fees on gross receipts from dry cleaning operations. Almost all states have established some type of fee based on solvent usage. Typically, perchloroethylene (PCE/Perc) is assessed a higher fee because of the risks posed, the costs for remediation, and the greater likelihood that contamination will or has occurred. Typically, the seller of the solvent is responsible for collecting and paying the fees to the state. In addition, some states require the facility to report the amount of solvent used. Most states include some type of deductible payment when claims are made or cleanups conducted. These deductibles can be based on a sliding scale based on the cost of the cleanup, a sliding scale based on FTEs of the facility, or can be a fixed, deductible amount.

1.2 Eligibility

Most states have very specific language, either in program rules or in the enabling statute, which defines the types of facilities or parties eligible for remediation funding. The keys to eligibility lie in the definitions of facility, solvent, eligible parties, etc. Most states include some combination of the following in the definition of facility:

- Active Dry Cleaning Facilities
- Inactive Dry Cleaning Facilities
- Coin-operated Dry Cleaning Facilities
- Active Wholesale Supply Facilities
- Inactive Wholesale Supply Facilities
- Adjacent/Contiguous Properties
- Extent of contamination

In addition, most states specifically exclude certain types of facilities from being eligible to participate in their programs for remediation funding. These exceptions may include the following:

- Linen & Uniform Supply Facilities
- Coin-operated dry cleaners
- Federal/Military/State Property
- NPL sites
- Sites under administrative orders
- Wholesale Solvent Suppliers
- RCRA permitted facilities

Many states require applicants to comply with certain pollution prevention or best management practices to receive and maintain eligibility. Compliance with these requirements should limit or prevent future releases of dry cleaning solvent to the environment. These requirements include such things as direct-coupled delivery of PCE and diking or other secondary containment devices.

1.3 Prioritization

There are basically two types of state dry cleaner remediation programs; those that provide reimbursement to parties who conduct cleanups at dry cleaner facilities, and those in which the state agency directs the cleanup at dry cleaning facilities. Both of these types of programs use priority systems for either determining the order in which cleanups occur and/or the order in which reimbursements occur. For some states, the priority rank assigned to a site relates to the order in which reimbursement occurs. Portions of the funds are allocated to high, medium and low risk sites, assuming that there will be more money requested than available in the fund. Some reimbursement programs wait until the completion of the site investigation to prioritize the sites. As site reimbursement applications are received, the sites are categorized and funds reimbursed based on the available funds in those categories.

States that direct the remediation use a scoring system to determine which sites they will address first. Regardless of the purpose of the prioritization system, most systems have common elements relating to the system itself. Most systems include looking at possible receptors, the media in which contamination occurs, the geologic conditions, the toxicity and concentration of the contaminants at a particular site/facility, and the likelihood of catastrophic consequences. In addition, some scoring systems look at historical factors including the operating history of the facility. Some state systems also look at cost effectiveness in determining the priority for a site/facility.

1.4 Implementation

Probably the biggest difference in state dry cleaner programs is in the actual implementation or operation of the programs. Some states have reimbursement programs where they may provide oversight of the activities conducted by contractors on behalf of the applicant. Other states have state lead programs where the state contracts directly with environmental consulting firms and direct the contractor's activities. In addition, some state programs report to independent boards or councils. One state has outsourced its program to a private firm.

1.5 Benefits

Participation in a state dry cleaner program has numerous benefits. In general terms, the beneficiaries of the program include active dry cleaning facility owner/operators, inactive dry cleaning facility owner/operators, real property owners, nearby real property owners, the public and the environment. Some states provide liability protection to the person/facility participating in the state program. The protection does not include federal liability and may not include third party or personal injury liability. State programs provide funding for cleanups through industry fees, which limits state costs. Property transfers including refinancing and site redevelopment can occur without lending institutions preventing transactions from occurring due to real or perceived liability issues.

Without state dry cleaner programs, most dry cleaners would not be able to address contamination with or without enforcement action by other regulatory agencies. Most dry cleaners would not have the financial resources necessary to remediate the environmental contamination found at a dry cleaner site. Also, most dry cleaning sites are relatively small and would generally not be a high priority for typical remediation programs. Therefore these sites would not be addressed and the plumes would continue to migrate. Most state programs are not just reactive to past contamination but have instituted pollution prevention/best management practices to reduce or eliminate future releases of solvent to the environment.

2. STATE COALITION FOR THE REMEDIATION OF DRY CLEANERS

The State Coalition for Remediation of Dry Cleaners (SCRD) was established in 1998 with support from the U.S. EPA Office of Superfund Remediation and Technology Innovation and the National Ground Water Association (NGWA). It is comprised of representatives of states with established dry cleaner remediation programs. Currently the member states are Alabama, Florida, Illinois, Kansas, Minnesota, Missouri, North Carolina, Oregon, South Carolina, Tennessee, Texas, and Wisconsin. The Coalition's primary objectives are to provide a forum for the exchange of information and the discussion of implementation issues related to established state dry cleaner programs; to share information and lessons learned with states without dry cleaner-specific programs; and to encourage the use of innovative technologies in dry cleaner remediation. SCRD has developed into a strong network of states with a vast amount of experience, training, and expertise on remediating contaminated dry cleaner sites across the country. The Coalition meets twice a year to exchange information, discuss new and innovative technologies, talk about what works and what does not, and share experiences of the respective programs. Sharing this vast knowledge is extremely valuable. In addition to the semi-annual meetings, the Coalition conducts conference calls approximately every six weeks. Each state has benefited from membership in SCRD. The day-to-day work of the Coalition is carried out by subgroups addressing three major areas: Program Development/Administration, Project Management/Technical Issues, and Outreach.

SCRD has a website with a storehouse of information available, including the names and contacts for each member state and represented state. There are also links to the website of each member state. The State Coalition for the Remediation of Dry Cleaner's website is located at http://www.drycleancoalition.org/. The SCRD website includes the following:

2.1 Dry Cleaner Site Profiles

The site profiles include site-specific details about the remediation of dry cleaner sites throughout the United States. The intent is to provide users, particularly state officials, with information that can help them in making more informed decisions related to the remediation of sites in their states. The profiles represent the first step in what is hoped will become a comprehensive reference. The data in each profile includes the site name, location, description, hydrology, ground water contaminate, remediation technology used, results achieved, costs, lessons learned, and a point-of-contact for further information. A search capability is included: users may enter key words and search all profiles simultaneously.

2.2 Conducting Contamination Assessment Work at Dry Cleaning Sites

This report, prepared by SCRD members, provides helpful information for States and others involved in conducting site characterization work at dry cleaning facilities. There are approximately 36,000 active dry cleaning facilities in the United States and an estimated 75 percent of these are probably contaminated. This includes commercial, industrial and coin-operated facilities where soil and groundwater have been contaminated by dry cleaning solvents. In addition, there are an unknown number of former dry cleaning sites that also are contaminated. Since dry cleaning facilities are located in urban areas, dry cleaning solvent contamination has impacted a large number of public water supply wells and threatens many other well fields. This report includes background information on dry cleaning operations, planning and conducting site assessment/characterization, and a range of assessment technologies available. It also includes examples of the assessment/characterization approaches used by several SCRD states.

2.3 Obstacles to Complete PCE Degradation During Reductive Dechlorination

This paper describes the accumulation of cis-1,2-DCE, which appears relatively common at sites undergoing reductive dechlorination enhanced by the addition of hydrogen releasing compound (HRC®), molasses, sodium lactate, vegetable oil, or some other organic carbon source. The paper was presented at the Battelle conference in the spring of 2002 and is based on data from a dry cleaner site in Florida. It provides lessons that can be used to develop more rigorous site qualification procedures and design criteria for the application of accelerated reductive dechlorination at other sites.

2.4 Reported Leaks, Spills and Discharges At Florida Dry Cleaning Sites

This document presents data collected by the Florida Dry Cleaning Solvent Cleanup Program on reported spills, leaks and discharges of dry cleaning solvent and solvent-contaminated wastes at 334 dry cleaning facilities and 14 dry cleaning wholesale supply facilities located within the State. This information provides those involved in soil and groundwater assessment and cleanup with a better understanding of dry cleaning equipment, operations and waste management practices, particularly former industry practices. This information can facilitate identification of the contaminant source areas.

2.5 Chemical Used in Dry Cleaning Operations

Information on "Chemicals Used in Dry Cleaning Operations" is available as a viewable/ downloadable document and in a searchable, on-line database. These resources are intended to help those engaged in the assessment and remediation of contaminated dry cleaner sites and to assist regulators conducting compliance inspections at dry cleaning facilities. The document provides current and historical information on the types of chemicals, solvents, pre-cleaning/spotting agents, garment treatments, and solvent and equipment maintenance materials used in dry cleaning operations. The database provides information on dry cleaning products/trade names (as listed on Material Safety Data Sheets), who manufactures or distributes them, what the products are used for in dry cleaning operations, what chemical ingredients each product contains, the CAS number for each chemical ingredient, any alternate name by which these chemicals are known, and any additional information available.

The Coalition is currently working on a document titled Components of State Dry Cleaner Programs. The document will explore the various approaches or options states have incorporated into their cleanup programs. The document can be used as a resource for states in the development stage or for states that are considering restructuring their programs.

3. TENNESSEE DRY CLEANER ENVIRONMENTAL RESPONSE PROGRAM

On June 13, 1995, the Tennessee General Assembly enacted a law establishing the Dry Cleaner Environmental Response Program (DCERP). The Act established a "fund" made up of annual registration fees collected from the owners/operators of dry cleaning facilities and surcharges on dry cleaning solvent. The fees and surcharges are deposited into the Fund and become the revenue stream that provides for reimbursement of environmental response activity costs and all the personnel and administrative expenses of the program. Annual registration of active dry cleaning facilities is mandatory while participation in the environmental response program is voluntary. The purpose of the program is to provide funding and oversight for the investigation and, where necessary, remediation of impacted sites (contaminated with solvents from dry cleaning operations). Facilities eligible to receive reimbursement of environmental response costs include active or abandoned dry cleaning facilities and in-state wholesale distribution facilities. In addition, impacted third parties can have access to the fund for reimbursement of environmental response costs. The statute requires environmental response actions and associated methods and techniques to be "cost effective, reasonable, and technically feasible." The statute (TCA 68-217-111) provides liability protection to eligible parties for any release under any other law or for third party claims, if the site is fund eligible and is in material compliance with the program requirements. The liability protection does not apply to personal injury claims or liability under federal laws.

The Fund will provide reimbursement for the reasonable cost of Department-authorized inspection, investigation, and remediation activities. The applicant is responsible for accepting responsibility for program deductibles up to an established maximum amount. The program can authorize reimbursement of up to $200,000 per year for releases from any individual facility. In order for response costs to be reimbursable, a Dry Cleaner Approved Contractor (DCAC) must perform the work and the work plan and associated cost proposal must have received written pre-approval from the DCERP. Program rules identify the requirements a firm must meet in order to be placed on the DCAC list. The DCERP evaluates contractor qualifications and publishes a list of approved contractors. DCACs enter into a contractual relationship with the applicant and not the DCERP or the Department. DCACs can apply for one or more of the following contractor categories: Facility Inspection, Site Investigation, and Remediation.

Environmental response activities are conducted in phases. The first phase is a Facility Inspection (FI) that includes a determination of Fund eligibility through records review and an on-site inspection to determine compliance with BMPs. The BMPs are intended to prevent or eliminate the possibility of future releases of solvent to the environment. The program provides facilities an "opportunity to cure" BMP deficiencies which impact Fund eligibility. The second phase is a Prioritization Investigation (PI) that determines the presence or absence of solvent impacts, establishes a priority ranking for the site, and determines whether additional investigation is needed. The next phase is a Solvent Impact Assessment (SIA) that determines the extent and nature of the contamination. After the SIA, a Remedial Alternatives Study (RAS) evaluates the technical feasibility, costs, and effectiveness of implementing various remedial approaches. The next phase includes the implementation of the selected remedial action. A completion letter will be issued to the applicant after all required activities are completed at the site.

Currently, 52 facilities/sites are in some phase of environmental site investigation or remediation. The facilities/sites consist of both active and inactive dry cleaning facilities and one solvent supplier. Eight facilities/sites have received completion letters that no further environmental activity is necessary. The DCERP has reimbursed owners/operators in excess of $4.1 million for site investigation/cleanup.

4. MINNESOTA DRY CLEANER PROGRAM

The "Dry Cleaner Environmental Response and Reimbursement Law" (MN Stat. §115B.47 to 115B.51) (Dry Cleaner Fund) was passed into law in 1995. The law was proposed by the dry cleaner industry through its professional organization, the Minnesota Cleaners Association and became law because of their efforts. Since that time, the responsibility to administer the Dry Cleaner Fund and oversee remediation at dry cleaner sites has been an on-going task of the Minnesota Pollution Control Agency (MPCA). The Minnesota Department of Revenue has the responsibility to collect the annual registration and product fees generated by this law.

With only a few exceptions, dry cleaning facilities that provided services to the general public are eligible to apply for reimbursement from the Dry Cleaner Fund. An owner or operator of a dry cleaning facility need not be a Minnesota business to use the Dry Cleaner Fund, but the site must be located in Minnesota. The Dry Cleaner Fund is intended to benefit past or present members of the dry cleaning industry. For this reason, economic development agencies, lenders, and municipalities who have never been in the dry cleaning business are not eligible for reimbursement using the Dry Cleaner Fund.

Currently there are 288 dry cleaners in Minnesota who are paying annual fees and product fees into the Dry Cleaner Fund. Annual registration fees are based on the number of employees working at a facility. For facilities with full-time equivalence (FTE) of less than five, the annual registration fee is $1,000; for five to ten FTEs, the annual registration fee is $2,500; and for more than ten FTEs, the annual registration fee is $4,000. The current product fees collected by chemical wholesalers from dry cleaners for tetrachloroethene (Perc) are $7.50 per gallon, for hydrocarbon-based dry cleaning solvents the fee is $1.40 per gallon, and for all other non-aqueous solvents the fee is $.0.35 per gallon. The Dry Cleaner Fund generates about

$650,000 annually, with about $400,000 paid out annually in remediation and administrative costs.

The MPCA must determine that a cleanup is necessary and the contamination must have resulted from current or former dry cleaning operations. Other criteria are that the owner or operator is responsible for the first $10,000 in costs, the owner or operator had to have complied with all state and federal laws and regulations at the time of the release, and the dry cleaning facility must not be one of the six specifically excluded facilities listed in the Dry Cleaner Law.

All reasonable costs associated with the investigation and cleanup of a dry cleaner site release are eligible once the MPCA staff has determined a cleanup is necessary. However, the following costs are not eligible for reimbursement: any legal fees or costs; interest on loans used to pay for investigations or cleanup; any costs determined by the MPCA staff to be beyond the scope of an adequate investigation and/or cleanup effort; costs associated with the cleanup of contamination which are not associated with the operation of the dry cleaner; and any costs for remediation which are associated with illegal actions at the time of the release.

The dry cleaner program has been in existence since July 1, 1995. Since that time, 16 sites have been addressed or are being worked on. Eight of those sites have been cleaned up and returned to productive use. The Dry Cleaner Fund has reimbursed dry cleaners over $3 million. Of that amount, one site in Southeastern Minnesota has accounted for over $1 million. Working closely with the dry cleaners and their consultants, the MPCA staff has been able to keep the costs to a minimum. The dry cleaning industry has been extremely supportive of the efforts to find inexpensive but effective remediation techniques. Also, Minnesota's involvement in SCRD has provided MPCA staff with a valuable resource and knowledgeable contacts, allowing staff to stay abreast of the most current technologies and remediation options

CHAPTER 27

PCBS IN THE DELAWARE: A THIRTY-ONE-YEAR TECHNICAL AND LEGAL ODYSSEY

Itzchak E. Kornfeld, Esquire
Oayes Environmental Solutions, Philadelphia, PA

Abstract: The Metal Bank Superfund Site, located in northeast Philadelphia, has been the subject of study and litigation for the past 31 years. The odyssey began in 1972 when at least 21,000 gallons of PCB-rich transformer fluid leaked from a UST, spilling into the Delaware River. The Delaware River is tidal in the vicinity of the Site.

A shallow embayment or mudflat forms the western portion of the Site. It consists of fine silts and clays, with some occasional gravel in the subsurface. The Property itself is composed of urban fill. Groundwater underneath the Site flows into the Delaware River . Depth to groundwater varies from 7'-16' feet.

Following the 1972 spill PCB contaminated oil was measured in mudflats soils at levels exceeding 1500 ppm. In 1977 PCB concentrations measured at 1,539 ppm. In 1989 at the termination of a six-year pump and treat operation, the concentrations measured 1540, the same as prior to the remediation. Surface soils at the Property also contained PCBs at concentrations up to 42 ppm; total SVOCs at concentrations up to 2,008 ppm; and lead concentrations up to 27,000 ppm (or 22.7% of the sample). SVOCs and PAHs had high positive correlations with PCBs.

Key words: PCBs, Superfund, regulation, court action, PAHs, SVOCs.

1. INTRODUCTION

The distribution of Polychlorinated Biphenyls ("PCBs") is informed by scientific inquiry. However, the effects and risks of PCB distribution are guided, indeed, underpinned, by legal principles. Thus, since 1976 when Congress essentially banned the use of PCBs, by passing section 6(e) of the Toxic Substances Control Act's ("TSCA"), PCBs have come under intense regulatory scrutiny: specifically in the areas of risk analysis, mass loading and provenance research. From a regulatory perspective much of the research into these chemicals has come from Superfund sites.

It is from the latter perspective that we embark on the present odyssey. The main character in this drama is what today is known as The Metal Bank Superfund site ("the Site" or "Metal Bank" a/k/a "the Cottman Avenue Property"). (USEPA 2003).

PCBs are a group of synthetic organic chemicals which were widely used in the United States because of their chemical stability and low reactivity. Each PCB molecule consists of a chlorinated biphenyl - two hexagonal rings of carbon atoms connected by a carbon-carbon bond - containing from one to ten chlorine atoms attached in various locations. There are 209 different types - or "congeners" -- of PCBs, based upon the number of chlorine atoms and their positions on the carbon rings. (Court Opinion2003) (For a more elaborate discussion of PCBs, *see* reference 3.)

This article discusses the thirty-one-year history of the Metal Bank Site: beginning with an oil spill into the Delaware River from Metal Bank's operations, in 1972, through "Phase One" - the liability phase of a CERCLA and RCRA case – which was tried the over the course of seven weeks, beginning on August 19, 2002. Part I describes the Site. Part II provides an overview of the Site's history. In part III I discuss the legal history of this case. Part IV addresses the nature of the contamination, which is followed by Part V, a discussion of the Defendants' remediation efforts. In Part VI I discuss some of the trial issues and the expert opinions.

2. THE METAL BANK SITE

The Metal Bank Site is located on the shores of the Delaware River in Philadelphia, Pennsylvania. It measures 11.6 acres in size and is part of an industrial area that was established on the city's waterfront in the post-Civil-War era. The property is situated at the foot of Cottman Avenue in the northeastern portion of the city. It is bordered by Milnor Street and Interstate 95 to the north and the Delaware River to the south. (U.S. EPA

Region III Jan. 2003) (*See* Figures 1, 2 and 3). The Site is divided into three areas.

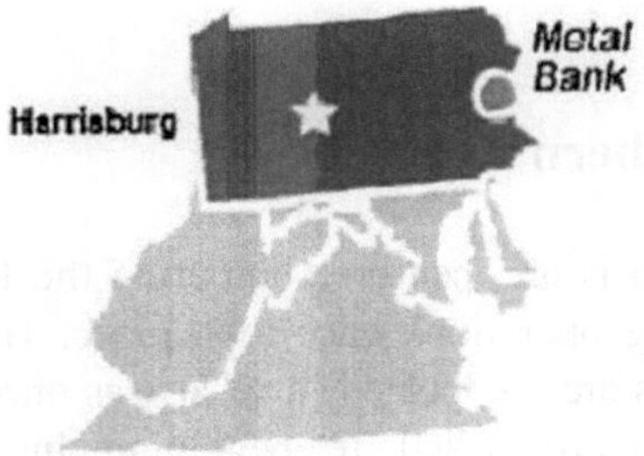

Figure 1. General Location of the Metal bank Site

Figure 2. Location of the Metal bank Site within the boundaries of Philadelphia, PA

2.1 The Courtyard Area

The Courtyard Area consists of an open area at the northern section of the property near Milnor Street. The buildings near the Courtyard Area were demolished in 1998.

2.2 The Southern Area

The Southern Area is an open area bordering the Delaware River where transformer-processing operations once took place. Historical aerial photos show that most of this area is located in what was once part of the Delaware River and was gradually filled in beginning in approximately 1950. Heterogeneous urban-fill material, most of which was placed before 1968, is about 15 feet thick. Its origin is unknown, but it contains construction debris, including chunks of concrete, brick, lumber, cloth and metal. (Court Opinion, 2003).

The Southern Area sits approximately 10 feet above Mean Sea Level, and most of it is located within the 100-year old floodplain. The outer edge of the Southern Area is steeply sloped, with large concrete blocks of material apparently placed for erosion control along approximately 550 feet of shoreline of the present-day Delaware River. This is the area known as the "upper rip rap." There is also a "lower rip rap" area sometimes referred to as the "beach." It consists of small heterogeneous fill material placed there as part of a clean-up response ordered by the United States Coast Guard following an "oil spill" in that area that occurred in 1972. (USEPA Record of Decision ("ROD"), December 1997).

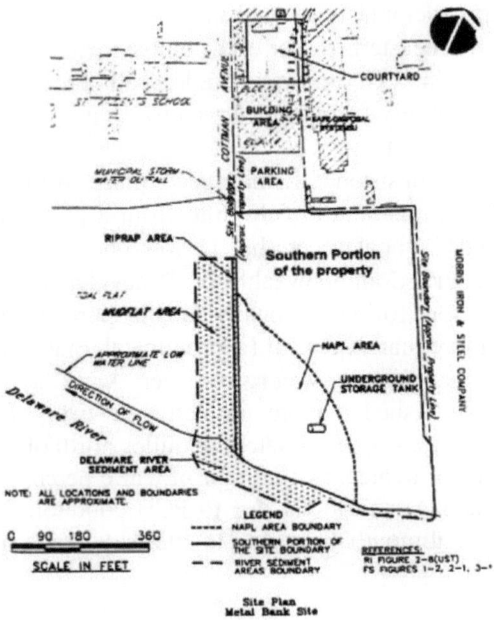

Figure 3. Plan View of the Metal bank Site, showing each of the Site's respective areas of concern

2.3 The Mud Flat and River Sediments Area

The River Sediments Area, located adjacent to the southern and western boundaries of the Property, includes both mudflat and river bottom. The Delaware River is tidal in the vicinity of the Site, with six to seven foot tides that reach maximum and minimum water levels every twelve hours. To the immediate west of the Property, the River forms a shallow embayment, which is completely submerged at high tide and which forms an exposed mudflat of five to seven acres in size at low tide. The mudflat consists primarily of fine silts and clays, with some occasional gravel in the subsurface, with the amount of gravel increasing closer to the border with the Metal Bank Site. The river bottom is composed of gravelly and sandy material, and it slopes gradually away from the Property. (Court Opinion, 2003).

The embayment is bordered to the north by St. Vincent's School - a former orphanage currently serving as a day-care center and an emergency shelter for at-risk children from the City of Philadelphia -- and to the west by the Quaker City Yacht Club, which serves as a boat launch for recreational boaters. A municipal combined stormwater/sewer outfall ("CSSO") owned by the City of Philadelphia is located at the foot of Cottman Avenue and empties into the northeastern corner of the embayment during and following periods of heavy rain. Groundwater underneath the Site flows into the Delaware River, generally from north to south. Depth of groundwater varies from seven to sixteen feet.

Recreational and subsistence fishing takes place in the Delaware River, approximately 200 yards south of the Site from a public access ramp that was formerly used for boat access but is now open only for fishing. On April 11, 2001, the Commonwealth of Pennsylvania issued a general statewide fish advisory for recreationally caught sport fish which advises the public to eat no more than one meal (approximately one half pound) of sport fish caught in the state's waterways per week. In addition, the Commonwealth has issued a more protective advisory for the Delaware River south of Yardley – approximately 20 miles north of the Site - advising the public to limit or avoid consumption of white perch, striped bass, carp, channel catfish, and American eel due to PCB contamination and to limit consumption of smallmouth bass due to mercury contamination. (Court Opinion, 2003).

3. SITE HISTORY

From 1968 to 1972, Metal Bank of America, Inc. ("MBA") recycled electrical transformers[6] purchased from utilities along the eastern seaboard for the reclamation of their copper cores and the recycling of their iron cases. As part of its recycling process Metal Bank drained oil from the used transformers on a concrete pad. The pad was angled towards the center and emptied into a 6,000 gallon underground storage tank ("UST"). The oil was periodically pumped out of the UST by a private contractor and was hauled from the Site and sold for industrial fuel purposes.

The operation called for dismantling the used electrical transformers, emptying them of their oily liquid, and pulling their copper cores. The reclamation operations were very sloppy. Moreover, the Site operators did not regard the oil as posing any health risks and were not careful about the

[6]An electrical transformer generally has an iron or steel casin g, within which a core of wound copper wire is immersed in a dielectric fluid or oil that is used for insulation and cooling purposes. Dielectric fluids used included mineral oil, as well as fluids consisting primarily of polychlorinated biphenyls ("PCBs "). From 1968 or 1969 until 1973, the scrap electrical transformers processed by Metal Bank frequently contained PCBs, polycyclic aromatic hydrocarbons ("PAHs") and other semi-volatile organic compounds ("SVOCs"), volatile organic compounds ("VOCs "), metals and other hazardous substances.

transformer draining and storage procedures. (Court Opinion). Oil dripped or spilled from leaking transformers as they were first unloaded at the Property. "Clam buckets" used to unload the transformers from the trucks, several at a time, sometimes ruptured transformers filled with oil that splashed onto the ground. The recycling operations led to oil releases in various locations on the property with the majority of the contamination in the vicinity of an underground storage tank which was used by Metal Bank to store the used transformer oil.

On the morning of August 3, 1972, the U. S. Coast Guard ("USCG") and inspectors from the Pennsylvania Department of Environmental Regulation ("PADER") responded to an oil spill in the Delaware River, approximately one-half mile to the west of the Metal Bank Site, in the vicinity of the Quaker City Yacht Club and traced its origin to the Cottman Avenue Property. (USEPA 1997). Numerous dead fish were reported. An inspection four days later revealed that the entire area near the bank of the Delaware River was saturated with oil, and areas were visible where oil had seeped through and poured over the bank of the Site into the river. (Court Opinion).

A Metal Bank employee observed oil floating in the river near the tank. He noticed that the level of oil in the tank would change at times when oil had neither been added nor pumped out. For example, if employees ceased work on Friday and capped the tank, they would return on Monday to find the tank overflowing if it had rained over the weekend.

When they then lowered a pump hose deep into the tank, they would pump water out of the bottom. The levels in the tank also rose when the tide was in, even if the tank was capped. (Court Opinion). The volume of oil observed entering the river at the Property indicated to the United States Coast Guard ("USCG") and to Pennsylvania Environmental inspectors that oil was flowing from the Site into the river, principally as a result of a leaking UST.

Following the Coast Guard's identification of the UST as the source of the oil spill into the Delaware River in August of 1972, Metal Bank had the tank inspected. There is no documentation of a leak being found or repaired. However, Metal Bank admitted in subsequent litigation with its insurance company, that the tank was ruptured and leaking if it had not been convinced by its own investigation that this was so. The Court reasoned that MBA would not have made such an admission if it was not true.

MBA maintained that the spill was composed of bunker "C" oil, and that its source was either from an upstream facility or from an oil barge. (USEAPA 1997 at p. 8). However, in Phase I of the litigation the Court found that "[b]y inference [and] by a preponderance of the evidence ... the tank did leak and was repaired during the tank inspection process." (Court Opinion at p. 7).

Subsequently, the USCG ordered Metal bank to perform numerous remedial measures, including laying booms along the river, cleaning up the spill and improving its housekeeping. Metal Bank also advised the USCG that it discontinued its transformer salvage operations. (USEPA 1997 at p. 8).

Following the oil spill the USCG collected samples from the Site, including samples of the soil and of the oil spill. The Coast Guard tested the oil using the best available technology of the time. Initially, it did not detect PCBs in the oil samples. In late 1972 and 1973, in response to recommendations made by the USCG, Metal Bank took limited actions to clean up its property. (Trial Court Opinion). It performed some surficial clean up of the southern portion of the property where the concrete pad was located, placed "booms" out to contain and collect oil in the river and along the shoreline, installed cylindrical caissons to capture the oil as requested by the USCG, and covered the ground with clean soil. However, Metal Bank did not undertake any efforts at that time to clean up subsurface contamination or to prevent subsurface oil from migrating into the river.

In the mid-1970s, in response to a USCG request that Metal Bank armor the shore, MBA installed the lower rip rap, consisting of smaller pieces of rubble, broken bricks and concrete. The lower rip rap was designed to stabilize the lower portion of the upper rip rap and to act as a buffer both for oil spills coming from the land toward the water and for oil that might come from the River toward the land. Thus, the lower rip rap likely was placed on top of oil that had not been recovered by the methodologies utilized by Metal Bank.

In September 1977, in response to continuing concerns, the USCG and other government agencies re-analyzed the 1972 samples using more recent and innovative analytical technology and detected the presence of PCBs in concentrations over 800 parts per million ("ppm"). Analyses of soils and liquids samples at the Site detected the presence of PCBs at levels up to 1579 ppm. Based on these findings, the USCG and EPA hired Roy F. Weston, Inc. ("Weston"), to define more fully the nature and extent of PCB contamination at the Site. Weston conducted an investigation and documented its findings in two reports dated October 1978 and March 1980. The 1978 Report showed that as many as 21,000 gallons of PCB-contaminated oil had pooled in the subsurface of the Metal Bank Site. The report concluded that this oil was releasing PCBs to the underlying groundwater and that PCBs from the Property were contaminating the Delaware River through oil and groundwater discharges. The 1980 report proposed a remediation plan.

At Metal Bank's request, Energy and Environmental Analysis, Inc. ("EEA") also investigated conditions at the Property. In its May 8, 1979 report, EEA estimated that the extent of the oil spill from the ruptured UST affected an area of approximately 75,000 square feet and contained about 11,700 to 46,750 gallons of oil, and involved 115 to 460 pounds of PCBs. The EEA report also estimated that groundwater transporting oil to the Delaware River moved at the rate of 17,053 gallons per day.

"Although these estimates assume that all of the oil came from Metal Bank's operations, even if a substantial portion came from upgradient sources as [the Metal Bank] defendants contend, the defendants' spills became commingled with all of the other oil, making any responsible defendant responsible for the entire clean up [under CERCLA's statutory scheme]." (Court Opinion at p. 8).

4. LEGAL HISTORY

On April 23, 1980, the United States, on behalf of the EPA, filed a civil suit against The Union Corporation, a New York Stock Exchange listed company, and two of the property's former owners, Irvin G. Schorsch, Jr. and his brother John B. Schorsch (collectively hereinafter "Metal Bank"). The Government sued under two statutes. The Resource Conservation and Recovery Act ("RCRA"), the Toxic Substances control Act ("TSCA"), which regulates PCBs, and also sought injunctive relief as well as costs of suit.

The case was assigned to Judge James T. Giles, who has presided over this case ever since. And in the interim has assumed the position of Chief Judge of the Federal District Court for the Eastern District of Pennsylvania.

In 1983, the EPA entered into a consent decree or Stipulation in which Metal Bank agreed to attempt to remediate the Site's contamination by constructing an oil recovery system to pump out and remove the oil.

On January 1989, pursuant to the Stipulation, Metal Bank petitioned the court to stop the oil recovery operation because it believed *that all* the oil had been recovered. In March of 1989 EPA collected samples from the monitoring wells on the Site. That sampling effort apparently showed that PCB-contaminated oil was still floating on the aquifer. (UEPA 1997 at p. 9). However, the Government did not agree that the Site no longer posed a Substantial hazard to human health or the environment. EPA monitoring in 1989 showed that despite eight years of groundwater pump and treat operations at the Site, a layer of PCB-contaminated oil at least three inches thick was still floating on the groundwater at some portions of the Site. (Court Opinion at p. 9). PCB concentrations measured in the oil layer were

1,539 ppm in 1977 prior to the oil recovery operation and almost the same, 1,540 ppm, in 1989 when the oil recovery operation was being terminated. On May 1991, EPA signed an Administrative Order by Consent with ten utilities, which had sent transformers to the Site and organized themselves as the Cottman Avenue PRP Group ("PRP Group"). Pursuant to the Administrative Order, the PRP Group conducted a Remedial Investigation/Feasibility Study ("RI") to define the nature, extent, and sources of contamination at the Site and to estimate the health and environmental risks associated with the contaminants at the Site. The RI Final Report, dated October 14, 1994, documented widespread contamination by PCBs, TPH, PAHs and other hazardous substances at the Site.

On June 12, 1989, the United States sought the Court's intervention in preventing a permanent shutdown of the recovery system. The Court denied that motion as moot, especially since the Government agreed with Metal Bank on a proposed order which extended the time in which EPA could conduct a final sampling effort until August 15. Those samples yielded an oil layer some 3 inches thick sitting on the groundwater in some portions of the Site. Curiously, the Government failed to pursue its displeasure with that ruling through the Court. Moreover, EPA ignored its remedies under the administrative process, which by 1989 was in full force, *e.g.*, EPA was initiating negotiations with the PRP group for performance of the RI.

Consequently, Metal Bank, believing that it had fulfilled the terms of the Stipulation and had settled the case, dismantled the oil recovery system, and under the Stipulation's terms negotiated with EPA to cover the Southern Area with fill and vegetated it. Additionally, Metal Bank repaired the fences around the Site and agreed to maintain them. Finally, the case remained on the suspense list until May 1998 when, after 28 years, the matter "was restored to the court's active trial docket in 1998, upon the Government's claim that remediation had failed or had not addressed all contamination concerns." (Court Opinion at 3).

In 1999, following reactivation of the case by its placement on the active trial list, the Government amended its complaint as follows: it dropped the TSCA claim; added a CERCLA count under that statute's §§ 107 and 113(b) to recover its cleanup costs, and under section 7003 of RCRA, 42 U.S.C. § 7003, which alleged that the Site posed an imminent and substantial endangerment to human health and the environment, and requested injunctive relief.

The case began to move. "Following extensive discovery and pre-trial proceedings, trial was phased as follows: Phase One would determine whether [the] defendants were liable and whether response costs were incurred by the Government; Phase Two would determine whether the

Government's response costs, if any, were reasonable and recoverable, as well as the scope of any further remedial action; and Phase Three would determine the liability of the third party defendants [including the utilities that contracted with MBA to purchase the transformers, capacitors, and other scrap metal and the City of Philadelphia]. Trial of Phase One commenced August 19, 2002" and lasted six weeks. On January 21, 2003, in an 84-page opinion Chief Judge James T. Giles ruled on Phase One of the trial that the former and current site owners — Union Corporation, Metal Bank of America Inc., and former Metal Bank owners and officers Irvin G. Schorsch, Jr. and John Schorsch — are liable for EPA's costs related to the cleanup of the Site. He also rejected each and every one of the Defendants' theories and arguments.

4.1 The Court Action

Notwithstanding, on December 30, 1982, EPA initiated an administrative action under its Superfund authority, proposing for listing on the National Priorities List ("NPL") and a Remedial Project Manager was assigned. The Site was formally added to the list on September 8, 1983, with a Hazard Ranking Score ("HRS") of 33.23. (Federal Register 1983).

To the author's best information this is the first Government CERCLA case where two parallel tracks, an administrative action and a lawsuit, moved forward at the same time.

Under the administrative powers in 1987 EPA identified another 10 potentially responsible parties (PRPs) - in addition to the four defendants - all of which were utility companies who supplied transformers to Metal Bank. Under the leadership of one PRP company, the PRPs formed a steering committee, which they named the "Cottman Avenue PRP Group", and signed an agreement with EPA in 1991 on how to investigate the site. Metal Bank declined to join this group. (EPA 2003). Thereafter, beginning in 1991, the PRP group conducted a Remedial Investigation and a Feasibility Study.

5. NATURE OF THE CONTAMINATION

The Metal Bank Site is contaminated with PCBs, PAHs, SVOCs, VOCs, petroleum hydrocarbons ("TPH"), metals, dioxins, furans, pesticides and other hazardous substances. The contamination extends across the surface and subsurface soils on the Site, in Delaware River sediment and in an adjacent embayment, known as the mud flat. Additionally, PCB-laden oil has been identified in the Site's groundwater, which may be seeping into the

Delaware River as a result of tidal wave movements underneath the Site. The levels and scope of contamination have been documented in the Remedial Investigation and Feasibility Study ("RI/FS"), the Record of Decision ("ROD"), the pre-design investigation ("PDI"), a Trench Study performed by the Metal Bank defendants in 2000, and mudflat data collected during 2002 by the Government's experts. (Court Opinion and ROD).

Furthermore, based on a risk analysis of these contaminants, EPA also found that recreational fishermen may be at risk from eating contaminated fish; future construction workers may be at risk from the PCBs in the oil beneath the site; as well as impacts on fish, bivalves and animals feeding in the mud flats.

In 1991, during the RI, a sample was collected from the southernmost – closest to the Delaware River – monitoring well, MW-6. Although EPA characterized the sample as located in the oil layer which was floating on the groundwater, that characterization may be incorrect. Nevertheless, the oil sample had a PCB concentration of 1,090 ppm. (ROD at p. 17). This sample, however, could not be replicated in the following years. In fact, EPA admits in the ROD that "[d]uring the 1992 sampling of MW6, the sampling technique used is likely to have disrupted any oil layer that may have been present. Only 7 ppb was detected in the oil layer while EPA's split detected 183 ppb. A bailer rather than a tube and low-yielding pump was used to purge the wells. Use of the bailer is likely to have dispersed any oil layer that may have been present and may have resulted in lowering the PCB levels in the oil. Previous PCB concentrations measured in the oil layer present at the Site are 1,539 ppm in 1977 prior to the oil recovery operation and 1,540 ppm in 1989 when the oil recovery operation was being terminated." (ROD at p. 17).

This author believes that even if *the bailer* disrupted or dispersed the oil layer, it is inconceivable that an oil sample would yield a result which is millions of percentage points lower, *i.e.,* 1539 ppm (1977) 1540 ppm (1989) 1,090 ppm (1991) 7 ppb and 183 ppb (EPA split) (1992). One inference which may be drawn from the precipitous drop in the concentration of PCBs over the course of the one year from 1991-92 is that the oil layer may have dissipated over the course of that year because the well was allowed to equilibrate. Alternatively, the decrease may be due to a lowering of the water table and the movement of oil through the porous urban fill that MW-6 is drilled through.

During the Remedial Investigation ("RI") in 1991, the PRP group's contractor drilled and installed 15 new monitoring wells as part of a hydrologic investigation. Of the 15 wells 7 were sampled in 1991 and all 15 were sampled in 1992.

"PCBs were detected in *groundwater samples* collected from monitoring wells MW6 and MW7. Concentrations of total PCBs in MW6 were *12.3* and *25.6* ppb during the 1991 and 1992 samplings, respectively. Total PCBs of 1.3 ppb were detected in MW7 during 1992 sampling." (USEPA 1997) (Emphasis supplied).

These groundwater concentrations of PCBs are in line with the 1992 value for MW-6 of 7 ppb. Thus, that value is most probably indicative of a sample taken in groundwater and not in any oil layer, if any. EPA characterizes the PCB-laden oil as a Non-Aqueous Phase Liquid or NAPL. It also characterizes PCBs as DNAPLs or Dense Non-Aqueous Phase Liquid as follows: "When PCBs are released into the environment through transformer oil spills, they usually bond to the oil and float on top of the water table because the specific gravity of oil (0.8) is lighter than water (1.0). This oil is known as the Light Non-Aqueous Phase Liquid ("LNAPL") layer. However, the specific gravity of PCBs themselves are heavier than water . . . and they sink forming a Dense Non- Aqueous Phase Liquid ("LNAPL") layer." (USEPA 1997). Nevertheless none of the investigations identified a DNAPL layer.

In 1995, pursuant to its Superfund administrative jurisdiction, EPA issued a proposed plan to clean up the Site by removing contaminated soils and collecting PCB-contaminated oil before it could be released into the Delaware River. EPA addressed comments from the general public as well as 22 volumes of comments submitted by Metal Bank, who argued against any cleanup.

EPA issued its formal decision, the Record of Decision or ROD, on how the site would be cleaned up in December 1997. In addition, EPA made minor changes to the ROD by issuing two Explanation of Significant Differences (ESDs) in September and December 2000. The major components of the ROD and subsequent ESDs include installing a trench to monitor and collect oil, installation of a sheet pile wall along a portion of the perimeter of the Metal Bank property adjacent to the Delaware River, excavation of PCB contaminated soil with concentrations of PCBs greater than 25 ppm and removal of oil, excavation of sediments with concentration of PCBs greater than 1 ppm in the immediate vicinity of the property in the Delaware River, and long-term biological and groundwater monitoring. In 1998, EPA issued a unilateral order to 13 PRPs (including ten utility companies and the Metal Bank defendants) to begin design and construction work. Metal Bank chose not to participate. The design work required by the order was completed by the utility companies in January 2003.

6. UNSUCCESSFUL REMEDIATION ATTEMPT

As noted previously, on December 13, 1983, the Court approved a Stipulation between the United States and Defendants that required Metal Bank to install and operate an oil recovery system until all recoverable oil was removed from the subsurface of the Site. As part of the consent decree the United States and the defendants agreed that the matter was settled, and that Metal Bank's remaining obligation was to complete the pump and recovery operation. Since the progress and/or completion of the remediation was a technical matter, the Court placed the matter in suspense, *i.e.*, no court action would be taken until the remediation was concluded. (Court Opinion at p. 9).

The Metal Bank Defendants hired an environmental consultant, who has a Ph.D. in chemistry education, to implement the system outlined by Weston in its 1980 report. However, he testified that the Weston plan was too costly and complicated. The consultant's system consisted of three recovery wells, several oil separation units, and several 55-gallon drums containing activated carbon to treat groundwater - which an EPA remediation engineer called a "Rube Goldbergish" contraption. The system removed most, but not all, of the subsurface oil at the Site. (Court Opinion at p. 9). Metal Bank's consultant argued that it was physically impossible to extract any more oil. He analogized the Site to an oil field, where, he asserted, only 33% of the oil was recoverable. However, this analogy is faulty at best. First, in oil fields the porosity is relatively uniform, whereas at the Site the fill's porosity is very heterogeneous. Second, the consultant did not consider that in oil field extraction operations, companies use secondary and tertiary recovery after primary recovery dwindles to a drip.

In the late 1980s, the Metal Bank Defendants shut down the system and dismantled it, placing approximately one to two feet of clean fill material over the surface of the Southern Area. (Court Opinion). After eight years of remedial efforts, the Defendants took the position that all feasible remediation had occurred and that the clean-up had been successful in that all remaining oil was permanently trapped and posed no risk of migration to the Delaware River or to the area of the embayment.

Pockets or layers of oil beneath the ground surface were found to contain PCBs at concentrations in the range of 520 ppm to 1,090 ppm. Courtyard Area soils at the ground surface and to a depth of approximately two feet were found during the RI to contain PCBs at concentrations up to 140 ppm.

6.1 Samples Collected During the Remedial Investigation, 1991 - 1993

Moreover, surface, subsurface sediments and groundwater were still contaminated. In its Opinion, the Court highlighted contamination that was found both on and off the Site. For example, during the RI, Southern Area surface soils were found to contain metals, PCBs and SVOCs. The levels are shown in Table 1.

Table 1. **Southern Area surface soils hazardous substances found during the RI to contain**

•arsenic at concentrations up to 6.8 ppm
•copper at concentrations up to 149 ppm;
•lead at concentrations up to 220 ppm;
•PCBs at concentrations up to 4.7 ppm;
•total SVOCs at concentrations up to 11.8 ppm.

Similarly, during the RI Southern Area subsurface soils were found to contain PCBs, pesticides, VOCs, SVOCs and metals. Table 2 shows the levels of these hazardous substances.

Table 2. **Southern area subsurface soils haza rdous substances found during the RI**

•PCBs at concentrations up to 42 ppm;
•the pesticide 4,4'-DDD at concentrations up to 11 ppm;
•total VOCs at concentrations up to 907 ppm;
•total SVOCs at concentra tions up to 2,008 ppm;
•arsenic at concentrations up to 21.1 ppm;
•lead at concentrations up to 227,000 ppm (or 22.7% of the sample)
•Mercury at concentrations up to 10.5 ppm.

Southern Area subsurface soils were also found to contain PCBs at concentrations up to 680 ppm, during the Pre-Design Investigation ("PDI"). Twenty-one sample locations contained PCBs in excess of 25 ppm; fifteen sample locations contained PCBs in excess of 50 ppm; and seven locations contained PCBs in excess of 100 ppm. A majority of the samples in excess of 100 ppm were located near the UST in the southwest corner of the Site.

In addition, PCBs, VOCs, SVOCs, metals and other hazardous substances found in Groundwater at the Site during the RI are shown in Table 3.

Table 3. Hazardous substances in Groundwater at the Site found during the RI

•PCBs at concentrations in water up to 25.6 parts per bil lion ("ppb"), and at a concentration in a floating oil sample of 1,090 ppm
•total VOCs at concentrations up to 5.6 ppm
•total SVOCs at concentrations up to 22.7 ppm
•total pesticides at concentrations up to 61 .3 ppb
•arsenic at concentrations up to 369 ppb in unfiltered samples,
• arsenic at concentrations up to 67 ppb in filtered samples
•chromium at concentrations up to 288 ppb in unfiltered samples
• chromium at concentrations up to 102 ppb in filtered samples
•lead at concentrations up to 1,382 ppb in unfiltered samples
• lead at concentrations up to 7.6 ppb in filtered samples
•mercury at concentrations up to 22.2 ppb in unfiltered samples
• mercury at concentrations up to 0.9 ppb in filtered samples

Table 4 shows the levels of PCBs, SVOCs and lead (Pb), found during the RI in Sediments in the Delaware River adjacent to the Site.

Table 4. Hazardous substanc es found in Sediments in the Delaware River adjacent to the Site during the RI

•PCBs at concentrations up to 6.8 ppm;
•total SVOCs at concentrations up to 244 ppm;
•lead at concentrations up to 2,030 ppm.

6.2 Samples Collected During the Pre-Design Investigation in 1999

On June 26, 1998, the EPA issued an Administrative Order for Remedial Design and Remedial Action to the Defendants and the members of the PRP Group. In accordance with the Administrative Order, the PRP Group conducted a Pre-Design Investigation ("PDI") to collect engineering data in support of the design for the remedy and to further determine the scope of contamination. (Court Opinion). The PDI final report, issued on January 21, 2000, confirmed the existence of a layer of oil floating on the groundwater table beneath the Southern Area and further delineated the extent of PCB contamination at the Site. (Court Opinion). A measurable layer of oil ranging in thickness from 0.125 inches to 5.75 inches was detected in the southwest corner of the property.

Sediments in the mudflat and Delaware River were found to contain PCBs at concentrations of up to 6.1 ppm during the PDI. Out of a total of 45 samples, 11 contained PCBs at concentrations exceeding 1 ppm. Sediments in the mudflat area were also found during a January 2002 sampling event to contain PCBs at concentrations up to 22.1 ppm.

In the Courtyard Area, PCBs were detected in two of eleven samples, one of them at a concentration of 8.2 ppm and the other at 190 ppm. In the three areas of concern delineated within the Southern Area, PCBs at concentrations above the EPA cleanup action level of 25 ppm were detected in 38 out of 231 subsurface borings .(Court Opinion at p. 11). *See* Figure 4.

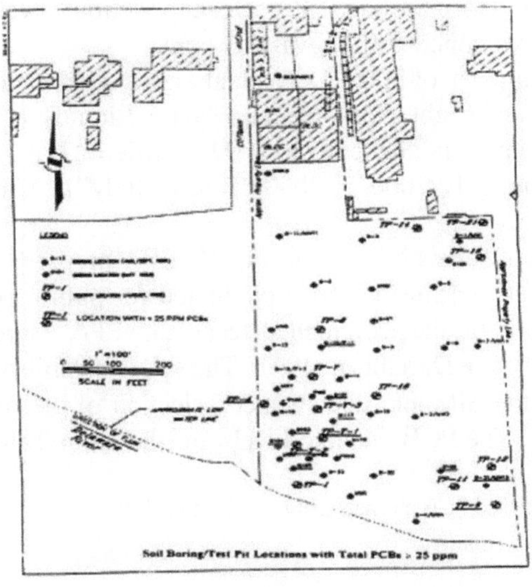

Figure 4. Remedial Investigation locations of soil borings/ test pits

The highest concentration, 680 ppm, was found in a soil sample taken from near the UST. Figure 5. shows the uncovered UST. Sediment samples taken from 45 locations in the River Sediments Area, including the mudflats had total PCB concentrations ranging from non-detect to 6.1 ppm.

Figure 5. Uncovering the UST during the 1999 Pre -Design Investigation, note the oily (dark) quality of the subsurface soil

As part of the PDI, the combined sewer system located underneath Cottman Avenue was inspected and seven sediment samples were collected from inlet pipes leading into the sewer. Only one of them had a PCB concentration higher than the EPA cleanup action level of 1 ppm for sediments, and that concentration was 1.3 ppm.

During the summer of 2000, Defendants conducted a field study to look for the presence of Light Non-Aqueous Phase Liquids ("LNAPL") at the Site, issuing a report in September, 2000, entitled "Data Report, Cottman Site Investigation, July 2000" ("2000 Trench Study"). (Court Opinion at p. 11).

Based on the results obtained in the RI, EPA prepared a Proposed Plan for remediation at the Site, which it circulated for public comment in 1995. Following review of the comments received, EPA issued a Record of Decision ("ROD") in December, 1997. The ROD established PCB cleanup action levels at the Site and documented selection of the remedial action to be implemented. The PCB cleanup levels are shown in Table 5.

Table 5. **ROD established PCB cleanup action levels at the Site**

10 ppm in surface soils
25 ppm in subsurface soils
1 ppm in mudflat sediments

6.3 Samples Collected During the Trench Study, 2000

As part of that study, the Metal Bank Defendants opened up seventeen (17) previously installed monitoring wells and piezometers to check for the presence of oil over the course of about four weeks. During this time, they found measurable oil in 10 of the wells and piezometers, with the thickest measurement being 15 inches of oil in one of the piezometers. Defendants also excavated five trenches using a trackhoe and observed oil in all of the trenches except for Trench number 3, which was not fully excavated due to the presence of a natural gas line. A layer of oil approximately four inches thick was found floating on the groundwater in one of the trenches.

In addition to measuring the oil thickness, the Metal Bank Defendants tested for PCBs in the groundwater and oil underneath the property, as well as in a groundwater "seep" discharging at the base of the upper rip-rap into the mudflat area. PCBs in groundwater were measured at levels up to 7.2 ppm; PCBs in oil were measured at levels up to 530 ppm; and PCBs in the groundwater seep were measured at a concentration of 0.22 ppb.

"In January 2002, EPA's consultants returned to the Site to sample mudflat sediments in the area on the western edge of the Site. Sediment samples yielded PCB concentrations up to 22.1 ppm, and samples of oily liquid that pooled in small holes excavated in the mudflat yielded PCB concentrations of up to 360 ppb." (Court Opinion at p. 12). Both the sediment and groundwater samples also showed significant levels of PAHs. All Government (EPA) witnesses present during the January 2002 sampling event described a strong, oily, petroleum smell coming from the shallow holes dug to gather sediment samples.

EPA's Senior Remedial Project Manager and two expert witnesses for the United States "all testified that they have seen oil sheens coming from the mudflats in the area near the rip-rap at the southwest corner of the Property. (T. Tr. at I - 120-121, II - 17, III - 6). All three testified that when they dug down a few inches into this area of the mudflats, they saw a dark, oily-looking and oil-smelling liquid percolate into the holes. (T. Tr. at I -- 113, II -- 18, III -- 6). The experts testified that they did not see or smell this oily material in areas of the mudflat further away from the southwest corner of the property. (T. Tr. at II-19, III - 6)". (Court Opinion at p. 14).

"At trial, although [the] defense ['s] expert [in fate and transport] did not agree that oil or oil residue is currently seeping from the Property through the lower rip rap, he opined that some oil could have been trapped by the lower rip rap installation and that the perceived oil sheens could be extractions of the oil deposit that are coming to the surface rather than seepage of oil trapped in subsurface areas of the Property bordered by the upper rip rap. (T. Tr. at IX - 141)." (Court Opinion at p. 8). However, the Court rejected this scenario and credited EPA's trail experts.

EPA's consultants again sampled mudflat sediments and groundwater on June 19, 2002. Analysis of these samples indicated the presence of dioxin-like PCB congeners, PAHs, volatile organic compounds ("VOCs"), semivolatile organic compounds ("SVOCs"), dioxins, and furans.

7. EXPERTS

In complex cases, such as this one, it is axiomatic that experts are an indispensable part of the fabric of a trial. In many cases the experts' theories are tested under the Supreme Court's 1993 *Daubert* decision. (U.S. Supreme Court, 1993; Kornfeld, 1992, 1994). Some examples of the Court's reference to and credit of certain experts have been previously cited. However, during the liability phase of this three phase case the Court was confronted with panoply of issues.

One of the main issues at trial was whether oil at the Site could or did migrate into the Delaware River or the mudflats. Chief Judge Giles found that although the Metal Bank's defense expert on soil and oil migration "did not agree that oil or oil residue is currently seeping from the Property through the lower rip rap, he opined that some oil could have been trapped by the lower rip rap installation and that the perceived oil sheens could be extractions of the oil deposit that are coming to the surface rather than seepage of oil trapped in subsurface areas of the Property bordered by the upper rip rap." (Court Opinion, at Transcript vol. IX – p. 141).

Moreover, the Court noted that "[s]heens of oil originating from the sediments adjacent to the Site were witnessed as recently as December 2001. (T. Tr. at IV - 17-18, 19-20). Defendants' trial experts opined that sheens present in the waters adjacent to the Site are biological in nature. The court credits the [Government's] witnesses who concluded that the sheens are oil. They testified that the sheens come from pockets of oil that can be produced upon digging shallow depths in the beach or lower rip rap, and that what looked and smelled like oil, is oil." (Court Opinion at p. 10).

Additionally, manufacturers of PCB-containing transformers used PCB Aroclor 1260 as the principal component of the transformers' dielectric fluid. The Court found that certain Aroclors were transported from the Metal Bank Property to other portions of the Site. For example, Judge Giles observed that "Aroclor 1260 has been found on-site, in the rip rap area, in the beach area and in the Delaware River sediments adjacent to the Site. Similarly, before 1971, manufacturers of PCB-containing capacitors used Aroclor 1242 as a principal component of capacitor dielectric fluid. Aroclor 1242 has also been found on-site, in the rip rap area, in the beach area and in the Delaware River sediments adjacent to the Site. Transformers manufactured during the period from approximately 1950 to 1970 contained polycyclic aromatic hydrocarbons ("PAHs"), including naphthalene, phenanthrene, and 2-methylnaphthalene. Naphthalene, phenanthrene and 2-methylnaphthalene are among the PAHs that have been found in the soils and sediments at the Site. (Court Opinion at p. 10).

7.1 Evidence of Fate and Transport

Another primary issue at trial was whether the PCBs at the Site had a similar chemical signature to those in the Delaware River and Mudflat sediments. Testing of PCBs requires an understanding of Aroclor mixtures. Each Aroclor mixture has a product number, the last two digits of which generally refer to the average percent of chlorine by weight. "Aroclor 1260," for example, contains 60% chlorine by weight. The individual congeners that make up Aroclors have different physical properties, including different rates of solubility, volatilization and degradation. These differences can affect each congener's persistence in the environment and its rate of migration, with higher chlorine congeners tending to be more stable in the environment. As PCB mixtures move through the environment, the absolute and relative concentrations of individual congeners change over time due to differences in their physical properties as well as differences in the rate of bioaccumulation by living organisms.

"The most common way to test for PCBs in the environment is to perform Aroclor testing by using a gas chromatograph instrument to detect the pattern of PCBs that is currently present in a sample. (T. Tr. at II - 53, 57-58). Where chromatograms, which are print-outs from the chromatograph, match the pattern of a known formula, the presence of a particular Aroclor can be confirmed. (T. Tr. at II - 53-57, 243-244). Aroclor testing has its shortcomings because as PCB mixtures are exposed to the environment, the concentration of individual congeners can change over time. (T. Tr. at II - 53-55, III - 115). In addition, the presence of a contaminant in a sample can interfere with analysis, causing inaccuracies in interpreting chromatograms. (T. Tr. at II - 61, 243-244). Although Aroclor testing is helpful for screening a site, it may be insufficient to characterize risk fully. (T. Tr. at III - 117)." (Court Opinion at p. 13).

The more accurate, but more expensive and less-frequently used means of measuring PCBs in the environment is congener testing. The United States' expert witness in the field of analytical chemistry and environmental engineering, reviewed chromatograms from various Aroclor testing at the Site, including chromatograms from the Remedial and Pre-Design

Investigations, and determined that PCB contamination at the Site has been consistently under-reported due to problems inherent in Aroclor analysis. "PCBs are known to be present in surface soils in the Courtyard Area at concentrations up to 190 ppm . . . in subsurface soils in the Southern Area at concentrations up to 680 ppm . . . in oil underneath the Southern Area at concentrations up to 530 ppm . . . and in the mudflat sediments at concentrations up to 22.1 ppm . . . ." (Court Opinion at p. 13).

7.2 The Impact of Polycyclic Aromatic Hydrocarbons

At trial the Government, unlike the Metal Bank Defendants, focused not only on PCBs but on the Site's entire suite of contaminants. For example, in its opinion the Court addressed a number of these contaminants, including *polycyclic aromatic hydrocarbons* ("PAHs") and made certain findings regarding this class of hazardous chemicals. These include the following.

PAHs are a group of semi-volatile organic compounds formed during the incomplete combustion of organic substances. PAHs are components of the raw material that is used to make various grades of oil, including mineral insulating oil. (Court Opinion at p. 13). "Research has shown that mineral insulating oil contains a high percentage of PAHs." (Court Opinion at p. 13, *citing* T. Tr. at II - 26). PAHs tend to persist in the environment for a long time, where they pose a risk to both humans and aquatic life. *(Op cit.).*

Exposure to high doses of PAHs can be acutely toxic to aquatic life. The harmful effects from lower doses may include immune-system dysfunction, organ failure and physiological impairment. In addition, some types of PAHs, including benzo (a) pyrene, are known to be carcinogenic. The National Oceanic and Atmospheric Administration ("NOAA") has determined the toxicity threshold level for PAHs (that is, the level at or above which an organism exposed to a hazardous substance would be expected to suffer a toxic effect) to be 44.8 ppm. The toxicity threshold for benzo (a) pyrene is 7 ppm.

"Total PAHs and benzo (a) pyrene have been measured in the mudflats adjacent to the Metal Bank Property at levels well above the toxicity thresholds. In January 2002, total PAHs were detected at concentrations as high as 869 ppm, and benzo (a) pyrene was measured at concentrations of up to 46.6 ppm." (Court Opinion at p. 13). In the early 1990's, the Delaware River Basin Commission ("DRBC") studied PAH concentrations in the Delaware River at locations both upstream and downstream from the Metal Bank Site. Samples taken from the mudflats at the Site had PAH concentrations ranging from ten to sixteen times greater than the concentrations reported by DRBC at other locations in the Delaware River.

7.3 Impact of Other Contaminants

The Court also noted that various metals - including arsenic, cadmium, copper, lead, and mercury - have been detected in the mudflats at the Site in concentrations above the toxicity threshold level. It also stressed that arsenic is of particular concern at the Site because it was detected in the mudflats at a concentration up to 290 ppm when the screening level (i.e., toxicity threshold) is only 8.2 ppm. Arsenic exposure at this level poses a threat to the aquatic ecosystem, and that dioxins and furans were also found at the Site. Similarly, the Court noted that dioxins and furans are halogenated aromatic hydrocarbons whose molecules, like PCBs, consist of two benzene rings with variable amounts of chlorine attached. Dioxins, the opinion found, "are naturally produced through the incomplete combustion of organic materials and are extremely toxic to human health . . . Furans are inadvertent contaminants present in PCB mixtures . . . ." (Court Opinion at p. 14).

7.3.1 Migration of Contaminants

The Court held that "[t]he evidence shows that contaminants from the Property have migrated, and are continuing to migrate, into the Delaware River and mudflats adjacent to the Property." (Court Opinion, p. 14). For example, it noted that "[t]housands of gallons of transformer oil are still present at the Site, and some of that oil is continuing to migrate from the Property into the River and mudflats. Oil has been observed in the subsurface of the Southern Area within the past two years." (Court Opinion, p. 14). Because of the heterogeneous nature of the fill and uncertainty as to the amount of oil that was originally released at the Site, it is difficult to quantify precisely how much oil remains. Oil has been visually observed in the mudflats.

Chief Judge Giles also noted that "Another way to measure oil is to look for Total Petroleum Hydrocarbons ("TPH"), which are derived from oil.... Concentrations of TPH in the range of 10,000 ppm and greater have been measured at the Property. (T. Tr. at II -- 35). The highest concentrations have been measured in the southwest corner of the Property, in the vicinity of the underground storage tank. (T. Tr. at II - 35). TPH has also been measured in the mudflats near the southwest corner in the range of 1,000 to 6,000 ppm. (T. Tr. at II - 36)". (Court Opinion at p. 14).

The evidence also demonstrated, the Court observed, "that contaminant concentrations decrease in a gradient from the southwest corner of the Property westward across the mudflats . . . While TPH concentrations in the mudflats are highest adjacent to the southwest corner of the Property . . .

TPH concentrations further away from the southwest corner, both toward the CSSO and towards the Yacht Club, are much lower (T. Tr. at II - 36-39). Similarly, in the mudflats, PCBs are found in the highest concentration in the area adjacent to the southwest corner of the Property, and the concentrations decrease with distance away from the southwest corner . . . Most PAH hotspots in the mudflats - those with total PAH concentrations of 40 ppm or greater - are located in the area of the mudflats adjacent to the southwest corner of the Property ...." (Court Opinion at p. 14 - 15).

Additionally, the Court noted that during the June 2002 mudflat sampling, the concentrations of PCBs, PAHs, volatile organic compounds, dioxins and furans detected in the samples from near the yacht club were from ten to one hundred times lower than the samples from near the Property.

Indeed, the Court found that in addition, the same group or suite of contaminants can be found on the Property and in the adjacent mudflats and river sediments. This suite includes PCBs, PAHs, dioxins, furans, and volatile organic chemicals. "The similarities between the contamination found in the two areas demonstrates by *a preponderance of the evidence* that the contamination in the mudflats came from the Site . . . The subsurface oil at the Property, groundwater flow and tidal actions account for transformer oil containing PCBs and other contaminants leaching into the mudflat . . . The lower rip rap installation was [placed] over oil residue that now has found pathways through the fill to the surface as sheens and to areas beneath the beach that are subject to disturbance. In addition, the historical evidence shows that there was an oil pathway to the embayment through or under the upper rip rap in 1972 when the underground oil leaked from the UST." (Court Opinion at p. 15).

The lower rip rap installation did not close off that pathway entirely, inasmuch as the contamination in the mudflats at the southwestern portion of the Property is consistent with the suite of contaminants found on, and under, the Site in the vicinity of the UST. The court credited the Government's expert witness in the field of analytical chemistry and environmental engineering, opinion that ". . . oil can find ways to migrate, regardless of whether or not you have a thick layer floating on the groundwater environment." That expert explained that where there are pockets of oil, "we just need to move little bits of it and carry those contaminants with it." Thus, "[t]here would be opportunities, that may be affected by tides . . . precipitation ... or drought conditions, that can cause some of that [oil thought to be trapped] to move." (Court opinion at 15).

Groundwater at the Property generally flows in a southwesterly direction from the Southern Area to the mudflats. Since the bulk of the Southern Area is constructed on artificial fill, the specific flow pathways for the groundwater and oil flow are difficult to characterize. "The presence of chunks of concrete, bricks, wire, pipe and other material in the heterogeneous fill may offer preferential pathways for the migration of contaminants . . . The court finds that the oil has found, and continues to find, pathways from the ground and groundwater through the rip rap, upper and lower and into the beach and mudflat areas. Defendants have failed to show that the pollution in the mudflats was caused exclusively by a source other than the Cottman Avenue Site." (Court Opinion at 15). The defendants argued that an upgradient lampblack factory was responsible for the TPH, PAHs, and other contaminants.

The Court addressed this argument in the following way. "A lampblack factory was in operation on a neighboring property from 1849 until 1970. Lampblack is a fine powdery material produced by burning low grade oils, creosote, coal tar, anthracene oil and crude oil. During operations, coal tar and crude oils were stored in tanks on the property. Due to the absence of waste product controls, discharges of lampblack and raw materials used to manufacture it, possibly may have occurred into the area surrounding the factory, although there are no records of spills. However, the court finds that this possibility does not account for the PCBs, PAHs, and petroleum hydrocarbons at the Site and in the adjacent mudflat and river sediments. The topography of the area makes it improbable that contaminants from the lampblack factory migrated to the Site. An historic channel running southeast past the lampblack property would have provided a natural pathway to the Delaware River for any discharges from the factory and would not account for contamination of the southwest portion of the Site or for groundwater contamination there under . . . ." (Court Opinion at 15 - 16).

With regards to the City of Philadelphia's combined storm and sewage outfall at the foot of Cottman Avenue, the Court made the following finding: "Similarly, the court finds that the CSSO is not a significant source of contamination to the mudflats . . . If it were, the PCBs and other contaminants would be distributed across the mudflats and would not be concentrated in the area adjacent to the southwest corner of the property . . . Given the overwhelming evidence tracing the contamination of the Metal Bank Site to its own operations, the court is unpersuaded by Defendants' argument that background chemicals (naturally occurring or other non-site related chemicals) account for the contamination at the Site." (Court Opinion at 16).

Moreover, Judge Giles noted that the Aroclor testing does not account for the likely presence of dioxin-like PCB congeners at the Site. For instance, limited sampling in the mudflat sediments adjacent to the Site in

June 2002 revealed the presence of dioxin-like PCBs along with Aroclor 1260. "The fact that Aroclor 1260 also occurs throughout the soils and LNAPL layer strongly suggests that dioxin-like PCBs, if tested for, would be found there as well." (Court Opinion at p. 20). Additionally, Aroclor testing tends to under-report PCB mixtures that have been exposed over time to weathering and degradation in the environment. The Aroclor testing of soils on the Property thus resulted in numerous "non-detect" samples, which, upon closer analysis by the Government's expert, were found to contain highly chlorinated PCB congeners.

7.3.1.1 Dioxins and Furans

As with PCBs, the data regarding dioxins and furans at the Site was also deficient and may have resulted in an underestimation of the risks from these highly toxic chemicals. Dioxins and furans, which may have been produced by the burning of PCB-contaminated cables that were recycled for their copper wiring at the Site, were detected at elevated levels in soils, mudflats, and in Corbicula clams at the Site. However, gaps in the existing data set caused one of the Government's experts to conclude that he could not reliably perform an assessment of the harm to and from the Site.

8. CONCLUSION

The Metal Bank of America Site is among the first listed Superfund sites. Although during the investigative phase, *i.e.,* the RI/FS, PDI and other studies, the Government and the Defendants major focus was on PCB contamination of the Site, at trial the Government concentrated its trial strategy not only on PCBs but PAH's VOCs, SVOCs and other contaminants. It is clear that across the span of the past 31 years, from a spill originating from the Metal Bank Property into the Delaware River in 1972, through the trial of Phase I, more and more data emerged which demonstrated two significant points. First, that PCBs, SVOCs, VOCs, metals and other contaminants at the Site exceed EPA's cleanup standards, and risk analyses were exceeded, in water and sediment in each of the four areas under scrutiny, the Courtyard, the Southern Area, the Mudflats and the sediments and water in the Delaware River. Second, that the owners and operators of the Site were responsible for at least two major spills, one of which was from a UST, used as a storage tank for PCB-laden oil, which was also the source of the other contaminants.

Nevertheless, from a legal point of view, the Government's two-prong approach, *i.e.,* filing suit and thereafter initiating an administrative action, is both duplicative and may work at cross-purposes. Since the Court has no jurisdiction over EPA's administrative action, the Agency can implement its remedy, among other initiatives, while the Court is still in the midst of trying that issue. And EPA has already stated that it may follow that course of action.

In an 84-page opinion, issued on January 22, 2003, the Court rejected the Defendants contention that the Site posed "no risk". Of course, CERCLA's liability scheme, as set forth in the statute's §§ 104 and 107, has no place for that argument. In fact, the statute has three specific elements. They are: (1) whether a person is an owner and/or operator of a vessel or a facility – which is defined broadly, in the statute; whether one is a person who at the time of disposal – which is broadly defined to include a spill – of any hazardous substance owned or operated any facility at which such hazardous substances were disposed of; (2) whether any person who by contract, agreement, or otherwise arranged for disposal or treatment, or arranged with a transporter for transport for disposal of hazardous substances; and (3) was there a release from the facility which required the Government to expend resources in response to the release.

The Court held that the Government met its burden in demonstrating that the Defendants fell within each of CERCLA's elements. Furthermore, the Court found that the contaminants migrated off site, to the Mudflats and the sediments and water in the Delaware River. In addition, the Court rejected the Defendants' contention that the 6,000 gallon UST that Metal Bank used to drain the transformers (that the owners and operators recycled for their copper cores and steel cases) did not leak. Moreover, Chief Judge Giles found that the Defendants' experts' opinions were unconvincing, because they looked at limited data sets or could not account for the occurrence of the contaminants.

In many respects this was a classic CERCLA liability case. The Defendants could not overcome their burden of proof. Moreover, they failed to support their theory that their operations were not the source of the widespread contamination. However, the case is non-conventional in at least two ways. First, the Defendants, rather than rest their case on the elements of CERCLA, posited a theory that they caused "no harm" – which is irrelevant under the statute. Second, this is the only case that I know of where the Government first filed suit and some three years later initiated a Superfund administrative action, both aimed at the same resolution, cleaning up the Site - essentially double dipping.

ACKNOWLEDGMENT

Mr. Kornfeld was a member of the defense trial team from 1999 until August 23, 2003, four days into the trial of *U.S. v. Union Corp. et al.*

The Author wishes to dedicate this article to the Honorable James T. Giles, Chief Judge of the Federal District for the Eastern District of Pennsylvania, who has shown great patience, a keen understanding of the tremendous volume of facts, history and technical material, and an extraordinary respect to counsel.

Mr. Kornfeld wishes to thank Betsy Lukens, Esquire, of EPA Region III's Office of Regional Counsel, for her generosity in supplying copies of exhibits showing recent sediment analysis; and Rick Greene of the Delaware Department of Natural Resources and Environmental Control, Office of Watershed Assessment, for stimulating discussions regarding fish-tissue contamination data.

REFERENCES

Agency for Toxic Substance and Disease Registry ("ATSDR"). 2003. HazDat Database ATSDR's Hazardous Substance Release and Health Effects Database. Located at http://www.atsdr.cdc.gov/hazdat.html, for *a list of contaminant descriptions and associated risk factors.*

Comprehensive Environmental Response Compensation and Liability Act ("CERCLA"), 42 U.S.C. § 9601, *et seq.* (1980), as amended by the Superfund Amendment and Reauthorization Act ("SARA") 1986. Popularly known as "Superfund".

Court Opinion in the case of United States v. Union Corp. et al. No. 80 -1589. January 22, 2003. Located *at* http://www.paed.uscourts.gov/documents/opinions/03D0023P.htm. "PCBs were manufactured in the United States between 1929 and 1977 as complex mixtures of individual congeners known as "Aroclors." (Court Opinion, *citing* Government Exhibit 725 at p. 1; Trial Transcript at II -50). Each Aroclor mixture has a product number, the last two digits of which generally refer to the average percent of chlorine by weight. "Aroclor 1260," for example, contains 60% chlorine by weight. (*Op cit.*). The individual congeners that make up Aroclors have different physical properties, including different rates of solubility, volatilization and degradation. (*Op cit.* at II-51). These differences can affect each congener's persistence in the environment and its rate of migration, with higher chlorine congeners tending to be more stable in the environment. (*Op cit.* at II- 1-52). As PCB mixtures move through the environment, the absolute and relative concentrations of individual congeners change over time due to differences in their physical properties as well as differences in the rate of bioaccumulation by living organisms. *(Op cit.,* Government Exhibit 725 at 1). The low reactivity and high chemical stability of PCBs made them useful for a number of industrial purposes, including as a constituent of insulating fluid used in electrical transformers and capacitors. (Gov. Ex. 725 at 1). "Askarel transformers," used a fluid containing 60 percent or more PCBs by weight. (*Op cit.* Tr. at II - 26-27, 67; T. Tr. at V - 170). "Mineral oil transformers" contained a grade of mineral insulating oil that was frequently contaminated with PCBs through recycling operations or reuse of contaminated mineral oil. (*Op cit.* Gov Ex. 527 at 1-2; Gov. Ex. 113 at ¶ 15; T. Tr. at I-67-71, II-27). PCBs can cause a variety of adverse health effects. They are classified as suspected human carcinogens and may damage the immune system, may cause developmental problems in children and impair reproductive systems. (*Op cit.* Tr. at III - 113; Gov. Ex. 488 at 15). PCBs have been shown to cause severe effects on exposed aquatic organisms and wildlife, including suppression of immune responses, impairment of reproduction and development, disruption of endocrine function, cancer, and organ enlargement and malfunction. (*Op cit.*Gov. Ex. 647 at 31). Most PCBs do not degrade very quickly in the environment, and they can persist for many years in sediments, where they can cause adverse effects, not only to individual organisms, but also to entire aquatic populations or ecosystems. (*Op cit.*Gov. Ex. 725 at 18). Court Opinion p. 12.

Daubert v. Merrell Dow Pharmaceuticals. 1993. Supreme Court Reporter , 113 p. 2786.

Earth Tech. 1994. Metal Bank/Cottman Avenue Remedial Investigation Report.

Federal Register. September 8, 1983. Vol. 48, 40673.

Itzchak E. Kornfeld. 1992. Comment to the Boundaries of Groundwater Modeling under the Law. Standards for Excluding Speculative Expert Testimony. Tort & Insurance Law Jour. 28, 59-68.

Itzchak E. Kornfeld. 1994. A Postscript on Groundwater Modeling: *Daubert,* "Good Grounds," and the Central Role of Cross -Examination. Tort & Insurance Law Jour. 29, 646-650.

Brian Marengo, P.E. 1998, revised 1999. A Review of PCB Loading Estimates to the Delaware Estuary from Combines Sewer Overflow (CSO). City of Philadelphia Water Department.

Ogden/HartCrowser. 1999. Pre-Design Investigation Report, Metal Bank NPL Site, Philadelphia, PA. 1999.

Resource Conservation and Recovery Act, codified at Tit. 42 § 6901 *et seq.*

Roy F. Weston, Inc. 1978. Investigation Report of Oil Spill into the Delaware River .

Tetra Tech EM Inc. 2002. Trip Report for the Metal Bank Site, Philadelphia, Pennsylvania. July 30, 2002.

Toxic Substances Control Act, codified at Tit. 15 § 2605(e). In passing TSCA's § 6(e) Congress specifically obliged EPA to regulate PCBs by establishing a legal presumption that PCBs pose an unreasonable risk. EPA's PCB regulations are located at 40 C.F.R. § 761.

USEPA (U.S. Environmental Protection Agency). 1997. Record of Decision ("ROD").

USEPA (U.S. Environmental Protection Agency). 2002. QASQA Laboratory Report, Metal Bank Superfund Site. U.S. Region III Office of Analytical Services and Quality Assurance, Fort Mead, Maryland. February 12, 2002.

USEPA (U.S. Environmental Protection Agency). 2003. Metal Bank of America (a/k/a Cottman Avenue) NPL Site. Current Site Information (NPL Pad). EPA ID# PAD046557096. Located *at* http://www.epa.gov/reg3hwmd/super/PA/metal -bank/.

This page intentionally left blank

CHAPTER 28

EXPERT OPINIONS IN ENVIRONMENTAL LITIGATION GATEKEEPING 10 YEARS AFTER *DAUBERT*

David G. Ries and Robert L. Burns, Jr., Esquire
Thorp Reed & Armstrong, LLP, Pittsburgh, PA

Abstract: Environmental litigation almost always involves scientific issues and expert opinions, often with multiple experts in different disciplines. In 1993, the United States Supreme Court issued its landmark decision in Daubert v. Merrell Dow Pharmaceuticals, Inc., 509 U.S. 579 (1993), which made significant changes in the standards for admissibility of expert opinions in federal courts. The Supreme Court established a gatekeeping requirement under which courts must screen expert opinions for reliability and exclude "junk science." The Court also established a new, more flexible test to be used in this process. These standards have now become better defined through their application by courts over the ten years since Daubert, including a number of environmental cases. A recent example of the application of these standards to an environmental case is Freeport-McMoran Resource Partners, L.P. v. B-B Paint Corp., et al., 56 F.Supp. 2d 823 (E.D. Mich. 1999), a private party CERCLA case in which one of the authors successfully argued a motion to exclude expert testimony for a group of 12 defendants, resulting in exclusion of the opinions and a judgment for the defendants. The expert in the case was a PhD. chemist and had served as an expert witness in over 180 cases. While admissibility of expert opinions is primarily the responsibility of attorneys, it is important for environmental professionals to understand the issues, both to assist attorneys and to take them into consideration in projects which may later result in litigation. This paper explores current standards for expert opinions in environmental litigation in both federal and state courts.

Key words: environmental litigation, expert opinions, *Daubert, Frye,* Federal Rules of Evidence

1. INTRODUCTION

In 1993, the United States Supreme Court issued its landmark decision in *Daubert v. Merrell Dow Pharmaceuticals, Inc.,* 509 U.S. 579, 113 S. Ct. 2786 (1993), which made significant changes in the standards for admissibility of expert opinions in federal courts. The Supreme Court established a gatekeeping requirement under which courts must screen expert opinions for reliability and exclude "junk science." The Court also established a new, more flexible test to be used in this process. An amendment to Federal Rule of Evidence 702[7], which became effective in December 2000, codified *Daubert's* general approach. These standards have now become better defined through their application by courts over the ten years since *Daubert,* including a number of environmental cases. Because environmental cases almost always involve scientific issues and expert opinions, *Daubert* and its progeny have had and will continue to have a major impact on environmental litigation. While admissibility of expert opinions is primarily the responsibility of attorneys, it is important for environmental professionals to understand the issues, both to assist attorneys and to take them into consideration in projects which may later result in litigation.

2. THE *DAUBERT* STANDARDS

Rule 702 of the Federal Rules of Evidence governs the admissibility of expert opinions. The rule, in its original form, provided:

> If scientific, technical, or other specialized knowledge will assist the trier of fact to understand the evidence or to determine a fact in issue, a witness qualified as an expert by knowledge, skill, experience, training, or education, may testify thereto in the form of an opinion or otherwise.

In *Daubert,* the Supreme Court held that district court judges have a "gatekeeping role" to determine evidentiary reliability of scientific evidence before it is admitted and redefined the requirements for admissibility of scientific evidence under Rule 702.

There are accordingly two distinct requirements enunciated in *Daubert:* (1) the procedural gatekeeping function for trial courts and (2) the newly defined substantive analysis for determining admissibility of scientific

[7] A link to the current version of the Federal Rules of Evidence is available at http://www.uscourts.gov/rules/newrules4.html. A link to the rules, with Advisory Committee Notes, is available at http://www.law.cornell.edu/rules/fre/overview.html.

evidence. The gatekeeping function requires a preliminary assessment by the district court of the proffered opinions before they are admitted:

> Faced with a proffer of expert scientific testimony, then, the trial judge must determine at the outset, pursuant to Rule 104(a), whether the expert is proposing to testify to (1) scientific knowledge that (2) will assist the trier of fact to understand or determine a fact in issue. This entails a preliminary assessment of whether the reasoning or methodology underlying the testimony is scientifically valid and of whether that reasoning or methodology properly can be applied to the facts in issue.

509 U.S. at 592-593, 113 S. Ct. at 2796. (Emphasis added).[8]

Admissibility is based on a two-step analysis in which the trial court determines (1) whether the proffered expert opinion reflects scientific knowledge, whether the findings are derived by the scientific method and whether the work amounts to good science (reliability), and (2) whether the proffered expert opinion is relevant to the task at hand (relevance)[9]. *E.g., Smelser v. Norfolk Southern Railway Co.,* 105 F.3d 299, 303 (6th Cir. 1997), *cert. denied,* 118 S. Ct. 67 (1997).

For the first step in this analysis, the Supreme Court provided a list of nonexclusive factors, which it characterized as "general observations," which a court should analyze in determining the reliability of scientific evidence:

1. Whether the scientific theory or technique "can be (and has been) tested";
2. Whether the scientific theory or technique "has been subjected to peer review and publication";
3. "The known or potential rate of error";
4. "The existence and maintenance of standards controlling the technique's operation"; and
5. "General acceptance" in the "relevant scientific community."

509 U.S. at 593-594; 113 S.Ct. at 2796-2797. (Emphasis added).

The Supreme Court noted that reviewing courts, in applying the *Daubert* analysis, must focus "solely on [the expert's] principles and methodology, not on the conclusions that they generate." 509 U.S. at 595, 113 S. Ct. at 2797.

In adopting the *Daubert* analysis, the Supreme Court held that the more restrictive approach to admissibility of experts of *Frye v. United States,* 293 F. 1013 (D.C. Cir. 1923), which had been followed in federal courts for years, was superceded by the adoption of the Federal Rules of Evidence.

[8] Fed. R. Evid. 104(a) governs determinations of preliminary questions of admissibility of evidence generally.

[9] Fed. R. Evid. 401-403 govern relevance generally.

Frye focused on "general acceptance" by scientists in the particular field of the proposed expert opinions:

> Just when a scientific principle or discovery crosses the line between the experimental and demonstrable stages is difficult to define. Somewhere in this twilight zone the evidential force of the principle must be recognized, and while courts will go a long way in admitting expert testimony deduced from a well-recognized scientific principle or discovery, the thing from which the deduction is made must be sufficiently established to have gained general acceptance in the particular field to which it belongs.

293 Fed. at 1014. Under this test, the underlying scientific principle(s) and methodology employed must be generally accepted by the relevant scientific community.[10]

Significantly, the Supreme Court viewed the *Frye* test as overly restrictive and replaced it with the more flexible *Daubert* analysis. Because the new test is more flexible, a wider range of expert opinions should be admissible after *Daubert* than was previously admissible under *Frye.* However, because of the procedural gatekeeping function articulated by the Supreme Court, a *Daubert* analysis has often resulted in exclusion of the proffered expert opinion. In addition, some courts have incorrectly applied *Daubert's* "general observations" almost as a mandatory checklist for admissibility rather than as part of a flexible analysis. Three surveys, in 1991, 1998 and 1999, found that Federal judges were more likely in 1998 than in 1991 to scrutinize expert opinions before trial and to then limit or exclude the proffered opinions. (Krafka, 2002)

3. APPLICATION OF *DAUBERT*

3.1 U. S. Supreme Court

The Supreme Court has decided *Daubert* issues twice since its original 1993 decision.[11] In *General Electric Co. v. Joiner, 522* U.S. 136, 118 S. Ct.

[10] Some courts, in applying *Frye,* have gone beyond this requirement and held that the methodology and the conclusion must both be generally accepted. E.g., *Thomas v. the West Bend Co., Inc.,* 760 A.2d 1174, 1179 (Pa. Super. 2000), *appeal denied,* 2001 WL 501438 (Pa. 2001).

[11] In addition, in *Weisgram v. Marley Co.,* 528, U.S. 440, 120 S. Ct. 1011 (2000), the U.S. Supreme Court held that a court of appeals may direct entry of judgment as a matter of law where it determines that evidence was erroneously admitted and that the remaining, properly admitted evidence is insufficient to support a verdict. The evidence which the

512 (1997), the Court held that the test for appellate review of admissibility of expert testimony is whether there was an abuse of discretion in admitting or excluding the expert testimony.[12]

Significantly, the Supreme Court also held that the gatekeeping function is not limited to evaluation of methodology, as it stated in *Daubert,* but also properly includes a review of the connection between the methodology and the expert's conclusions:

> But conclusions and methodology are not entirely distinct from one another. ... [N]othing in either *Daubert* or the Federal Rules of Evidence requires a district court to admit opinion evidence that is connected to existing data only by the *ipse dixit* of the expert. A court may conclude that there is simply too great an analytical gap between the data and the opinion offered.

522 U.S. at 146, 118 S.Ct. at 519.

Most recently, in *Kumho Tire Co., Ltd. v. Carmichael*, 526 U.S. 137, 119 S. Ct. 1167 (1999), the Supreme Court held that the *Daubert* analysis applies to all expert testimony based on "technical or other specialized knowledge" and is not limited to scientific opinions. With respect to technical opinions, however, the *Daubert* factors have to be adjusted to fit the facts of the particular case, with the goal of testing the reliability of the proffered expert opinion. 526 U.S. at 149-151, 119 S. Ct. at 1175. The Court noted that the *Daubert* factors are intended to be "helpful, not definitive" and that district courts have "considerable leeway in deciding in a particular case how to go about determining whether particular expert testimony is reliable." 526 U.S. at 151-152, 119 S. Ct. at 1175-1176. The specific *Daubert* factors are to be applied by a district court "... where they are reasonable measures of the reliability of expert testimony." Id.

3.2 Courts of Appeals and District Courts

The Advisory Committee Note to the recent amendment to Rule 702 contains a list of additional factors which federal courts have applied in determining the reliability of expert opinions:

Eighth Circuit found was erroneously admitted was expert testimony which the court found to be inadmissible under *Daubert.* The admissibility of the expert testimony was not an issue before the Supreme Court.

[12] Where the issue is whether or not the district court applied the proper legal standard under the Federal Rules of Evidence, review is plenary. Review of the district court's application of the correct legal standard is for abuse of discretion. *E.g., Tormenia v. First Investors Reality Co., Inc.,* 251 F.3d 128, 134-135 (3d Cir. 2000).

1. Whether experts are "proposing to testify about matters growing naturally and directly out of research they have conducted independent of the litigation, or whether they have developed their opinions expressly for purposes of testifying." *Daubert v. Merrell Dow Pharmaceuitcals, Inc.*, 43 F.3d 1311, 1317 (9th Cir. 1995) (on remand).
2. Whether the expert has unjustifiably extrapolated from an accepted premise to an unfounded conclusion. *See General Elec. Co. v. Joiner,* 522 U.S. 136, 146 (1997) (noting that in some cases a trial court "may conclude that there is simply too great an analytical gap between the data and the opinion proffered").
3. Whether the expert has adequately accounted for obvious alternative explanations. *See Claar v. Burlington N.R.R.,* 29 F.3d 499 (9th Cir. 1994) (testimony excluded where the expert failed to consider other obvious causes for the plaintiff's condition). *Compare, Ambrosini v. Labarraque,* 101 F.3d 129 (D.C. Cir. 1996) (the possibility of some uneliminated causes presents a question of weight, so long as the most obvious causes have been considered and reasonably ruled out by the expert).
4. Whether that expert "is being as careful as he would be in his regular professional work outside his paid litigation consulting." *Sheehan v. Daily Racing From, Inc.*, 104 F.3d 940, 942 (7th Cir. 1997). *See Kumho Tire Co. v. Charmichael,* 119 S.Ct. 1167, 1176 (1999) (*Daubert* requires the trial court assure itself that the expert "employs in the courtroom the same level of intellectual rigor that characterizes the practice of an expert in the relevant field").
5. Whether the field of expertise claimed by the expert is known to reach reliable results for the type of opinion the expert would give. *See Kumho Tire Co. v. Carmichael,* 119 S.Ct. 1167, 1175 (1999) (*Daubert's* general acceptance factor does not "help show that an expert's testimony is reliable where the discipline itself lacks reliability, as, for example, do theories grounded in any so-called generally accepted principles of astrology or necromancy."); *Moore v. Ashland Chemicals, Inc.,* 151 F.3d 269 (5th Cir. 1998) (en banc) (clinical doctor was properly precluded from testifying to the toxicological cause of the plaintiff's respiratory problem, where the opinion was not sufficiently grounded in scientific methodology); *Sterling v. Velsicol Chem. Corp.,* 855 F.2d 1188 (6th Cir. 1988) (rejecting testimony based on "clinical ecology" as unfounded and unreliable).

Advisory Committee Note to 2000 Amendment to Rule 702.

Applying these additional factors as part of a *Daubert* analysis, the Sixth Circuit and Ninth Circuit have held that particularly close scrutiny is required where expert opinions have been developed for litigation, rather than in the regular practice of the expert's profession:

> These factors are to assist the court in determining "whether the analysis undergirding the experts' testimony falls within the range of accepted standards governing how scientists conduct their research and reach their conclusions." *Daubert* (on remand), 43 F.3d at 1316. The Ninth Circuit has added another factor to assist the court in its inquiry: "whether the experts are proposing to testify about matters growing naturally and directly out of research they have conducted independent of the litigation, or whether they have developed their opinions expressly for purposes of testifying" because the former "provides important, objective proof that the research comports with the dictates of good science." Id. at 1317.

Smelser, 105 F.3d at 303 (quoting from *Daubert* on remand).

The Ninth Circuit's opinion in *Daubert* (on remand) explains this additional special scrutiny for litigation opinions:

> One very significant factor to be considered is whether the experts are proposing to testify about matters growing naturally and directly out of research they have conducted independent of litigation, or whether they have developed their opinions expressly for purposes of testifying. . . . in determining whether proposed expert testimony amounts to good science, we may not ignore the fact that a scientist's normal workplace is the lab or in the field, not the courtroom or lawyer's office.
>
> If the proffered expert testimony is not based on independent research, the party proffering it must come forward with other objective, verifiable evidence that the testimony is based on scientifically valid principles.

Daubert v. Merrell Dow Pharmaceuticals, Inc., 43 F.3d 1311, 1317-1318 (9th Cir. 1995); *cert. denied,* 516 U.S. 869, 116 S. Ct. 189 (1995). (Quotation omitted). (Emphasis added).

Under this approach, the party seeking to have the expert opinion admitted has the burden of showing that the expert's conclusions are "based on sound science" which requires an "objective, independent validation of the expert's methodology." *Smelser,* 105 F.3d at 303.

Experience alone may qualify a witness to testify as an expert if reliability of the opinion is established. The Advisory Committee Note to the 2000 Amendment recognizes that an expert may be qualified based on experience:

> Nothing in this amendment is intended to suggest that experience alone – or experience in conjunction with other knowledge, skill, training or education – may not provide a sufficient foundation for expert testimony. To the contrary, the text of Rule 702 expressly contemplates that an expert may be qualified on the basis of experience. In certain fields,

experience is the predominant, if not sole, basis for a great deal of reliable expert testimony. *See, e.g. United States v. Jones,* 107 F.3d 1147 (6th Cir. 1997) (no abuse of discretion in admitting the testimony of a handwriting examiner who had years of practical experience and extensive training, and who explained his methodology in detail); *Tassin v. Sears Roebuck,* 946 F.Supp. 1241, 1248 (M.D.La. 1996) (design engineer's testimony can be admissible when the expert's opinions "are based on facts, a reasonable investigation, and traditional technical/mechanical expertise, and he provides a reasonable link between the information and procedures he uses and the conclusions he reaches"). *See also Kumho Tire Co. v. Carmichael,* 119 S.Ct.1167, 1178 (1999) (stating that "no one denies that an expert might draw a conclusion from a set of observations based on extensive and specialized experience.").

A recent district court case which analyzed an expert opinion based solely on the expert's knowledge and experience is *Pappas v. Sony Electronics, Inc.,* 136 F. Supp. 2d 413 (W.D. Pa. 2000), a case involving the cause of a fire in a residence. In *Pappas,* the court excluded an electrical engineer's testimony concerning the cause of the fire where the engineer's opinions were based solely on his knowledge and experience and he failed to establish the reliability of his methodology. In rejecting the proffered opinion, the court observed:

> ... In short, Brugger's opinion is based on nothing more than his training and years of experience as a fire investigator and engineer. While there may be cases when experience and training alone provide an adequate foundation for an expert opinion under Rule 702, this is not one of them. *See, e.g., Oddi,* 234 F.3d at 157 [FN16]
>
> FN16. Before expert testimony grounded solely in knowledge and experience can be admissible under Rule 702, the expert must still prove that his methodology is reliable. In such cases, the expert may do so by: 1) discussing his experience and knowledge in detail; 2) explaining the methods he has used in the past; 3) indicating the success or failure that he has enjoyed in employing these methods; and 4) testifying about how he used the same methods in the investigation at issue. At the *Daubert* hearing, Brugger did not present this kind of detailed testimony.
>
> 136 F. Supp. 2d at 425.

While *Daubert* was intended to make admissibility of expert testimony more flexible, it is often better known for excluding unreliable expert opinions as "junk science." This is the case because district courts have been giving proffered expert testimony close scrutiny under the gatekeeping procedure which *Daubert* established. A survey of federal judges by the Federal Judicial Center in 1998 reported that "[j]udges were more likely to scrutinize expert testimony before trial and less likely to admit expert testimony in 1998 than in 1991." (Johnson, 2000) However, according to the Advisory Committee Note to Rule 702, acceptance of expert opinions, rather than exclusion, has generally been the result in cases applying *Daubert:*

> A review of the case law after *Daubert* shows that the rejection of expert testimony is the exception rather than the rule. *Daubert* did not work a "seachange over federal evidence law," and "the trial court's role as gatekeeper is not intended to serve as a replacement for the adversary system." *United States* v. *14.38 Acres of Land Situated in Leflore County, Mississippi,* 80 F.3d 1074, 1078 (5th Cir. 1996).

4. ADVISORY COMMITTEE NOTE TO 2000 AMENDMENT TO RULE 702

The Federal Judicial Center has prepared a *Reference Manual on Scientific Evidence,* now in its second edition, to assist federal judges in managing scientific evidence. A copy of the manual is included in *Moore's Federal Practice.* It is also available on the Internet at www.fjc.gov, under "Publications." This manual is an important reference for attorneys proffering and opposing scientific evidence and for expert witnesses because it is an official publication used by federal judges.

Another helpful reference is *Scientific Evidence Review, Monograph No. 4 (1999),* published by the American Bar Association Section of Science and Technology. This 466-page book includes a general analysis of *Daubert* and its progeny, as well as discussion of their application in each of the federal courts of appeals and each of the states. In 2003, the American Bar Association published *Scientific Evidence Review – Admissibility of Scientific Evidence in the Courtroom, Monograph No. 6,* which continues the analysis.

Finally, the Advisory Committee Note to the 2000 Amendment to Rule 702 contains an overview of *Daubert* and the cases applying it.

On some occasions, federal courts have employed court-appointed expert witnesses, either to advise the court on expert issues or to actually testify at trial. Federal Rule of Evidence 706 governs court-appointed experts. The *Reference Manual on Scientific Evidence* (Second Edition) contains a discussion of court-appointed experts at pages 59 through 63, both under Rule 706 and under district courts' inherent powers.

In the fall of 1998, the American Academy for the Advancement of Science started a pilot program, *Court Appointed Scientific Experts,* to assist federal judges in obtaining independent scientific and technical experts. The project is designed to help judges locate highly qualified scientists and engineers to serve as experts for the courts, rather than for the litigants. Information on the project is available at www.aaas.org/spp/case/case.htm.

5. AMENDMENT TO FEDERAL RULE 702

Rule 702 was amended, effective December 1, 2000, to codify the general approach adopted by the Supreme Court in *Daubert.* The original language of Rule 702 remains the same. Three specific requirements (in bold below) have been added to the rule:

Rule 702. Testimony by Experts

> If scientific, technical, or other specialized knowledge will assist the trier of fact to understand the evidence or to determine a fact in issue, a witness qualified as an expert by knowledge, skill, experience, training, or education, may testify thereto in the form of an opinion or otherwise, if (1) the testimony is based upon sufficient facts or data, (2) the testimony is the product of reliable principles and methods, and (3) the witness has applied the principles and methods reliably to the facts of the case.

This amendment "affirms the trial court's role as gatekeeper and provides some general standards the trial court must use to assess the reliability and helpfulness of proffered expert testimony." It does not attempt to codify the specific *Daubert* factors enunciated by the Supreme Court as "general observations." Advisory Committee Note to 2000 Amendment to Rule 702.

6. BASES FOR EXPERT OPINIONS

An important consideration in admissibility and sufficiency of expert opinions is the grounds upon which an expert may base his or her opinion. The Federal Rules of Evidence provide substantial flexibility in this area. Rule 703 provides:

> The facts or data in the particular case upon which an expert bases an opinion or inference may be those perceived by or made known to the expert at or before the hearing. If of a type reasonably relied upon by experts in the particular field in forming opinions or inferences upon the subject, the facts or data need not be admissible in evidence in order for the opinion or inference to be admitted. ...

There are, however, limits to this flexibility, and courts have held that it may not be used to permit a testifying expert to serve as "a mouthpiece" for another nontestifying expert. This rule requires careful consideration in preparation of an environmental case.

A good example of the application of this rule in an environmental case is *Dura Automotive Systems of Indiana, Inc. v. CTS Corp.*, 285 F.3d 609 (7th Cir. 2002), a CERCLA contribution case involving groundwater contamination. A company which was required by EPA to clean up contamination near its plant sought reimbursement from the owner of a nearby plastics manufacturer.

Based on *Daubert,* the District Court excluded the company's sole expert witness, a hydrogeologist, because he admitted that he was not an expert in the hydraulic models QuickFlow and SLAEM, and sought to testify concerning the results of modeling performed by professional groundwater modelers in his firm.[13]

In affirming the exclusion of the opinions, the Court of Appeals noted that construction of the models was an "iterative process" and that the "soundness of the underlying judgment" of the modelers was at issue and it was, accordingly, proper to prohibit the hydrogeologist from testifying concerning it.

The Court concluded:

> Had Durn wanted to use SLAEM and QuickFlow to determine the current capture zone of the Elkhart Field, we might well have a different case; such use might be quite routine. Dura wanted to use these models to determine the capture zone twenty years ago. The affidavits make clear that adapting the models to that use required a host of discretionary expert judgments for the [modelers], not [the hydrogeologist], to make.

285 F. 3d at 615.

When an environmental dispute is in litigation, it is important to carefully document all of the underlying tasks for an expert opinion (sampling plan, sample collection, lab analysis, modeling, etc.) and to make sure that a foundation can be established for its admissibility. This may require

[13] The company later sought to add opinions by the modelers, but the court excluded them because it was six months beyond the deadline under the expert disclosure rules.

testimony of multiple experts if they exercised their own professional judgment in the process.

7. *DAUBERT* IN THE STATE COURTS[14]

Before the Supreme Court's *Daubert* decision, most states applied the *Frye* test or similar tests to determine admissibility of expert testimony. State courts are now facing the issue of whether to adopt *Daubert* or stay with their established tests for admissibility of expert opinions. Some states have adopted the *Daubert* test (*e.g.,* Connecticut and Texas)[15] while others continue to employ the *Frye* test, either because they have declined to adopt *Daubert* or have deferred decision on the issue (*e.g.,* Illinois, Minnesota and New York).[16] Still other states have their own tests which differ from both *Daubert* and *Frye* (*e.g.,* South Carolina and Wisconsin).[17] This section reviews the approaches recently taken by Pennsylvania, Michigan, and Massachusetts in deciding this issue. Pennsylvania and Michigan courts have continued to use the *Frye* test, although Michigan has adopted a *Daubert* approach by statute for personal injury and property damages cases. Massachusetts courts have generally adopted the *Daubert* approach.

7.1 Pennsylvania Courts

In 1998, Pennsylvania adopted Pennsylvania Rule of Evidence 702 which is modeled on Federal Rule 702, with some variation. The Comment to Pa.R.E. 702 expressly notes:

> Adoption of Pa.R.E. 702 does not alter Pennsylvania's adoption of the standard in *Frye v. United States,* 293 F. 1013 (D.C. Cir. 1923), which requires scientific evidence to have "general acceptance" in the relevant scientific community. ... Pennsylvania courts have not yet decided whether the rationale in *Daubert* supercedes or modified the *Frye* test in

[14] The authors acknowledge the assistance of Jennifer A. Schnore, an associate at Thorp Reed & Armstrong, LLP, in updating the state law materials.

[15] *See, State v. Porter,* 698 A.2d 739 (Conn. 1997); *cert. denied,* 118 S. Ct. 1384 (1998); *E.I. duPont de Nemours and Co., Inc. v. Robinson,* 923 S.W. 2d 549 (Tex. 1995).

[16] *See, People v. Miller,* 670 N.E. 2d 721 (Ill. 1997), *cert. denied,* 117 S.Ct. 1338 (1997) (deferring decision); *Goeb v. Tharaldson,* 615 N.W. 2d 800 (Minn. 2000); *People v. Wesley,* 633 N.E. 2d 451 (N.Y. 1994).

[17] *See, State v. Dinkins,* 462 S.E. 2d 59 (S.C.1995) (state's own reliability and relevance test); *State v. Peters,* 534 N.W. 2d 867 (Wis. Ct. App. 1995) *review denied,* 537 N.W.2d 572 (Wis. 1995) (limited relevancy and qualification test; reliability is determined by the fact finder).

Pennsylvania. *Commonwealth v. Crews,* 536 Pa. 508, n.2, 640 A.2d 395 (1994).

Although the Pennsylvania Superior Court and the Pennsylvania Commonwealth Court (both intermediate appellate courts) have held that the *Frye* test, rather than the *Daubert* test, continues to apply in Pennsylvania, *Wack v. Farmland Industries, Inc.,* 744 A.2d 265 (Pa. Super. Ct. 1999); *Checcio v. Frankford Hospital,* 717 A.2d 1058 (Pa. Super. 1998); *Blum v. Merrell Dow Pharmaceuticals, Inc.,* 705 A.2d 1314 (Pa. Super. 1997), *aff'd,* 764 A.2d 1 (2000); *McKenzie v. Westinghouse Electric Co.,* 674 A.2d 1167 (Pa. Commw. 1996) *allocatur denied,* 547 A.2d 733 (Pa. 1997), the Pennsylvania Superior Court recently limited the application of *Frye* to instances where theories of "novel" scientific methodology were presented, rather than whenever "science enters the courtroom". *Trach v. Fellin,* 817 A.2d 1102, 1110 (Pa. Super. Ct. 2003). The *Trach* court held that:

> As the *Frye* court so elegantly stated, [ ] 'While courts will go a long way in admitting expert testimony deduced from a well-recognized scientific principle or discovery, the thing from which the deduction is made must be sufficiently established to have gained general acceptance in the particular field in which it belongs.' *Frye,* 293 F. at 1103. In this single, simple sentence, the *Frye* court recognized that the essence of admissibility is general acceptance: that a principle or discovery can fall by the wayside as science advances is just another way of saying it is not generally accepted. We therefore conclude that we are merely stating the law in Pennsylvania when we state they *Frye* applies only to novel science.

Id. at 1110.

Importantly, the Pennsylvania Supreme Court is currently deciding *Grady v. Frito-Lay,* which may determine whether Pennsylvania will continue to employ evidence standards established in *Frye* or if Pennsylvania will adopt the *Daubert* test. 789 A.2d 735 (Pa. Super. 2001) (en banc), *allocatur granted,* 800 A.2d 294 (Pa. 2002) (oral arguments were held in March 2003). When it granted the Petition for Allowance of Appeal, the court instructed the parties to address the effect of *Frye* and *Daubert* on the sole issue of whether the Superior Court correctly applied the law when it reversed the decision of the trial court to exclude the testimony of Dr. Beroes Ph.D., P.E., and associate professor emeritus of chemical engineering at the University of Pittsburgh. *Grady,* 800 A.2d 294 (Pa. 2002). Id. Dr. Beros created and conducted an experiment that allegedly proved that a Dorito chip could tear the walls of the esophagus. *Grady,* 789 A.2d at 738-39.

Nonetheless, until the court renders an opinion on *Grady,* if proceeding in a Pennsylvania state court and the methodologies of novel expert scientific evidence are at issue, counsel should argue their positions under the *Frye* standard, or perhaps, alternatively, under both *Frye* and *Daubert.*

7.2 Michigan Courts

Michigan courts also continue to generally apply the *Frye* test rather than the Daubert test for the admissibility of expert testimony. Michigan refers to its test as the *Davis-Frye* test. *See, People v. McMillan,* 539 N.W.2d 553, 555 (Mich.App. 1995) (declining to apply *Daubert* absent a ruling by the Michigan Supreme Court displacing *Davis-Frye*).

The Michigan Court of Appeals decision in *Nelson v. American Sterilizer Co.,* 566 N.W.2d 671, 673-674 (Mich.App. 1997), however, appears to be a departure from this general rule. There, the court placed no express reliance on the *Davis-Frye* test. Rather, in applying Michigan Rule of Evidence 702, it cited to *Daubert* and applied a broad standard of reliability based on the holdings in numerous federal cases.

In 1999, the Michigan Court of Appeals again addressed the issue in *Anton v. State Farm Mutual Automobile Insurance Co.,* 607 N.W.2d 123, 127 (Mich.App. 1999). Though the court cited heavily to the opinion in *Nelson,* it maintained that it was applying the *Davis-Frye* test for admitting expert testimony. Id. The court expressly declined to address the continued applicability of the *Davis-Frye* test under Michigan law. Id. at 127. However, the court noted that Michigan Rule of Evidence 702, "unlike its federal counterpart, incorporates a 'recognized' standard for the admissibility of scientific evidence." Id. 607 N.W.2d at 127 n.3. Michigan Rule of Evidence 702 begins with "If the court determines that recognized scientific, technical or other specialized knowledge will assist the trier of fact ..." (Emphasis added.) This language appears to refer to a variation of the "general acceptance" required under the *Frye* test.

It is notable that, in 1996, the Michigan Legislature passed a law outlining the factors which a court must consider in determining whether to admit expert testimony in actions for the death of a person or for injury to a person or property. See M.C.L. § 600.2955(1) (1996). In *Greathouse v. Rhodes,* the Michigan Court of Appeals noted that, in passing § 600.2955(1), the Legislature apparently intended to codify the holding in *Daubert.* 618 N.W.2d 106 (Mich. App. 2000), *rev'd on other grounds,* 650 N.W.2d 351 (Mich. App. 2001).

7.3 Massachusetts Courts

Massachusetts employed the *Frye* "general acceptance" test until the Massachusetts Supreme Judicial Court decided *Commonwealth v. Lanigan,* 641 N.E. 2d 1342 (Mass. 1994). In *Lanigan,* the defendant appealed his conviction for rape and indecent assault and battery on the ground that the trial court erroneously admitted incriminating DNA evidence. Id. at 1344-45. The court acknowledged that it traditionally applied a *Frye* analysis when determining the admissibility of expert testimony based on scientific knowledge, but that it had applied *Frye* flexibly. Id. at 1348. The court, faced with the Commonwealth urging that it adopt *Daubert* and abandon *Frye,* held the following:

> We accept the basic reasoning of the *Daubert* opinion because it is consistent with our test of demonstrated reliability. We suspect that general acceptance in the relevant scientific community will continue to be the significant, and often the only, issue. We accept the idea, however, that a proponent of scientific opinion evidence may demonstrate the reliability of validity of the underlying scientific theory or process by some other means, that is, without establishing general acceptance.

Id. at 1349.

In *Commonwealth v. Senior,* the court reaffirmed that in the *Lanigan* case, it adopted *Daubert* "in part." 744 N. E.2d 614, 618 (Mass. 2001). The *Senior* court held that "[u]nder the *Daubert-Lanigan* standard, although general acceptance is a relevant factor, it is not the 'essential ingredient.' Indeed, the ultimate test is the reliability of the theory or process underlying the expert's testimony." Id. (citations omitted).

Hence, Massachusetts has not specifically adopted the *Daubert* four-part test for evaluating expert testimony, but it has relied on its principles extensively when reviewing the admissibility of expert scientific testimony.

8. *DAUBERT* IN FEDERAL ENVIRONMENTAL CASES

The *Daubert* standards have been applied by federal courts in a number of environmental cases in the last few years. For the most part, the courts have applied a straightforward *Daubert* analysis. Some recent examples include:

A. Courts of Appeals

- *Dodge v. Cotter Corp.*, 328 F.3d 1212 (10th Cir. 2003) (Fourth appeal arising from cases seeking damages under CERCLA and tort claims based on contamination from a uranium mine; reversed district court's rulings on admissibility of expert opinions because the "Findings lack the degree of specificity that would allow us to determine whether the district court properly applied the relevant law;" district court opinion contained no discussion of geologist's conclusion that it was proper to use a general text for establishing baselines for metals where the area at issue was mineral rich and geologically varied.)
- *Stevenson v. E.I. DuPont DeNemours and Co.,* 327 F.2d 400 (5th Cir. 2003) (Claim by neighboring landowners that petrochemical plant contaminated their property and affected their health; challenge to air modeling expert and expert opinions on soil sampling; *Daubert* analysis involves challenge to admissibility of evidence; does not change court's role in reviewing sufficiency of evidence which has been admitted.)
- *U.S. v. Dico, Inc.*, 266 F.3d 864 (8th Cir. 2001) (CERCLA case involving TCE contamination in groundwater; affirmed admission of hydrogeologist's opinions despite defendant's contention that expert ignored other sources of TCE and failed to consider other data, model was not basis for opinion on source of TCE, expert did consider other data and factual basis for expert opinion is an issue of credibility, not admissibility.)
- *NutraSweet Co. v. X-L Engineering Co.,* 227 F.3d 776, 787-790 (7th Cir. 2000) (Suit by property owner alleging that plaintiff's property was contaminated by improper disposal of VOCs on the adjoining property; the court affirmed the district court's conclusion that analysis of historical photographs is a well-accepted technique so as to bear a sufficient indicia of reliability; the photographic analysis was used by plaintiff's environmental expert to confirm opinions based on solvent degradation (speciation), chemical chromatography and groundwater migration analysis which were not challenged on *Daubert* grounds.)
- *St. Martin v. Mobil Exploration & Producing US, Inc.*, 224 F.3d 402 (5th Cir. 2000), *reh. denied,* 234 F.3d 31 (5th Cir. 2000). (Action by property owners against oil companies for damage to marshes from alleged failure to maintain spoil banks on canals; ecology expert qualified to testify based on years of experience and personal observations; *Daubert* factors are nonexclusive and need not be rigidly applied in every case.)
- *Gussack Realty Co. v. Xerox Corp.,* 224 F.3d 85 (2d Cir. 2000) (Groundwater contamination case based on CERCLA and state torts; defendant offered opinion that involved property was upstream and

upgradient; plaintiff offered expert opinion that groundwater flowed toward plaintiff's property, in a different direction than surface water, based on review of defendant's expert's analysis and state agency data; admissibility affirmed – experts need not conduct their own tests, may rely on their interpretation of data collected by others.)

- *U.S. v. Cunningham,* 194 F.3d 1186 **(11th Cir. 1999)** (Criminal RCRA case in which the defendant was convicted of conspiracy and illegal transportation and disposal of hazardous waste; affirmed trial court's exclusion of defense expert witness who based his opinion on an unproven test method, was unfamiliar with applicable regulations and disagreed with EPA's regulatory determination that barium was a hazardous waste.)
- *Kalamazoo River Study Group v. Rockwell Int'l. Corp.,* 171 F.3d 1065 **(6th Cir. 1999)** (Two-site case where PCB's released at one site were alleged to have traveled through a ditch and river to a second site; affirmed exclusion of expert opinion, where district court found that expert's opinion was based on "speculation, conjecture and possibility" and that "the inadequate factual basis makes [the] affidavit scientifically unreliable"). *See also* district court opinion at 3 F. Supp. 2d 815 (W.D. Mich. 1997).
- *Burns Philip Food, Inc. v. Cavalea Continental Freight, Inc.,* 135 F.3d 526 **(7th Cir. 1998)** (Claim of contamination of land from diesel fuel on adjacent property; affirmed exclusion of environmental expert's opinions which district court found to be based on inadequate investigation and testing).

B. District Courts

- *Seneca Meadows, Inc. v. ECI Liquidating, Inc.,* 121 F. Supp. 2d 248, 252-254 (W.D.N.Y. 2000) (Landfill case against generator defendants; court denied summary judgment where there were conflicting affidavits as to whether a defendant's wastes were hazardous; the defendant did not challenge plaintiff's expert's methodology, but claimed factual errors underlying his analysis; the court found that the opinions were admissible and that the fact issues would be best addressed by cross-examination and presentation of contrary evidence at trial.)
- *Freeport-McMoran Resource Partners L.P. v. B-B Paint Corp., et al.,* 56 F. Supp. 2d 823 (E.D. Mich. 1999) (Excluding expert opinions and granting summary judgment in a waste transshipment case – discussed below).
- *State of New York v. Almy Brothers, Inc.,* 1998 WL 57666, 46 E.R.C. (BNA) 1339 (N.D.N.Y. 1998) (Third-party contribution claim against prior owner-operator of a milk processing facility; third-party plaintiff

admitted that she had no direct evidence of disposal of hazardous substances by dairy operator; district court held that opinions of a geologist and environmental engineer, both with substantial experience in remediation, concerning the likelihood of contamination from operation of a milk processing facility and absence of contamination from plaintiff's operations was sufficient to preclude summary judgment.)

- *U.S. v. SCA Services of Indiana, Inc.*, 1995 WL 569634 (N.D. Ind. 1995) (Waste disposal site; expert's opinion excluded based on lack of foundation for conclusion that all products in the involved class contain the same hazardous constituents).
- *Thomas v. FAG Bearings Corp.*, 846 F. Supp. 1382 (W.D. Mo. 1994) (A CERCLA case in which the defendant claimed that third party defendants contributed to TCE contamination in groundwater; court excluded expert opinions which it found to be "concocted of impermissible bootstrapping of speculation upon conjecture"; speculation that contamination entered groundwater and, if it did, that it traveled to involved area; "no information available to say with any degree of certainty that contaminants went from point 'A' to point 'B'").

While most federal environmental cases have applied a straightforward *Daubert* analysis, two cases contain language which suggests that a more liberal standard may apply to admissibility of expert opinions in environmental cases because of the remedial nature of the environmental laws. *B. F. Goodrich v. Betkowski,* 99 F.3d 505 (2d Cir. 1996), *cert. denied,* 118 S. Ct. 2318 (1998), a CERCLA case involving disposal at a landfill, has language which supports this approach. The court notes that: "[e]nvironmental science, like epidemiology 'is ill suited to lead a fact finder toward definitive answers, dealing as it does in statistical probabilities.'" In effect, the court seems to say that environmental science, like epidemiology, is a "soft science" and that a more liberal analysis should apply to admissibility of expert opinions in these fields.

If this is what the court intended, it is not supported by *Daubert* and a majority of the cases decided under it. The Supreme Court's decisions in *Daubert* and *Joiner* both involved epidemiology issues. Further, numerous circuit court and district court cases have applied a standard *Daubert* analysis to epidemiology issues. As the cases above demonstrate, most courts have also applied a standard *Daubert* analysis to environmental cases. A close review of the expert opinions and supporting bases in *B. F. Goodrich* suggests that it could have been decided the same way under a standard *Daubert* analysis. The court found that the expert's opinions were supported by sufficient research, including EPA publications.

In *F. P. Woll & Co. v. Fifth & Mitchell Street Corp.,* 1999 U.S. Dist. LEXIS 874, 48 E.R.C. (BNA) 1362 (E.D. Pa. 1999), the district court followed *B. F. Goodrich* and used similar language concerning the application of *Daubert* in environmental cases.

9. FREEPORT-MCMORAN/FOREST WASTE: A CASE STUDY

Freeport-McMoran, 56 F. Supp. 2d 823, *supra,* is a recent example of the application of *Daubert* to proffered expert opinions in an environmental case. The court excluded the proposed expert testimony and granted summary judgment for defendants.[18]

The case is a CERCLA contribution action involving a claim of transshipment of waste from one National Priority List ("NPL") site (the Berlin & Farro site) to a second NPL site (the Forest Waste site). The plaintiff was a generator at the Forest Waste site which settled with EPA and agreed to participate in remediation of the Forest Waste site as a part of a group of Potentially Responsible Parties (PRPs). The defendants were generators who were alleged to have sent drummed waste to the Berlin & Farro site, a waste incineration facility. Defendants settled with EPA for contamination at the Berlin & Farro site.

Plaintiff claimed that some of the drummed waste which defendants sent to the Berlin & Farro site was not incinerated there but, instead, was transshipped to the Forest Waste site. EPA and the PRP group at Forest Waste decided not to bring claims against the Berlin & Farro generators relating to the Forest Waste site.

The plaintiff had no evidence that any of defendant's drummed wastes were disposed of at the Forest Waste site. Instead, plaintiff attempted to prove indirectly that defendants' drummed waste was transshipped from the Berlin & Farro site to the Forest Waste site for disposal.

Plaintiff's theory was as follows:

1. Each defendant arranged for the disposal of drummed waste at the Berlin & Farro site;
2. Each defendant's drummed waste that arrived at the Berlin & Farro site contained solid materials;
3. After the liquids were drained or pumped from each defendant's drums at Berlin & Farro, residual solids remained;

[18] D.G. Ries, one of the authors of these materials, argued the *Daubert* motion for a defense group of 12 defendants.

4. Each defendant's drums containing residual solids were transshipped from Berlin & Farro to the Forest Waste site.
5. The residual solids remaining in each defendant's drums which were transshipped to the Forest Waste site contained hazardous substances like those present at the Forest Waste site.

Plaintiff originally retained the expert to review each defendant's waste stream to express opinions as to CERCLA hazardous substances which they contained. In order to develop plaintiff's transshipment theory, plaintiff later requested the expert to provide an opinion as to the presence of a solid component in each defendant's waste.

Plaintiff's expert had a Ph.D. in chemistry, taught at a small college for several years, and then worked at EPA in the fields of CERCLA and RCRA for several years. He then became a litigation consultant and served as an expert witness in over 183 cases, primarily for the federal government.

The expert studied the available materials about each defendant's waste streams and expressed opinions which were essentially carbon copies as to each of the defendants:

1. That each defendant's waste contained some solids;
2. That the solids would settle by gravity, to the bottom of the drums in which they were stored;
3. That some residual solids would remain in the drums when the overlying liquids were poured or decanted at Berlin & Farro; and
4. That the residual solids contained hazardous substances of the type found at the Forest Waste Site.

The evidence established that every effort was made to completely drain the drums at Berlin & Farro by pouring out their contents or pumping them out. Only drums which still contained too much solid material after this process for the drums to be recycled were disposed of onsite or at one of several landfills. Most drums were sold to drum recyclers.

There was no evidence that any defendant's drums contained too much solid material for the drums to be recycled. The expert did not express any opinions on quantities of solid materials in any defendant's drums.

What was missing from the expert's analysis, *inter alia,* was any fact evidence or opinion that drums which contained some unspecified quantities of residual solids would be transshipped to Forest Waste, rather than being recycled.

After the close of discovery, including a two-day deposition of plaintiff's expert, defendants filed a motion to strike the expert opinions, along with motions for summary judgment. The district court granted these motions. In excluding the proffered expert opinions, the court found:

- The proposed opinions were "utterly lacking in any indicia that would establish any of the *Daubert* factors";

- The only source for substantiation of the expert's theories was his "own experience"; and
- The opinions were unreliable because they were based on a misunderstanding of the facts.

56 F. Supp. 2d at 832-837.

The expert's opinions might have been scientifically reliable and admissible if he had limited his analysis to his original task –comparing the hazardous substances found in each defendant's waste streams to the hazardous substances found at the site. However, he went well beyond this task. He attempted to express carbon-copy opinions as to the solids in each defendant's wastes, based on limited information, which varied as to each defendant, with no observation or testing of the wastes and without an adequate scientific or factual basis. When asked if there was a set of information about each defendant's wastes which he sought in order to perform his analysis, he responded that there was not and that he relied on the information which plaintiff's counsel elected to give him. When asked if there are any published or generally accepted standards for his analysis, he stated that there may be, but he could not point to them and did not refer to them.

10. CONSIDERATIONS FOR ENVIRONMENTAL PROFESSIONALS

As noted in the Introduction, it is important for environmental professionals who may testify in litigation to understand the general principles of admissibility of expert opinions. While offering and challenging expert opinions is primarily the responsibility of attorneys, environmental consultants often get involved before attorneys in matters that may later result in litigation. In such cases, they can provide the foundation for litigation through documented, scientifically reliable work in accordance with the litigation standards discussed above. Providing such a foundation is often an issue of documentation as well as the nature and methodologies of the services. Generally accepted standards and methodologies and generally accepted treatises and reference materials should be employed and their use should be documented. Litigation may require more rigorous and more detailed analysis than general environmental protection services. Each task must be analyzed in light of how it will fit into a courtroom presentation. It is particularly important that subcontractors and other service providers, like labs, know that a matter may be headed for litigation and that special attention is addressed to predicates for admissibility, particularly, documentation and personal observation or supervision by qualified individuals. Qualifications of persons performing each task should be reviewed to be sure that they will be sufficient for litigation, especially if they could be called to testify on behalf of a client. This is particularly important for professionals who exercise their own professional judgment and reach their own expert opinions rather than simply performing support functions such as data collection.

Once a matter reaches litigation, environmental professionals who are familiar with these principles can provide significant assistance to attorneys in offering and challenging expert opinions. In situations where environmental consultants are retained for litigation, they should follow the same standards and methodologies normally used for their work, with even more rigorous analysis, if appropriate.

11. CONCLUSION

In federal courts, admissibility of expert opinions is now governed by Fed. R. Evid. 702 which requires that the expert opinion must be based on (1) sufficient facts and data, (2) reliable principles and methods, and (3) reliably applied principles and methodology. Reliability is measured by a flexible analysis described in *Daubert* and the subsequent cases applying *Daubert.* This analysis serves both to exclude "junk science" and unreliable opinions and to permit admission of reliable opinions which have not yet reached general acceptance. Expert opinions must also be relevant to the issues in dispute. Similar considerations apply in state courts which have adopted *Daubert.*

In states which continue to apply the *Frye* analysis, admissibility of expert opinions requires general acceptance of the expert's principles and methodology in the relevant scientific community. Some courts also require general acceptance of the conclusions as well as the methodology. Finally, some jurisdictions apply the *Frye* test to all expert opinions while others limit its application to novel scientific evidence.

REFERENCES

Cynthia H. Cwik and John L. North, *Scientific Evidence Review,* Monograph No. 4 (American Bar Ass'n 1999)

Cynthia H. Cwik and John L. North, *Scientific Evidence Review – Admissibility of Scientific Evidence in the Courtroom,* Monograph No. 6 (American Bar Ass'n 2003)

Federal Judicial Center, *Reference Manual on Scientific Evidence (2d. edition)* (2000)

Molly T. Johnson et al., *Expert Testimony in Federal Civil Trials, A Preliminary Analysis* (Federal Judicial Center 2000,). Available at www.fjc.gov, under "Publications."

Carol Krafka, et al., *Judge and Attorney Experiences, Practices and Concerns Regarding Expert Testimony in Federal Civil Trials* (Federal Judicial Center 2002) Available at www.fjc.gov, under "Publications."

This page intentionally left blank

PART VIII: REMEDIATION

CHAPTER 29

EVALUATION OF TWO ORGANOCLAYS, CLINOPTILOLITE, AND HYDROXY-APATITE AS SORBENTS FOR HEAVY METAL REMOVAL FROM WATER

George R. Alther[1], Fred D. Tillman, Jr.[2] and James A. Smith[2]
[1]Biomin Inc.,Ferndale, MI, and [2]University of Virginia, Charlottsville, VA

Abstract: A series of column tests were performed with four sorbent media to determine their effectiveness for the removal of heavy metals from water. The sorbents are an organoclay , an organoclay/anthracite blend, clinoptilolite (a zeolite), and hydroxy-apatite. Heavy metals studied were Ni, Cr, Cu, Cd, and Zn. Four column tests were conducted using an influent solution containing all four metals (one test for each sorbent). Four column tests were conducted using an influent solution containing only Zn (on e for each sorbent), and two column tests were conducted using an influent solution containing only Ni (with clinoptilolite and hydroxy-apatite only). For the Zn-hydroxy-apatite experiment, the inflow solution also contained 200 mg/L Ca^{++} to simulate hard, water. Aside from ion-selective resins, hydroxy-apatite is the only sorbent that is believed to retain its sorption capacity for metals in hard water. Metal effluent breakthrough (data indicated that the organoclay and the organoclay/anthracite blend removed all heavy metals, but only to a limited extent, with effluent metal concentrations equal to 50% (of the influent concentration appearing after 2 to 3 pore volumes of flow. These sorbents have the unique capability of removing both heavy metals and nonpolar organic pollutants from water simultaneously. Hydroxy-apatite demonstrated the greatest removal of metals from solution. Pore volumes of flow required to produce effluent concentrations equal to 50% of the influent concentration ranged between 50 (Ni) and 160 (Cu). Although hydroxy -apatite performed best for metal removal, it should be noted that clinoptilolite is superior to most media for ammonia removal.

Key words: heavy metals, organoclay, anthracite, hydroxy-apatite, clinoptilolite, zeolite

1. INTRODUCTION

Sorption of heavy metals by sorbents such as bentonite, organically modified bentonite (organoclay), clinoptilolite (zeolite), and hydroxy-apatite, has been studied previously using batch experiments (Alther, 1998; Alther, 2000; Alther, 2001; Alther and Ellis, 2001a,b). Cation-exchange capacities of standard sorbent media typically range between 70 meq/100 g for bentonite to 600 meq/100 g for synthetic zeolites. Organoclays typically have much lower cation exchange capacities than bentonite owing to the exchange of relatively large, organic quaternary ammonium cations onto the clay that are not easily displaced by ammonium cations. Mumpton (1981) presents the results of several studies on the sorption capacity of zeolites.

Organoclays are particularly effective at removing relatively non-polar organic pollutants such as oil, polychlorinated biphenyls, polycyclic aromatic hydrocarbons, chlorinated solvents, and gasoline hydrocarbons from water (Alther, 2002). When used as a pretreatment for activated carbon systems, Organoclays can increase the life of the activated carbon by 7 to 10 times. However, little is known about the ability of organoclays to remove heavy metals from solution. In addition, there are relatively little data that quantify the sorption of metals to organoclays, zeolites, and hydroxy-apatite in flow-through, column reactors (which mimic field-treatment systems).

Given the strict regulatory requirements on heavy-metal discharges to the environment, it is important to develop cost-effective sorbents that can treat metal-contaminated waters (or in some cases, water contaminated with both metals and organic pollutants). In this study, we evaluate the suitability of four sorbents (an organoclay, an organoclay/anthracite blend, a zeolite (clinoptilolite), and hydroxy-apatite for the removal of heavy metals from water. Experiments are performed with combinations of metals and with single metals in flow-through columns to better assess real world performance and possible kinetic sorption limitations.

2. EXPERIMENTAL MATERIALS AND METHODS

Initially, all five metals were combined in one stock solution, with later experiments involving two of the metals individually. For all experiments, a column constructed of 30-inch-long (76.2 cm) by 3-inch-diameter (7.62 cm) poly-vinyl-chloride (PVC) and fitted with flexible couplings, reducing bushings and NPT thread-to-5/8" hose fittings was used. Two layers of circular type-304 stainless steel woven cloth (20x20 mesh) were placed inside each end-fitting to suspend the sorbent inside the column and prevent sorbent material from entering the tubing. A 55-gallon (208.2 L) self-supporting polyethylene tank was used to prepare and hold the inflow metal solution. A Master-Flex peristaltic pump was used to force the aqueous-metal solution up through the column to displace void-space air and ensure maximum contact with the sorbent material. Figure 1 shows a picture of the experimental column system.

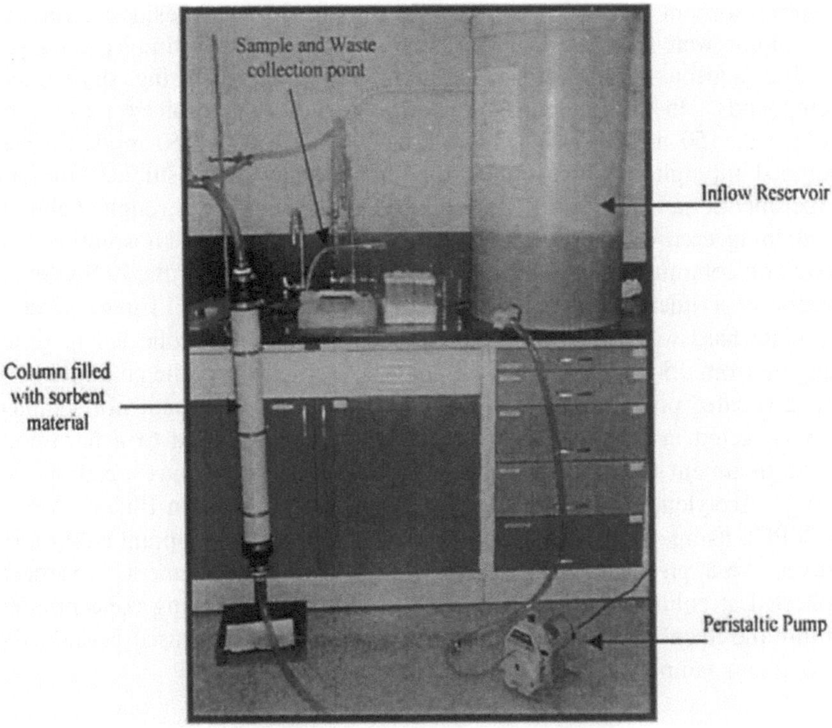

Figure 1. Photograph of experimental setup showing PVC column, peristaltic pump, inflow reservoir and waste collection carboy. Aqueous -metal solution is pumped up through the column in order to displace pore -space air and prevent short -circuit flow.

Each packed column was back-flushed with clean water several times prior to each experiment in
order to remove fines associated with the sorbent. After the fines were removed, the physical properties of each packed column were measured. For the organoclay/anthracite column, the porosity, pore volume, sorbent mass, experimental flow rate, and water residence time in the column were 0.377, 1.3 L, 2.9 kg, 160 mL/min, and 8.3 min respectively. For the organoclay column, the porosity, pore volume, sorbent mass, experimental flow rate, and water residence time in the column were 0.359, 1.2 L, 2.9 kg, 156 mL/min, and 8.0 min respectively. For the clinoptilolite column, the porosity, pore volume, sorbent mass, experimental flow rate, and water residence time in the column were 0.329, 1.4 L, 2.6 kg, 72 mL/min, and 20 min respectively. For the hydroxy-apatite column, the porosity, pore volume, sorbent mass, experimental flow rate, and water residence time in the column were 0.42, 1.3 L, 2.9 kg, 160 mL/min, and 8.3 min respectively.

The aqueous-metal inflow solution was prepared using dry-reagent compounds. In the first set of four experiments, a reservoir containing 60 mg/L Cd, 150 mg/L Cr, 120 mg/L Cu, 60 mg/L Ni and 60 mg/L Zn was pumped through columns containing the sorbents to be studied. In later experiments, a 785 mg/L Zn solution was pumped through columns containing each of the four sorbents. Lastly, a 437 mg/L Ni solution was tested on columns of clinoptilolite and hydroxy-apatite. For the Zn/hydroxy-apatite experiment, 200 mg/L Ca^{++} was also added to the influent water to simulate hard water. In each experiment, samples were collected at times ranging from 15 minutes to several hours at the outflow of the column in 50-mL graduated polypropylene tubes and then capped. Effluent not sampled was collected in 5-gallon (18.9 L) plastic carboys and sent to a hazardous waste treatment facility. Samples were analyzed for aqueous-metal species using Acetylene-Air Flame/Atomic Adsorption (Perkin-Elmer Model 5100PC), using single-element hollow cathode lamps. Two-point calibration curves were prepared from stock solution for each element. Samples collected at column outflow were diluted with water to bring concentration within the linear range of instrument. Check samples were used periodically throughout sample analysis to ensure quality control.

3. RESULTS

Figure 2 presents the breakthrough curves for each metal for the organoclay/anthracite column. The vertical axis is relative concentration, defined as the ratio of the inflow concentration to the outflow concentration.

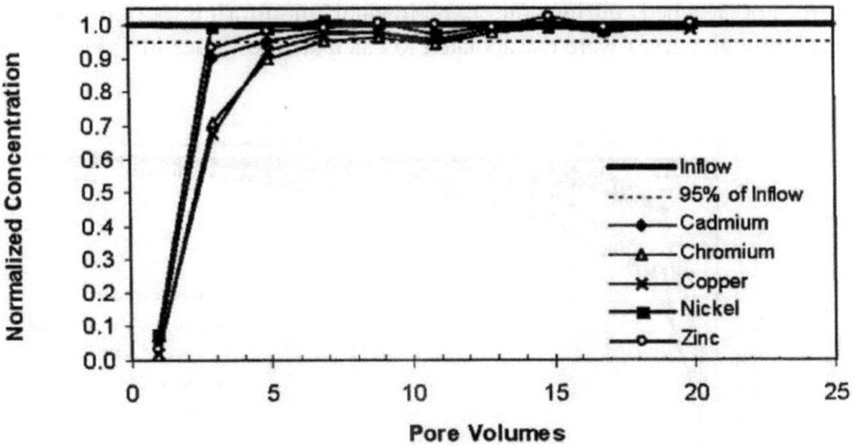

Figure 2. Breakthrough curves of individual metals from an inflow solution containing 60mg/L Cd, 150 mg/L Cr, 120 mg/L Cu, 60 mg/L Ni and 60 mg/L Zn through a column containing organoclay /anthracite.

The horizontal axis is pore volumes of flow. Figures 3-5 present similar curves for each of the other three sorbents. Figure 6 presents breakthrough curve data for Zn (as a single solute) on all four sorbents. Figure 7 presents breakthrough curve data for Ni on hydroxy-apatite with 200mg/L Ca^{++} and clinoptilolite. In all cases, the breakthrough curves are approximately S-shaped (or are approaching an S-shape), which is characteristic of advective/dispersive transport through a water-saturated porous medium. The time for complete breakthrough varies with both metal type and sorbent, indicating that some metals are sorbing to a greater extent than others. In the absence of sorption, the relative solute concentration (defined as the influent concentration divided by the effluent concentration) is expected to equal 1 at 1 pore volume of flow. For all experiments, the relative concentrations of 0.5 correspond to pore volumes of 2 or greater, indicating that there is measurable sorption of every metal to every sorbent.

Table 1 presents summary information about the column experiments with all five metals present in the inflow solution. Table 2 presents summary information for Zn (as a single solute) transport through all four sorbent media. Table 3 presents summary information for Ni transport through clinoptilolite and hydroxy-apatite (with 200 mg/L Ca^{++} as the co-solute). These data were calculated from the breakthrough curves shown in Figures 2-5. We define breakthrough as the point where the relative concentration equals 0.95. For (he transport of some metals through the clinoptilolite and hydroxy-apatite columns, relative concentrations of 0.95

were not reached during the experimental monitoring period, so the breakthrough curves were extrapolated to calculate the values in Table 1.

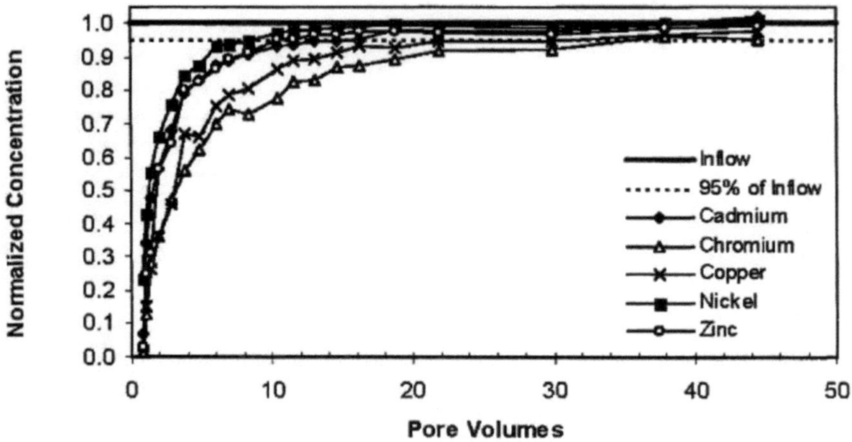

Figure 3. Breakthrough curves of individual metals from an inflow solution containing 60 mg/L Cd, 150 mg/L Cr, 120 mg/L Cu, 60 mg/L Ni and 60 mg/L Zn through a column containing organoclay.

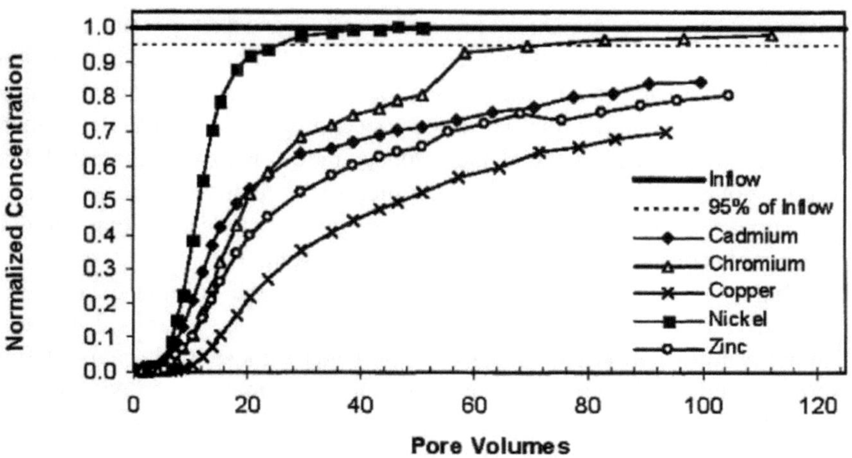

Figure 4. Breakthrough curves of individual metals from an inflow solution containing 60 mg/L Cd, 150 mg/L Cr, 120 mg/L Cu, 60 mg/L Ni and 60 mg/L Zn through a column containing clinoptilolite.

Table 1. Column breakthrough results presenting estimated adsorption capacity of filtration media to each metal species. Column inflow solution contained 60 mg/L Cd, 150 mg/L Cr, 120 mg/L Cu, 60 mg/L Ni and 60 mg/L Zn.

Sorbent	Metal	Inflow mass to breakthrough[a] (mg)	Mass sorbed to breakthrough[a] (mg)	Sorbed/ Incoming %	Mass Sorbed/ Mass of Sorbent (mg/kg)	% by weight
Organoclay/ Anthracite	Cd	857	107	12.5%	37	0.0037%
	Cr	2,143	343	16.0%	118	0.0118%
	Cu	1,714	267	15.6%	92	0.0092%
	Ni	857	62	7.2%	21	0.0021%
	Zn	857	85	9.9%	29	0.0029%
	Total	6,428	863	13.4%	297	0.0297%
Organoclay	Cd	2,242	209	9.3%	71	0.0071%
	Cr	5,605	879	15.7%	299	0.0299%
	Cu	4,484	496	11.1%	169	0.0169%
	Ni	2,242	136	6.1%	46	0.0046%
	Zn	2,242	185	8.3%	63	0.0063%
	Total	16,815	1,907	11.3%	648	0.0648%
Clinoptilolite	Cd	12,394	2,632	21.2%	752	0.0752%
	Cr	33,052	4,794	14.5%	1,369	0.1369%
	Cu	24,789	9,871	39.8%	2,820	0.2820%
	Ni	12,394	814	6.6%	233	0.0233%
	Zn	12,394	3,640	29.4%	1,040	0.1040%
	Total	95,023	21,751	22.9%	6,213	0.6213%
Hydroxy-Apatite	Cd	23,487	13,977	59.5%	5,315	0.5315%
	Cr	62,633	40,588	64.8%	15,434	1.5434%
	Cu	46,974	29,053	61.8%	11,048	1.1048%
	Ni	23,487	4,643	19.8%	1,766	0.1766%
	Zn	23,487	10,765	45.8%	4,094	0.4094%
	Total	180,069	99,026	55.0%	37,656	3.7656%

[a] Breakthrough is measured 95% of inflow solution except for Cd, Cr, Cu and Zn on both clinoptilolite and hydroxy-apatite which were extrapolated to 95% breakthrough from available data.

Table 2. Estimated sorptive capacity for each sorbent material to an aqueous inflow solution containing 785 mg/L of Zn.

Sorbent	Inflow mass to breakthrough[a] (mg)	Mass sorbed to breakthrough[a] (mg)	Sorbed/ Incoming (%)	Mass Sorbed/ Mass of Sorbent (mg/kg)	% by weight
Organoclay/Anthracite	10,916	2,249	21%	677	0.068%
Organoclay	12,719	2,771	22%	950	0.095%
Clinoptilolite	119,678	32,374	27%	8,959	0.896%
Hydroxy-Apatite	243,642	69,069	28%	28,717	2.872%

[a] 95% breakthrough estimated by linear extrapolation of measured data for organoclay/anthracite, clinoptilolite and hydroxy-apatite sorbents.

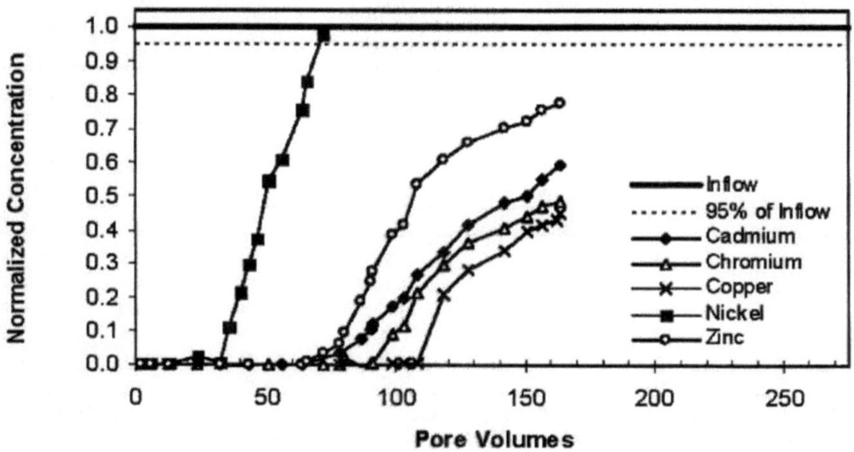

Figure 5. Breakthrough curves of individual metals from an inflow solution containing 60 mg/L Cd, 150 mg/L Cr, 120 mg/L Cu, 60 mg/L Ni and 60 mg/L Zn through a column containing hydroxy -apatite.

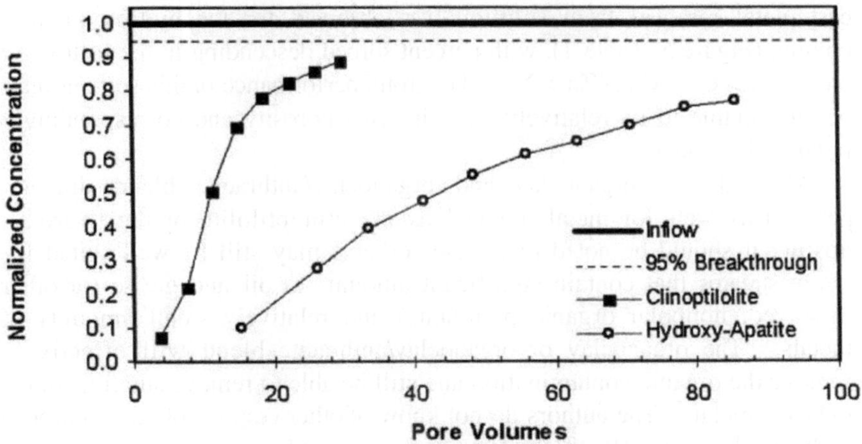

Figure 6. Breakthrough curves from an inflow solution containing 785 mg/L of Zn through columns containing the listed sorbent materials.

4. DISCUSSION

Considering first the multiple-metal-sorption experiments (Figure 2 and Table 1), several important trends can be identified. For the organoclay/anthracite blend, it appears that sorption of all five metals is similar and relatively weak. Sorption increases for the pure organoclay (Figure 3, Table 1), with percent sorbed descending in the following order: Cd > Cr > Cu > Zn > Ni. The greater sorption of the metals to the organoclay compared to the organoclay/anthracite blend suggests that the organoclay is a better metal sorbent than the anthracite. Replacing the anthracite in the blend with organoclay therefore results in a better sorbent for the metals. Sorption of the metals is further increased when the sorbent is changed to clinoptilolite, with percent sorbed descending in the following order: Cu > Cr > Zn > Cd > Ni. Unlike organoclays or anthracite, zeolites like clinoptilolite have polar (charged) surfaces that can effectively sorb metals or charged metal-hydroxides, and this feature of the sorbent likely explains its improved performance relative to the organoclay or organoclay/anthracite blend. It may be possible to regenerate the clinoptilolite column with a **3% $NaCO_3$** solution. Other zeolites, such as chabazite, may have a higher removal capacity than clinoptilolite, but these zeolites are more expensive and therefore may not be of practical use. The best metal removal from solution was observed for the hydroxy-apatite column (Figure 5, Table 1), with percent sorbed descending in the following order: Cu > Cr > Cd > Zn > Ni. The strong performance of this sorbent may be attributable to its relatively large internal porosity and correspondingly polar surface area.

Although the organoclay and organoclay/anthracite blend did not perform as well for metal removal as the clinoptilolite or the hydroxy-apatite, it should be noted that these sorbents may still be well suited for waste streams that contain significant amounts of oil and grease (or other dissolved, nonpolar organic pollutants) and relatively small amounts of metals. The organoclay or organoclay/anthracite blend will effectively remove the organic contamination and still be able to remove small amounts of heavy metals. The authors do not know of other comparable dual-purpose sorbents that can effectively treat such a mixed-contaminant wastewater stream.

In considering the transport of Zn as a single solute through the four porous media (Table 2 and Figure 6), the four sorbents show similar relative sorption behavior, e.g. sorption decreases in the following order: Hydroxy-apatite > clinoptilolite > organoclay > organoclay/anthracite. However, the magnitude of sorption, expressed on a percentage weight basis (Table 2), increases for every sorbent compared to the multiple-metal experiments (Table 1). This indicates that sorption of the metals is competitive. Each sorptive media will likely have better metal removal efficiency when only one or two metals are present in the waste stream compared to multiple metals (assuming everything else is equal). Similar conclusions can be drawn for Ni sorption (as a single solute) to clinoptilolite and hydroxy-apatite (Table 3 and Figure 7). Exposing the hydroxy-apatite to 200 mg/l of Ca^{++} seems to have had no deleterious effect on the media's performance, although we have not performed a single-solute Ni/hydroxy-apatite experiment in the absence of Ca^{++}. Such a test would be required to positively confirm this observation. It is a well-known fact that zeolites, clays, and synthetic ion-exchange resins lose substantial capacity for the removal of metals in hard water. Hydroxy-apatite may not have this limitation.

Table 3. Estimated sorptive capacity for each sorbent material to an aqueous inflow solution containing 437 mg/L of Ni.

Sorbent	Inflow mass to breakthrough[a] (mg)	Mass sorbed to breakthrough[a] (mg)	Sorbed/ Incoming (%)	Mass Sorbed/ Mass of Sorbent (mg/kg)	% by weight
Clinoptilolite	18,039	6,213	34%	1,711	0.171%
Hydroxy-Apatite	62,233	25,945	42%	10,057	1.006%

[a] 95% breakthrough estimated by linear extrapolation of measured data for both clinoptilolite and hydroxy-apatite sorbents.

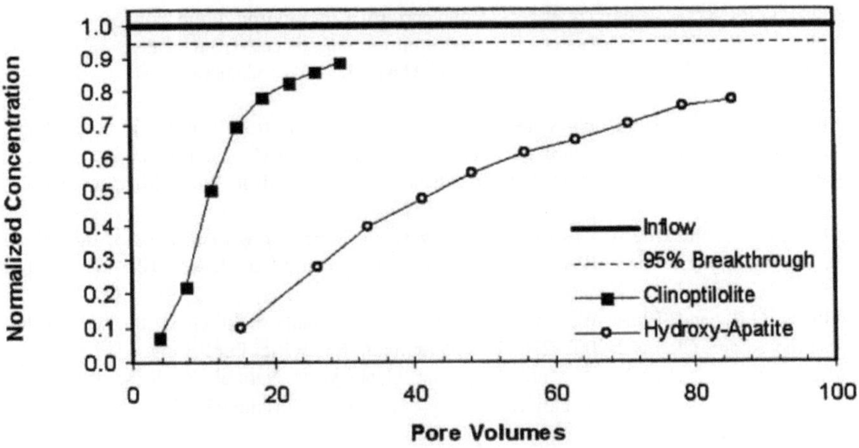

Figure 7. Breakthrough curves from an inflow solution containing 437 mg/L of Ni through columns containing clinoptilolite and hydroxy-apatite.

5. CONCLUSIONS

Four sorbents were studied for the removal of heavy metals from water. All four sorbents showed measurable metal sorption when studied in column experiments with inflow solutions containing all five metals and in a limited number of single-solute column experiments. Sorption of Zn and Ni to clinoptilolite and hydroxy-apatite exhibited competitive sorption, and it is likely that competitive sorption effects exist with the other sorbent-metal combinations. Hydroxy-apatite performed best in the experiments, and appears to be an effective sorbent for heavy metals. Its performance did not appear to be hindered by the presence of hardness (200 mg/L Ca^{++}). In multi-metal column sorption experiments, each sorbent preferentially sorbed different metals, although in general, Cu and Cr uptake was strongest and Ni uptake was weakest. Although the organoclay and organoclay/anthracite blend generally did not perform as well as the clinoptilolite and the hydroxy-apatite, these sorbents may be well suited for removal of oil and/or other nonpolar organic pollutants in combination with small quantities of heavy metals.

REFERENCES

Alther, G.R., 1998. Organoclays boost performance of activated carbon and enhance zero discharge option. Environ. Technol. 40 -43.

Alther, G.R., 2000a. Maximize water cleanup performance. Environ. Protection. 37 -41

Alther, G.R., 2000b. Organoclays remove humic substances from water. In: Humic Substances. 341 p. (Ghabbou r, A.A., and Davis, G., Eds.) Cambridge, Royal Society of Chemistry.

Alther, G.R., 2001. Organoclays remove organics and metals from water. In: Proceedings, Soils, Sediments and Groundwater. Amherst, University of Massachusetts.

Alther, G.R., 2002. Using organoclays to enhance carbon filtration. Waste Management. 22, 507-513.

Alther, G.R., and Ellis, W., 2001a. Efficient wastewater cleanup with clay -based flocculants. In: Proceedings, AESF/EPA Conference for Environmental Excellence. 127-131. Orlando, AESF.

Alther, G.R. and Ellis, W., 2001b. Bentonite flocculants help clean up wastewater. Proceedings, Cleantech 2001. 236 -240. Flemington, NJ, Witter Publishing.

Mumpton, F. (Ed.), 1981. Mineralogy and Geology of Natural Zeolites, Review in Mineralogy, Vol. 4. Mineralogical Society of America, Washington, DC. 225 p. Chelsea, MI, Bookcrafter, Inc.

CHAPTER 30

EX-SITU TREATMENT OF DENSE NON-AQUEOUS PHASE LIQUIDS USING CALCIUM OXIDE (QUICK LIME)

Wm. Gordon Dean
WRS Infastrucrture & Enviornment Inc., Tallahassee, FL

Abstract: Open excavation was selected to remove dense, non -aqueous phase liquids (DNAPLs) identified by previous assessment activities at the FDOT Fairbanks Disposal Pit in Gainesville, Florida. The DNAPLs resulted from the disposal of asphalt testing residue and consisted primarily of trichloroethene and 1,1,1-trichloroethane. Presumptive and visual evidence of DNAPLs had been found in three areas of the site. Excavation of all three areas was conducted and DNAPLs were observed as predicted.

The original plan called for ex-situ vacuum extraction of all contaminated soils. All potentially contaminated soils were to be processed through a rotary trommel, staged in 100 cubic yard piles, and sampled for analytical testing. Due to the increasing clay content of the deeper soils, the trommel could not be used for most of the potentially contaminated soils. Quick lime was mixed into the soils at an approximate 5% ratio to improve the soil handling characteristics. The combination of the heating and chemical reaction between the soils and the lime removed the contaminants to below the leachability soil cleanup target levels, and all contaminated soils were ultimately treated by mixing with quick lime. The soils treated with quick lime were returned to the excavation to stabilize the slopes after confirmation soil samples verified the treatment goals had been reached. Approximately 10,000 cubic yards of soil were treated with quick lime, saving several million dollars as compared to off -site disposal.

Key words: DNAPL, calcium oxide, quick lime, ex-situ soil treatment

1. SITE BACKGROUND

The Fairbanks Disposal Pit (FDP) is an abandoned sand borrow pit, approximately 10 acres in size, in Fairbanks, Alachua County, Florida. The FDP is owned by the Florida Department of Environmental Protection (FDEP) and operated by the Florida Department of Transportation (FDOT). FDOT purchased the site in 1956 as a borrow source for sand and clay for road construction. Beginning in 1956, the site was used to dispose of waste and topsoil collected from roadsides, demolition debris, and construction debris. Waste generated by the FDOT's Bureau of Materials Research (BMR) Laboratory was also collected and disposed at the FDP. Disposal activities continued until 1982. No records exist to document the quantity of hazardous wastes disposed during site operations.

The wastes derived from the BMR Laboratory included chemicals used in laboratory testing. A major portion of their work is to test asphalt paving materials to determine if they meet FDOT specifications. From 1956 until approximately 1961, carbon tetrachloride was the solvent used to dissolve asphalt samples. In 1961, carbon tetrachloride was replaced by the solvent 1,1,1-trichloroethane. In 1980, the Abson recovery process began to be used to dissolve the asphalt samples. Benzene was the solvent used in this process until 1981, when trichloroethene was substituted.

Between 1956 and 1978, waste solvents were recycled after distillation in a solvent still. The still was closed in 1978 when it was identified as a fire hazard. Information regarding disposal of waste from the distillation unit could not be found. During removal activities conducted in 1983, a total of 1,046 drums were excavated from the pit area. In addition to the initial drum removal in 1983, further excavation at the site in 1990 removed 345 drums. An additional 35 drums were removed during closure activities in 1994 and 1995. In addition to disposal of drummed wastes, liquid wastes were reportedly poured directly into trenches on site.

WRS Infrastructure & Environment, Inc. (WRS) was retained by FDOT in January 1994 to complete the physical closure activities in response to a CERCLA Administrative Order. Following soil excavation activities, WRS was tasked with designing and installing groundwater treatment systems to address groundwater contamination identified in four aquifers that underlie FDP. Groundwater treatment systems for all four aquifers were installed and started up between December 1995 and April 1996. In 1998 it was noted that the influent concentration in two surficial groundwater recovery trenches closest to the 10-acre site were not decreasing, suggesting a source of contamination was still present. A Geoprobe™ investigation was conducted in 1999 in an effort to assess possible sources contributing to the dissolved phase contaminant concentrations observed in the recovery trenches. WRS initiated direct-push investigations on the modified RCRA cap, downgradient of the cap, and along the horizontal recovery trenches. Additional assessments using a membrane interface probe, a Rotasonic™ drill rig, and extensive soil sampling by the University of Waterloo were conducted in 2000. Presumptive evidence of dense, non-aqueous phase liquids (DNAPL), defined as groundwater concentrations greater than 1 % of the solubility limit, was found in four areas under the northern half of the 10-acre site. Concentrations as high as 1,200,000 μg/l of trichloroethene were detected and DNAPL was visually identified in three of these areas. The DNAPL was located on top of, and diffused into, a clay layer located approximately 35 to 40 feet below land surface.

Based on the findings of additional assessment activities, FDOT elected to proceed with physical removal of soils that were possibly acting as continuing contaminant sources. WRS's scope of work included the following proposed remediation activities:

- Removal of the clean overburden (land surface to 30 feet below land surface)
- Dewatering using horizontal dewatering trenches and sump pumps
- Excavation of potentially contaminated soils below a depth of 30 feet
- Power screening these soils using a rotary trammel to remove any debris
- Sampling and segregating clean and contaminated soils
- Ex-situ vacuum extraction of contaminated soils in an existing, lined storm water pond
- Addition of 5% calcium oxide (quick lime) as needed to improve the soil handling characteristics of clayey soils that were expected to be encountered at depth.

The goal of the remediation was to remove all soils exceeding the leachability soil cleanup target levels in Chapter 62-777, Florida Administrative Code. The target levels for the contaminants of concern were benzene (7 μg/kg), 1,1-dichloroethene (60 μg/kg), trichloroethene (30 μg/kg), and 1,1,1-trichloroethane (1,900 μg/kg).

2. QUICK LIME PROPERTIES

Calcium oxide (quick lime) is a white crystalline solid manufactured by heating (calcining) limestone, dolomite, coral, oyster shells, or chalk (which are all mainly calcium carbonate) to drive off carbon dioxide. More than 90% of the lime produced in the United States is used for basic or industrial chemistry. The primary uses are (Kirk-Othmer 1991a, b; Budavari, 1986):

- steel manufacture
- metallurgy
- air pollution control
- water and sewage wastewater treatment
- cement and mortar
- chemical manufacturing
- glass and paper manufacturing
- sugar refining

Calcium oxide is also used frequently in the construction industry for soil drying, soil modification, and soil stabilization. Quick lime reacts with water to produce heat (as high as 800° C) and calcium hydroxide (hydrated lime). When applied to wet soils or mud, the reaction consumes approximately 30% water by weight; removes additional water as steam; and provides dry solids for stability. The most dramatic soil modifications occur in clay soils of moderate to high plasticity. Modification occurs because the calcium cations supplied by the lime replace the cations normally present on the surface of the clay mineral. The clay surface mineralogy is permanently altered and the clay soils and changed to silts, producing the following benefits (Little, 1999):

- Plasticity reduction
- Reduction in moisture-holding capacity (drying)
- Swell reduction
- Improved stability

Highly plastic clays can be immediately and permanently changed to low plasticity silts using quick lime. This was the reason quick lime was proposed to amend the soil-handling properties of the clays expected to be found at depth at the FDP site.

From an environmental standpoint, calcium oxide is benign. The US Food and Drug Administration has affirmed that calcium oxide is generally recognized as safe as a direct human food ingredient. It may be used in food with no limitation other than current good manufacturing practice. However, working with quick lime does present some difficulties. It reacts with any moisture, including sweat, eyes, and lung tissue, making adequate personnel protection essential. It is also incompatible with many organics, halogens, acids, and combustible materials.

3. SOIL REMEDIATION USING QUICK LIME

Excavated soil from below 32 feet at the FDP site demonstrated increasing clay content with depth. As the clay content increased, the rotary trommel became unable to effectively process the soils. As proposed in the original work plan, quick lime was added to improve the materials handling. The objective was to modify the clays into non-cohesive silts that could be processed by the trommel. Quick lime was mixed with the soils with a large road tiller to amend the soil for treatment. Approximately 5% by volume quick lime was added to the clayey soils. The reaction of the quick lime with the soil moisture generated significant amounts of heat. In addition, quick lime chemically reacts with many organics. Because the contaminants of concern were volatile organics, WRS sampled the soils after addition of the quick lime to determine the effect, if any, of the quick lime addition. Analytical testing of the soils after mixing with the quick lime demonstrated that the target levels had been achieved and that no further treatment was required. Perimeter air sampling did not detect any chlorinated solvents during the treatment process. Leaching tests were also conducted on the treated soils to determine if the quick lime was somehow masking the chlorinated solvents during the soil analyses. No contaminants were detected in the leachate samples. It was concluded that the soil remediation resulted from volatilizing the chlorinated compounds by the heat of reaction between the quick lime and water, possibly combined with a direct chemical reaction with either the halogens or the organic compounds.

The soil-processing procedure was then modified by staging all potentially contaminated soils in one large pile in the processing area. Soils containing visible DNAPL were treated in the excavation by mixing in quick lime with a backhoe prior to removal. All potentially contaminated soils were placed in the lined soil treatment area in 18-inch lifts. The quick lime was delivered in 2,000 pound supersacks which have a discharge chute built into the bottom. A backhoe was used to pick up the supersacks and spread the quick lime over the soils to be treated. The quick lime was then mixed into the soils using a large road tiller. The treated soils were stockpiled and sampled. A total of 569 soil samples were sampled from the soil processing area. All soils were ultimately classified as clean although some soils required additional treatment before a clean status was achieved. Soil analytical results initially categorized as contaminated (above the leachability soil cleanup target levels), per on-site mobile laboratory results, were re-processed with quick lime. Clean soil was hauled to the clean soil stockpile. No soils exceeding the leachability standards left the processing area.

Soils from below approximately 35 feet, soils previously placed in the lined pond, and sediments from the dewatering equipment were processed in this manner. Approximately 11,500 cubic yards of contaminated soils were treated with the quick lime.

Once the planned excavation was completed, floor and wall samples were taken in the excavation to confirm that all contaminated or potentially contaminated material had been removed and that only clean, undisturbed material remained. If the confirmation soil samples indicated values exceeding the leachability levels, an additional ten foot by ten foot area around the location was excavated an additional two feet deep. These two foot intervals continued until the confirmation samples were below leachability levels. As the excavation activities continued, it was apparent that source areas would need to be excavated much deeper than the initial 37 feet. The final excavation depth reached 49 feet below land surface in one area of the site. The excavation boundaries were also extended further west and north based on the confirmation sample results.

4. COST

The delivered cost of the quick lime was approximately $200 per 2,000-pound supersack. Using a 5% by volume mixture resulted in a material cost of approximately $10 per cubic yard. The multiple soil-handling steps dictated by the site constraints added approximately $20 per cubic yard in handling and processing costs. The total cost to treat the contaminated soils to the leachability target levels was approximately $30 per cubic yard, or a total of $350,000 for the approximately 11,500 cubic yards of treated soils. These costs did not include the cost of clean overburden removal, dewatering, or treatment of the dewatering discharge. Compared to an off-site transportation and disposal cost of $250 to $300 per ton, a savings of $2.5 to $3 million was realized. Any potential liability from the off-site disposal was also eliminated.

5. CONCLUSIONS

Calcium oxide (quick lime) was used to successfully remediate 11,500 cubic yards of soils contaminated with chlorinated solvents and DNAPLs for a unit cost of approximately $30 per cubic yard. The soils were successfully treated to the leachability soil cleanup target levels in Chapter 62-777, Florida Administrative Code and to non-detectable concentrations in most cases. The treatment mechanism was most likely a combination of heat to vaporize the volatile contaminants and a direct chemical reaction with the quick lime. Further research is needed to identify the exact mechanisms. The cost savings compared to off-site transportation and disposal was $2.5 to $3 million and any potential liability from the off-site disposal was eliminated.

REFERENCES

Aldrich Chemical, Inc. MSDS Safety Information for Calcium Oxide; February 10,1992

Ash Grove Cement Co. MSDS for Calcium Oxide

Budaveri, S; Calcium Oxide; The Merck Index., 11th Edition, 1989

Calcium Oxide – Crops; National Organic Standards Board Technical Advisory Panel Review; Compiled by OMRI for the USDA National Organic Program; April 4, 2002

Fisher Scientific, Inc. MSDS for Calcium Oxide

Hazardous Substance Fact Sheet for Calcium Oxide; New Jersey Department of Health and Senior Services

Kirk-Othmer, Kroschwitz, Howe-Grant, eds; Calcium Compounds; Wiley Interscience Publ.; 1991a

Kirk-Othmer, Kroschwitz, Howe-Grant, eds; Vol. 15 Lime and Limestone, Encyclopedia of Chemical Technology, 4th edition; Wiley Interscience Publ.; 1991b

Lime Dries Up Mud; National Lime Association

Lime in Soil, National Lime Association

Little, Dallas N.; Evaluation of Structural Properties of Lime Stabilized Soils and Aggregates, Volumes 1 through 4; National Lime Association, January 5, 1999

Mallinckrodt Baker, Inc. MSDS for Calcium Oxide

National Fire Protection Association Fire Protection Guide on Hazardous Materials; 7th ed.

Petry, Thomas M.; Final Project Report - Investigation of Agents of Practical Use to Stabilize Slopes and Erosion Along the I -55 Corridor South of Sikeston, Missouri; MODOT Project No. RI 99-030; April 19, 2001

Soil Acidity and Liming – Extension Internet Inservice Training ; Clemson University Extension Service, Department of Crop and Soil Environmental Science

Using Lime for Soil Stabilization and Modification; National Lime Association; Updated March 2001

This page intentionally left blank

CHAPTER 31

NAPL CONTAINMENT USING IN SITU SOLIDIFICATION

James R. Greacen[1] and Edward J. Walsh[2]
[1]The RETEC Group, Concord, MA and [2]Lyme Properites, LLC, Cambridge, MA

Abstract: Dense non-aqueous phase liquid (DNAPL) at a depth of 20 feet below grade covered a three-acre area of a former manufactured gas plant site, preventing site closure and inhibiting development of the property. As part of site closure, soil and DNAPL were stabilized *in situ* using a crane-mounted, 10-foot-diameter auger capable of injecting a 7% cement grout mixture. The auger was advanced approximately 22 feet below grade to extend beneath the DNAPL and key the stabilized mass into the underlying clay layer. In total, 2,256 overlapping columns were advanced over an area covering approximately 144,000 square feet. To demonstrate that performance requirements were met, uncured samples were collected, cured in the laboratory, and analyzed for NAPL saturation. Samples of the soil and DNAPL that had cured *in situ* were also collected to allow direct observation of stabilized material and to provide additional supporting documentation of the permanence of the remedy. The solidification remedy allowed site closure to proceed, and the construction of two 250,000-square-foot life science buildings is nearing completion. The buildings are scheduled for occupancy in 2004.

Key words: DNAPL, LNAPL, in situ, solidification, NAPL saturation

1. INTRODUCTION

Containment technologies are often used to prevent the migration of nonaqueous phase liquid (NAPL) when removal is not a viable option. Hydraulic containment is often used in cases where light nonaqueous phase liquid (LNAPL) is present and require remediation. Containment of both LNAPL and dense nonaqeous phase liquid (DNAPL) is typically accomplished using sheet pile walls or slurry walls that surround the NAPL. *In situ* solidification is a method of mixing soil and adding a reagent mixture to create a solidified monolith in the subsurface. It may be used to surround NAPL to construct a barrier wall or used within the NAPL to eliminate the occurrence of NAPL altogether.

In Massachusetts, a waste-disposal site cannot be permanently closed without, at a minimum, eliminating the occurrence of NAPL at thicknesses greater than one-half inch. At a site in Cambridge, *in situ* solidification was used to eliminate the occurrence of DNAPL and allow a development to proceed in a parcel that had remained dormant for over 30 years.

2. BACKGROUND

From circa 1850 to 1960, Cambridge Gas and Light Company operated a manufactured gas plant (MGP) in Cambridge, Massachusetts. Since then, the ten-acre property remained undeveloped due in part to the significant volumes of tars and oils left behind by the former industrial operations, including DNAPL covering a three-acre area. As much as four feet of DNAPL lay atop a confining clay unit at a depth of 20 feet below grade. Lesser volumes of LNAPL were also present. Despite its location in a thriving urban area, the property remained undeveloped for over 30 years since the decommissioning of the MGP.

Lyme Properties, a developer of biotech office and lab space as well as other commercial properties, hired RETEC to evaluate the costs of remediating the property as part of property development. Various remediation scenarios were combined with potential development plans to assess whether remediation of the property was economically feasible. Among the technologies reviewed were excavation with off-site disposal, excavation with on-site solidification, *in situ* solidification, and NAPL extraction with a variety of enhancements including steam stripping, surfactants, and high vacuum extraction. The excavation options were rejected due to the costs of treatment or disposal of the NAPL and soil. The recovery technologies were rejected due to a combination of cost, time to complete the remedy, and the uncertainty of meeting the regulatory criteria.

In situ stabilization was selected as a remedy that could be completed in a relatively short time, and would result in a cost that could be factored into the development.

3. TECHNOLOGY DESCRIPTION

In situ solidification using soil augers is an established technology widely used in geotechnical construction to improve the strength of soil. Soil mix walls constructed using similar equipment and procedures have been widely used for structural support on recent projects in the Boston area associated with the Central Artery or "Big Dig" construction project. Stabilization also has a long history of successful use in the remediation of chemical contaminants in soil. In this case, the primary objective of solidification was to eliminate and permanently immobilize NAPL as required to achieve a permanent remedy under Massachusetts regulations. The technology in this application was used to solidify the NAPL rather than prevent leaching of constituents from the soil. Nevertheless, by reducing the hydraulic conductivity of the material, solidification has the additional benefit of reducing the amount of groundwater coming into contact with the MGP residuals.

4. SITE PREPARATION

Many of the subsurface structures of the former MGP required removal prior to stabilizing the soil and NAPL. Building foundations, pipes, portions of subsurface tanks, and large quantities of debris were excavated and shipped offsite to allow the solidification equipment to be advanced to the target depths and adequately mix the soil in place. Removal of subsurface debris and 8,500 cubic yards of soil were also required to maintain existing grades by offsetting the volume of material injected during the solidification process and any "swelling" of the soil resulting from the solidification process.

5. FULL-SCALE EQUIPMENT

The soil mixing equipment consisted of a 150-ton crane, mixing platform, 70-foot Kelly bar, auger assembly, fume hood, vapor treatment system, and grout plant. The auger assembly consisted of a 10-foot diameter auger equipped with ports to inject a reagent mixture into the subsurface. The mixing platform that provided the power to rotate the auger was mounted on the crane and was capable of generating 450,000 ft-lb of torque. A vapor-collection shroud placed at the ground surface over the auger collected vapors released during the mixing process and directed them to a granular activated carbon-vapor treatment system (Figure 1).

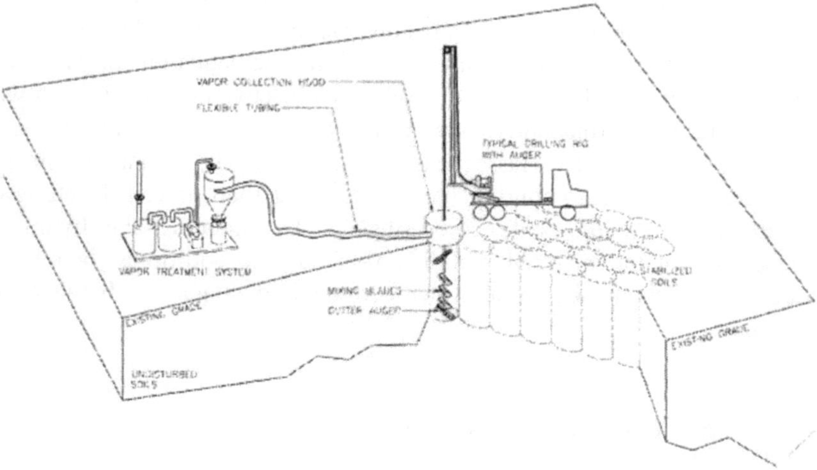

Figure 1. Conceptual overview of *in situ* solidification

6. FIELD PILOT TESTING

During the first three days in the field, the solidification equipment was tested by performing a series of "test runs" using a variety of grout mixtures evaluated as part of a treatability study. During the test runs, any necessary modifications to the equipment and grout mixture were performed. The initial test runs were performed using a reagent mixture of approximately 10% cement kiln dust (CKD) by weight of soil, and 2% bentonite by weight of CKD. Several days later, it was observed that the mixed columns were not curing quickly enough meet the solidification schedule. Due to the slow curing of the CKD, a reagent mix consisting of 5% Portland cement by weight of soil was used as the solidification reagent. This mixture successfully eliminated the presence of NAPL, leaving a wet, non-cohesive column of soil and grout mixture; however, the soil columns did not cure sufficiently over a 24-hour periods to allow moving equipment onto them to stabilize adjacent soil columns. As a result, the reagent mixture was modified to use a Portland cement ratio of 7% by weight of the soil, and 1-2% bentonite by weight of cement. Although the 5% cement mixture successfully eliminated the NAPL and hardened over a one-week period, the 7% mix also eliminated the NAPL and cured sufficiently within 24 hours to allow the solidification equipment to move more freely around recently-stabilized areas.

7. FIELD APPLICATION

The soil and NAPL were stabilized *in situ* using a crane-mounted, 10-foot diameter auger that injected the cement grout mixture. The auger was advanced approximately 22 feet below grade to extend beneath the DNAPL and key the stabilized mass into the underlying confining clay layer. The auger was first advanced to the target depth and the soil mixed as the auger was moved up and down through the soil column. During the mixing process, the reagent mixture was added to the soil. The soil columns were overlapped to ensure full coverage, with no unstabilized areas between columns. Once the reagent mixture cured, a stabilized monolith of hardenedsoil remained. During nine months of field application, 2,256 overlapping columns were advanced over an area covering approximately 144,000 square feet.

8. SAMPLE COLLECTION

Monitoring was performed during solidification operations to ensure that performance requirements were met. Samples were collected both immediately after solidification (uncured samples) and after the concrete-based reagent mixture had set (cured samples). Uncured stabilized samples were collected to allow the laboratory to place the soils into precast forms for NAPL saturation analysis. Samples of the soil and NAPL that had cured *in situ* were also collected to allow direct observation of stabilized material and to provide additional supporting documentation of the permanence of the remedy.

9. SAMPLE ANALYSIS

The following qualitative procedures were used for documenting the absence of "free" NAPL:

visual observation of the samples in the field during the collection of the uncured samples;

visual observations of the centrate emanating from samples of the stabilized material after they have been spun for an hour in a centrifuge at 1,000 Gs following ASTM Method D425; and

visual observations of the recovered cores of stabilized material.

A separate quantitive test for the absence of NAPL was conducted by analysis of the cured samples for NAPL saturation and NAPL saturation capacity. Design performance is achieved if the in-place NAPL saturation is less than the NAPL saturation capacity of the stabilized soil. Hydraulic conductivity testing was used to measure the stabilized area's resistance to groundwater flow and to evaluate the relative homogeneity of the stabilized mass.

A total of 52 samples were analyzed following a 28-day curing period. A summary of the saturation sample test results is provided in Table 1. Post treatment NAPL saturation indicates the concentration of total NAPL within the soil matrix of the stabilized soil following treatment, while NAPL saturation capacity indicates the concentration of total NAPL the stabilized soil is capable of holding. All 52 samples demonstrated that the stabilized material has additional capacity to hold NAPL in the residual (immobile) phase.

Table 1. Stabilized Soil Data

	Bulk Density (g/cc)	*Porosity*	*Post Treatment NAPL Saturation (%Pore Volume)*	*NAPL Saturation Capacity (%Pore Volume)*	*Hydraulic Conductivity (cm/sec)*
Minimum	1.07	37.7	<0.1	11.2	9.07×10^{-8}
Maximum	1.63	55.5	14.9	52.6	3.06×10^{-5}
Average	1.33	46.6	3.7	25.5	2.70×10^{-6}

The permeability testing results indicate that the hydraulic conductivity of the stabilized material ranges from 3.06×10^{-5} cm/sec to 9.07×10^{-8}

cm/sec, demonstrating several orders of magnitude reduction in permeability.

10. PROJECT STATUS

In situ solidification was performed over a three-acre plume of DNAPL to meet the Massachusetts requirement of eliminating the occurrence of NAPL to achieve permanent site closure. By combining the remediation with site development, remediation of the NAPL was accomplished in less than one year at a cost of approximately $80 per cubic yard, or about half of the estimated cost for excavation and offsite treatment/disposal. Two buildings that overlie the stabilized soil are nearing completion. One building is a 247,000 square foot building that will be used as office, retail, and biotech space. The second building is a 273,000 square foot "green building" that will be the corporate headquarters for Genzyme, a biotech company.

CHAPTER 32

ANAEROBIC BIODEGRADATION AND BIOTRANSFORMATION USING EMULSIFIED EDIBLE OILS

M. Tony Lieberman[1], Nicholas L. Lindow[2], Dr. Robert C. Borden[2], and Gary M. Birk[3]

[1]Solutions Industrial & Environmental Services, Inc.,Raleigh, NC, [2]North Carolina State University,Raleigh, NC, [3]EOS Remediation, Inc., Raleigh, NC

Abstract: Once in groundwater, chlorinated solvents, nitrate, perchlorate and dissolved metals from acid rock drainage pose serious water quality threats. Options for clea nup of these contaminants vary, but many have high capital costs that need long -term maintenance, continuing attention, and are not appropriate for lightly contaminated areas or where the source area is poorly defined. The introduction of emulsified edible oil substrate (EOS®) to the aquifer has been shown to accelerate anaerobic biodegradation and biotransformations of these widespread contaminants (U.S. Patent #6,398,960).

Emulsified edible oils have been shown to be very effective as a long-lasting, natural time-release, organic substrate. Microcosm and field data have confirmed their longevity in aquifers for over three years without re -application as well as their ability to stimulate the desired biological activity. Degradation of chlorinated ethenes and ethanes can result in formation of daughter products and non -toxic end products, but the outcome is very site - specific. By contrast, perchlorate and nitrate are rapidly removed from groundwater matrices given the proper environmental conditions for growth of generally ubiquitous perchlorate - reducing and nitrate-reducing microorganisms. The biological creation of reducing conditions in acidic rock drainage streams can induce precipitation of dissolved metals and neutralized waste streams. The EOS® microemulsion has been shown effective for promoting these outcomes.

Key words: Edible Oil Substrate, Anaerobic Biodegradation, Anaerobic Biotransformation, Chlorinated Solvents, Perchlorate, Nitrate Reduction, Acid Rock Drainage

1. INTRODUCTION

A novel, low-cost technology has been patented (U.S. Patent #6,398,960) for delivering a low solubility, slowly degradable substrate to the subsurface to enhance the *in situ* biotransformation of a variety of groundwater contaminants including chlorinated solvents, perchlorate, nitrate, and dissolved metals. Chlorinated solvents are a frequently encountered problem in groundwater at Department of Defense (DoD) facilities, textile manufacturing plants, asphalt-testing labs and dry cleaning establishments. Anaerobic reductive dechlorination has been shown to be an efficient microbial means of transforming more highly chlorinated species to less chlorinated species. Groundwater contamination by perchlorate (ClO_4^-) has recently been identified as a major environmental issue, especially for the DoD. In many cases, perchlorate has entered groundwater through the release and/or disposal of ammonium perchlorate (AP), a strong oxidant that is used extensively in solid rocket fuel, munitions and pyrotechnics. Recent studies have shown that millions of gallons of drinking water aquifer material have been impacted by perchlorate, which is an inhibitor of thyroid uptake of iodine in humans. USEPA has recently released a draft risk assessment that could result in a drinking water limit for perchlorate of 1 μg/L.

High nitrate concentrations in groundwater are a major problem in agriculture, livestock, poultry–producing areas and elevated nitrate concentrations are also a major problem associated with land disposal of treated sludge from domestic wastewater treatment plants. High nitrate concentrations in groundwater are a health concern and can lead to water quality problems in surface water. The US Geological Survey found that nitrate concentrations exceeded the drinking water standard in over 15% of the wells sampled indicating that millions of acres are contaminated with excess nitrate. The principal effects of nitrate contamination arise as a result of the conversion of nitrate to nitrite in the oral cavity and/or stomach. Nitrite oxidizes hemoglobin to methaemoglobin, which interferes with the transport of oxygen by the blood, a condition known as methaemoglobinaemia.

The effect of acid rock drainage (ARD) on groundwater is a significant and growing problem for the mining industry. There are between 100,000 and 500,000 abandoned mines in the United States along with the many thousands of active mines. Currently there are 50 mines listed on the Superfund National Priorities List and the Environmental Protection Agency (EPA) estimates there are 2.2 million acres of metals-polluted surface water resulting from mining operations in just eight States. Estimated costs for control of ARD in just the State of Pennsylvania alone, ranges between 5 and 15 billion dollars (PADEP, 1997).

Because of the magnitude of these problems, it is vitally important to continue development of effective control measures to manage these environmental and economic problems. Control technologies must be inexpensive to implement with little required maintenance, because many of these sites are large and located in remote areas. Edible Oil Substrate (EOS®) is one of a handful of technologies being evaluated for use within the United States and overseas to address these pressing groundwater issues.

2. EDIBLE OIL SUBSTRATE (EOS®) PROCESS- U.S. PATENT #6,398,960

The EOS® process uses a patented soybean oil-in-water emulsion to enhance *in situ* biodegradation in groundwater by bringing a biodegradable organic substrate in the form of an edible oil into contact with the contaminants. This substrate serves as a carbon source for cell growth and as an electron donor for energy generation. To achieve this, the EOS® emulsion is injected through permanent wells or temporary direct-push points. Water is subsequently injected to distribute and immobilize the oil. Once in the aquifer, the oil slowly biodegrades over time providing a slow continuous source of dissolved organic carbon (i.e., fermentation products) to support biodegradation of the target contaminants.

EOS® can be injected into "hot spots", throughout the plume or as a permeable reactive barrier using conventional wells or direct-push injection points. All materials used in the process are Generally Recognized As Safe (GRAS), food-grade materials (21 CFR 184.1400) which typically facilitate obtaining regulatory approval for *in situ* application. The amount of EOS® injected into the subsurface is determined based on the concentrations of the target compounds, the concentrations of various biodegradation and geochemical parameters, and the geologic and hydrogeologic conditions. The following sections discuss laboratory and field results that have been obtained that illustrate the potential effectiveness of EOS® on perchlorate, nitrate, heavy metals and chlorinated solvent contamination in groundwater.

3. PERCHLORATE

Perchlorate is highly soluble in water and does not readily sorb to mineral surfaces. A variety of studies have shown that microorganisms from a wide variety of aquifers can anaerobically biodegrade perchlorate when supplied with appropriate organic substrates and related amendments. A growing volume of information has been developed demonstrating that a wide diversity of microorganisms can degrade perchlorate to chloride and oxygen (Coates et al. 1999; Coates and Pollock 2003). Perchlorate-reducing organisms are widespread in the environment (Coates et al., 1999; Logan, 2001; Wu et al., 2001) and can use a variety of different organic substrates (e.g. acetate, propionate, lactate, etc.) as electron donors for perchlorate reduction (Herman and Frankenberger 1998; Coates et al., 1999). The specific biochemical pathways necessary for perchlorate reduction are not well known, but there is good evidence that the pathway proceeds via the following sequence (Logan et al., 2000):

ClO_4^- (perchlorate) → ClO_3^- (chlorate) → ClO_2^- (chlorite) → Cl^- (chloride) + O_2

The effect of EOS® ingredients on perchlorate biodegradation has been studied in two separate microcosm studies. Initially, triplicate microcosms were constructed using sediment and groundwater from the Neuse River Waste Water Treatment Facility in Raleigh, NC. Each bottle was constructed to contain 100 mL of groundwater, 130 mg/L perchlorate, 50 mg of the target substrate to be evaluated, and 50 mL of moist sediment. Laboratory analysis showed that the perchlorate concentration was reduced to below the analytical detection limit.

Microcosms were established using aquifer material and groundwater from a perchlorate-impacted site in Maryland. Microcosms were constructed with 175 mL of groundwater, 1,000 mg/L of the target substrate to be evaluated, and 50 mL of moist sediment. Microcosms were prepared with live and killed controls and the following treatments: (1) no sediment, no carbon control; (2) live control with no added carbon; (3) killed control; (4,5,6) three separate emulsified oils and (7) emulsified oil plus bioaugmentation culture. Perchlorate was rapidly degraded in all bottles that received emulsified oils (Figure 1). After 37 days, perchlorate was below detection in all substrate-amended bottles. A pilot scale project at the Maryland site is currently being used to demonstrate the effectiveness of EOS®, injected to form a permeable reactive barrier, for promoting *in situ* biodegradation of perchlorate. The use of edible oil substrate is shown to be an effective, inexpensive remediation approach with wide applicability.

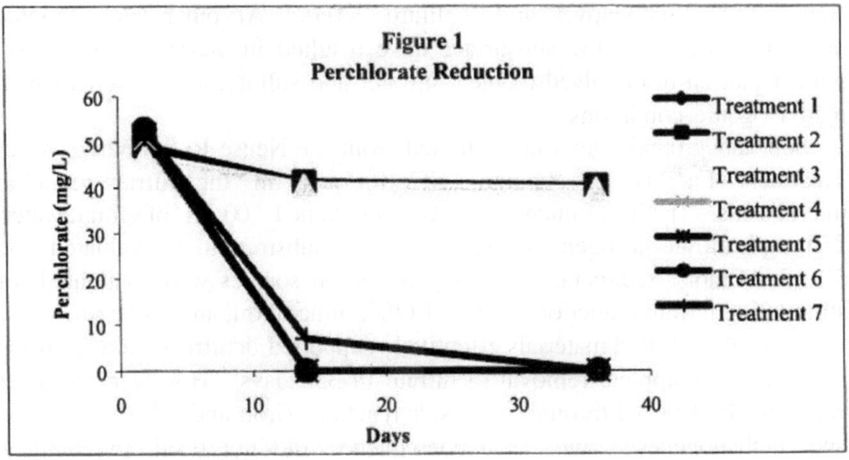

Figure 1. Effect of different carbon sources on perchlorate reduction in batch microcosms containing sediment and groundwater from a site in Maryland.

4. NITRATE REDUCTION (DENITRIFICATION)

There is extensive data on biological nitrate reduction under ambient conditions (ITRC 2002, Martin et al., 1999). Nitrate, like perchlorate, is very mobile, nonsorptive and resistant to biodegradation in many aquifers. While denitrifying bacteria are ubiquitous in soils and groundwater, denitrification can be very slow due to the lack of a carbon energy source for bacterial growth. However, in aquifers with higher levels of naturally occurring or anthropogenic organic carbon, nitrate is rapidly removed.

Nitrate is stable in aerobic environments. However, under anoxic conditions, denitrification can reduce nitrate into nitrogen gas.

Oil + NO_3 (Nitrate) → NO_2 (Nitrite) → NO (Nitrous Oxide) → N_2 (Nitrogen Gas)

The denitrifying bacteria responsible for this process are facultative anaerobes that can use nitrate as a terminal electron acceptor in their metabolic pathways. EOS® acts as the electron donor and carbon source. Denitrifying bacteria are sensitive to pH levels and prefer neutral conditions. For example, studies have documented extensive nitrate removal during groundwater flow through organic-rich riparian buffers prior to surface water discharge (Fennessy and Cronk 1997; Martin et al. 1999; Borin and

Bigon 2002; Schoonover and Williard 2003). At other sites, organic materials released to the subsurface have resulted in anaerobic conditions with depletion of dissolved oxygen, nitrate, and sulfate, and the creation of methanogenic conditions.

Soil and groundwater were collected from the Neuse River Waste Water Treatment Facility in Raleigh, NC for use in the nitrate-reducing microcosms. Triplicate microcosms each contained 100 mL of groundwater, 250 mg/L nitrate-nitrogen, 50 mg of the target substrate to be evaluated, and 50 mL of moist sediment. A variety of carbon sources were evaluated for stimulating nitrate reduction, such as EOS®, mineral oil, and molasses.

Most of the tested materials effectively supported denitrification resulting in partial to complete removal of nitrate in 350 days. However, molasses was quickly depleted through other side reactions (iron and sulfate reduction and methanogenesis) and, as a consequence, did not result in complete nitrate removal.

Based on the microcosm results, several carbon sources were selected for further study in intermittent flow columns. The columns were operated for two weeks without any substrate to establish steady-state conditions. One column was treated with various oil emulsions, and one column was left untreated as a control. No inoculum was required due to the natural abundance of denitrifying bacteria.

Figure 2 shows nitrate and dissolved iron concentrations versus time in the effluent of each column. Within one week of treatment, nitrate was removed to below detection and has remained below detection for four months. Nitrite was never detected in the column effluents. Both columns produced significant amounts of dissolved iron indicating highly reducing conditions. Dissolved oxygen in the effluent dropped to non-detect for both columns. The effluent pH for both treated columns remained at a steady 6.5. These bench-scale studies successfully demonstrated the potential application of the patented emulsified oils into groundwater systems contaminated with nitrate.

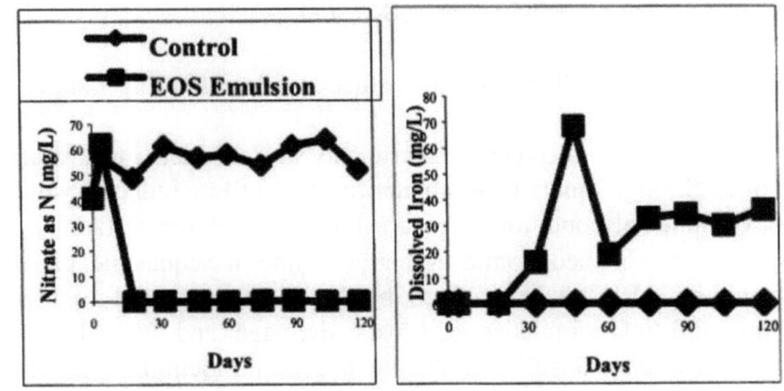

Figure 2. Nitrate and dissolved iron concentrations in the effluent of EOS® emulsion and untreated columns

5. ACID ROCK DRAINAGE - METALS BIOTRANSFORMATION

Coal and hard rock metal mining (including gold, copper, lead and zinc) often results in extensive surface and groundwater contamination. Coal and other important metal ores are found associated with sulfide deposits. During mining operations, these ores and related sulfide minerals [pyrite (FeS_2), pyrrhotite (FeS), chalcopyrite ($CuFeS_2$) and enargite (Cu_3AsS_4)] are exposed to oxygen and water, resulting in the formation of large amounts of sulfuric acid and dissolution of heavy metals including iron (Fe), manganese (Mn), copper (Cu), cobalt (Co), eadmium (Cd), nickel (Ni), and zinc (Zn). The low pH, high sulfate and metals concentrations of surface waters can be very detrimental to aquatic life.

Anaerobic bioremediation processes may be used to reduce sulfate and immobilize heavy metals in the subsurface if a carbon and energy source is available to drive iron and sulfate reduction. Hydrogen and low molecular weight organic acids can be produced from the fermentation of a variety of organic substrates including compost, manure, sugars, oils, and organic rich sediments. Naturally occurring iron-reducing bacteria (IRB) and sulfate-reducing bacteria (SRB) use the hydrogen (and organic acids) to reduce ferric iron (Fe^{3+}) and sulfate (SO_4^{2-}) to pyrite (FeS_2). Shown below is a representative reaction for iron and sulfate reduction using hydrogen as terminal electron acceptor. These reactions consume H^+ causing a decline in acidity and increase in pH.

$$Fe^{3+} + H^{+} + 2\ SO_4^{2-} + 7.5\ H_2 \rightarrow FeS_2 + 8\ H_2O$$

Similar reactions will immobilize other metals including copper, lead, and zinc.

An initial series of microcosm experiments was conducted to evaluate the ability of EOS® ingredients to stimulate iron and sulfate reduction under the harsh environmental conditions typical of acid rock drainage (ARD). Coal mine spoils were obtained from a former coalmine in Sequatchie Valley, TN for use in the ARD studies. Each bottle was constructed in triplicate containing 190 mL simulated acid rock drainage and 25 mL of moist sediment. An adapted sulfate-reducing bacterial inoculum was obtained from a Successive Alkalinity Producing System (SAPS) pond in Pennsylvania. Live and killed controls were compared with coal mine spoils with and without added carbon amendments.

Figure 3 shows the observed variation in dissolved sulfate, dissolved iron and pH in the different microcosms. Values shown are the average of triplicate incubations. There was no significant change in the sulfate, iron or pH levels in the killed control, acidic live control and neutralized live control. The data show that the EOS® ingredient resulted in 100% reduction in sulfate and increase in pH to over 6 in both the acid and neutralized treatments. Sulfate reduction was somewhat more rapid in the neutralized bottles. However, the low pH of the microcosms did not significantly impact treatment in any bottle. Iron was removed from the acid oil treatments at an approximate ratio of 1 mole iron per two moles sulfur indicating iron is precipitated as an iron disulfide with the same chemical composition as pyrite.

The microcosm results indicate that EOS® can provide a very effective treatment for ARD. To further evaluate this process, six columns were packed with coal mine spoils similar to the material used to construct the ARD microcosms. The columns were allowed to equilibrate with an influent solution consisting of 20 mM $FeSO_4$ acidified to pH~3 or neutralized FeSO4 (pH~ 5). Four columns (2 acidic and 2 neutralized) received a one-time treatment of EOS®, and a microbial inoculum from the ARD microcosms. One acidic and one neutralized column remained untreated as no added carbon controls.

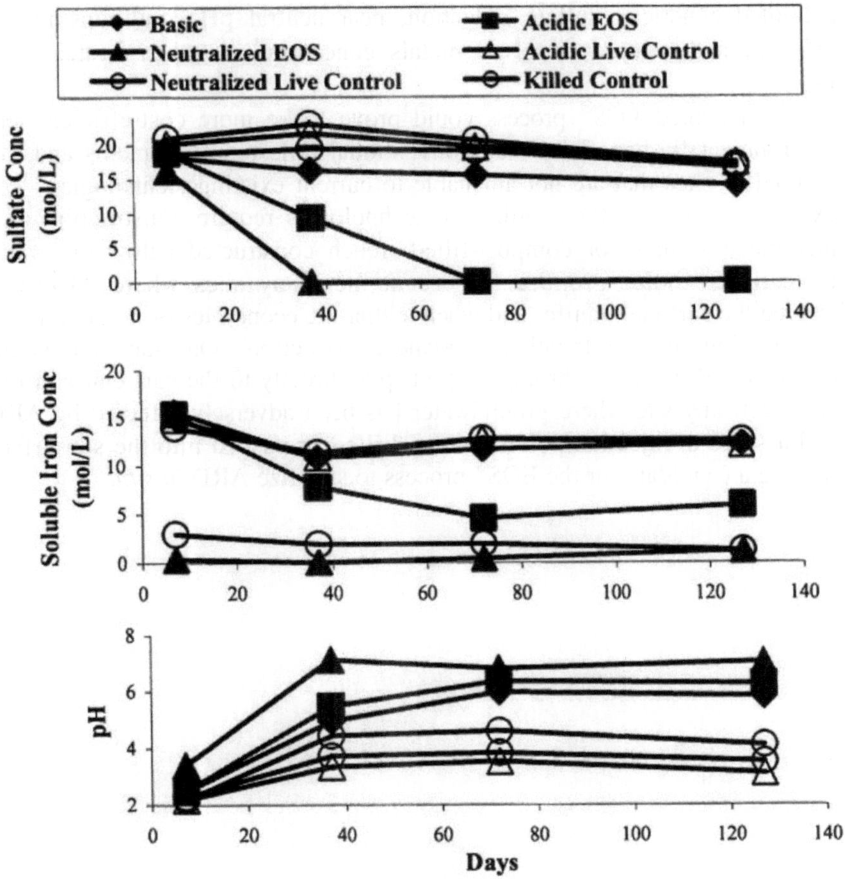

Figure 3. Variation in dissolved sulfate, iron, and pH in ARD microcosms

Figure 4 shows monitoring results from the columns receiving an acidic ARD influent (pH ~ 2.6 to 2.8). In all live carbon amended columns, there is a dramatic reduction in sulfate and increase in pH. There was also a dramatic drop in dissolved iron. Copper and zinc remained below detection in the effluent of the carbon-amended columns, but increased to high levels in the untreated column. The only metal with a lower removal efficiency was manganese. This is believed to be due to the relatively high aqueous solubility of manganese sulfide. As in the microcosms, the low pH did not significantly inhibit the bacterial consortia responsible for sulfate reduction.

The total reduction of metals in the effluent was approximately 750 mg/L, about half that of sulfate. Again the metals removal was limited by the availability of reduced sulfide in the effluent solution. The effluent pH increased to near neutral conditions in all treated columns. The columns exhibited beneficial sulfate reduction, near neutral pH conditions in the effluent, and reduced dissolved metals concentrations when treated with EOS®.

The patented EOS® process could prove to be more cost-efficient and environmentally beneficial than conventional limestone treatments and can be used at sites that are not amenable to current existing technologies. For example, current ARD-stabilization technologies require construction of a man-made wetland or compost-filled trench constructed below the water table. EOS® holds promise for use at the many sites where ARD has entered the surficial aquifer and where either the economics or practicality of construction prevents trench or wetland construction. One such site would be where ARD leaches through a spoils pile directly to the surficial aquifer. In effect, any site where groundwater has been adversely affected by ARD and a series of injection points can be drilled or pushed into the subsurface may be a candidate for the EOS® process to stabilize ARD *in situ*.

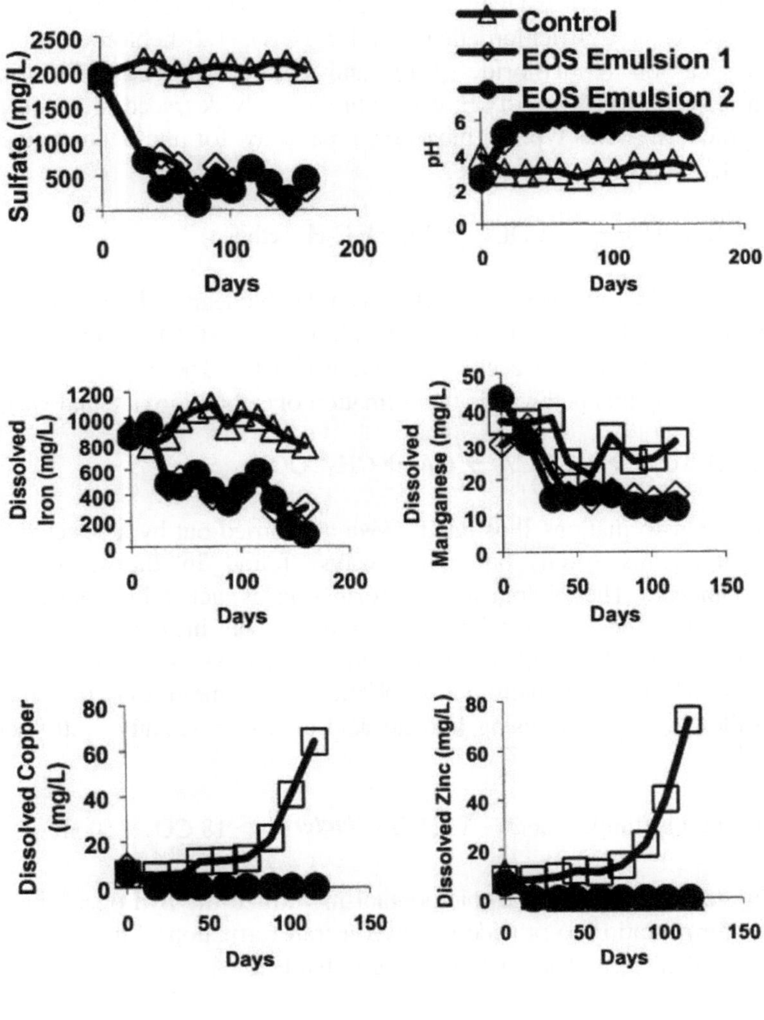

Figure 4. Sulfate, pH, and heavy metals response to EOS® treatment in AMD columns

6. CHLORINATED SOLVENTS

Chlorinated aliphatic hydrocarbon solvents amenable to *in situ* anaerobic bioremediation include tetrachloroethene (PCE), trichloroethene (TCE), *cis*-

1,2-dichloroethene (*cis*-DCE), vinyl chloride (VC), 1,1,1-trichloroethane (1,1,1-TCA), 1,1,2-trichloroethane (1,1,2-TCA), 1,2-dichloroethane (1,2-DCA), carbon tetrachloride (CT), and chloroform (CF). Chlorinated ethenes, such as PCE and TCE, can be biologically degraded into non-toxic end products. The typical breakdown pathway for these compounds is illustrated below.

$$\text{PCE} \rightarrow \text{TCE} \rightarrow \text{1,2-DCE} \rightarrow \text{VC} \rightarrow C_2H_4 \text{ (ethene)}$$

In the same manner, 1,1,1-TCA can be biodegraded to intermediate compounds including 1,1-DCA and chloroethane (CA), which then can abiotically transform to acetic acid (CH_3COOH). The result of complete degradation by this pathway is the formation of carbon dioxide and water.

$$\text{1,1,1-TCA} \rightarrow \text{1,1-DCA} \rightarrow \text{CA} \rightarrow CH_3COOH$$

The transformations illustrated above are carried out by a selected set of microorganisms often, but not always, found in the contaminated environment. The biological transformation is generally considered to involve two steps. After introduction of the oil into the groundwater, biodegradation of the oil results in removal of oxygen and production of hydrogen (H_2). These microbial metabolic transformations are illustrated in the following equation using linoleic acid as a representative fatty acid in EOS®:

$$C_{18}H_{32}O_2 \text{ (linoleic acid)} + 34\ H_2O \text{ --}\textit{Bacteria}\text{--> } 18\ CO_2 + 50\ H_2$$

In step two, specialized microorganisms utilize the hydrogen generated by oil fermentation to provide energy for transformation of the CAHs. For chlorinated ethenes, this reaction is illustrated below:

$$C_2HCl_3 \text{ (TCE)} + 3\ H_2 \text{ --}\textit{Bacteria}\text{--> } C_2H_4 \text{ (ethene)} + 3\ Cl^- + 3\ H^+$$

As a result of an increasing number of laboratory and field studies showing that chlorinated solvents can be biodegraded by naturally occurring microorganisms, the wealth of project knowledge about the effectiveness and limitations of this approach continues to grow. When provided with an appropriate organic substrate, indigenous microorganisms can readily convert an aquifer to conditions suitable to promote reductive dechlorination (Ellis et al., 2000; Lee et al., 1998). However, as simple this may seem, each project site is different and these differences can dramatically influence the results.

A pilot study evaluating the use of EOS® for controlling chlorinated solvent migration was implemented at Altus Air Force base, OK. Historical releases of degreasing agents resulted in a chlorinated solvent plume with TCE concentrations reaching **78,000 μg/L** in the source area (US EPA 2002). Geology at the site consists of reddish-brown, moderately plastic, sandy clay to a depth of 15 ft, underlain by fractured clayey shale with occasional gypsum layers. Most groundwater flow and contaminant transport appears to occur through a series of weathered shale fractures. A line of six permanent 2-inch PVC wells spaced 5 ft apart and screened from 8 to 18 ft below ground surface, was installed perpendicular to groundwater flow approximately 250 feet downgradient from the source area. EOS® was injected into these wells to form a permeable reactive barrier.

TCE concentrations dropped immediately after injection, as illustrated by the data from injection well TS-IW-3 (Figure 5A). Although the concentrations of total ethenes [TCE, *cis*-1,2-dichloroethene (*cis*-DCE), vinyl chloride (VC), ethene and ethane] initially decreased, these temporary reductions were likely due to dilution and/or sorption to the oil. As Figure 5A illustrates, approximately 7.5 months after injection, the concentration of total ethenes (molar concentration) was more than 90 percent of the pre-injection TCE concentration. This demonstrates that dilution/sorption was no longer significant and that the observed reductions in contaminant concentrations were due to biodegradation. Over the 13-month interval since EOS® injection, TCE has declined from **9.9 μM/L (1,300 μg/L)** to below the detection limit (BDL) in the center injection well.

Similar results were observed in monitoring well TS-MW-5, 20 feet downgradient of the barrier (Figure 5B). Emulsion reached TS-MW-5 immediately after injection, as evidenced by a rise in TOC to 2200 mg/L one day after application. Post-injection monitoring over 13 months has shown that TCE decreased from **12.6 μM/L (1,660 μg/L)** to BDL and *cis*-DCE from 9.3 to **0.75 μM/L** (900 to **73 μg/L**). There has been a concomitant increase in VC from 7.0 to **28.3 μM/L** (440 to **1,770 μg/L**) and ethene from 0.25 to **18.2 μM/L** (6.9 to **510 μg/L**). The increase in total ethenes (molar concentration) in this well may be a result of enhanced desorption/dissolution as dissolved TCE is removed through enhanced reductive dechlorination. Although TOC has substantially decreased from the starting concentration, the continuous downgradient migration of dissolved TOC from areas closer to the injection barrier would be expected to support additional reduction of VC to ethene and ethane.

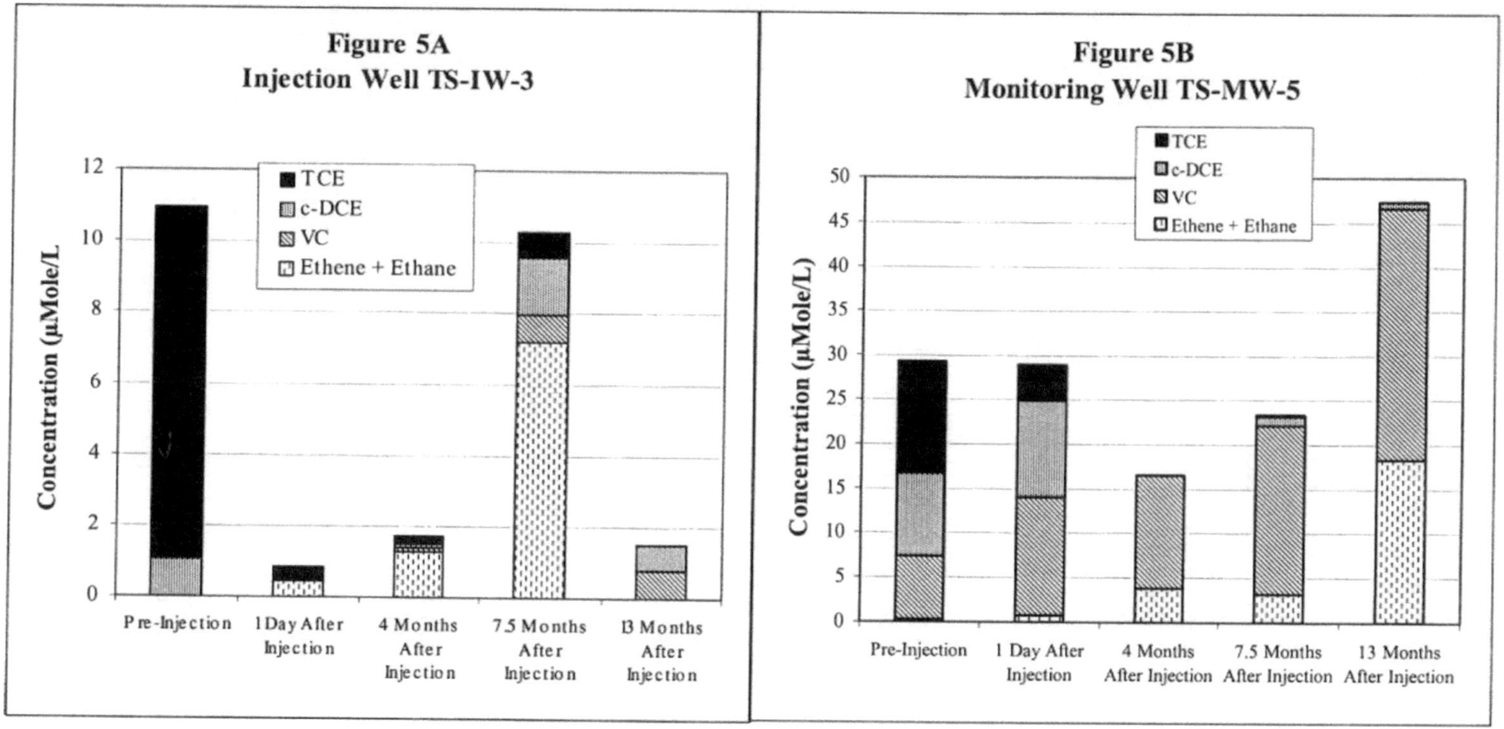

Figure 5. Concentration of TCE and biodegradation daughter products in Injection Well TS-IW-3 (Figure 5A) and Monitoring Well TS-MW-5 (Figure 5B) before and after injection of EOS®.

7. CONCLUSIONS

The use of anaerobic biodegradation for cleanup of many chlorinated solvents in groundwater has become a widely recognized and accepted technology. As a result of an increasing number of laboratory and field studies showing that chlorinated solvents can be biodegraded by naturally occurring microorganisms, the wealth of project knowledge about the effectiveness and limitations of this approach continues to grow. Interest and research into this technology has also resulted in potential for expanded application for treating other significant groundwater issues including perchlorate, nitrate and heavy metal contamination.

The criteria for selection of the optimal substrate for remediation of these contaminants include: 1) the ability to effectively support and promote anaerobic reductive biodegradation; 2) the ability to be delivered to the subsurface cost-effectively; 3) longevity of residual substrate *in situ;* 4) not creating adverse secondary water quality effects on the aquifer; and 5) the cost of substrate and its application.

Using injectable substrates eliminates the need for installation and maintenance of aboveground treatment equipment. Following injection of substrate, the entire cleanup occurs *in situ* relying on a combination of dissolution of substrate, contaminant migration, and subsurface retardation factors to control the rate of degradation.

The patented EOS® process is the result of continuing research addressing this important problem. As shown in this paper, edible oils can support reductive biodegradation in laboratory and column studies. In particular, EOS® has been shown to be inexpensive and capable of enhancing the desired biotransformations processes, while providing extended longevity in the subsurface. (EOS Remediation, Inc. (http://www.EOSRemediation.com) is licensed to sell the EOS® product.

REFEERENCES

Borin, M. and Bigon, E. 2002. Abatement of **NO_3-N** concentration in agricultural waters by narrow buffer strips. Environ. Poll. 117, 165-168.

Coates, J.D., Michaelidou, U., Bruce, R,A., O'Connor, S.M., Crespi, J.N., and Achenbach, L.A.. 1999. Ubiquity and diversity of dissimilatory (per)chlorate -reducing bacteria. Appl. Environ. Microbiol. 65 (12), 5234-5241.

Coates, J.D and Pollock, J. 2003. Potential for in situ bioremediation of perchlorate in contaminated environments. Presented at: In Situ and On -Site Bioremediation, the Seventh International Symposium, Orlando, FL, June 2003.

Ellis, D.E., Lutz, E.J., Odom, J.M., Buchanan, R.J., Bartlett, C.L., Lee, M.D., Harkness, M. R., Deweerd, K.A. 2000. Bioaugmentation for accelerated in situ anaerobic bioremediation. Env. Sci. Tech. 34, 2254-2260.

Fennessy, M.S. and Cronk, J.K. 1997. The effectiveness and restoration of potential riparian ecotones for the management of nonpoint source pollution, particularly nitrate, Crit. Rev. Environ. Sci. Tech. 27(4), 285-317.

Herman, D.C. and Frankenberger, Jr., W.T. 1998. Microbial -mediated reduction of Perchlorate in groundwater. J. Environ. Qual. 27, 750-754.

Interstate Technology and Regulatory Council (ITRC) *In Situ* Bioremediation Team. August 2002. Technical/Regulatory Guidelines: A systematic approach to *in situ* bioremediation in groundwater including decision trees on *in situ* bioremediation for nitrates, carbon tetrachloride and perchlorate. August 2002.

Lee, M.D., Odom, J.M., and Buchanan, R.J. 1998. New perspectives on microbial dehalogenation of chlorinated solvents : Insights from the field. Ann. Rev. Microbiol. 52, 423-452.

Logan, B.E. and Koenigsberg, S.S. 2000. The potential for *in situ* perchlorate degradation. In: 2nd International Conference on Remediation of Chlorinated and Recalcitrant Compounds, Monterey CA.

Logan, B.E. 2001. Assessing the outlook for perchlorate remediation . Env. Sci. Technol. 35 (23), 482A - 487A.

Martin, T. L., Kaushik, N.K., Trevors, J.T., and Whitely, H.R. 1999. Review: denitrification in temperate climate riparian zones. Water, Air and Soil Pollution, 111, 171 -186.

National Institute for Public Health and Environmental Protection (RIVM), "Integrated Criteria Document: Nitrate Effects." Appendix to Report No 758473012, The Netherlands: 1989.

PADEP (Pennsylvania Department of Environmental Protection), 1997. A Status Report on the Environmental Legacy of Coal Mining in Pennsylvania, Harrisburg, PA.

Schoonover, J.E. and Williard, K.W.J. 2003. Ground water nitrate reduction in giant cane and forest riparian buffer zones. J. Amer. Water Res. Assoc. 39(2), 347-354.

US EPA. 2002. Food-grade oil injections used to stimulate reductive dechlorination. In Technology News and Trends, Issue 2, page 2.

US Patent #6,398,960. Borden, R.C. and Lee, M.D. Method for remediation of aquifers. June 4, 2002

Wu, J., Unz, R.F., Zhang, H., and Logan, B.E. 2001. Persistence of perchlorate and the relative numbers of perchlorate - and chlorate -respiring microorganisms in natural waters, soils and wastewater. Bioremediation Journal5(2),119-130.

PART IX: RISK BASED CLEANUP

CHAPTER 33

A REVIEW OF FUZZY SET THEORETIC APPROACHES AND THEIR APPLICATIONS IN ENVIRONMENTAL PRACTICE

Venkatesh Uddameri
Texas A&M University-Kingsville, Kingsville, TX

Abstract: Fuzzy set theory deals with characterizing and quantifying vague and imprecise information and as such can be of high value in environmental practice. While statistical randomness arises due to lack of adequate data and variability in a system, fuzziness incorporates imprecision arising from lack of complete information, conflicting viewpoints or theories, incomplete or incorrect models of physical phenomenon and subjective decisions made during analysis. Clearly, fuzziness is quite prevalent in environmental practice and must be quantified for risk-informed decision making. While fuzzy logic approaches offer several advantages, lack of understanding and training among practitioners is seen as a significant bottleneck in routine utilization of fuzzy based schemes. Conventional mathematical approaches, such as establishing relationships between inputs and outputs, regression, calculus, and hypothesis testing can all be carried out on fuzzy variables. The literature on fuzzy set -theoretic approaches is rich and a significant amount of research is being carried out to identify solutions to new problems and refine existing approaches. Several theoretical contributions suitable for use in environmental practice have been identified in this review. In addition, selected example environmental applications where fuzzy logic based schemes have been employed have also been documented. It is hoped that environmental practitioners and regulators will appreciate the utility of fuzzy logic in characterizing imprecision that plagues environmental analysis and utilize its functionality in routine risk, remedial, regulatory and other applications.

Key words: Imprecision, Vagueness, Mathematical models, Decision making, RBCA

1. INTRODUCTION

Environmental analysis, especially risk-based decision making, is often plagued by uncertainties in data, imprecision in available modeling tools and methodologies and vagueness in understanding of the underlying scientific and technical underpinnings. These limitations affect regulatory and policy decisions that are to be made at the site, and must therefore be quantified and communicated during the risk-based decision making process. Statistical approaches such as the monte-carlo simulation are most commonly employed to characterize parametric uncertainties during risk-assessments (e.g., Batchelor et al., 1998). However, in addition to statistical randomness associated with input parameters, regulatory decision-making must also reconcile with inherent uncertainties arising due to differences between site-characteristics and the chosen mathematical model. Further, other subjective traits such as the risk preferences of the decision makers, effects of conflicting and confounding scientific theories must also be quantified for effective risk-informed decision-making.

The concept of fuzzy sets was first developed by Lotfi Zadeh to deal with imprecise or vague phenomena (Zadeh, 1965). Unlike conventional set theory, wherein an event either belongs or does not belong to a set, fuzzy sets allow for an event to have partial memberships in one or more sets. This feature allows effective reconciliation of conflicting descriptions of a same event by different stakeholders. In addition, fuzzy sets can often be described using linguistic descriptors such as *"high", "medium"* and *"low"* and as such are invaluable to communicate information to a wide range of stakeholders in an easy to understand and intuitive manner. Given these and other benefits, fuzzy logic based approaches are increasingly being proposed for environmental approaches (e.g., Lee et al., 1994; Guymonn et al., 1999; Freissinet et al., 1999; Uddameri, 2003a).

Despite a growing number of research publications, the utilization of fuzzy logic based approaches in routine environmental practice is fairly minimal. The major causes limiting the utility of fuzzy logic approaches in environmental practice include – a lack of training in basics of fuzzy logic among environmental practitioners, non-availability of easy to use software programs that implement fuzzy logic solutions and limited regulatory acceptance. Most papers on fuzzy logic tend to be highly mathematical in nature and as such may not be readily comprehensible. In addition, the applications of theoretical concepts are often not illustrated in many publications, thereby limiting the utility of such approaches in routine environmental practice. The overall goal of this paper is to provide a cursory overview of fuzzy logic and discuss various environmental applications in which fuzzy logic concepts have been used. It is hoped that this review will provide practitioners with a starting point to explore the utility of fuzzy logic in their own applications.

2. BACKGROUND

2.1 Fuzzy Set

Fundamental to the study of fuzzy logic is the concept of fuzzy set. As an example, consider a set of measured nitrate concentrations in a water-well to be [1.0, 8.0, 12, 9.9] ppb. If the allowable nitrate concentration is 10 ppb, the data can be divided into two sets: Allowable: [1.0, 8.0, 9.9]; and Non-allowable: [12]. However, given the limitations in sampling and measurement techniques, it is quite likely that the measured value of 9.9 ppb may not be different from the allowable standard of 10 ppb and will probably be considered non-allowable by many environmental practitioners. In addition, many practitioners may consider the value of 8.0 ppb to be only *somewhat allowable.* Linguistic descriptors such as *somewhat allowable* cannot be accounted for in traditional sets (or crisp sets), but can be easily handled in fuzzy sets. A fuzzy set for a variable, say nitrate concentrations, often consists of two or more sub-sets that span the domain of interest, which is referred to as the *universe of discourse,* and is plotted on the x-axis. The elements of the fuzzy set can belong to one or more fuzzy subsets that span the universe of discourse. The extent to which the element belongs to a fuzzy subset is characterized by the degree of membership, which is plotted on the y-axis. For example, the value of 8.0 in Figure 1 has a membership of 0.8 in fuzzy set *not allowable* and a membership of 0.2 in the fuzzy set *allowable* indicative of the fact that it is only *somewhat allowable.* Fuzzy subsets often overlap with one another, causing elements to have membership in more than one subset. The extent of overlap characterizes the degree of fuzziness in the variable. If the subsets don't overlap, the elements will belong to a single subset just like a conventional (crisp set). The relationship between the degree of membership and the domain of the fuzzy subset is called the membership function. Fuzzy sets are mathematically characterized by this membership function. They can be of any arbitrary shape, although, triangular, trapezoidal or bell-shaped sets are most commonly used in practice (Cox, 1995). It is important to note than more than one element in the subset can have the same degree of membership and it is not necessary for the fuzzy set to have one or more elements that have a degree of membership of unity or zero (extreme values). Fuzzy sets are commonly represented using **α–cuts.** The **α–cut** simply represents the range of values the fuzzy set assumes when the degree of membership is equal to α.

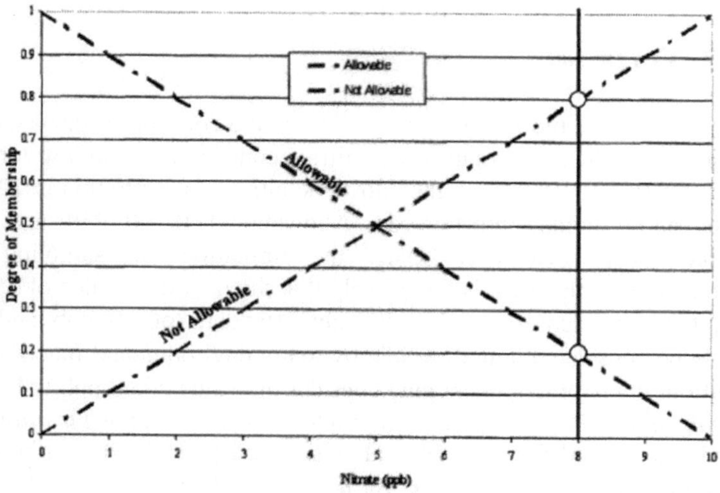

Figure 1.

2.2 Fuzzy Number

A fuzzy number is a special form of a fuzzy set, where the membership function is convex (i.e., has an increasing and a decreasing part), and at least one element of the set has a degree of membership of unity. Just like fuzzy sets, fuzzy numbers can be represented by membership functions of arbitrary shapes using **α-cuts**. However, triangular and trapezoidal fuzzy numbers are most commonly used in many applications (e.g., Kaufmann and Gupta, 1988). A triangular fuzzy number can be represented in many ways (see Figure 2), and is considered symmetric if the midpoint of the domain has a membership degree of unity.

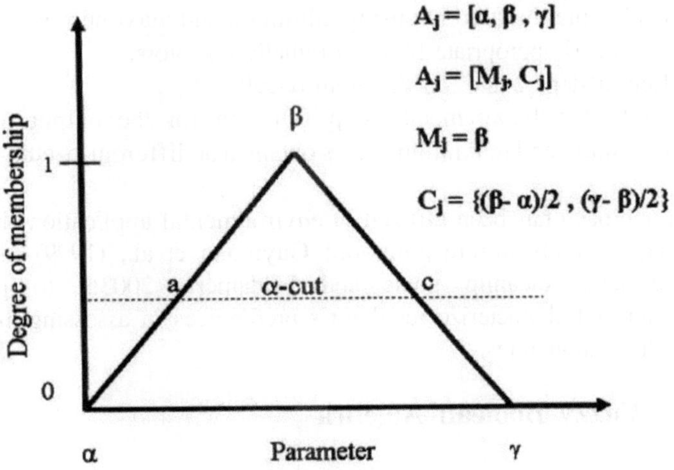

Figure 2.

3. MATHEMATICAL ANALYSIS WITH FUZZY SETS AND NUMBERS

Several different mathematical analyses, such as arithmetic and boolean operations, differential and integral calculus, regression analysis, hypothesis testing, correlation analysis and optimization can all be performed using fuzzy variables and are discussed below:

3.1 Fuzzy Arithmetic

Fuzzy arithmetic provides appropriate mathematical formalisms to add, subtract, multiply and divide fuzzy numbers. Using these operators, other mathematical functions can be solved using conventional mathematical formulas. A generalized **α-cut** procedure to perform algebraic calculations in a fuzzy framework is presented in Dubois and Prade, (1988) and entails the following steps:

1. Select a value of membership function **(α-cut)**.

2. For each input select the minimum and maximum value corresponding to the selected α-cut.
3. Calculate the minimum and maximum value for the output, by evaluating the function using minimum and maximum values for the input and appropriate fuzzy arithmetic operators.
4. Repeat steps 2 and 3 at different α-cuts.
5. Build the fuzzy membership function for the output using the minimum and maximum values obtained at different α-cuts.

This approach has been utilized in environmental applications by Lee et al., (1994) to study nitrate pollution; Guymonn et al., (1999) to develop screening level cleanup goals and Uddameri (2003a) to propagate imprecision and characterize regulator's preferences in assessing monitored natural attenuation rates.

3.2 Fuzzy Boolean Algebra

Boolean algebraic operators such as AND, OR, NOT can be used to manipulate fuzzy sets as well as fuzzy numbers. These operations are used extensively in fuzzy rule-based systems or fuzzy controls to map relationships between inputs and outputs using a set of IF-THEN rules. Mathematical approaches to carry out these Boolean operations have been presented by Zadeh (1965), Mamdani and Assilian, (1975), Sugeno (1985) and others. MATLAB fuzzy logic toolbox provides convenient tools to carry out fuzzy Boolean operations (Mathworks, 2002), other similar software is commercially available as well. Fuzzy rule-based systems have been widely used for a variety of environmental applications to assess nitrate pollution, rank pesticides, compute groundwater recharge and to control wastewater treatment systems. (e.g., van der Werf, 1998; Lee et al., 1995; Coppola et al., 2001; Chen et al., 2001).

Development of fuzzy rule-based systems entails the following four steps: 1) In the fuzzification step, the universe of discourse of the inputs and outputs are divided into two or more overlapping fuzzy subsets. Typically an odd number of subsets are used and while these subsets can be of any arbitrary shapes, triangular, trapezoidal or other regular geometries are commonly adopted. 2) A rule-base connecting fuzzified inputs to fuzzy output set is developed, and forms the backbone of the fuzzy rule-based system. Fuzzy rule-bases consist of a set of IF-THEN rules (e.g., IF SOIL = *CLAY* and WATER-TABLE = *DEEP* THEN LEACHING FACTOR is *LOW*). Here SOIL, WATER-TABLE are fuzzy inputs and LEACHING FACTOR is a fuzzy output of interest. Expert knowledge is often gleaned to develop pertinent fuzzy rules. However, fuzzy rules can also be generated from data using approaches suggested by Wang and Mandel (1991) and Nozaki et al., (1997). The number of rules in conventional rule-based systems increase exponentially with increased number of inputs. This phenomenon, referred to as combinatorial explosion, is often a limiting factor in development of expert systems (Cox, 1995). However, fuzzy expert systems work on the principle of parallel rule-firing and are known to produce fairly reliable and robust predictions even when all the rules are not specified (Kosko, 1994). 3) When inputs are presented to the fuzzy rule-based system, all pertinent rules are fired simultaneously, and the output fuzzy set (which is an aggregate of all the rules is generated). The development of the output fuzzy set entails Boolean operations such as implication and aggregation, which are performed using mathematical constructs provided by Zadeh, (1965); Sugeno, (1985) and Mamdani and Assilian, (1975). 4) The output fuzzy set is transformed into a single number during the defuzzification step as, a crisp output, such as an allowable concentration, is most often of interest than a fuzzy set. There are several possible approaches to defuzzification. However, the centroid method is most commonly used in many applications owing to its simplicity (Cox, 1995).

3.3 Fuzzy Regression

Bivariate and Multivariate regression are commonly used statistical techniques to relate outputs to inputs and in some instances identify trends between the variables. However, in many environmental applications, the exact nature of the statistical model may not be known with certainty and the selected functional form is not precise. Even when the functional form is known with certainty, there may not be sufficient data to obtain reliable parameter estimates for the regression coefficients. These factors tend to induce fuzziness into the regression scheme, which is distinct from statistical randomness. Also, different decision makers may feel differently about the adopted regression scheme leading to fuzziness in the decision-making process that is based on the regression scheme. Conventional statistical techniques cannot capture these varied preferences of the decision makers. Fuzzy regression on the other hand provides mechanisms to incorporate the beliefs of the decision maker (e.g., Wang and Tsaur, 2000; Uddameri, 2003b).

There are several different approaches to carrying out fuzzy regression; Chang and Ayubb, (2001) have summarized some these procedures. Early development of fuzzy regression was based on the minimization of the vagueness criterion proposed by Tanaka et al., (1982) and essentially entails solving a linear programming model aimed at maximizing the compatibility between the regression model and the data (i.e., minimize the vagueness in the model). While the minimization of vagueness is intuitive and easy to implement, it is sensitive to outliers and does not include the concept of least squares. Least squares based fuzzy linear regression schemes have been proposed by Diamond (1988), Savic and Pederwicz, (1991) and Wang and Tsaur, (2000). The approaches of Savic and Pederwicz (1991), is a two step procedure where the midpoints of the fuzzy coefficients are obtained using ordinary least squares, and the width of the regression coefficients are obtained via minimization of the vagueness criterion. The approach proposed by Diamond (1988) utilizes an iterative procedure to minimize the residual errors and is rather difficult to implement. The recent approach presented by Wang and Tsaur, (2000) provides a simpler algorithm for fuzzy least squares estimation that is based on nonlinear optimization. While most algorithms focus on estimating parameters for linear regression models, fuzzy nonlinear regression is also possible and has been discussed by Bardossy, (1990). Fuzzy regression has been utilized in environmental applications by Bardossy and Duckstein, (1990); Bardossy, (1990), Uddameri (2003b) and Uddameri and Kuchanur (2003) among others.

3.4 Fuzzy Optimization

Environmental decision-making often requires an optimal reconciliation between competing and conflicting objectives. The prescribed remedial measures must minimize risks, maximize the usefulness of available resources and at the same time be cost effective (i.e., minimize costs). In addition, specified corrective action schemes are also constrained by regulatory and logistic requirements. Linear and nonlinear programming schemes are widely used to identify optimal measures given a set of constraints in risk-based decision-making situations. However, crisp linear and nonlinear programming schemes cannot capture succinctly the different values and needs of the various stakeholders, and is also limited by uncertainties arising due to lack of perfect knowledge during the decision making process.

Fuzzy optimization provides an approach to quantify, uncertainties arising from lack of perfect information and conflicts caused by differing values and needs of various stakeholders. Typically, fuzziness can arise in the objectives governing the decision-making process and/or manifest in the constraints under which the decision must be made. Mathematical approaches have been developed to handle situations where fuzziness is only in the constraint set or when both the objective function and the constraint set are both fuzzy. The problem of fuzzy optimization was first addressed by Bellman and Zadeh (1970) and since has been studied by a variety of authors (e.g., Zimmermann, 1978; Tanaka and Asai, 1984; Rommelfanger, 1996; Liu, 2001).

From a mathematical standpoint, fuzzy optimization schemes utilize fuzzy set theoretic concepts to convert the fuzzy optimization problem into a crisp optimization problem that can be solved easily. Fuzzy optimization problems where the constraints are fuzzy (i.e., the constraints have to be met only in an approximate sense) and the objectives are optimized to a certain aspiration levels can be transformed into crisp linear and nonlinear programming models and solved using available software (e.g., Zimmermann, 1990). Maeda, (2001) and Zhang, et al., (2003) present solutions to fuzzy optimization schemes where fuzziness is only associated with the objective function. On the other hand, Liu (2001) has tackled the problem involving fuzziness in the constraints alone.

Fuzzy optimization has had a fair amount of use in water resources management problems (e.g., Slowinski, 1987; Teegavarapu and Simonovic, 2000). In addition, fuzzy optimization has been used to regulate air pollution, manage water quality, assess maximum pollution loadings and optimize solid waste management practices (e.g., Sommer and Pollatschek, 1978; Lee and Wen, 1996; Lee and Wen, 1997; Huang et al., 2001).

3.5 Fuzzy Hypothesis Testing

Statistical hypothesis testing aims to compare the statistical moments (such as mean, standard deviations) of two or more treatments to evaluate if the samples in the different treatments are drawn from the same population. Conventional statistical tests make rigorous assumptions about the hypothesis (e.g., the two means are exactly equal in some statistical sense). On the other hand fuzzy hypothesis tests do not make rigorous assumptions with regards to the hypothesis. Fuzzy hypothesis, such as the two means are nearly equal, are more suited when assumptions underlying the statistical measures cannot be met. In addition, fuzzy hypothesis testing becomes necessary when the collected data have imprecise meanings and can be modeled as fuzzy sets.

Approaches to fuzzy hypothesis testing have been presented by Saade and Schwarzlander, (1990); Watanabe and Imaizumi, (1993); Grezegorzewski, (2000); Taheri and Behboodian, (2001); Grezegorzewski, (2001). However, the approaches of Watanabe and Imaizumi (1993) and Grezegorzewski, (2001) appear to be pragmatic. It is important to recognize that fuzzy hypothesis testing does not lead to a crisp "acceptance" or "rejection" but only indicates the degree of rejection of null hypothesis. The test-statistic is often a fuzzy number and must be defuzzified to obtain a crisp acceptance or rejection (Grezegorzeweski, 2001). Hypothesis testing on fuzzy numbers can also be carried out by computing the fuzzy statistic of interest for different treatments, and testing the defuzzified values using conventional statistical approaches. However, the final results are often extremely sensitive to the defuzzification procedure and as such can lead to misleading values (e.g., Grezegorzewski, 2001). Mathematics of fuzzy hypothesis testing is not as well developed as conventional hypothesis testing, and this may be one of the reasons why the use of fuzzy hypothesis testing has not been reported in the environmental literature.

3.6 Fuzzy Correlation Analysis

Correlation is a statistical technique where the joint relationship between two variables of interest is examined with the aid of a measure of interdependence between the variables. As an example, correlation analysis can be used to establish a relationship between the turbidity and phosphorus pollution. In some instances, only qualitative characteristics of the variables are available that can, however, be ranked according to some quality. In such cases, rank correlation is used to establish correlations between the variables.

Correlations between "crisp" valued variables can be calculated using correlation coefficient or spearman rank correlation coefficient. The correlation coefficients assume a value between –1 to 1 and the variables exhibit high correlations when the correlation coefficient is close to –1 (negative correlation) or 1 (positive correlation).

Chiang and Lin (1999) extended the concept of statistical correlation to fuzzy numbers. Chaudhury and Bhattacharya, (2001), demonstrate that rank correlation can be used for fuzzy numbers, if the membership values for the elements in the fuzzy set are known. The fuzzy numbers are ranked on the basis of their membership functions prior to the application of the spearman rank correlation. Chaudhury and Bhattacharya (2001) also present an alternative metric for correlating fuzzy numbers. Liu and Kao (2002) have developed a methodology where the correlation coefficient is a fuzzy number. Partial correlation coefficient is another measure of correlation that removes (partials out) the effects of other variables. Chiang and Lin (2000) have developed a fuzzy counterpart to the partial correlation coefficient. However, the author is not aware of an environmental application where fuzzy correlation measures have been used.

3.7 Calculus Operations with Fuzzy Variables

As integral and differential operations can be written as algebraic operations, fuzzy arithmetic can be used to carry out calculus operations on fuzzy variables (e.g., Kaufmann and Gupta, 1985). Fuzzy rule-based systems have also been used to approximate the behavior of differential equations (Leephakpreeda, 2002). Analytical solutions for $\mathbf{N}^{th}$ -order fuzzy differential equation have been presented in frequency domain (Yue et al., 1998a) and time-domain (Yue and Guangyuan, 1998) and for a system of first-order fuzzy differential equations (Yue et al., 1998b). However, these approaches have been noted to be of little use, as the solution might exist for only an arbitrarily short time (Buckley and Feuring, 1999a). Buckely and Feuring (2000) have proposed an alternative approach to solving first-order differential equation (initial-value problem) and have presented some solutions to elementary partial differential equations (Buckley and Feuring, 1999b). Babolian et al., (2003) provide numerical solutions for fuzzy differential equations using the Adomian method.

Dou et al., (1995) combined fuzzy set theories with finite-difference calculus to estimate groundwater heads under uncertainty. This approach was also extended by them to simulate transient groundwater flow under imprecision (Dou et al., 1997a) and also to model solute transport in groundwater (Dou et al., 1997b). Schulz and Huwe (1997) assessed the impacts of parametric imprecision in an unsaturated water flow model using fuzzy set theoretic approaches and Zou and Lung (2000), combined Eulers method with fuzzy set theory to model imprecision in a dynamic phosphorous models. However, more rigorous mathematical approaches such as those suggested by Buckely and Feuring (2000) do not appear to have been utilized in environmental applications.

4. SUMMARY AND CONCLUSIONS

The objective of this study was to provide a brief overview of fuzzy set theory and demonstrate its utility in environmental practice. Fuzzy logic aims to deal with vague and imprecise information and can effectively reconcile between conflicting and subjective preferences that arise during the decision-making process. It is important to recognize that fuzziness is different from statistical uncertainty. Statistical randomness is data-driven and is characterized using probabilities (either frequentist or subjective) of occurrence of an event. On the other hand, fuzzy logic is based on the rather subjective concept of possibility, and characterized using a membership function that accumulates evidence for membership in a set of events (Cox, 1999). Statistical randomness arises due to lack of sufficient measurements, while fuzziness can be due to lack of complete information, conflicting viewpoints or theories, incomplete or incorrect models of physical phenomenon and subjective decisions made during analysis.

Conventional mathematical analysis such as regression, mathematical modeling, optimization and hypothesis testing can all be performed using fuzzy set-theoretic approaches. The literature on fuzzy set-theoretic approaches is rich and a significant amount of research is being carried out to identify solutions to problems and refine existing approaches. Several theoretical contributions suitable for use in environmental practice have been identified in this review. In addition, example environmental applications where fuzzy logic based schemes have been employed have also been documented. While this review is by no means complete, it expectantly provides a good starting point to explore the utility of fuzzy set theoretic approaches in environmental practice. It is hoped that environmental practitioners and regulators will appreciate the utility of fuzzy logic in characterizing imprecision that plagues environmental analysis and utilize its functionality in routine risk, remedial, regulatory and other applications.

ACKNOWLEDGMENTS

Support for this work from the NSF Center for Excellence in Science and Technology – Research on Environmental Sustainablility in Semi-Arid Coastal Areas (CREST-RESSACA; Cooperative Agreement # HRD-0206259) at Texas A&M University-Kingsville (TAMUK) and from Frank Dotterweich College of Engineering - TAMUK is gratefully acknowledged.

REFERENCES

Babolian, E., H. Sadeghi, et al. (2003). Numerically Solution of Fuzzy Differential Equations by Adomian Method. Applied Mathematics and Computation In press.

Bardossy, A. (1990). Note on Fuzzy Regression. Fuzzy Sets and Systems 37: 65-75.

Bardossy, A., I. Bogardi, et al. (1990). Fuzzy Regression in Hydrology. Water Resources Research 26(7): 1497-1508.

Batchelor, B., J. Valdes, et al. (1998). Stochastic Risk Assessment of Sites Contaminated by Hazardous Waste Sites. Journal of Environmental Engineering 124(4): 380 -388.

Bellman, R. E. and L. A. Zadeh (1970). Decision-Making in a Fuzzy Environment. Management Science 17(4): 141 -164.

Buckley, J. J. and T. Feuring (1999). Comments on Frequency/Time Domain Methods for Solutions of N-Order Fuzzy Differential Equations by Zhang, Wang and Liu. Fuzzy Sets and Systems 105: 185-186.

Buckley, J. J. and T. Feuring (1999). Introduction to Fuzzy Partial Differential Equations. Fuzzy Sets and Systems 105: 241-248.

Buckley, J. J. and T. Feuring (2000). Fuzzy Differential Equations. Fuzzy Sets and Systems 110: 43-54.

Chang, Y-H. and B. M. Ayyub (2001). Fuzzy Regression Methods - A Comparative Assessment. Fuzzy Sets and Systems 119: 187-203.

Chaudhuri, B. B. and A. Bhattacharya (2001). On Correlation between Two Fuzzy Sets. Fuzzy Sets and Systems 118: 447-456.

Chen, W. C., N. B. Chang, et al. (2001). Advanced Hybrid Fuzzy -Neural Controller for Industrial Wastewater Treatment. Journal of Environmental Engineering 127(11): 1048-1059.

Chiang, D.-A. and N. P. Lin (1999). Correlation of Fuzzy Sets. Fuzzy Sets and Systems 102: 221-226.

Chiang, D.-A. and N. P. Lin (2000). Partial Correlation of Fuzzy Sets. Fuzzy Sets and Systems 110:209-215.

Coppola, E. A., L. Duckstein, et al. (2002). Fuzzy Rule -Based Methodology for Estimating Monthly Groundwater Recharge in a Temperate Watershed. Journal of Hydrologic Engineering 7(4): 326-335.

Cox, E. (1995). Fuzzy Logic for Business and Industry , Charles River Media, Inc. Diamond, P. (1988). Fuzzy Least Squares. Information Science 46: 141-157.

Dou, C., W. Woldt, et al. (1997). Numerical Solute Transport Simulation Using Fuzzy Sets Approach. Journal of Contaminant Hydrology 27: 107-126.

Dou, C., W. Woldt, et al. (1995). Steady state Groundwater Flow Simulation with Imprecise Parameters. Water Resources Research 31(11): 2709 -2719.

Dou, C., W. Woldt, et al. (1997). Transient Ground-Water Flow Simulation using Fuzzy Set Approach. Ground Water 35(2): 205-215.

Dubois, D. and P. Prade (1988). Possibility Theory: An Approach to Computerized Processing of Uncertainty, Plenum Press.

Freissinet, C., M. Vauclin, et al. (1999). Comparison of First-Order Analysis and Fuzzy Set Approach for Evaluating Imprecision in a Pesticide Groundwater Pollution Screening Model. Journal of Contaminant Hydrology 37: 21 -43.

Grzegorzewski, P. (2000). Testing Statistical Hypothesis with Vague Data. Fuzzy Sets and Systems 112: 501-510.

Grzegorzewski, P. (2001). Fuzzy Tests - Defuzzification and Randomization. Fuzzy Sets and Systems 118:437-446.

Guyonnet, D., B. Come, et al. (1999). Comparing Two Methods of Addressing Uncertainty in Risk Assessments. Journal of Environmental Engineering 125(7): 660-666.

Huang, G. H., N. Sae-Lim, et al. (2001). An Interval-Parameter Fuzzy Stochastic Programming Approach for Municipal Solid Waste Management and Planning. Environmental Modeling and Assessment 6: 271 -283.

Kaufmann, A. and M. Gupta (1985). Introduction to Fuzzy Arithmetic - Theory and Applications. New York, Van Nostrand Reinhold Company.

Kaufmann, A. and M. Gupta (1988). Fuzzy Mathematical Models in Engineering and Management Science. New York, North-Holland Publishers.

Kosko, B. (1994). Fuzzy Systems as Universal Approximators. IEEE transactions on Computers 43(11): 1329-1333.

Lee, C.-S. and C.-G. Wen (1996). River Assimilative Capacity via Fuzzy Linear Programming. Fuzzy Sets and Systems 79: 191-201.

Lee, C.-S. and C.-G. Wen (1997). Fuzzy Goal Programming Approach for Water Quality Management in a River Basin. Fuzzy Sets and Systems 89: 181-192.

Lee, Y. W., M. F. Dahab, et al. (1994). Fuzzy Decision Making in Ground Water Nitrate Risk Management. Water Resources Bulletin 30(1): 135-147.

Lee, Y. W., M. F. Dahab, et al. (1995). Nitrate Risk Assessment using Fuzzy Set Approach. Journal of Environmental Engineering 121(3): 245-256.

Leephakpreeda, T. (2002). Novel Determination of Differential -Equation Solutions: Universal Approximation Method. Journal of Computational and Applied Mathematics 146: 443-457.

Liu, S-T. and C. Kao (2002). Fuzzy Measures for Correlation Coefficient of Fuzzy Numbers. Fuzzy Sets and Systems 128: 267-275.

Liu, X. (2001). Measuring the Satisfaction of Constraints in Fuzzy Linear Programming. Fuzzy Sets and Systems 122: 263-275.

Maeda, T. (2001). Fuzzy Linear Programming Problems as Bi -Criteria Optimization Problems. Applied Mathematics and Computation 120: 109-121.

Mamdani, E. H. and A. Assilian (1975). An Experiment in Linguistic Synthesis with a Fuzzy Logic Controller. International Journal of Man -Machine Studies 7: 1-13.

Mathworks (2002). MATLAB 6.5.

Nozaki, K., H. Ishibuchi, et al. (1997). A Simple but Powerful Heuristic Method for Generating Fuzzy Rules from Numeric Data. Fuzzy Sets and Systems 86: 251 -270.

Rommelfanger, H. (1996). Fuzzy Linear Programming and Applications. European Journal of Operations Researc h 92: 512-527.

Saade, J. J. and H. Schwarzlander (1990). Fuzzy Hypothesis Testing with Hybrid Data. Fuzzy Sets and Systems 35: 197-212.

Savic, D. and W. Pedrycz (1991). Evaluation of Fuzzy Regression Models. Fuzzy Sets and Systems 39: 51-63.

Schulz, K. and B. Huwe (1997). Water Flow Modeling in the Unsaturated Zone with Imprecise Parameters using Fuzzy Approach. Journal of Hydrology 201: 211 -229.

Slowinski, R. (1986). A Multicriteria Fuzzy Linear Programming Method for Water Supply System Development Planning. Fuzzy Sets and Systems 19: 217-237.

Sommers, G. and M. A. Pollatschek, Eds. (1978). A Fuzzy Programming Approach to an Air Pollution Regulation Problems. Progress in Cybernetics and Systems Research: General System Methodology, Mathematical Systems Theory, Fuzzy Systems,
Biocybernetics, and Theoretical Neurobiology, Vol III. Washington, D.C., Hemisphere.

Sugeno, M. (1985). Industrial Applications of Fuzzy Control , Elsevier Science Publishing Co.

Taheri, S. M. and J. Behboodian (2001). A Bayesian Approach to Fuzzy Hypothesis Testing. Fuzzy Sets and Systems 123: 39-48.

Tanaka, H. and K. Asai (1984). Fuzzy Linear Programming with Fuzzy Numbers. Fuzzy Sets and Systems 13: 1-10.

Tanaka, H., S. Uejima, et al. (1982). Linear Regression Analysis with Fuzzy Model. IEEE Transactions on Syst. Man Cybernetics (SMC-2): 903-907.

Teegavarapu, R. S. and S. Simonovic (1999). Modeling uncertainty in reservoir loss functions using fuzzy sets . Water Resources Research 35(9): 2815 -2825.

Uddameri (2003a). Estimating Natural Attenuation Rate Constants Using a Fuzzy Framework. Ground Water Monitoring and Remediation 23(3): 105-111.

Uddameri, V. (2003b). Estimating Field-Scale Longitudinal Dispersivity using Fuzzy Linear Regression. Environmental Geology accepted.

Uddameri, V. and M. Kuchanur (2003). Fuzzy QSARs for Predicting Kow of Persistent Organic Pollutants. Chemosphere in press.

van der Werf, H. M. G. and C. Zimmer (1998). An Indicator of Pesticide Environmental Impact Based on A Fuzzy Expert System. Chemosphere 36(10): 2225 -2249.

Wang, H.-F. and R.-C. Tsaur (2000). Insight of a Fuzzy Regression Model. Fuzzy Sets and Systems 112: 355-369.

Wang, L.-X. and J. M. Mendel (1992). Generating Fuzzy Rules by Learning from Examples. IEEE Transactions on Syst. Man Cybernetics 25(2): 353-361.

Watanabe, N. and T. Imaizumi (1993). A Fuzzy Statistical Test of Fuzzy Hypothesis. Fuzzy Sets and Systems 53: 167-178.

Yue, Z. and W. Guangyuan (1998). Time Domain Methods for the Solutions of N-Order Fuzzy Differential Equation. Fuzzy Sets and Systems 94: 77-92.

Yue, Z., W. Guangyuan, et al. (1998). Frequency Domain Methods for the Solutions of N - Order Fuzzy Differential Equation. Fuzzy Sets and Systems 94: 45-59.

Yue, Z., Q. Zhong, et al. (1998). Solving Processes for a System of First-Order Fuzzy Differential Equations. Fuzzy Sets and Systems 95: 333-347.

Zadeh, L. (1965). Fuzzy Sets. Information Control 8: 338-353.

Zhang, G., Y.-H. Wu, et al. (2003). Formulation of a Fuzzy Linear Programming Problems as Four-Objective Constrained Optimization Problems. Applied Mathematics and Computation 139: 383-399.

Zimmermann, H. (1990). Fuzzy Mathematical Programming. Encyclopedia of Artificial Intelligence - Second Edition Vol 1., New York, John Wiley and Sons.

Zimmermann, H.-J. (1978). Fuzzy Programming and Linear Programming in Several Variables. Fuzzy Sets and Systems 1: 45-55.

Zou, R. and W.-S. Lung (2000). Uncertainty Analysis for a Dynamic Phosphorous Model with Fuzzy Parameters. Water Quality and Ecosystem Modeling 1: 237-252.

This page intentionally left blank

CHAPTER 34

RISK-BASED MANAGEMENT STRATEGIES AND INNOVATIVE REMEDIES FOR SURFACE WATER PROTECTION: A CASE STUDY

S. Grant Watkins, P.G.; Kristin Durocher Wandland; Brian Ray
ENSR Corporation

Abstract: Releases of chlorinated aliphatic hydrocarbons have impacted soils and groundwater at a South Carolina facility. Plume discharges into nearby tributaries have contaminated surface waters with up to 4,600 μg/L of trichloroethene. Hydrogeologic studies, flow modeling, and pilot tests were performed to develop surface water protection remedies.

Part of the groundwater plume migrates off-site to the northeast and discharges into an underground storm water pipe and into a pond. Studies found that >99 percent of chlorinated hydrocarbons in the pond and downstream tributary originate from infiltration of contaminated groundwater into piping upstream of the pond. A surface water protection plan was developed using risk -based, innovative treatment approaches. The plan proposed plume containment by the subsurface pipe and *in-situ* treatment of water in the pipe. Risk -based components of this remedy included fencing the pond and developing wildlife - and human health-based standards for the pond and its downstream tributary. Successful pilot testing was conducted using air diffusion and ozone oxidation for water treatment inside the pipe.

Remedies for off-site surface waters west and south of the facility utilize *in-situ* enhanced reductive dechlorination treatment barriers around the tributaries. Trichloroethene was reduced in surface water and in the aquifer by 43% -99% in these areas.

Key words: Chlorinated hydrocarbons, trichloroethene, surface water remediation, air diffusion, ozone treatment, reductive dechlorination

1. INTRODUCTION AND BACKGROUND

The project site is a former manufacturing plant located in western South Carolina. Historical operations caused releases of trichloroethene (TCE) and other chlorinated aliphatic hydrocarbons (CAHs) to soils and groundwater beneath the facility. The complex structural and hydrogeologic conditions of the underlying saprolite-fractured rock aquifer system allowed the groundwater plumes to migrate in multiple directions away from the property. These plumes migrate off-site and discharge into three separate unnamed first-order streams at distances up to 1,500 feet from the facility. Concentrations of TCE as high as 4,600 micrograms per liter (μg/L) were detected in surface water northeast of the site (i.e., the northeast tributary). Trichloroethene concentrations up to 1,800 μg/L were detected in springs that feed the southern tributary. A separate plume from a waste pit source is encroaching on the western tributary, where very low concentrations of CAHs are periodically detected in surface water.

The state regulatory agency required that a surface water protection plan (SWPP) be submitted for impacted water bodies. To support the SWPP, surface water studies, groundwater-to-surface water flow modeling, contaminant fate modeling, and treatability pilot tests were conducted. These studies concluded that each of the three impacted surface water bodies has unique physical and hydrologic conditions and highly variable plume distribution in the adjacent aquifer. Additionally, exposure risks are different for each area based on accessibility to the surface water, presence and quality of ecological habitats, and beneficial use of each water body in the watershed. ENSR determined that a single cleanup remedy was not applicable to all three areas and developed an approach that uses innovative, risk-based cleanup methods and management approaches for the contaminated surface water. The following sections describe procedures used to develop and implement the SWPP.

2. REGIONAL GROUNDWATER FLOW AND PLUME MIGRATION

Comprehensive hydrogeologic studies have been conducted to determine plume distribution and subsurface controls on plume migration (Watkins et. al., 2002). The site is located on a topographic ridge that forms a drainage divide between two watershed sub-basins. A regional hydrologic divide traverses this area in a northwest to southeast direction (see Figure 1). The underlying saprolite-fractured rock aquifer is a single groundwater system comprised of three hydro-stratigraphic zones. CAHs have impacted each of the three zones of the aquifer system.

Shallow groundwater flow diverges from the site in multiple directions toward tributaries on both sides of the drainage divide. Figure 1 shows a groundwater potentiometric map and total volatile organic compound (VOC) plume map for the shallow unit. Shallow groundwater is hydraulically connected to local streams as evidenced by the equipotential lines and the presence of VOCs in surface water where the groundwater plume discharges.

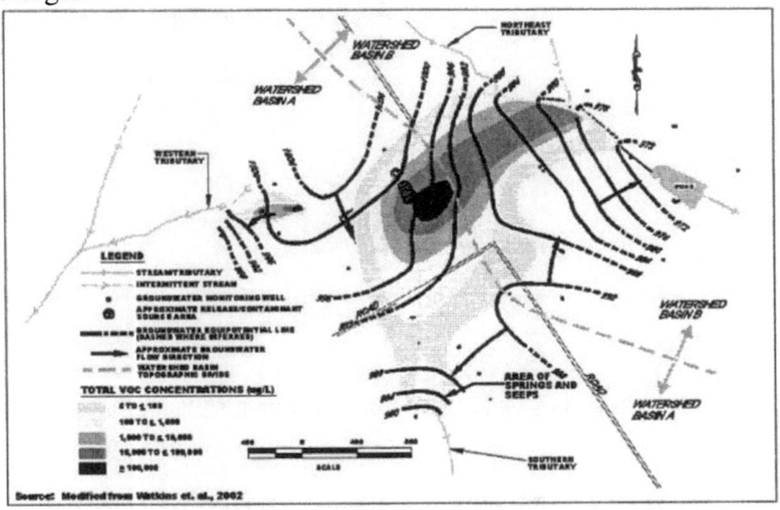

Figure 1. Potentiometric and VOC Plume Map: Saprolite Groundwater

Figure 2 shows a potentiometric map and total VOC plume for groundwater in the partially weathered rock and shallow fractured bedrock (i.e., the intermediate zone). Equipotential lines and vertical gradients between the shallow and intermediate zones indicate that groundwater in the intermediate zone also is hydraulically connected to tributaries and streams in this area. Regional groundwater flow in deeper fractured rock is toward the southeast and crosses the watershed divide without apparent influence from or hydraulic connection to the headwaters of local tributaries.

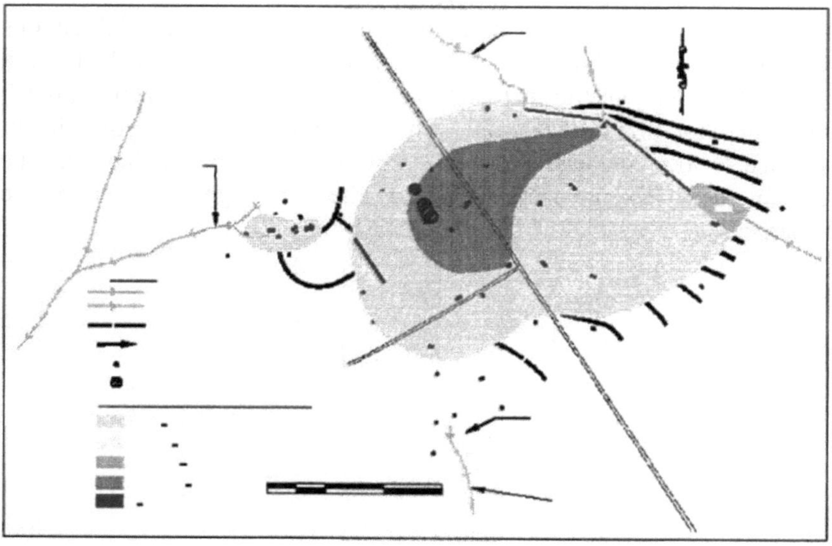

Figure 2. Potentiometric and VOC Plume Map: PWR/Shallow Bedrock

3. SURFACE WATER STANDARDS

Under State regulations, unclassified water bodies share the classification of the stream to which they are a tributary (SC R61.68). Each of the three tributaries described above flows into Class FW (freshwater) streams. Class FW water bodies are suitable for primary and secondary recreation, industrial and agricultural use, drinking water (following conventional treatment), fishing, and survival and propagation of a balanced and indigenous community of aquatic organisms.

South Carolina has developed surface water standards (WQS; SC R.61-68) for many constituents for the protection of human health and aquatic or terrestrial life. The State also has VOC screening values for the protection of human health by ingestion of water and aquatic organisms. VOCs, however, are generally not bioaccumulative. None of the VOCs found in the groundwater plumes are included in either the USEPA (1999) Persistent Bioaccumulative and Toxic (PBT) Chemical list or the USEPA (1995) Great Lakes Water Quality Initiative list of bioaccumulative compounds of concern (BCCs).

South Carolina does not have aquatic life screening values for VOCs. Federal sources were searched for alternative screening values. Chronic (long-term exposure) screening values were found for TCE and many of the CAHs in USEPA (2001) Region 4 guidance and USEPA (1996) EcoTox thresholds. Additional chronic level screening values (LOELs; lowest observed effect levels) were cited from USEPA (45 FR 79328).

Screening values were developed by ENSR for the protection of terrestrial organisms. Since none of the VOCs in the plume are bioaccumulative, values were developed based on the direct ingestion of surface water by wildlife. The amount of surface water ingested directly and indirectly by wildlife species and toxicity reference values (TRVs) for the VOCs were used to estimate whether or not site concentrations have the potential to pose an adverse effect to wildlife from the ingestion of surface water.

4. NORTHEAST TRIBUTARY CHARACTERIZATION AND CLEANUP GOALS

An unnamed stream (the northeast tributary) is located approximately 1,000 feet east-northeast of the site. When the adjacent off-site property was being developed in the early 1990's, the county engineering agency required that storm water management and sediment erosion controls be constructed on the property. To address this requirement, a 1,300-foot section of the tributary was partially rerouted and channeled underground through a 60-inch diameter concrete drain pipe and box culvert system. The piping system transports surface water below ground along the trace of the former stream valley. The northeast tributary emerges from the buried culvert pipe and flows into a 1-acre storm water retention pond (Figure 3). Surface water flows from the retention pond through a spillway and pipe and continues east-southeast along the original stream course. The man-made storm water piping system and storm water retention pond are the most significant hydrologic features affecting the selection of a surface water remedy. Although the pond was fenced as part of the SWPP remedy, access to the downstream tributary is unrestricted. Access to water in the buried pipe is limited to two box culverts, a drop inlet and the pipe inlet and outlet.

After VOCs were found in water within one of the box culverts and in the retention pond, ENSR initiated a surface water characterization, flow/mixing zone modeling study, and risk evaluation of this area. Stream seepage studies were performed and hydrologic gauging measurements were conducted in the northeast tributary, piping and adjacent piezometers to determine the groundwater-surface water interactions. Surface water sampling and flow measurements were conducted throughout the culvert pipe and northeast tributary.

Hydrologic studies concluded that the northeast tributary is a gaining stream and receives groundwater discharge from shallow and intermediate depths of the aquifer. It was also discovered that the bevel-and-socket joints on the concrete culvert piping are loosely sealed and are not watertight, allowing infiltration of groundwater through the pipe joints. Substantial quantities of groundwater infiltrate those portions of the pipe that are below the water table and the pipe sections that intercept the plume contain high concentrations of VOCs at the joint seeps. The pipe is sloped such that more surface area of the pipe is submerged below the water table in the down slope direction, thereby allowing a progressively greater influx of groundwater with distance downstream. Table 1 and Figure 3 summarize the water flow and sampling results at various locations in the pipe, pond and northeast tributary.

Table 1. Water Flow and Sampling Results in Northeast Tributary

Location	Distance Along Pipe (ft)	Water Flow (gpm)	Flow By Section (gpm)	TCE (ug/L)	Total VOC's (ug/L)
Upstream Station GP-C	NA	23	NA	<1	2.8
Closed Box Culvert #1	0	41	+18	<1	1.3
*Pipe Joint 45	120	49	+8	11.8	11.8
*Pipe Joint 9	450	58	+9	17,900	17,900
Open Box Culvert #2	555	60	+2	4,600	4,633
*Pipe Joint 63	590	139	+79	3,800	8,576
Drop Inlet (SW-A)	1,155	233	+94	2,400	2,444
Pipe Discharge Into Pond	1,290	291	+58	660	678
Pond Exit	1,680	305	+14	420	482
Downstream Tributary	2,175	NM	NA	210	222

*Water quality samples collected from seeps in pipe joints (not from flowing water) at these locations

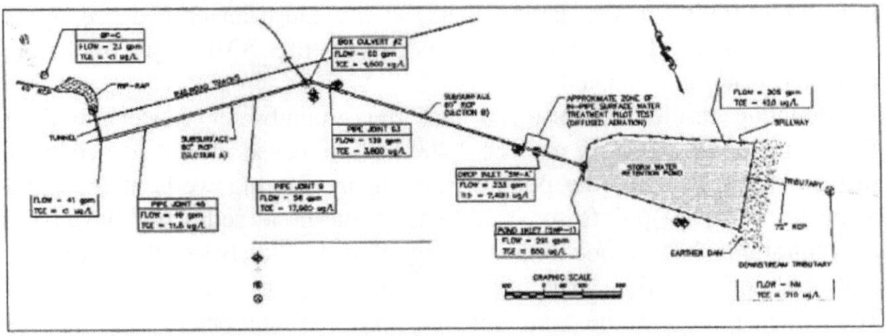

Figure 3. Northeast Tributary Storm Water Pipe and Pond System With Sampling Results

4.1 Flow and Dilution Modeling of Northeast Tributary

ENSR performed contaminant fate and flow modeling using GTC3M (NCDENR, 1996) to estimate acceptable concentrations of VOCs in the aquifer surrounding the pond that would not cause exceedance of surface water standards after mixing and dilution occurs. Model scenarios were run for TCE; the other VOCs detected in surface water at the pond are below South Carolina surface water maximum contaminant levels (MCLs) for human health exposures, which is the most conservative assumption for cleanup.

The G3CTM model uses a two-step calculation that uses advection, longitudinal dispersion, chemical retardation, and *in situ* chemical biodegradation via first order decay process as inputs for the first-step calculation. This part of the model utilizes the ODAST (One-Dimensional Analytical Solute Transport) program as described in Groundwater Transport: Handbook of Mathematical Models, the American Geophysical Union Water Resources Monograph Number 10 (Javandel et. al, 1984) to simulate contaminant fate and transport in the aquifer.

The second step in the model (uses a linear mixing equation to calculate the maximum ratio C/C_0 (groundwater concentration to maximum source concentration) at the plume's leading edge from the surface water body. To simulate mass loading only from the aquifer, model runs assumed that water discharging into the pond via the culvert pipe does not contain TCE. The 7Q10 flow of the pond was assumed to equal the sum of surface water and groundwater discharge into the pond. Using this calculation, the maximum TCE concentration in the aquifer that will not cause an exceedance of the surface water standard after mixing occurs at the 7Q10 flow (or actual measured flow) is estimated.

Modeling results demonstrated that the groundwater concentration of TCE at the pond boundary must be 2,400 µg/L to cause an exceedance of the surface water MCL in the pond. Existing monitoring wells at the pond boundary and well points screened beneath the pond sediments have TCE concentrations that are one to three orders of magnitude below this estimated threshold concentration. The model also predicted that future groundwater TCE concentrations at the pond boundary would remain one to two orders of magnitude below this threshold concentration. Therefore, direct discharge of groundwater into the pond is not expected to cause surface water MCLs to be exceeded now or in the future.

Mass loading calculations were performed to evaluate contaminant contribution from various sources in the system. The VOC contribution from groundwater discharging directly into the retention pond was determined to range from 0.014 pounds per day (lbs/day) to 0.0017 lbs/day. The VOC mass discharging from the culvert pipe system directly into the pond was determined to be 1.8 lbs/day, or approximately 99% of the total contaminant loading on the pond. Contaminant discharge from the aquifer directly to the pond is insignificant and controlling the VOCs exiting the culvert pipe would provide the greatest level of cleanup in the pond and its downstream tributary.

4.2 Remediation Goals for Northeast Tributary

In the northeast tributary, development of ecological screening values was complicated by underground channelization of the stream and the presence of the storm water retention basin. The man-made stormwater retention pond and downstream provide an area where humans and ecological organisms could be exposed to VOCs in the surface water. Institutional controls (i.e., fencing) were used to eliminate human exposure to the pond. Wildlife and aquatic life screening values were therefore deemed appropriate remedial goals for the pond. However, the State later acknowledged the stormwater retention pond as an in-stream treatment system requiring no specific remedial goals.

Human exposure could not be eliminated in the northeast tributary downstream of the pond. The remedial goal for this section of the northeast tributary was determined to be the human health-based water+organisms criterion, the most stringent of the applicable standards. The underground segment of the northeast tributary does not provide viable habitat for human, or aquatic and terrestrial wildlife receptors. Therefore, no remedial goal was determined for this section.

5. CHARACTERIZATION AND CLEANUP GOALS OF WESTERN AND SOUTHERN TRIBUTARIES

5.1 Characterization and Plume Impact Modeling of the Western Tributary

The western tributary is a discharge receptor for a separate, small groundwater plume originating from a former waste pit on this part of the property. CAHs and minor concentrations of several aromatic hydrocarbons exist primarily in the shallow and intermediate-depth aquifer zones adjacent to the tributary to a maximum depth of 55 feet. A shallow monitoring well at the edge of the tributary contains 140 µg/L of TCE. Historically, very low concentrations of CAHs have been detected twice in surface water.

A hydrogeological study was conducted around the western tributary. Stream seepage studies were performed and hydrologic gauging measurements were conducted in the stream, streambed well points, and adjacent piezometers and monitoring wells to determine the local groundwater-surface water interactions. The western tributary is intermittent at its headwaters and becomes perennial near the intersection with the plume. There is a strong upward flow gradient at the leading edge of the groundwater plume, and groundwater seeps are common in this area during the rainy season. The elevation where perennial flow occurs in the stream varies by season based on water table changes.

ENSR performed a simple contaminant fate and transport model of the dissolved CAH plume using the BIOCHLOR® model (Newall et. al., 2000) to support a remedial action plan for the waste pit source area. Target cleanup goals for groundwater at the discharge boundary of the tributary and for surface water were conservatively set as the human health-based MCL. The model predicted that if TCE concentrations were reduced to 250 µg/L at the contaminant source area, cleanup goals at the plume boundary could be maintained by natural attenuation of the plume. The remediation strategy developed for this area proposed installation of an *in-situ* treatment barrier to protect surface water until such time that the plume was stabilized by engineered treatment systems in the source area.

5.2 Characterization of the Southern Tributary

A detailed hydrogeologic investigation and stream study were also performed at the off-site southern tributary. The southern tributary contains a series of active, perennial springs that form the headwaters of the stream. Within the past forty years, a series of earthen dams were built along the tributary to create three small ponds. These dams were partially removed later to reestablish normal stream flow. Remnants of the dams and organic soils associated with the former ponds remain in the area.

Springs that feed the former location of the upper pond on the southern tributary are contaminated with CAHs. One of the springs had surface water TCE concentrations of 1,800 μg/L during the past two years. CAHs above the MCL concentrations are also found in surface water of the stream up to 450 feet downstream of the springs. The stream is fully accessible throughout this distance. The adjacent aquifer is impacted by CAHs migrating south from the manufacturing facility. In the area of the springs, the vertical extent of CAHs is found at relatively shallow depths of 45 feet.

5.3 Remediation Goals for the Western and Southern Tributaries

In the western and southern tributaries, potentially viable aquatic and wildlife habitat, and unrestricted access by humans required remedial goals to be developed that would be protective of all receptors.

6. SURFACE WATER TREATMENT PILOT TEST: NORTHEAST TRIBUTARY

A diffused aeration and ozonation pilot test was conducted on water flowing in the buried culvert pipe upgradient of the storm water retention pond on the off-site property east of the manufacturing facility. The test was conducted at drop inlet location "SW-A" located closest to the storm water retention pond (see Figure 3). Objectives of the pilot test were to determine the feasibility and design criteria of a diffused aeration/ozonation system as a method to reduce dissolved VOC concentrations in water flowing through the culvert pipe. An interim goal was to evaluate whether direct treatment of water in the subsurface pipe would be sufficient to remove VOCs entering the storm water retention pond to levels commensurate with the exposure risk at the pond and its downstream tributary. The pilot test used diffused aeration to volatilize VOCs, followed by an ozone injection stage for enhanced oxidation of VOCs.

6.1 Pilot Test Equipment

Water aeration and monitoring systems were constructed and placed in and around the culvert pipe as depicted in Figure 4. The aeration tubing consisted of 0.5-inch diameter polypropylene diffusion tubing installed in ten 50-foot sections inside the pipe parallel to the water flow. The tubing is manufactured with small diameter slits to provide a fine bubble size (0.2 to 3 mm) based on varying injection pressures. The tubing has a weighted keel on the bottom and is secured to metal support rack.

The end of each air diffusion tube leading to the air blower was connected to a section of non-perforated polypropylene tubing, which was attached to a 2-inch manifold pipe and then to a regenerative pressure blower. An ozone generator capable of producing 9 grams of ozone per hour at 10 cubic feet per minute (cfm) supplied air was used as an on-site ozone supply.

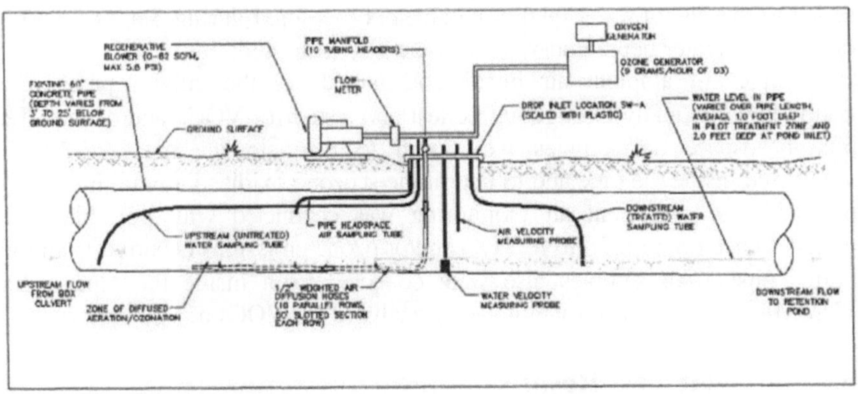

Figure 3. Pilot Test Schematic For Northeast Tributary

6.2 Pilot Test Procedures

The pilot test was conducted in three stages. Stage 1 used atmospheric air only to produce an air diffusion treatment zone in the culvert pipe. Air diffusion caused dissolved VOCs to be volatilized from the water as they were transported through the bubble-saturated water treatment zone. Air only was injected at rates ranging from 50-57 cfm for a period of 21 hours during Stage 1 of the test.

Stages 2 and 3 of the test used an air/ozone mixture at two different concentrations for a total period of 27 hours. Ozone is a powerful gaseous oxidant and disinfectant that is commonly used to treat water supplies and wastewater. Ozone is twelve times more soluble in water than oxygen, but its instability creates a short life span and it quickly breaks down to oxygen. The ozone/air mixture was injected at ozone concentrations ranging from 23.6 parts per million (ppm) in Stage 2 to 59.9 ppm (Stage 3).

As shown in Figure 4, water sampling was conducted at locations upstream and downstream of the in-pipe aeration zone. Dedicated sampling tubes were installed inside the pipe to collect water samples without entering the pipe during the tests. Samples of the water flowing through the pipe culvert and in the pond were collected at specific intervals of the test to determine the treatment efficiency as the water passed through the aeration zones and to evaluate the longevity of ozone dissolved in the water. Parameters analyzed for the water samples included total VOCs, pH, dissolved oxygen, turbidity, oxidation-reduction (redox) potential, and conductivity. Dissolved ozone was added to this list of water quality parameters measured during Stages 2 and 3 of the test. Water depth and velocity were also measured.

Samples of ambient air inside and outside of the culvert pipe were collected periodically throughout the test and tested for VOCs and ozone The pipe headspace air samples were used to estimate the mass of VOCs volatilized inside the pipe and to determine if ozone oxidized any of the vapor-phase VOCs. Ambient air monitoring was conducted with portable field meters to determine if fugitive VOC and/or ozone emissions occurred from the buried pipe. Air samples also were collected from inside the pipe using SUMMA® canisters and quantitatively analyzed for VOCs at a laboratory.

6.3 Pilot Test Results

Table 2 summarizes pilot test data and graphically depicts the water treatment efficiencies achieved during the test. Baseline water sampling before air injection began indicated that dissolved TCE concentrations were 8.3% lower in the downstream sample location (SW-A "Downstream"), which had a TCE concentration of 1,100 μg/L, than in the upstream sample location (SW-A "Upstream": TCE=1,200 μg/L). Furthermore, an 11.8% difference in TCE was measured between the SW-A "Downstream" sample location and the pond inlet sample (SWP-1; TCE=970 μg/L) before air injection began, further demonstrating dilution of the contaminated water as it travels downstream in the pipe. These results confirm earlier surface water studies that showed a natural reduction of VOC concentrations in the pipe water between the second box culvert and the pond due to influx of cleaner groundwater, an effect that was factored into the test results. TCE concentrations in the untreated water at SW-A "Upstream" varied throughout the test, suggesting that contaminant concentrations in the pipe water are dynamic. Water flow rates measured at SW-A were relatively consistent during the test.

During Stage 1 air injection, the total decrease in TCE concentrations between the two sampling points was 38.1%, of which approximately 30% of the reduction was attributed to the air diffusion volatilization. The air/water sparging ratios varied from 3.5:1 to 4.9:1 during Stage 1 testing. Dissolved oxygen concentrations in water increased an average of one unit between the upstream and downstream locations.

Table 2. Summary of Diffused Aeration and Ozonation Water Treatment at Northeast Tributary

SW-A "Upstream"			Ozone Injection Concentration (ppm)	SW-A "Downstream"			TCE Reduction Between Sampling Points (%)	Net Treatment Zone Removal Efficiency
Date/ Time	Elapsed Time (hrs)	TCE (ug/L)		Date/ Time	Elapsed Time (hrs)	TCE (ug/L)		
2/26/2001 12:40	0	1200	No Injection	2/26/2001 12:30	0	1100	8.3	NA
2/27/2001 8:30	19.83	2100	0	2/27/2001 8:07	19.62	1300	38.1	29.8
2/27/2001 14:10	25.49	1800	23.6	2/27/2001 14:10	25.67	1100	38.9	30.6
2/28/2001 12:20	47.65	1600	59.9	2/28/2001 12:20	47.84	830	48.1	39.8

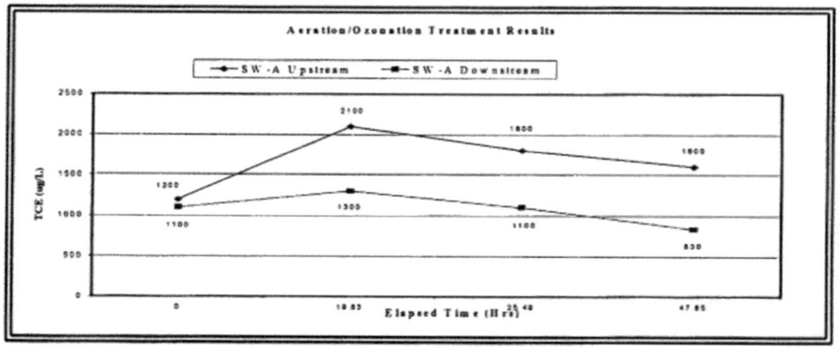

Adding 23 ppm of ozone to the water and reducing the air/water ratio to 3.5:1 slightly increased the TCE removal by a negligible 0.8% in Stage 2. The total TCE reduction between the two sampling points was 38.9% during Stage 2, of which 8.0% was attributed to natural dilution effects. Ozone concentrations in water and ambient air inside the pipe increased slightly during Stage 2. Ozone was not detected in ambient air above background levels at ground surface or in the pipe outlet at the pond (SWP-1).

For test Stage 3, the ozone injection concentration was increased to 59.9 ppm and an air/water ratio of 3.2:1 was maintained. The higher ozone concentrations increased the treatment efficiency approximate 10% over the lower ozone concentration used in test Stage 2, producing a net 40% reduction of TCE in the pipe water above the natural dilution effects (i.e., the total TCE reduction was 48%). Dissolved ozone concentrations in water did not increase proportional to the increased ozone injection rate. Measurable increases in ozone concentrations were observed in headspace air inside the culvert pipe during Stage 3. These results suggest that the ozone contact and retention time in water was low and most of the ozone injected at higher concentrations was escaping into the pipe headspace.

Water samples were collected from two locations in the retention pond to determine if quality of water exiting the storm water retention pond into the tributary had improved during the pilot test. Surface water samples were also collected at the pond outlet at the beginning of the pilot test and three days later at the end of the test. The pond outlet sample collected at the beginning of the pilot test showed **260 μg/L** of TCE, while the sample collected at the end of the test contained only **15 μg/L** of TCE. The magnitude of apparent TCE decrease (94.2%) between the two samples collected at the pond outlet is not commensurate with the volume of pond water exchanged (estimated as one third of the pond volume) or the comparatively lower VOC reductions in the culvert pipe. Other factors may have contributed to the lower TCE concentration measured in the pond at the end of the test.

7. SURFACE WATER REMEDIES AT THE WESTERN AND SOUTHERN TRIBUTARIES

Potential surface water protection remedies considered for the western and southern tributaries focused on *in-situ* treatment technologies that have low operating costs and minimal infrastructure requirements that might impede use of the off-site properties. Because biologically enhanced reductive dechlorination (ERD) had been successfully pilot tested in another part of the site, this technology was selected for application in a reactive barrier design to contain the groundwater plumes and to protect surface waters of these tributaries. In both treatment areas, commercially available Hydrogen Release Compound® (HRC®) was injected to create an *in-situ* biological ERD zone where groundwater redox is lowered and reductive dechlorination of trichloroethene is enhanced. Lactic acid, the active ingredient in HRC® gel, acts as an electron donor and the initial metabolic reaction that occurs produces hydrogen. In turn, the free hydrogen is further used by a different group of indigenous microbes to drive the microbial redox reaction that dechlorinates TCE and other chlorinated compounds. Details of each injection/treatment area are described below.

7.1 Western Tributary Reactive Treament Barrier

A total of 42 HRC® injection borings were installed in the leading edge of the plume upgradient of the western tributary. Locations of the HRC® injection borings, superimposed on the VOC plume map, are shown on Figure 5. The injection borings were installed in three groups to form a series of parallel ERD treatment barriers located upgradient of the tributary. This design is intended to intercept and treat the leading edge of the VOC plume before it discharges into the tributary or migrates further off-site.

The HRC® was injected in two different phases over a period of one year. The first phase of injections began near the tributary and the next phase of injections were installed further upgradient closer to the plume source. Total depths of HRC® borings around the western tributary varied from 43 feet below ground surface to 52 feet, with an average depth of 45 feet for most borings. Attempts to probe and inject HRC® below depths of 52 feet were unsuccessful due to probe refusal and high soil pressures.

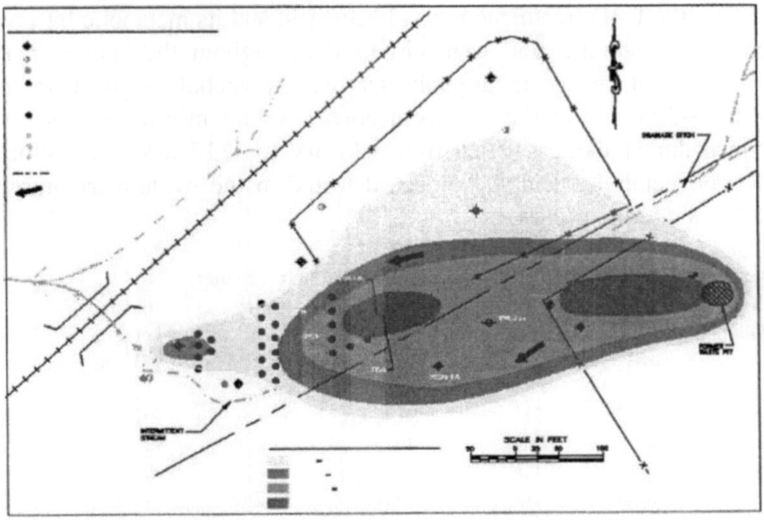

Figure 5. Total VOC Plume and ERD Barrier Location at Western Tributary

7.2 Southern Tributary Reactive Treatment Barriers

A total of 23 HRC® injection borings were installed for an ERD pilot test around the springs that form the headwaters of the southern tributary. HRC® injection borings were installed in semi-circular rows upgradient of each major spring to form the treatment barriers. The HRC® injection rows were spaced 12 feet apart, with the closest row being 25 feet from a spring.

The maximum HRC® injection depths around the springs of the southern tributary were 20 feet. Deeper injections were not used for pilot tests in this area because the highest concentrations of groundwater VOCs occur in the upper 25 feet of the aquifer and because the springs receive groundwater discharge primarily from this shallow zone of aquifer.

7.3 ERD Barrier Results

Table 3 summarizes aquifer ERD treatment results around the western tributary. In this treatment area, TCE reductions of 45.9% to 99.8% have been observed at the monitoring wells and piezometers that monitor the aquifer upgradient of the tributary. The compound cis-1,2-dichloroethene (cis, 1,2-DCE), a dechlorination byproduct of TCE, increased in most wells and has subsequently declined slowly over longer periods of time at most of these wells. Reduction of total chlorinated VOCs has varied from 31.8% to 67.5% in the ERD treatment zone. Lactic acid and its metabolic by-products acetic and propionic acid were dispersed throughout the barrier treatment zone. The net change in groundwater redox potential ranged from Δ-121 mV to Δ-325 mV in the treatment zone. Actual minimum redox values obtained during the test varied from –23 mV to –117 mV. No chlorinated VOCs or metabolic acids have been detected in the western tributary since HRC® injections occurred.

Table 3. Groundwater ERD Treatment Results at Western Tributary

Well Location	Initial TCE (ug/L)	*Final TCE (ug/L)	% TCE Reduction	Initial Total VOC's (ug/L)	*Final Total VOC's (ug/L)	% VOC's Reduction
MW-27s	140	26	81.4	190	101	46.8
MW-27i	14	2.8	80	20	8.6	57
MW-29s	29	1.3	95.5	62	42.3	31.8
MW15s	12,000	16	99.8	18,257	11,092	39.2
**MW-15i	3,700	2,000	45.9	12,400	7,100	42.7
SP-PZ-3	900	2	99.7	8,131	4,136	49.1
SP-PZ-4	21,000	29,000	None	40,425	54,153	None
SP-PZ-7	4,500	800	82.2	6,200	2,014	67.5
PZ99-1A	630	190	69.8	10,130	1,090	59.6

* Final Groundwater quality results 12 to 18 months after HRC injections occurred

** Maximum TCE and total VOC reductions occurred after 4 months of testing for well MW - 15i

Table 4 summarizes surface water treatment results associated with the ERD barriers installed around the major springs in the southern tributary. Because only one shallow piezometer/well point is located within the southern tributary ERD treatment zone, effectiveness of this treatment method was qualitatively determined by monitoring changes in water quality at the adjacent seeps and springs. During the 18-month period after HRC® was injected upgradient of the springs, TCE concentrations in surface water were reduced within a range of 57.9% to 87.4%. Total chlorinated VOC reductions of 59.6% to 84.2% occurred during this period. Metabolic acids were not detected in surface water samples and no geochemical changes were observed in surface water.

Table 4. Surface Water and Groundwater ERD Treatment Results at Southern Tribuatary

Sampling Location	Initial TCE (ug/L)	*Final TCE (ug/L)	%TCE Reduction	Initial Total VOCs (ug.L)	*Final Total VOCs (ug/L)	%VOCs Reduction
Surface Water						
Spring SP-2	95	40	57.9	120.8	43.6	63.9
Spring SP-3	990	400	59.6	990	400	59.6
SW-E	130	42	67.7	134.2	48.4	63.9
SW-F	38	4.8	87.4	38	6	84.2
Groundwater						
**SP-1	<1	8.8	**NA	1.8	11.8	**NA

*Final groundwater quality results 12 to 15 months after HRC injections occurred.

**TCE concentration was 500 ug/L at SP -1 one year before test. The "initial" or "final" result may be anomalous.

8. SUMMARY

The regulatory agency ruled that direct treatment of surface water in the culvert pipe of the northeast tributary would be allowed under a National Pollutant Discharge Elimination System (NPDES) permit. It was further ruled that the culvert pipe and the pond collectively will be considered components of the surface water "collection and treatment system", which eliminates the requirement to treat surface water to the human health MCLs within these structures. Human health MCLs will apply to surface water in the downstream tributary at the pond outfall where unrestricted exposure could occur. Although not required by the regulatory agency, the surface water wildlife standards for the storm water pond will become a self-imposed cleanup goal to be used in conjunction with fencing to restrict access to the pond.

The air diffusion pilot test demonstrated that dissolved VOCs were effectively removed from water in the culvert pipe even when moderately low air/water ratios were applied. Addition of 59.9 ppm of ozone to the water increased VOC removal efficiency only by 10%, which was not the degree that was expected or that would justify the cost of using ozone in a full-scale application. The treatment efficiency of ozone oxidation was likely limited by the minimal contact time of the ozone in the shallow pipe water at the test location. An expanded air diffusion pilot test was initiated in October 2003, and a full-scale diffusion system will be installed in 2004. This system will have 130 linear feet of treatment zone and will use higher air/water ratios ranging from 10:1 to 14:1 in the pipe to increase volatilization effectiveness.

Evaluations for full-scale applications of *in-situ* ERD treatment barriers and possible enhancements to these systems are planned for the western and southern tributaries. This technology appears to have the greatest potential for use as a sole remedy near the western tributary. Surface water impacts are more severe in the southern tributary and an aggressive ERD program, possibly combined with supplemental plume treatment and control technologies, will be required to achieve the more stringent surface water cleanup goals in this tributary.

REFERENCES

Javendel, I., Doughty, C., Tsang, C.F., 1984. Groundwater Transport: Handbook of Mathematical Models, Water Resources Monograph Series, Volume 10, Am. Geophys. Union, Washington, D.C.

NCDENR, 1996. G3CTM Groundwater Contaminant Transport Model. North Carolina Department of Environment and Natural Resources, Division of Water Quality, Groundwater Section, Raleigh, North Carolina.

Newell, C.J., C. E. Aziz, and J. Gonzales, 2000, BIOCHLOR Natural Attenuation Decision Support System, U.S. Environmental Protection Agency, Center for Subsurface Support, Ada, OK. USEPA Document EPA/600/R-22/008.

South Carolina Water Classifications and Standards. Chapter 61, R.61 -68, June 22, 2001.

USEPA, 1980. Water Quality Criteria. Federal Register Notice Number 45 FR 79328.

USEPA, 1995. Final Water Quality Guidance for the Great Lakes System, 40 CFR Parts 9, 122, 123,131, and 132. 60 FR 15366 -15425.

USEPA, 1996. EcoTox Thresholds. Eco Update Publication Volume 3 Number 2; 9345.0-12FSI. United States Environmental Protection Agency Office of Solid Waste and Emergency Response. EPA 540/F-95/038. January, 1996.

USEPA, 1999. Persistent Bioaccumulative and Toxic (PBT) Chemicals; Final Rule. 40 CFR Part 372. 64 FR 58665 -58753.

USEPA, 2001. Supplemental Guidance to RAGS: Region 4 Bulletins, Ecological Risk Assessment. Originally published November 1995. Website version last updated November 30, 2001: http://www.epa.gov/region4/waste/ots/ecolbul.htm

Watkins, S.G., Banks, P.A., and Billington, E.D., 2002. Hydrogeological Factors That Influence Chlorinated DNAPL Migration: A Case Study in a Piedmont Saprolite -Fractured Bedrock Terrain. In Proceedings of the 2002 Petroleum Hydrocarbons and Organic Chemicals in Groundwater: Prevention, Assessment and Remediation. National Ground Water Association. NGWA Press.

PART X: SITE ASSESSMENT

CHAPTER 35

APPLICATION OF PASSIVE VAPOR DIFFUSION SAMPLERS AT ENGLAND AIR FORCE BASE, LOUISIANA

Tom Dragoo[19], John Hicks[1], Don Vroblesky[2], and Rafael Vazquez[3]
[1]Parsons,Denver, CO, [2]USGS, Columbia, SC, and [3]Air Force Center for Environmental Excellence, San Antonio, TX

Abstract: Vapor sample collection using passive vapor diffusion samplers (PVDSs) represents a relatively new technology that employs passive sampling me thods for monitoring volatile organic compounds (VOCs). The sampler is buried in the bottom sediment of a surface -water body and is used to detect VOCs in groundwater discharging to surface water. Results of a field-scale PVDS application performed at a site at the former England Air Force Base (AFB), Alexandria, Louisiana between March and April of 2003 are presented. One objective for the PVDS application at England AFB was to determine if contaminated groundwater is discharging from Solid Waste Management Unit (SWMU) Number (No.) 41 into the adjacent Big Bayou and/or its tributary, the Unnamed Drainage Canal. A horizontal contaminant profile was developed by placing PVD samplers in the bottom sediment of Big Bayou and the Unnamed Drainage Canal, and analyzing the collected vapor samples for VOCs. The resulting data will aid in evaluating the fate and transport of contaminated groundwater emanating from SWMU No. 41. A second objective of this project was to evaluate the performance of the PVDSs and to provide guidance to the Air Force on the use of this technology to delineate contaminated groundwater discharge areas in surface water bodies. PVDS results showed detections of VOCs along the southern embankment of the Big Bayou;

and all PVDSs containing at least 0.9 micrograms per liter (ug/L) of *cis*-1,2-dichloroethene (DCE) were clustered along about 60 feet of the southern shoreline immediately downstream from shallow well S41L027MW, which contained the highest detected *cis*-1,2-DCE concentrations during the June 2002 groundwater sampling event.

Key words: PVDS, Surface-water body, Groundwater, Horizontal contaminant profile

1. INTRODUCTION

This report provides the results of the application of Passive Vapor Diffusion Sampler (PVDS) technology at the former England Air Force Base (AFB), Louisiana. England AFB was selected for the application of PVDS to assess whether groundwater, contaminated with chlorinated solvents, emanating from Solid Waste Management Unit (SWMU) Number (No.) 41 is discharging into the adjacent Big Bayou and/or its tributary (the Unnamed Drainage Canal) (Figure 1). The application was performed by Parsons for the Air Force Center for Environmental Excellence, Science and Engineering Division (AFCEE/ERS). The objectives of the PVDS application were to:

- Determine if contaminated groundwater is discharging from SWMU No. 41 into the adjacent Big Bayou and/or its tributary, the Unnamed Drainage Canal; and
- Evaluate the performance of the PVDSs and provide guidance to the Air Force on the use of this technology to delineate contaminated groundwater discharge areas in surface-water bodies.

PVD sampling is a relatively new technology designed to utilize passive sampling techniques. The approach and information regarding PVD sampling is largely drawn from the US Geological Survey (USGS) document, *Guidance on the Use of Passive-Vapor-Diffusion Samplers to Detect Volatile Organic Compounds in Ground-Water-Discharge Areas, and Example Applications in New England* (USGS, 2002). A PVD sampler consists of an empty 40-milliliter (mL) glass vial enclosed in two low-density polyethylene (LDPE) bags. The sampler is buried in the bottom sediment of a surface-water body with the goal of detecting VOCs discharging to surface water.

Each PVDS is attached to a wire surveyor flag or other marker for sampler identification, location, and recovery. PVD samplers are deployed in suspected groundwater discharge zones beneath surface-water bodies by burying them approximately 0.6 to 1.5 feet into bottom sediments. In some circumstances, this deployment may involve the use of divers or other means of deep-water emplacement, but often the samplers can be installed using a shovel or hand auger. PVD samplers should be allowed to remain in place until sufficient time has elapsed for the samplers to equilibrate and for pore-water to recover from environmental disturbances caused by sampler deployment. Typically, for sandy sediment, this equilibration period is two weeks. Longer equilibration times may be required for less permeable material. Factors affecting PVDS deployment and data interpretation include: the location and lithologic heterogeneity of the discharge zone, the orientation of the surface-water body relative to groundwater flow, the offshore distance of the contaminant-discharge zone, the potential temporal nature of the discharge-zone location and discharging concentrations, potential degradation of volatile organic compounds (VOCs) by bed-sediment microorganisms, and subsurface stream flow.

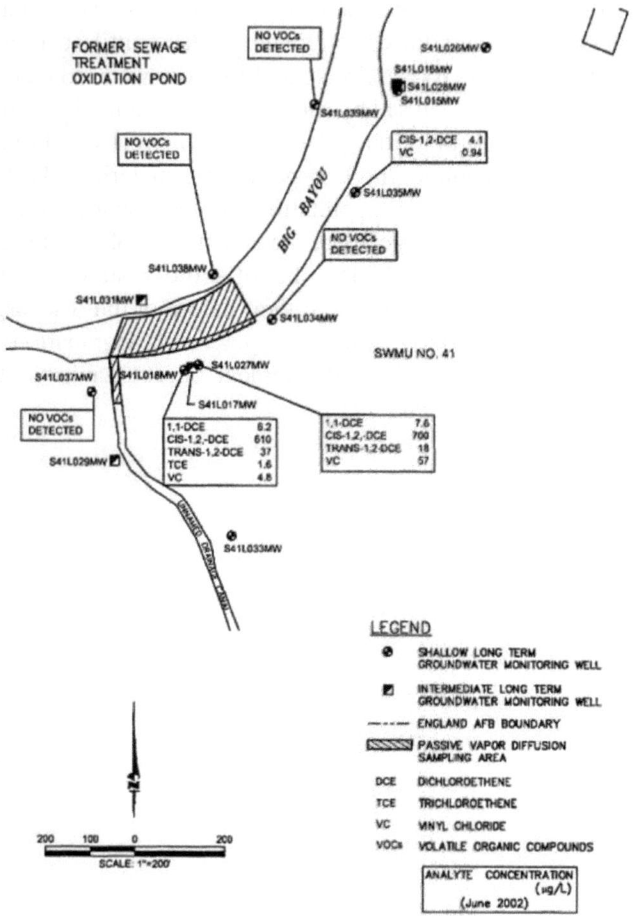

Figure 1. Passive vapor diffusion sampling area with results from the 2002 groundwater sampling event

2. SCOPE AND DESCRIPTION OF FIELD ACTIVITIES

PVDS application at England AFB required two mobilizations to the site: one to deploy PVD samplers in designated locations, and a second to retrieve the samplers and submit them to a fixed-base laboratory for analysis of VOCs by US Environmental Protection Agency (USEPA) Method 8260B (i.e., using a Gas Chromatograph/Mass Spectrometer [GC/MS]). The PVD samplers were installed in March 2003, and remained in place for nearly six weeks before being retrieved in April 2003. This report summarizes PVDS field activities and results.

2.1 Deployment Mobilization

The PVD samplers were deployed from March 10 through 13, 2003 by a two-man field crew. During this phase, several tasks were completed including piezometer installation, stream gauge installation, control line installation, and PVDS assembly and deployment.

During the PVDS deployment mobilization, various site conditions were monitored and recorded to help determine the following:

- Surface-water elevations and the actual locations of the wetted perimeters of the Big Bayou and the Unnamed Drainage Canal
- Surface-water elevations compared to expected normal conditions for that seasonal time-frame
- Appropriate PVDS deployment locations
- Groundwater discharge areas
- Proper retrieval mobilization schedule

In order to obtain this information, various types of equipment were installed at the site, as discussed in the following subsections.

2.1.1 Piezometer Installation

During the deployment mobilization, three piezometers were installed at the site to confirm that the sampled portions of the Big Bayou and the Unnamed Drainage Canal were gaining reaches (i.e., groundwater was discharging to surface water). Piezometer locations are shown on Figure 2. Piezometer (PZ) 1 and PZ2 were installed in the sediment of Big Bayou at depths of 5 feet and 3.5 feet below top of sediment (bts), respectively. PZ3 was installed in the Unnamed Drainage Canal at a depth of 1.5 feet bts. The PZs consisted of ¾-inch-diameter steel conduit pipe. The PZs were installed with a carriage bolt loosely placed into the bottom of the pipe to avoid plugging the pipe with sediment as it was driven into the sediment. Once the pipe was driven to its desired depth, a piece of steel rebar was placed down through the pipe to knock out the carriage bolt, allowing groundwater to flow into the pipe.

Subsequent water-level readings from two of the piezometers (PZ1 and PZ3) indicated that they were not operating properly; therefore, on 12 April 2003, four additional PZs (PZ4 through PZ7) were installed by England AFB contractor personnel. Locations of PZ4 through PZ7 are shown on Figure 2. PZ4 was installed to replace PZ1 and is located one foot west of PZ1. PZ5 was installed to provide supplemental data in another portion of the site, and is located 25 feet northwest of PZ1. PZ6 and PZ7 were installed to replace PZ3 and are located 1 and 2 feet south of PZ3, respectively. PZ4 and PZ6 consist of ¾-inch steel conduit pipe and were installed in the same manner as PZ1 through PZ3, described above. PZ5 consists of ¾-inch polyvinyl chloride (PVC) casing, a PVC drive point, and a 1-foot long PVC screen interval. PZ7 consists of ¾-inch galvanized steel pipe, a stainless-steel drive point, and a 1-foot-long stainless-steel screen interval. PZ4 through PZ7 were installed in the sediments of Big Bayou and the Unnamed Drainage Canal at an approximate depth of 3 feet bts.

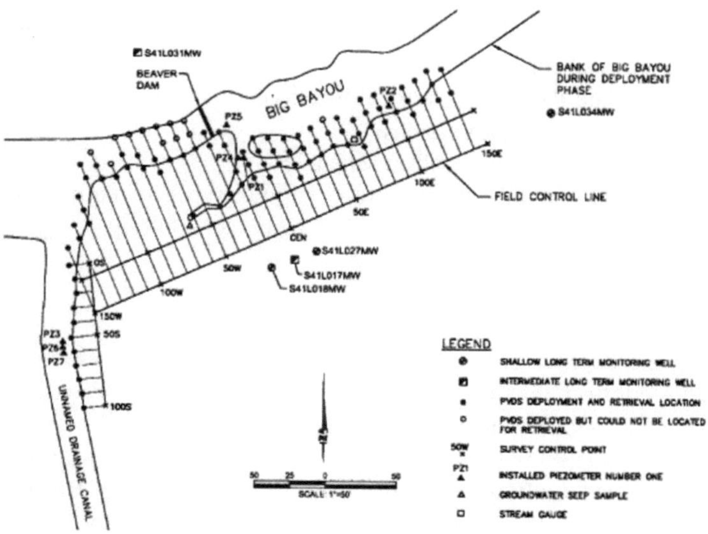

Figure 2. Passive vapor diffusion sampler deployment and retrieval locations.

2.1.2 Stream Gauge Installation

During the deployment phase, a stream gauge was also installed at the site to enable rapid monitoring of the water level in the Bayou and facilitate decision-making on the proper timing of the retrieval mobilization. The stream gauge location is shown on Figure 2. The stream gauge consists of a 2-inch wide by 6-foot long porcelain enamel, rust-resistant water-mark gauge that is graduated to the hundredth of a foot an marked at every foot and every tenth of a foot.

2.1.3 Control Line Installation

Two 300-foot-long and one 100-foot-long control lines were installed at the site during the deployment phase. These control lines were used as reference points and allowed for uniform placement of PVD samplers. Additionally, these control lines could be used as future reference points if additional or confirmation sampling is required at certain locations. Locations of these control lines are illustrated on Figure 2. The two 300-foot-long control lines were installed in an east/west orientation, parallel to each other and adjacent to Big Bayou. The 100-foot-long control line was installed in a north/south orientation adjacent to the Unnamed Drainage Canal. The locations of the control lines were surveyed by a licensed local surveyor.

2.1.4 PVDS Assembly and Deployment

Each PVDS consisted of an uncapped, 40-mL vial placed into a zip-lock bag (inner bag) and then placed into a second zip-lock bag (outer bag) (Figure 3). The PVDS was then tied to an 18-inch-long wire surveyor flag with zip ties, with the uncapped vial opening facing downward. Some PVD samplers were constructed with two uncapped vials, each having its own inner bag and both placed into the same outer bag, to facilitate collection of duplicate field samples.

Figure 3. Passive vapor diffusion samplers following retrieval.

Deployment of the PVD samplers occurred from March 11 through 13, 2003. PVDS deployment locations are illustrated on Figure 2. Every PVD sampler was connected to a 50-pound-test fishing line that extended from the control line to the PVDS. The fishing line was installed to aid in relocating the PVDSs during the retrieval phase.

For shallow-water deployment (i.e., water up to two feet deep), samplers were installed with a shovel by inserting the shovel into the sediment and pushing forward to create an opening between the back of the shovel and the sediment. In most cases, when the shovel was removed, the inserted PVDS was buried by natural collapse of sediments; however, in some instances backfilling of sediments was necessary.

In deep-water deployment (i.e., water deeper than two feet), a drive-point assembly was used. The drive-point assembly consisted of a 72-inch-long, 1¾-inch outside-diameter (OD), 1 5/8-inch inside-diameter (ID) electrical conduit outer pipe; a 74-inch long, 1½-inch OD, 1 3/8-inch ID electrical conduit inner pipe; and an 80-inch-long, 7/8-inch OD, 3/4-inch ID PVC pipe. A 2-inch OD pipe cap was attached to the top of the 74-inch-long pipe, and a 2-inch-long steel point was flush mounted to the bottom of the same pipe. The outer and inner pipes were driven into the bottom sediments by simply pushing down on the assembly. At the desired depth of installation, the inner pipe was removed, leaving a hole in the sediment extending about 2 inches deeper than the bottom of the outer pipe. The surveyor flag end of the PVDS was then pushed into the bottom of the PVC pipe until it was stopped by the volatile organics analysis (VOA) vial. The PVC pipe and attached sampler were then inserted and pushed through the outer pipe into the sediment. The sampler was held in place by the PVC pipe as the outer pipe was removed. Sediments then collapsed around the sampler, and the PVC pipe was removed from the sediment.

The target PVDS deployment depth was 12 inches bts at all locations including both shallow and deep water deployment. Again, PVD samplers were installed with the uncapped vial opening facing downward. Each sample was assigned a unique sample identification number identifying where the PVDS was deployed.

2.2 Site Monitoring

In between the deployment and retrieval of the PVD samplers, various site conditions were monitored to aid in establishing an appropriate retrieval schedule. Monitoring was focused on confirming that the sampled reaches of the surface water bodies were gaining (i.e., receiving groundwater discharge) for a sufficient time period prior to sampler retrieval. These activities included monitoring climatic conditions, surface-water levels using the stream gauge, and groundwater levels using the piezometers.

Base contractor personnel periodically visited the site to visually assess water levels and site conditions. These visits consistently revealed that water levels were below expected normal levels. Additionally, data from the Southern Regional Climate Center were monitored for cumulative daily rainfall. Throughout the deployment mobilization, the daily rainfall data showed slightly less than normal accumulation. Prior to and during the retrieval mobilization, the daily rainfall data showed significantly lower than normal accumulation.

Surface-water levels were also monitored because high surface-water levels may be indicative of losing conditions (i.e., surface water recharging the underlying groundwater system). Surface-water levels were monitored throughout the project by recording water level readings from the installed stream gauge. Site monitoring of PZs was also necessary to help identify whether the surface water bodies at the site (Big Bayou and Unnamed Drainage Canal) were losing (surface water discharging into groundwater) or gaining (groundwater discharging into surface water). Surface water losses or gains may cause temporal or cyclic changes in concentrations of VOCs that could affect interpretations of the horizontal extent and concentrations of VOCs in groundwater (USGS, 2002). A losing reach of surface water typically corresponds to lower groundwater levels in the piezometer (relative to surface water levels); a gaining reach typically corresponds to higher groundwater levels (relative to surface water levels). Therefore, in order to collect PVDS data representative of groundwater conditions, the PVD samplers had to be retrieved following a period during which the surface water bodies were gaining.

Following stabilization, groundwater potentiometric head measurements within the PZs and surface water levels outside the PZs were collected and compared to determine whether the vertical hydraulic gradient was upward (gaining reach) or downward (losing reach). Data from PZ2 and PZ4 indicate an upward vertical hydraulic gradient between the deployment and retrieval events. Water levels inside and outside of PZ3, installed in the Unnamed Drainage Canal, suggest the presence of losing conditions at that location (i.e., groundwater was being recharged by surface water); this condition may have been caused by the raising of the water level behind the beaver dam that was constructed between the sample deployment and retrieval events (see Section 2.3.2).

It appears that the remaining PZs may have needed more time to stabilize before they displayed accurate groundwater potentiometric conditions at the site. PZ5 and PZ7 were dry. The water level in PZ6 was slowly increasing over time, and it is not known if it had reached static equilibrium by April 22. When PZ1 was initially installed, water did not readily accumulate, indicating that it may have been plugged or installed in a very low-permeability zone. Therefore, clean water was added to this piezometer until it was completely full in an attempt to see if the water level declined until static equilibrium was reached. The water level in PZ1 did not decline more than 1 foot during the measurement period, suggesting that this piezometer was plugged.

2.3 Retrieval Mobilization

The PVD samplers were retrieved from April 21 through 22, 2003 by a two-man field crew. Site conditions (i.e., piezometer and stream gauge readings) also were monitored as the PVD samplers were retrieved. In addition, one groundwater seep sample was collected at the site as described in Section 2.3.3.

2.3.1 PVDS Retrieval

Retrieval of PVD samplers began on the morning of April 21 and ended April 22, 2003 (Figure 3). Retrieval of a PVDS consisted of excavating it from the sediment, inspecting it, capping and labeling the PVDS, then placing it on ice for shipment to the laboratory. A total of 13 of the 98 PVD samplers deployed could not be found and therefore were not retrieved; the locations of these 13 samples are shown on Figure 2. Collected PVD samplers were shipped to the laboratory for analysis of VOCs.

2.3.2 Site Conditions

Upon returning to the site for the retrieval phase, a beaver dam was discovered in the Big Bayou that had not been present during the deployment mobilization. The beaver dam was located approximately 20 feet upstream (west) of PZ5 (Figure 2). As a result, there were different water elevations above and below the beaver dam that varied by approximately one foot. In order to ease retrieval of PVD samplers and avoid inducing losing stream conditions downstream of the dam, the dam was left in place until all of the downstream PVD samplers were retrieved. Subsequently, the dam was breached to allow water in the Big Bayou to flow freely downstream. This resulted in a fairly rapid rise in water levels downstream from the dam, but upstream water levels showed very little decrease. After the dam was breached, the remaining PVD samplers located upstream of the dam and in the Unnamed Drainage Canal were retrieved.

2.3.3 Groundwater Seep Sample

During the retrieval of the PVD samplers, a groundwater seep was discovered that appeared to be emanating from SMWU No. 41. On April 22, 2003 a six-inch-wide by eighteen-inch-deep pit was excavated, just upgradient of the beginning of the seep, to collect a groundwater sample. The location of the groundwater seep sample is illustrated on Figure 2. This sample was put on ice and shipped to the laboratory for analysis of VOCs by USEPA Method 8260B.

3. RESULTS

3.1 PVDS

98 PVD samplers were deployed at the site; however, only 85 were retrieved and analyzed for VOCs. The distribution of detected chlorinated VOCs is illustrated on Figure 4. As shown on this figure, chlorinated compounds were detected at 14 locations at concentrations ranging from 0.075 ug/L to 4.5 ug/L for *cis*-1,2-DCE. *cis*-1,2-DCE was the most prevalent compound detected at the site, and was present in 12 of 14 locations where chlorinated VOCs were detected.

All of the PVD samplers with detected concentrations of chlorinated VOCs were located at or near the southern embankment of Big Bayou and along the eastern embankment of the Unnamed Drainage Canal. Furthermore, all samplers containing at least 0.9 ug/L of *cis*-1,2-DCE were clustered along about 60 feet of the southern shoreline immediately downstream from shallow well S41L027MW, which contained the highest detected *cis*-1,2-DCE concentrations during the June 2002 groundwater sampling event. As described in Section 2.3.2, the water level in the Big Bayou was artificially raised approximately 1 foot due to construction of a beaver dam approximately 20 feet west (upstream) of PZ5 following PBDS deployment. It is possible that the higher water levels in the Bayou upstream of the dam (including in the Unnamed Drainage Canal) may have caused some vertically downward migration of surface water in this area and impacted PVDS results; there were no analyte detections in PVD samplers installed in the bayou upstream of the dam. There were, however, two detections located in the Unnamed Drainage Canal. There were no piezometers installed in the bayou upstream of the dam to confirm the nature of surface-water/groundwater interactions in this reach. However, as described in Section 2.2, water-level data from piezometer PZ3, installed in the Unnamed Drainage Canal, suggest that losing conditions may have been present at that location.

3.2 Groundwater Seep Sample

On April 22, 2003 one groundwater seep sample was collected and analyzed for VOCs. Analytical results for this sample are shown on Figure 4. The only VOC detected in the groundwater seep sample was *cis*-1,2-DCE at a concentration of 0.075 μg/L. The location of this sample is illustrated on Figure 3.1.

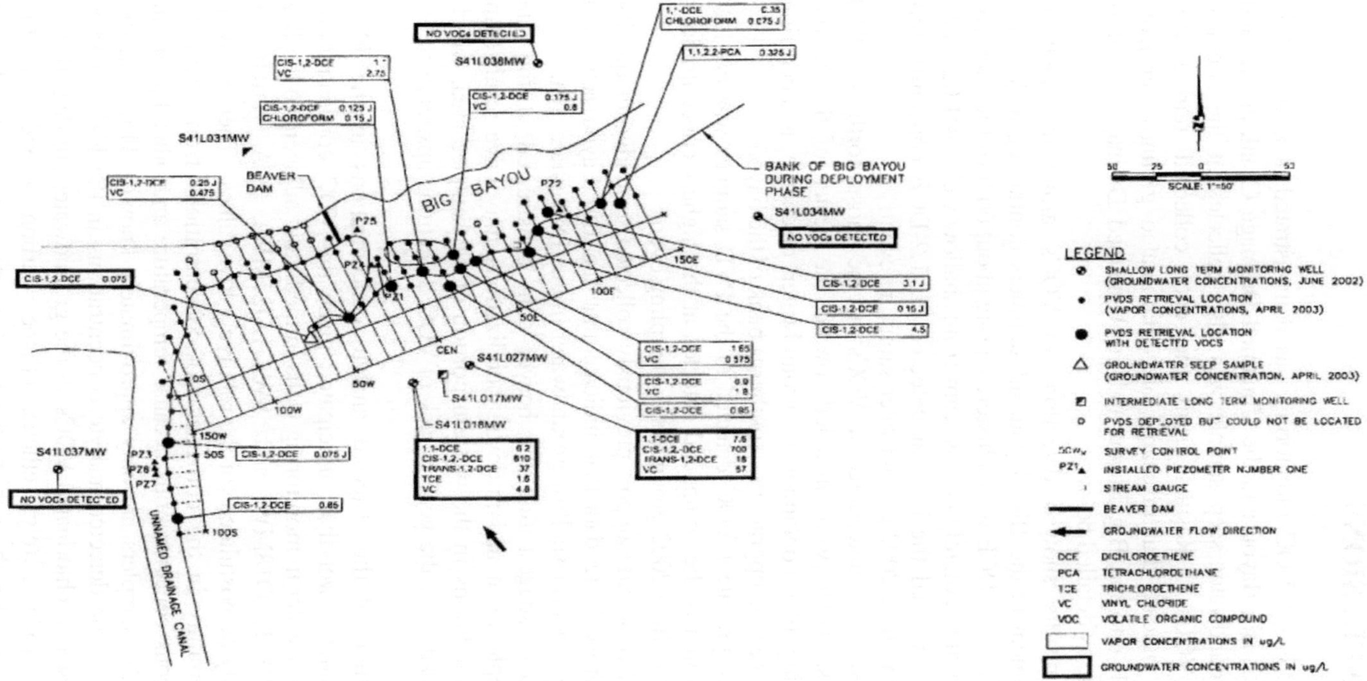

Figure 4. Passive vapor diffusion sampler results. Results are also provided for the 2002 groundwater sampling event and the groundwater seep sample

4. CONCLUSIONS

Use of PVDSs for VOC monitoring was demonstrated at a total of 85 locations in both Big Bayou and the Unnamed Drainage Canal, at England AFB. One groundwater seep sample was also collected at the site and analyzed for VOCs. Results from the data collected during this demonstration indicate that discharge of contaminated groundwater from SWMU No. 41 into the Big Bayou and the Unnamed Drainage Canal is likely occurring, as detailed below:

- PVDS results showed detections of VOCs along the southern embankment of the Big Bayou; all samplers containing at least 0.9 ug/L of *cis*-1,2-DCE were clustered along about 60 ft of the southern shoreline immediately downstream from shallow well S41L027MW, which contained the highest detected *cis*-1,2-DCE concentrations during the June 2002 groundwater sampling event.
- There were no detections of VOCs in the most northerly PVD samplers, which were installed toward the center of the bayou, suggesting that contaminated groundwater discharge is occurring mainly near the upgradient (southern) bank of the bayou.
- The most prevalent VOC detected in the PVD samples was *cis*-1,2-DCE, which is the compound detected at the highest concentrations during the June 2002 groundwater sampling event.
- The PVDS results support previously collected groundwater quality and hydrogeologic data that indicate that the Big Bayou acts as a discharge area for shallow groundwater in this immediate area.

Available data suggest that the beaver dam constructed at the site between the deployment and retrieval mobilizations may have resulted in losing stream conditions in the western portion of the sampled area, which also may be a cause for the general lack of VOC detections upstream of the dam.

The magnitude of the detected analyte concentrations in the PVD samples was much lower than anticipated given the VOC concentrations detected in groundwater in monitoring wells immediately adjacent to the Big Bayou (i.e., SW41LO18MW and SW41LO27MW). A variety of explanations may be postulated for the relatively low values. There is some degree of uncertainty in the conversion of concentrations from vapor to equivalent aqueous concentrations because temperature and other factors in the vicinity of the samplers are only approximately known. However, the low magnitude of the detected analyte concentrations in the PVD samples may also suggest that chlorinated VOCs in the groundwater are degraded to non-toxic end products (e.g., ethene, ethane, carbon dioxide) as the groundwater approaches the surface water. This may occur via reductive dechlorination if the sediment environment beneath the Bayou is highly organic, anaerobic, and reducing. Alternatively, *cis*-1,2-DCE and especially vinyl chloride can be directly oxidized in aerobic environments. Another potential explanation for the low magnitude of the PVDS detections is that the VOC concentrations in the discharging groundwater may become diluted due to contact with surface water beneath the stream bed.

Overall, results from this PVDS application indicate that this sampling methodology can facilitate identification of groundwater VOC plume discharge into surface-water bodies. It should be noted that the cross-gradient (i.e., eastern and western) boundaries of the VOC plume discharge area were not precisely delineated by this sampling effort. Trace-level VOC concentrations were detected in two of the eastern-most samples collected (Figure 4), and plume discharge in the western portion of the sampled area may have been masked by the effects of the beaver dam. However, the lack of *cis*-1,2-DCE detections in the eastern-most five rows of the sampling grid indicate that discharge of chlorinated ethenes to the Bayou diminishes in this area.

REFERENCES

US Geological Survey (USGS). 2002. Guidance on the Use of Passive -Vapor-Diffusion Samplers to Detect Volatile Organic Compounds in Ground -Water-Discharge Areas, and Example Application in New England. Water -Resources Investigations Report 0 2-4186.

This page intentionally left blank

CHAPTER 36

ULTRA-FAST FIELD GAS CHROMATOGRAPHY FOR SITE CHARACTERIZATION AND FIELD MONITORING

Michael A Marando
GEI Consultants, Inc.

Abstract: The zNose™ Model 4200 analyzer is a field gas chromatograph (GC) capable of analyzing volatile organic compounds (VOCs) and semivolatile organic compounds (SVOCs) to part-per-billion (ppb) levels in less than two minutes. The analyzer is manufactured by Electronic Sensor Technology of Newbury Park, CA and has a surface acoustic wave (SAW) detector that can detect and quantify the mass of VOC and SVOC compounds to picogram (10^{-12} **gram)** levels. This capability allows for real-time decisions to be made in the field and allows site managers to be more effective in conducting site assessments by reducing time and costs. The zNose™ can also produce VaporPrints™, a visual representation of the constituents in the sample that allows for identification of mixtures of chemicals. VaporPrints™ can be generated for complex mixtures of chemicals to assist in environmental forensics. Using the zNose™ for screening SVOC contaminated soil samples and for screening contaminated groundwater samples is discussed.

Key words: VOC, SVOC, Site Remediation, Site Assessment, Gas Chromatography, zNos

1. INTRODUCTION

Accurate chemical data generated in the field can facilitate environmental investigations and remediations. Field analytical screening tools can generate these data to assist investigators and remediation managers. This capability allows for real-time decisions to be made in the field. Investigators can be more effective in conducting assessments by obtaining field data that can assist them in selecting more appropriate samples for laboratory analysis. Site managers can more effectively use their analytical budgets to obtain more useful data. In addition, rapid monitoring of process technologies can be performed to enhance process control for meeting remediation goals. Currently some field techniques for analyzing organic compounds include immunoassay, photoionization detectors, flame ionization detectors, and field portable gas chromatographs.

The zNose™ Model 4200 Analyzer (Figure 1) is a field portable gas chromatograph manufactured by Electronic Sensor Technology of Newbury Park, CA. Volatile organic compounds (VOCs) and semi-volatile organic compounds (SVOCs) can be analyzed with the zNose™ to part-per-billion (ppb) levels in less than two minutes. The zNose™ uses a surface acoustic wave (SAW) detector that detects and quantifies the mass of VOC and SVOC compounds to picogram (10^{-12} gram) levels. The zNose™ can also generate VaporPrints™, a visual representation of the constituents in the sample that allows for identification of mixtures of chemicals. This article discusses how the zNose™ operates and some environmental applications for the zNose™. The article also discusses using the zNose™ to screen non-aqueous phase liquid (NAPL) samples and to screen soil contaminated with pentachlorophenol (PCP).

Figure 1. The zNose™ Model 4200 Analyzer, Photo courtesy of Electronic Sensor Technology

2. THE ZNOSE™ MODEL 4200 ANALYZER

The zNose™ operates in a two-step process (Figure 2). In the first step, a pump draws vapors through a heated inlet onto a loop trap filled with Tenax adsorbent. Alternatively, a solvent sample of up to one microliter can be injected into the port of the instrument for analysis while the pump is collecting vapors. After the sample has been concentrated, a valve switches the loop trap to the inject position to begin the second step of the analysis. In the inject position the loop trap is placed in line with the helium carrier gas. The Tenax is heated to 250°C, the analytes are desorbed from the Tenax, and the analytes are separated along a capillary column one meter in length.

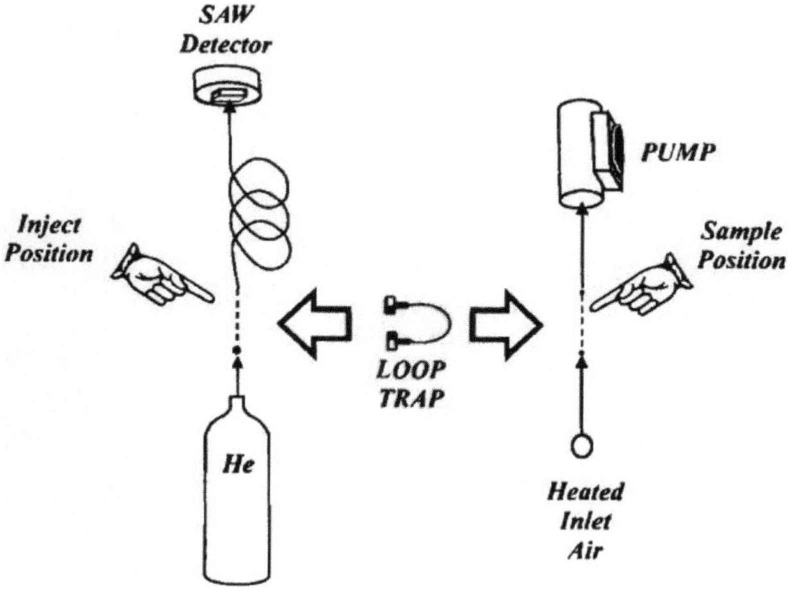

Figure 2. Simplified zNose™ Operation Diagram, Diagram courtesy of Electronic Sensor Technology

The column can operate at a constant temperature (isothermal operation) or can ramp as high as 18°C per second. Instead of using an oven to heat the column, as is the case with typical gas chromatography, heating is performed electrically by increasing the current along the outside of the column. As the compounds are eluted from the column, they are focused onto the SAW detector.

SAW detectors are used in electronic noses for the detection of specific compounds. SAW detectors work by deposition and are typically coated with polymers that adsorb specific compounds. Unlike SAW detectors in many electronic noses, the SAW detector in the zNose™ is uncoated, and is 1000 times more sensitive than coated SAW detectors (Staples et al., 1998).

The SAW detector initially resonates at a fixed vibration frequency. As analytes elute from the column they condense onto the detector and change the vibration frequency. As the mass of the detector increases, the vibration decreases proportionally. The zNose™ measures these changes and produces a chromatogram based on the difference in vibration change, or the change in mass of the detector. Figure 3 is a chromatogram of benzene, toluene, ethyl benzene, and m- and o-xylene (BTEX). Figure 4 is a chromatogram of ten polycyclic aromatic hydrocarbons (PAHs) and PCP. The zNose™ reports either a mass or a concentration based on a user calibration.

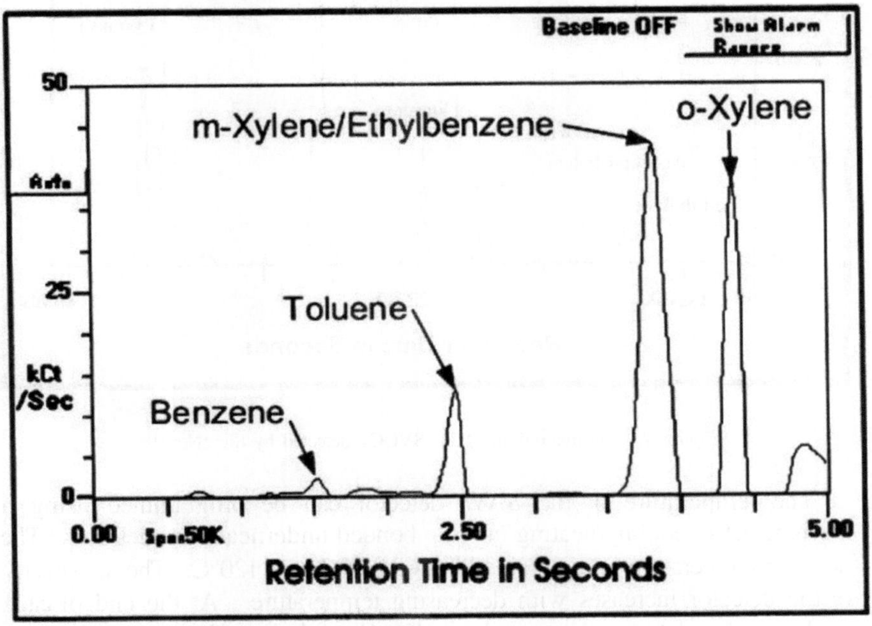

Figure 3. Chromatogram of BTEX compounds detected by the zNose™

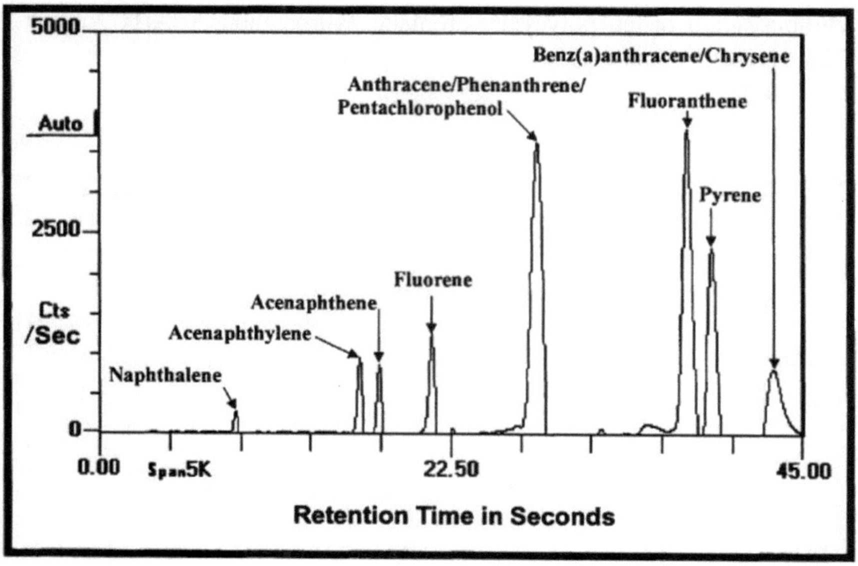

Figure 4. Chromatogram of 11 SVOCs detected by the zNose™

The temperature of the SAW detector can be programmed using a thermoelectric cooling/heating module bonded underneath the detector. The detector temperature is maintained between 0°C and 120°C. The sensitivity of the detector increases with decreasing temperature. At the end of each analysis, the detector is heated to 130°C to boil off compounds from previous injections.

The output of the zNose™ can also be represented as a VaporPrint™, which is a polar plot of the analytes present in the sample. This image may be represented as the direct output of the SAW detector or as the converted chromatogram. VaporPrints™ allow the spectrum of analytes to be visualized and allows for easier recognition. Figure 5 is an example of a VaporPrint™ that was generated from groundwater contaminated with No. 2 Fuel Oil.

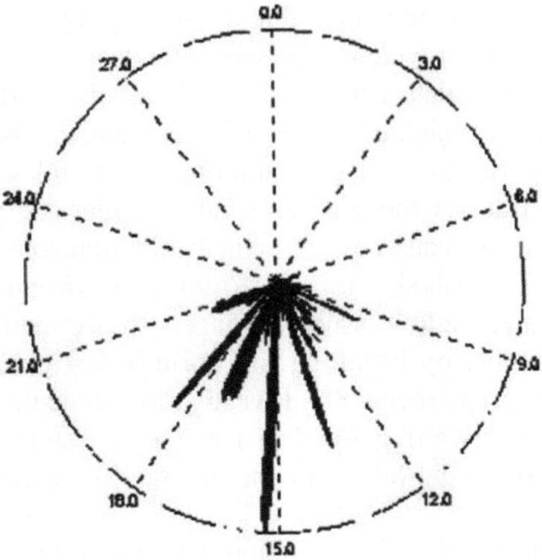

Figure 5. Headspace VaporPrint™ of No. 2 Fuel Oil in Groundwater

2.1 Applications

The zNose™ has been used by both commercial and government sectors since 1996 for trace analysis and fingerprint recognition. The zNose™ is used in the food and beverage industry to monitor quality control. Researchers who study the physiology of plants, bacteria, and animals have found the zNose™ to be very useful (Staples, 2003). The zNose™ has also been validated by the White House Office of National Drug Control Policy for the detection of cocaine and heroin vapors (Staples, 1999). The zNose™ is currently being tested at Dugway Proving Ground for the detection of chemical warfare agent (Business Wire, 2002), and is being used in the chemical industry for the detection of airborne chemicals for worker safety (Staples, 2003).

The zNose™ has also been used to a more limited extent in environmental applications. The instrument has been tested by the US EPA for the detection of chlorinated VOCs in groundwater (EPA, 1998a) and PCBs in soil (EPA, 1998b). The zNose™ has also been used at Lawrence Livermore Laboratories for the detection of VOCs in groundwater (Williams, et. al, 1997). The Electric Power Research Institute (EPRI) has also evaluated the zNose™ for air monitoring at remediation sites.

The zNose™ can be beneficial to investigators and remediation managers. If VOC or SVOC specific data are beneficial in the field, if the contaminant concentrations are less than 1 part per million (ppm), and if visual or olfactory senses are not reliable in identifying contaminated media, then the zNose™ could provide useful screening data.

Using the zNose™, data can be obtained rapidly in the field and site investigations can be expedited. For soil gas surveys, the zNose™ has a built-in pump to purge the soil gas sampling point and sample the vapors without having to transfer the soil vapor into a tedlar bag prior to analysis. VaporPrint™ analysis can be performed simultaneously to obtain a fingerprint of the analysis for performing environmental forensics. Obtaining this data can allow project managers to use their sample analysis budget more effectively by installing soil borings, test pits, and monitoring wells in more appropriate locations. Investigation strategies can be adjusted in the field instead of waiting for data from a laboratory. For very large sites, a GPS stamp can be added to the zNose™ data file to better track sampling locations.

The zNose™ can facilitate remediation activities by providing managers with field data to determine the limits of excavation. Also, if on-site remediation is performed, pre-treatment and post-treatment samples can be screened and the data used to optimize the treatment process and evaluate treatment effectiveness.

Because the cycle time for each analysis is typically two to three minutes, quality control samples can be run often. Several analyses from each sample can be analyzed to verify the precision of the data. Continuing calibration runs can be performed frequently without taking much time away from analysis of samples. Blanks can be run often to verify that cross-contamination is not occurring. The speed of the analysis also allows troubleshooting to be expedited because the operator can quickly identify problems.

3. EXAMPLES OF REMEDIATION USES

At a property in Massachusetts, steps were taken to contain No. 2 Fuel Oil that had migrated onto the site from an adjacent property. While excavating, a second area of oil contamination was discovered on the property and cross-gradient to the area contaminated with No. 2 Fuel Oil. This second area of contamination was of particular concern because it may have represented a new release. To evaluate this second area, headspace vapors from groundwater samples collected from each area were screened with the zNose™. Figure 5 is a headspace VaporPrint™ for a sample of the known No. 2 Fuel Oil contamination. Figure 6 is a headspace VaporPrint™ for a sample collected in the second area of oil contamination. The two VaporPrints™ look very similar and appear to have a similar composition.

A laboratory fingerprint analysis later confirmed that both groundwater samples had the same contaminant. The second area of oil contamination was therefore considered part of the known No. 2 Fuel Oil release and did not represent a new reporting condition.

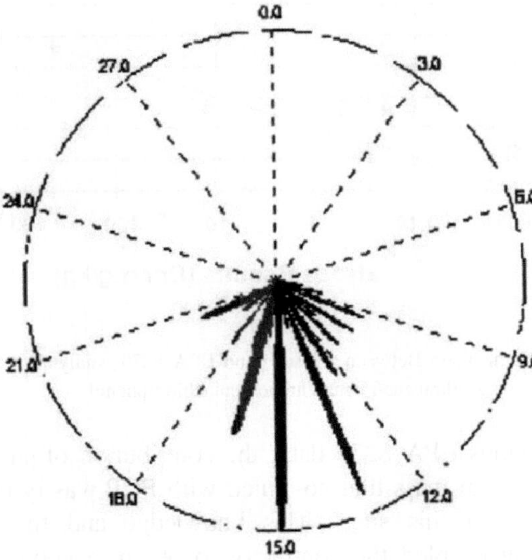

Figure 6. Headspace VaporPrint™ of Unknown Sample

On behalf of a remediation contractor, the zNose™ was used to monitor the effectiveness of a chemical oxidation treatment process in treating pentachlorophenol (PCP) and other SVOCs in soil. The contractor was interested in monitoring the treatment effectiveness in real-time so they could optimize their process and expedite remedial operations. The contractor provided eight samples to be analyzed for phenanthrene and PCP using the zNose™. Split samples were sent to a laboratory for EPA 8270 analysis. A methylene chloride extraction was used to extract the SVOCs from the 10 grams of soil.

Figure 7 shows the comparison between the EPA 8270 and zNose™ results. The diagonal line on the chart shows where the EPA 8270 results and the zNose™ are equal. In some cases, multiple zNose™ analyses were performed on a sample that had one EPA 8270 analysis. The data show a good correlation between the zNose™ and the EPA 8270 results, demonstrating that the zNose™ is a useful field screening tool for PCP, phenanthrene, and anthracene.

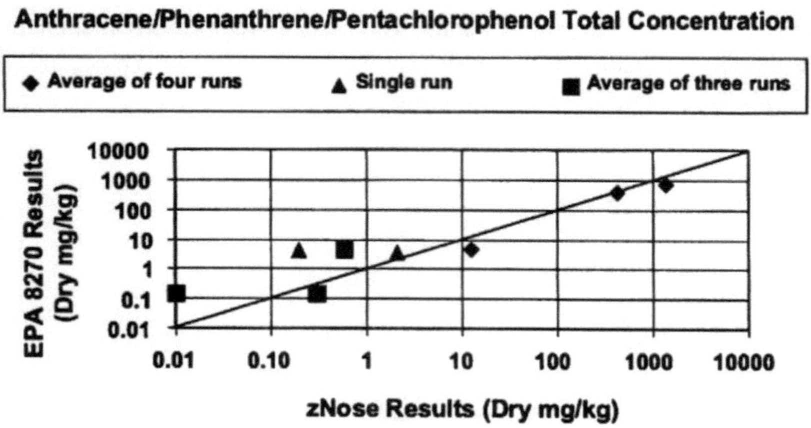

Figure 7. Comparison Between zNose™ and EPA 8270 Analysis Results for Anthracene/Phenanthrene/Pentachlorophenol

Based on previous EPA 8270 data, the contribution of phenanthrene and anthracene to the total mass that co-eluted with PCP was typically 10% for the contamination at this site. This knowledge and the results of the zNose™ screening enabled the contractor to obtain a good estimate of the PCP concentration. The remediation contractor has since mobilized the zNose™ to a remediation site where remedial activities are currently underway.

4. CONCLUSION

The zNose™ Model 4200 Analyzer is a useful field tool for obtaining chemical specific VOC and SVOC data for groundwater, soil gas, soil, and air. These data can help expedite field operations and facilitate real-time optimization of treatment processes to save time and money. The zNose™ can also generate VaporPrints™ to provide a field fingerprinting tool to assist in environmental forensics. Phenanthrene and PCP concentrations measured in soil samples using the zNose™ and by US EPA laboratory testing methods were similar.

ACKNOWLEDGMENTS

The author expresses his appreciation to Dr. Edward Staples and Ken Zeiger of Electronic Sensor Technology for their assistance, Ash Jain of the Electric Power Research Institute for providing preliminary funding, and Paul Maus from Soil Savers, Inc. for additional funding.

REFERENCES

Business Wire.com webpage. October 16, 2002.

Staples, E.J., Matsuda, T., and Viswanathan, S. 1998. "Real Time Environmental Screening of Air, Water, and Soil Matricies Using a Novel Field Portable GC/SAW System." Environ. Strategies for the 21ST Century, Asia Pacific Conference, April 8-10, 1998. Singapore.

Staples, E.J., 1999. "The Electronic Nose – A New Law Enforcement Tool." 1999 Office of National Drug Control Policy Technology Symposium, March 8-10, 1999. Washington, D.C.

Staples, E.J., 2003. Personal Communication.

USEPA (U.S. Environmental Protection Agency). 1998a. "Environmental Technology Verification Report. Field Portable Gas Chromatograph. Electronic Sensor Technology, Model 4100." EPA/600/R-98/141. November 1998.

USEPA (U.S. Environmental Protection Agency). 1998b. "Environmental Technology Verification Report. Portable Gas Chromatograph/Surface Acoustic Wave Detector. Electronic Sensor Technology, 4100 Vapor Detector." EPA/600/R -98/114. August 1998.

Williams, D., Staples, E.J., Watson, G., McGuire, D., Pappas, G. 1997. "Use of a Surface Acoustic Wave Detector Incorporated into a Field Portable Rapid Gas Chromatograph to Detect Chlorinated Hydrocarbons from DNAPL Contaminated Ground Water." Fifth Symposium on Field Screening Methods for Hazardous Wastes and Toxic Chemicals. January 29-31, 1997. Las Vegas, NV.

This page intentionally left blank

CHAPTER 37

GEOSTATISTICAL MODELING AND MAPPING OF SEDIMENT CONTAMINANT CONCENTRATIONS

Kandiah Ramanitharan, Laura J Steinberg, Gerhard Piringer
Tulane University

Abstract: This paper demonstrates the application of geostatistical modeling techniques to the quantification of Aroclors and heavy metals in Hudson and Duwamish river sediments, respectively. The objective was to compare models and modeling parameters between contaminants and sites, as well as to investigate a modeling modification that may accomodate curved river segments. Gaussian, exponential and spherical variograms were used to model micro-scale spatial correlations, and ordinary kriging and simple kriging were used as interpolation techniques. The major macro -anisotropy direction paralleled the river flow at both sites. The macro-anisotropy ratio was comparable between Aroclors, but varied between the metals. Comparable results for Aroclors could be due to similar physico - chemical properties governing fate and transport processes. Generally, spherical and exponential variograms fitted the data better than Gaussian variograms at both sites. With the exception of comparable micro-anisotropy parameters between the Aroclors, variogram parameters varied between contaminants and sites. Subdividing curved river segments and modeling the reaches separately did not improve predictions. Expanding the study to other types of water bodies and research into the use of co -kriging promises further insights.

Key words:Contaminated Sediments, River, Geostatistics, Hotspots, GIS

1. INTRODUCTION

In the remediation of contaminated aquatic bottom sediments, sampling and quantification of contaminants are important steps. Accuracy and cost of contaminant quantification, in part, depend on the number of samples and the quantification technique used. Hence, choosing the appropriate mathematical approach to contaminant quantification becomes an important part of the remediation process, and it also influences the sampling strategy.

Geostatistics is a mathematical tool that can assist in contaminant quantification. It studies spatial correlations between categorical or continuous attributes – such as concentrations – that are observed at fixed locations. A spatial correlation can be separated into a macro-scale correlation and a micro-scale correlation. The macro-scale relationship (trend) can be modeled with statistical or deterministic interpolation techniques. Geostatistical techniques extract micro-scale spatial correlations from the data and express them in variograms, and then use this information in interpolation techniques such as kriging.

Geostatistics has been extensively used in other geo-environmental disciplines. For example, research in subsurface hydrology (ASCE, 1990a; ASCE, 1990b; and Kirandis, 1995) on kriging applications in groundwater studies includes work on parameter distribution mapping and sampling designs. McKenna and Poeter (1995), Wingle (1997), and Wingle et al. (1999) explored ways to reduce subsurface hydrology uncertainty using geostatistics. In the field of surface water hydrology, Cressie and Majure (1997) developed a spatio-temporal model for predicting the distribution of livestock waste (in the form of nitrate concentration) in the upper North Bosque watershed. Qian (1997) compared universal kriging and Bayesian kriging in the estimation of the area affected by phosphorus runoff in an Everglades wetland. Genton and Furrer (1998) used a variogram-kriging model for predicting the spatial distribution of rainfall in Switzerland. De Oliveira and Ecker (2002) presented a Bayesian geostatistical model to map total nitrogen hotspots in Chesapeake Bay.

Although geostatistics has been used to analyze other types of environmental data, few studies have used geostatistics in research on contaminated aquatic bottom sediments. In some of these studies, geostatistics was used to map the distribution of contaminated concentrations (Furrer and Genton, 1999; Steinberg and Ramanitharan, 2000; Steinberg and Ramanitharan, 2001). Butcher (1996) used co-kriging as a tool to incorporate screening and laboratory PCB concentration data in the Hudson River bottom sediment study.

In water bodies with highly convoluted shapes, euclidian distance may not properly represent the distance between two distantly separated data points in variogram calculations. Without knowing the sediment transport path, there is no exact way to measure the "true" distance between two data locations. However, for practical purposes, the variogram models may be improved by dividing a long river stretch into a series of reaches, where the euclidian distance approximates the sediment transport path length. Each reach is then separately modeled and the results are combined.

The objective of this paper was to study macro-scale and micro-scale spatial concentration components; to study different variogram model-kriging techniques; to compare effective models and model parameters between the two river sites and between contaminants; and to investigate a modeling modification that may accomodate highly convoluted river segments.

2. STUDY SITES AND DATA

2.1 Study area

This research uses data from two contaminated sediment sites. Heavy metals were the contaminants at the Duwamish River site, and PCB mixtures (Aroclors) were studied at the Hudson River site.

Duwamish River

At each of 300 locations, concentrations of three heavy metals – chromium, cadmium, and nickel – were measured in 1998 along an eight-kilometer reach of Duwamish River, WA (Figure 1). The data sets were taken as part of the LODRIV98 study and were extracted from the SEDQUAL 4.3 database (Washington State Department of Ecology, 2002).

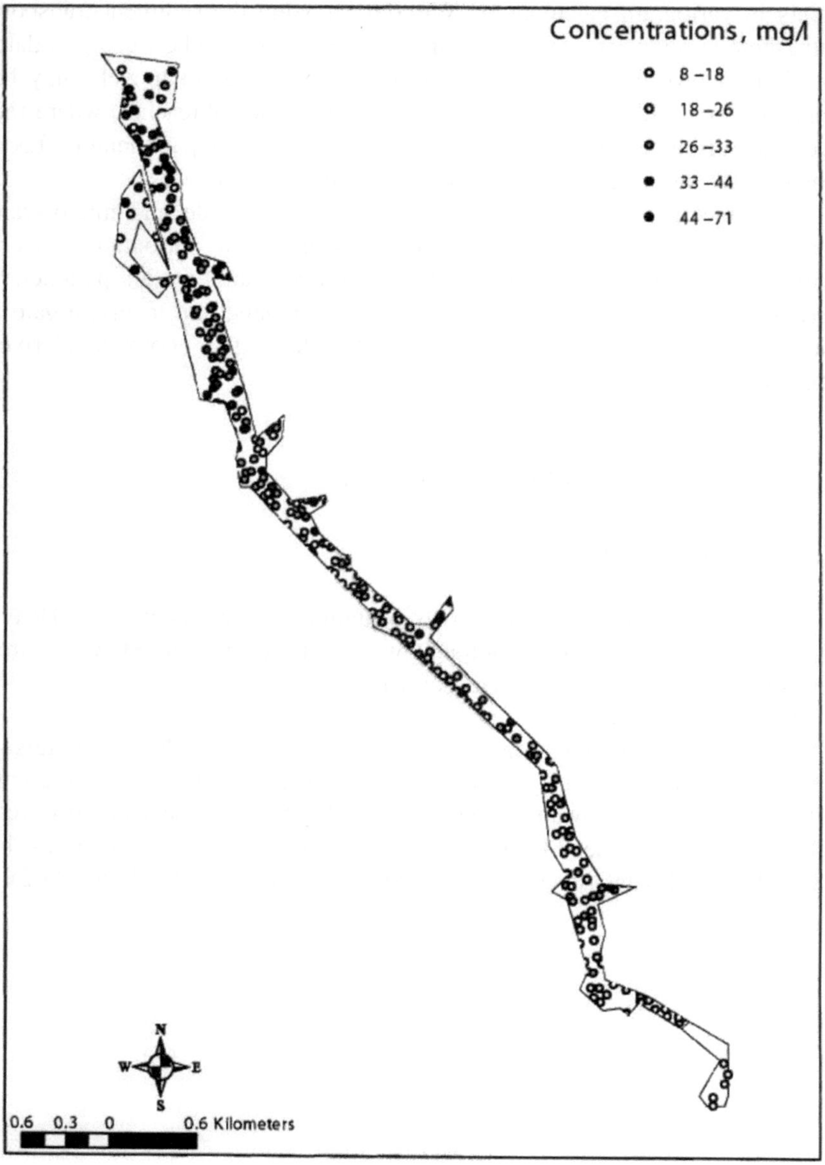

Figure 1. Chromium data in Duwamish River

Hudson River

At each of 116 locations, concentrations of three commercial PCB mixtures – Aroclors 1242, 1254, and 1260 – were measured in 1984 along a five-kilometer reach of the Hudson River, NY (Figure 2). The data sets were collected by New York Department of Health and extracted from the database for the Hudson River PCBs Reassessment RI/FS, Release 5.0 (USEPA, 2000).

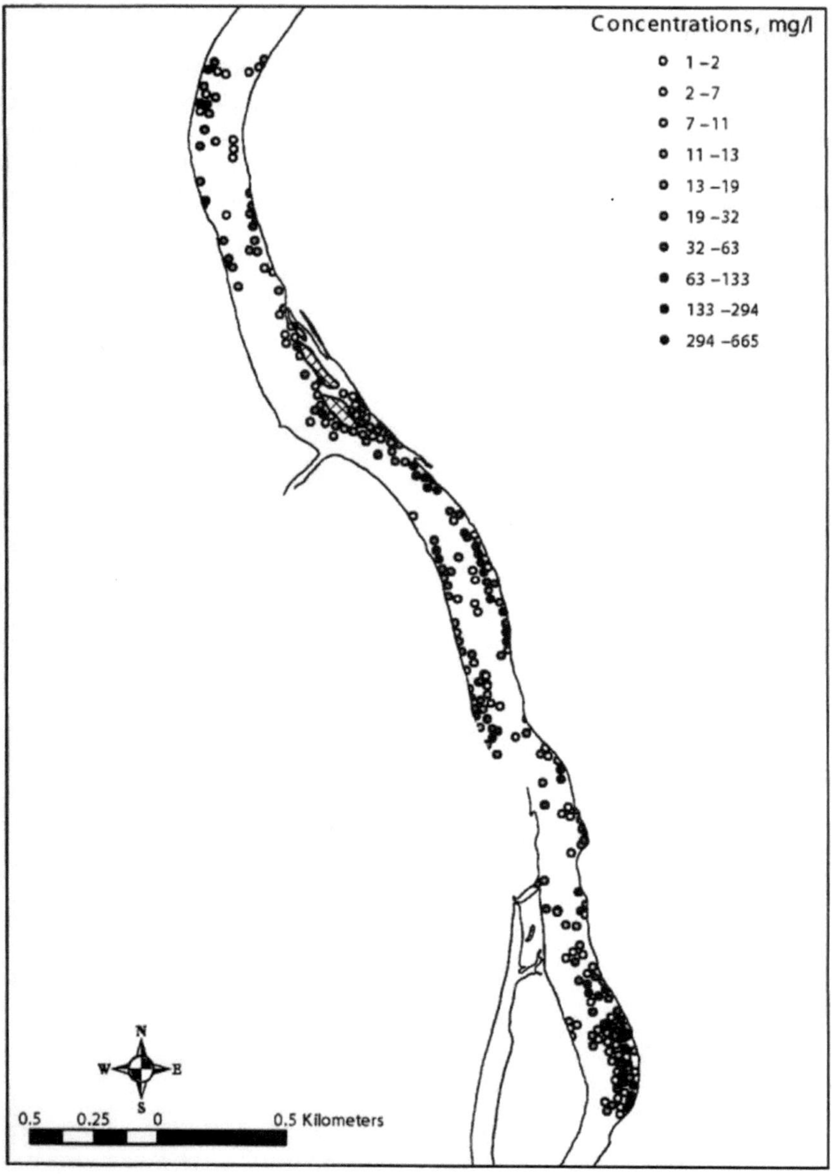

Figure 2. Aroclor 1242 data in Hudson River

3. THEORY

3.1 Variograms

When a set of spatial data for an attribute is mapped, geostatistics considers how the relationship between data varies with the nearness of their locations. This proximity can statistically be quantified with the introduction of spatial covariance functions, correlograms, or by their counterparts, semivariograms (for convenience and in this work, semivariograms are commonly called variograms). An experimental variogram is obtained from spatial data and is subsequently fitted to a theoretical variogram, which is a conditionally negative definite function. In this study, three theoretical variograms (spherical, exponential and Gaussian) were used to fit the experimental variograms. To select the best variogram model, cross-validation statistics were used after kriging. In general, a theoretical variogram is defined by three parameters, its nugget, sill, and range (Figure 3). Nuggets provide an estimate for measurement errors and micro-scale variations. Sill defines the maximum variance that can be reached, and range defines the distance at which the sill is reached. Partial sill is defined as the difference between the sill value and the nugget value. Variogram ranges will vary depending on the direction from which data are correlated to a given location's data. Thus, the micro-anisotropy ratio is the ratio between the highest range value and the lowest range value. The direction in which the largest range occurs is indicated by the micro-anisotropy angle.

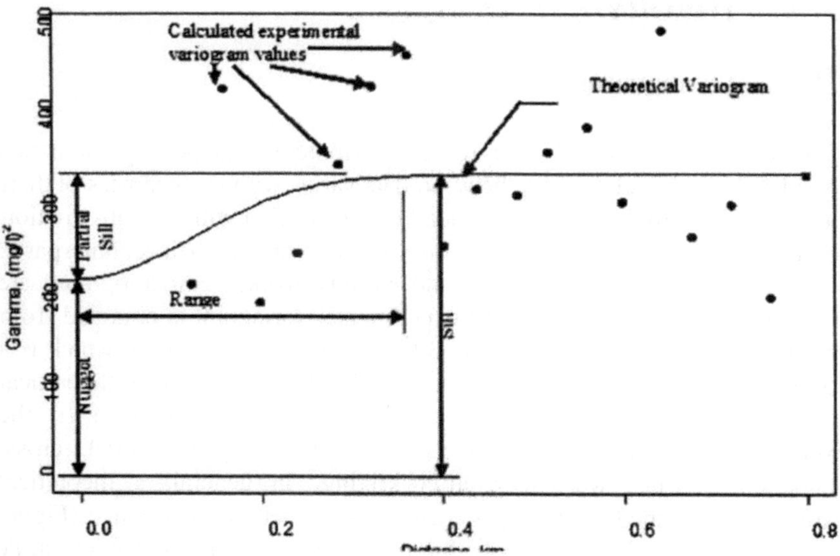

Figure 3. Typical omnidirectional variogram. Gamma = variogram value, Distance = Km

3.2 Kriging

In a geostatistics context, kriging can be defined as a generalized linear regression technique used with a variogram model for spatial data interpolation. In this study, the spatial data are the measured contaminated sediment concentrations.

The linear regression estimator $Z^*(\underline{u})$ is the unknown concentration to be estimated by kriging. It is defined as

$$Z^*(\underline{u}) - m(\underline{u}) = \sum_{\alpha=1}^{N(\underline{h})} \lambda_\alpha \left[Z(\underline{u}_\alpha) - m(\underline{u}_\alpha) \right] \tag{1}$$

where $\lambda_\alpha(\underline{u})$ is the weight assigned to the measured concentration $z(\underline{u}_\alpha)$, which is interpreted as a realization of the random variable $Z(\underline{u}_\alpha)$. The means $m(\underline{u})$ and $m(\underline{u}_\alpha)$ are the expected values of the random variables, $Z(\underline{u})$ and $Z(\underline{u}_\alpha)$ (Goovaerts, 1997). The mean $m(\underline{u})$, is used to represent the large-scale variation (trend) in the data. The objective of minimizing the estimation variance, $\sigma^2_E(\underline{u})$ can be stated as,

$$\sigma^2_E(\underline{u}) = Var\ \{Z^*(\underline{u}) - Z(\underline{u})\}, \tag{2}$$

under the constraint

$$E \; \{Z * (\underline{u}) - Z(\underline{u})\} = 0 \;. \tag{3}$$

Two types of kriging techniques, simple kriging (SK) and ordinary kriging (OK), were used in our case studies.

Simple kriging (SK)

$$m(u) = m, \; known, \quad \ldots \underline{u} \; ? \; A \tag{4}$$

In simple kriging, the mean **m(u̲)** is assumed to be a known constant throughout the region of concern, A.

Ordinary kriging (OK)

$$m(u') = constant, \; but \; unknown, \quad \ldots \underline{u'} \; ? \; W(\underline{u}) \tag{5}$$

In ordinary kriging, the mean is allowed to fluctuate globally, but assumed constant within the local neighborhood **W(u̲)**.

A moving elliptical window was used to select and restrict the data for trend fitting. Hence, this ellipse is used to define the "macro-scale anisotropy" property of the data.

4. RESULTS AND DISCUSSION

4.1 Results

Duwamish River

Chromium data were normally distributed. Trend visualization showed a generally increasing first order polynomial trend along the river. Thus, the ordinary kriging technique was employed rather than simple kriging. To fit the local means, the two macro anisotropy parameters were determined by iteratively optimizing a moving elliptical shape window. For all three heavy metals, the major macro anisotropy direction was roughly parallel to the flow direction, while anisotropy ratios were different.

The full set of chromium data (299 data locations) was divided into a model calibrating subset (224 data randomly selected locations) and a model validating subset (remaining 75 data locations). Both subsets were used in a first experiment to compare the performance of exponential, spherical and Gaussian variogram models. In a second experiment designed to improve predictions for convoluted river stretches, the stretch was subdivided into three reaches, each with its own calibration and validation data sets. Three different variogram models were fitted for each reach, and the results were compared to results obtained from using only one variogram model for the whole river stretch.

In both experiments, based on cross-validation statistics, it was found that the exponential and spherical models performed slightly better than the Gaussian model (Table 1). Figure 4 shows as a typical result the mapped predictions from the single-stretch spherical variogram model. However, when reach-specific variogram models were compared to each other in the second experiment, all root mean square errors (RMS) were within 1% of the average of the three RMS errors. Many of the single-stretch model parameters were comparable to those of reach two. Overall, no improvement was achieved by modeling the river stretch as three separate reaches. The cross validation RMS error for calibration data of the single model was 6.382 mg/l, and the combined RMS error for modeling with the three individual reach models was 6.405 mg/l. RMS values for the validation set were 9.118 mg/l and 9.370 mg/l, respectively. Micro anisotropy angles, as well as sill, nugget, and range values did not show any similarities between the three metals studied.

Table 1. Fitted variogram models for chromium concentrations in the Duwamish case study. Ordinary kriging was used. Anisotropy angles are given in degrees, ranges and axis lengths are in meters

Data Set	**Macro Anisotropy Parameters**			**Variogram Model**	**Micro Anisotropy Parameters**			**Cross Validation Statistics (Calibration-Set)**	
	Major Axis	**Minor Axis**	**Angle**		**Major Axis**	**Minor Axis**	**Angle**	**Mean Error (mg/l)**	**RMS Error (mg/l)**
Single Stretch	4000	2500	135	Spherical	1363	692	343	-0.141	6.382
				Exponential	1363	718	340	-0.1248	6.403
				Gaussian	1363	658	342	-0.1681	6.409
Reach 1	4000	1000	160	Spherical	2371	733	347	0.06978	6.854
				Exponential	1778	1778	n/a	0.1568	6.894
				Gaussian	948	901	2	-0.004005	7.065
Reach 2	6000	1000	135	Spherical	1304	645	316	-0.269	6.944
				Exponential	1067	596	311	-0.3397	6.757
				Gaussian	1126	547	316	-0.2801	6.959
Reach 3	3000	1000	345	Spherical	949	206	353	-0.02071	4.786
				Exponential	948	188	352	-0.02782	4.795
				Gaussian	948	196	353	-0.01128	4.852

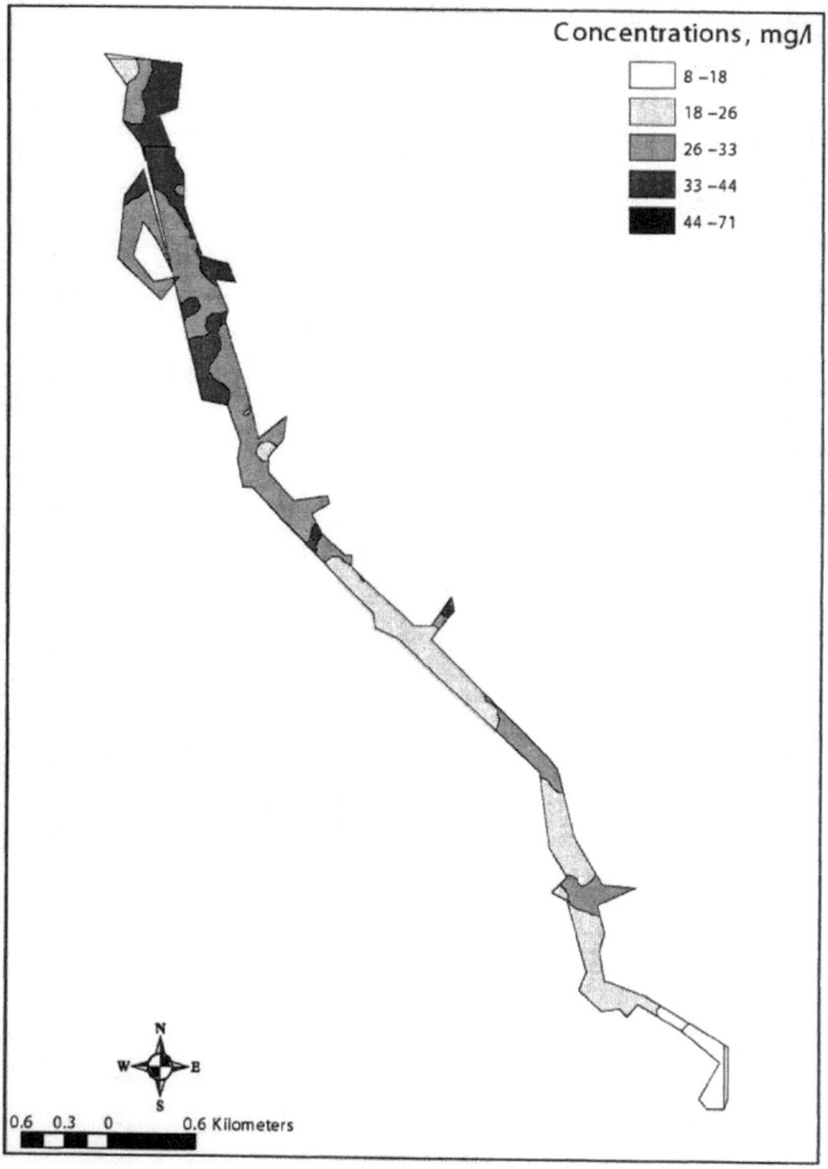

Figure 4. Chromium concentration mapping at the Duwamish site – single-stretch spherical variogram model

Hudson River

Extremely high concentrations of the three studied PCB mixtures – Aroclor 1242, Aroclor 1254, and Aroclor 1260 – were not clustered, but spread out along the river stretch as individual spikes. Both a normal score transformation (NST) and a log transformation were used to transform the data into a form suitable for fitting with the variogram models. Normal-score-transformed data and log-transformed data were combined with simple kriging and ordinary kriging, respectively.

When fitting log-transformed data to models combined with ordinary kriging, the major macro anisotropy direction was parallel to the river flow direction regardless of the Aroclor modeled. While macro anisotropy ratios had varied between the metal contaminants at the Duwamish river site, they were comparable for all three Aroclors at the Hudson river site. While using NST-SK models, regardless of the type of the Aroclor, the major axis of micro anisotropy was found to be parallel to the river flow (Table 2).

As in the Duwamish case study, two experiments were performed with the Hudson river data. In the first experiment, variogram models were fitted to the different Aroclor data sets and then compared. In the second experiment, the whole Hudson river stretch was divided into an upper and a lower reach, and three different variograms were fitted to the data from each reach.

Since Aroclor concentrations did not vary much with the location on the river stretch, NST-simple kriging (NST-SK) models were also implemented. NST-SK-exponential variogram models yielded the lowest RMS value for Aroclor 1242 and Aroclor 1254 models. In general, either a spherical or an exponential model performed better than a Gaussian model (Table 2), with the exception of the Aroclor 1260 data set. As a typical result, Figure 5 shows Aroclor 1242 concentrations mapped from single-stretch exponential model predictions. For these data, the Gaussian model was only insignificantly better than the spherical model. Similar to macro anisotropy parameters obtained from log-transformed data, micro anisotropy angles and ratios were comparable between Aroclors modeled with the residuals present after removing trend. The angles were parallel to the river flow, and the ratios all had a value of approximately six, with the exception of the Aroclor 1260 data set. However, sill, nugget and range values did not show any similarities between Aroclors.

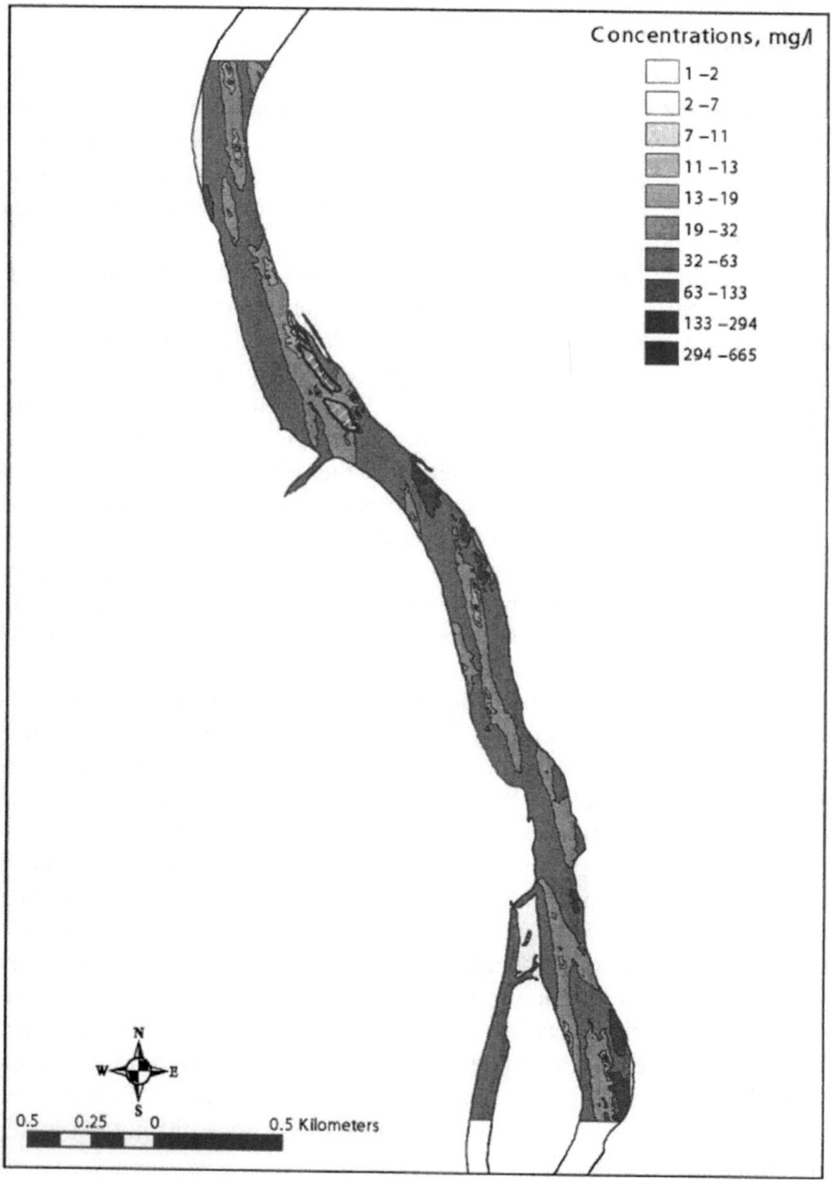

Figure 5. Aroclor 1242 concentration mapping at Hudson site – single-stretch exponential variogram model. Cross-hatched areas designate islands

When the individual reach models and the whole-stretch model were compared for Aroclor 1242 data, the whole-stretch model yielded the better prediction. The single NST-SK-exponential variogram model was found to perform best, followed by the combined NST-SK-exponential variogram model.

Table 2. Fitted single models (NST-Simple kriging) for Aroclor concentrations in Hudson case study. Anisotropy angles are given in d egrees

Aroclor	Data Set	Variogram Model	Micro Anisotropy Parameters		Cross Validation Statistics (Calibration-Set)	
			Ratio	Angle	Mean Error (mg/l)	RMS Error (mg/l)
1242	Single Stretch	Spherical	6.39	349	-0.09683	61.4
		Exponential	6.50	349	0.01196	60.46
		Gaussian	6.53	349	0.04109	61.26
1254	Single Stretch	Spherical	5.46	349	-0.1231	3.235
		Exponential	6.25	351	-0.1162	3.184
		Gaussian	6.25	349	-0.1097	3.259
1260	Single Stretch	Spherical	3.20	333	-0.0234	4.583
		Exponential	3.28	333	0.1278	4.617
		Gaussian	3.79	333	-0.1598	4.582
1242	Upper Reach	Spherical	3.15	352	1.207	39.15
		Exponential	4.55	334	0.6286	38.48
		Gaussian	3.54	356	0.9836	39.19
	Lower Reach	Spherical	5.39	346	-1.303	72.55
		Exponential	5.75	345	0.2138	70.9
		Gaussian	5.75	345	-0.8994	71.72

4.2 Discussion

Trend results

Contaminant data from both river bottom sediments exhibited macro anisotropy. While Aroclor data had to be transformed to dampen the effect of concentration spikes, macro-anisotropy angles could be compared between the two sites, since the angles are not affected by a data transformation. The major anisotropy axis was parallel to the river flow at both sites. Both the Duwamish and Hudson river sites are near-straight and have high length-to-width ratios. Sediment transport at these two sites occurs mainly in the flow direction. This may explain the dominant influence of the flow direction on the macro-anisotropy directions. On the other hand, the other parameter of spatial trend, the macro anisotropy ratio, differed from one metal to the next at the Duwamish river site, probably demonstrating the effect of contaminant-specific properties on the sediment distribution or differences in the location of metal sources.

Variogram results

Exponential and spherical variograms produced better results for the Hudson and Duwamish rivers. Exponential or spherical variograms indicate linearly changing correlations (Goovaerts,1997). Micro-spatial correlations of sediment concentrations – and thus the variogram models that describe them – can be expected to relate to variations in local conditions. However, in the absence of detailed information about the conditions at the studied sites, further research is needed to confirm and explain the better fit of exponential and spherical variogram models.

Micro-anisotropy parameters associated with the variograms demonstrate the influence of the contaminants and their sources. In contrast to the macro anisotropy angles, which were mainly determined by the river flow direction, micro anisotropy angles were found to vary from metal to metal, but not between Aroclors, as were micro anisotropy ratios. Anisotropy parameters may have been the same for all Aroclors because they share the same contaminant source and have similar chemical properties. In contrast to the Aroclor contaminants, heavy metals vary in their physico-chemical characteristics. Other variogram parameters varied between contaminants at both sites.

Effect of Subdividing the River Segment

At both sites, whole-stretch variogram models performed better than the individual reach models. Since the studied river stretches are almost straight, subdividing the stretch into reaches did not improve the concentration prediction, but reduced the data points available to fit the models. If a river is highly curved – in contrast to the two river sites studied here – subdividing the stretch may provide better predictions, if data density in each reach is high enough. A low number of data pairs in a reach may increase model inaccuracy and predictions may not improve over a single-stretch model. However, the results of this study demonstrated that the effectiveness of subdividing a convoluted river stretch depends on the extent of convolution, the number of data available, and the spread of the data points.

5. CONCLUSIONS

Geostatistical modeling of sediment contaminants showed that the major macro anisotropy directions are dominated by the shape and boundary of the river stretch studied. Macro anisotropy ratios, as well as micro spatial correlations may be influenced mainly by the contaminant's physico-chemical properties, in addition to local sediment characteristics. However, more in-depth research using larger data sets is needed to confirm these findings.

At both river sites, either spherical or exponential variogram models tended to provide the best predictions. Similarity of micro anisotropy parameters between the organic Aroclor contaminants may allow improved predictions by applying co-kriging techniques. However, results for heavy metals demonstrated that, in general, micro anisotropy parameters and other variogram parameters can be expected to be contaminant-specific and have to be determined individually.

While the contaminant concentration of highly convoluted rivers may be mapped with a series of regionalized variogram models, this technique yields no improvement for river stretches that are only slightly curved.

The study presented here was limited to rivers only, but has been expanded to other types of water bodies by the authors. In future research efforts, the geostatistical modeling approach for aquatic sediment contamination mapping should be compared to deterministic modeling approaches such as Inverse Distant Weight (IDW) models. This study was restricted to univariate geostatistical modeling. However, using secondary and soft data may considerably improve the predictions. Therefore, the potential benefits of using additional techniques like kriging with a drift, co-kriging, and Bayesian and fuzzy krigings should be investigated.

ACKNOWLEDGEMENT

The Authors want to thank Martin Payne, Washington State Department of Ecology, for the Duwamish data, and David Kluesner, EPA, for the Hudson river data. The first author wants to thank the Department of Civil & Environmental Engineering, Tulane University for financially supporting him during this study.

REFERENCES

ASCE Task Committee on Geostatistical Techniques in Geohydrology of the Ground Water Hydrology Committee of the ASCE Hydraulics Division. 1990a. Review of geostatistics in geohydrology I: basic concepts. *ASCE Journal of Hydraulic Engineering,* 116(5). 612-632.

ASCE Task Committee on Geostatistical Techniques in Geohydrology of the Ground Water Hydrology Committee of the ASCE Hydraulics Division. 1990b. Review of geostatistics in geohydrology II: applications. *ASCE Journal of Hydraulic Engineering,* 116(5). 633-658.

Butcher, J. B. 1996. Co-kriging to incorporate screening data: Hudson River sediment PCBs. *Water Resources Bulletin,* 32(2). 349-356.

Cressie, N. and Majure, J. J. 1997. Spatio -temporal statistical modeling of livestock waste in streams. *Journal of Agricultural, Biological, and Environmental Statistics* , 2(1). 24-47.

De Oliveira, V. and Ecker, M.D. 2002. Bayesian hot spot detection in the presence of a spatial trend: application to total nitrogen concentration in Chesapeake Bay. *Environmetrics,* 13(1). 85-101.

Furrer, R. and Genton, M. G. 1999. Robust spatial data analysis of Lake Geneva sediments with S+SpatialStats. *Systems Research and Info. Systems* , 8. 257-272.

Genton, M. G. and Furrer, R. 1998. Analysis of rainfall data by simple good sense: is spatial statistics worth the trouble? *Journal of Geographic Information and Decision Analysis,* 2(2). 11-17.

Goovaerts, P. 1997. *Geostatistics for Natural Resources Evaluation* . Oxford Univ. Press, New York. 483 pp.

Kitanidis, P. K.. 1995. Recent advances in geostatistical inference on hydrogeological variables. *Reviews of Geophysics* 33 Supplement, U.S. National Report to International Union of Geodesy and Geophysics 1991-1994. 1103-1109.

McKenna, S. A and Poeter, E. P. 1995. Field example of data fusion in site characterization. *Water Resources Research,* 31(12). 3229-3240.

Qian, S. S. 1997. Estimating the area affected by phosphorus runoff in an everglades wetland: a comparison of universal kriging and bayesian kriging. *Environmental and Ecological Statistics,* 4. 1-29.

Steinberg, L. J. and Ramanitharan, K. 2000. Spatial modeling of contaminated sediments in Puget Sound. In: *WEFTEC 2000,* CD-ROM, Water Environment Federation; Anaheim.

Steinberg, L. J. and Ramanitharan, K. 2001. *Improved Methods for Spatial Modeling of Contaminated Aquatic Sediments.* Final Report, RCS Program, Louisiana Board of Regents.

Stumm, W. and Morgan, J. J. 1996. *Aquatic Chemistry.* John Wiley & Sons, Inc. New York. 1022 pp.

USEPA (U.S. Environmental Protection Agency). 2000. *Database for the Hudson River PCBs Reassessment RI/FS. Release 5.0,* CD-ROM, TAMS Consultants, Inc. Bloomfield, NJ.

Washington State Department of Ecology. 2002. *Sediment Quality Information System – SEDQUAL 4.3,* CD-ROM, The Sediment Management Unit (SMU) of the Toxics Cleanup Program, Olympia, WA. July 2002.

Wingle, W. L. 1997. Evaluating subsurface uncertainty using modified geostatistical techniques. *PhD Dissertation.* #T-4595. Colorado School of Mines, Golden, CO. 202 pp.

Wingle, W. L., Poeter, E. P., and McKenna, S. A. 1999. UNCERT: Geostatistics , uncertainty, analysis and visualization software applied to groundwater flow and contaminant transport modeling. *Computers & Geosciences,* 25. 365-376.

This page intentionally left blank

CHAPTER 38

A PASSIVE DIFFUSION BAG SAMPLER DEMONSTRATION AT MULTIPLE US DEPARTMENT OF DEFENSE INSTALLATIONS

John Tunks, John Hicks, Javier Santillan, Rafael Vazquez
Parsons; Air Force Center for Environmental Excellence (AFCEE)

Abstract: Groundwater sample collection using passive diffusion bag samplers (PDBSs) represents a relatively new technology that employs passive sampling methods for monitoring volatile organic compounds (VOCs) in groundwater. The potential benefits and cost savings associated with using PDBSs for long -term monitoring are significant, as no purge waters are generated, and labor requirements for sampler installation and retrieval are minimal. Results of a field-scale PDBS demonstration performed at 14 Department of Defense installations between May 2001 and May 2002 are presented. The primary objective of the PDBS demonstration was to assess the effectiveness of the PDBS method by comparing groundwater analytical results for VOCs obtained using the current (conventional) sampling method with results obtained using the PDBS method. The comparison of the conventional and diffusion sampling results will allow assessment of the appropriateness of implementing diffusion sampling for VOCs at each sampled well.

Key terms: PDBS, No-purge, Diffusion, Passive, Bag, Sampling, Long-term monitoring

1. INTRODUCTION

This report presents the results of a passive diffusion bag sampling (PDBS) demonstration consisting of groundwater monitoring of volatile organic compounds (VOCs) at 14 United States (US) Department of Defense (DoD) installations. The demonstration was performed by Parsons for the Air Force Center for Environmental Excellence, Science and Engineering Division (AFCEE/ERS). The objectives of the PDBS demonstration were to: develop vertical profiles of VOC concentrations across the screened intervals of the sampled monitoring wells; assess the effectiveness of the PDBS method by statistically comparing groundwater analytical results for VOCs obtained using the current (conventional) sampling method (i.e., micropurge/sample or 3-casing-volume purge/sample) during regularly scheduled long-term monitoring (LTM) events with results obtained using the PDBS method; and compare the costs of PDB and conventional sampling.

Diffusion sampling is a relatively new technology designed to utilize passive sampling techniques that eliminate the need for well purging. Specifically, a diffusive-membrane capsule is filled with purified water, sealed, suspended in a well-installation device, and lowered to a specified depth below the water level in a monitoring well. Over time, the VOCs in the groundwater diffuse across the membrane, and the VOC content of the water inside the sampler reaches equilibrium with groundwater in the well casing. The sampler is subsequently removed from the well, and the water in the diffusion sampler is transferred to a sample container and submitted for laboratory analysis of VOCs. Benefits of diffusion sampling include reduced sampling costs and reduced generation of investigation-derived waste.

Once a diffusion sampler is placed in a well, it remains in place until chemical (i.e., VOC) equilibrium is achieved between the water in the well casing and the water in the diffusion sampler. Depending on the hydrogeologic characteristics of the aquifer, the diffusion samplers can reach equilibrium within 3 to 4 days (Vroblesky, 2001). Groundwater samples collected using the diffusion samplers are generally thought to be representative of water present within the well during the previous 24 to 72 hours.

PDBS relies on the natural flow of groundwater through a well screen, and therefore the results obtained from this method will not always be comparable to results obtained using conventional sampling methods which induce groundwater flow into a well by creating a hydraulic gradient through well purging. A variety of factors can cause mixing in an unpumped well, such as vertical flow, convection, and diffusion. In a mixed well, the PDBS will represent the mixture or the zone of inflow. In addition, contaminant stratification sometimes is observed in unpumped wells (Vroblesky and Peters, 2000). In this case, the PDBS may represent the depth interval at which the sampler is placed. PDB samples are not susceptible to matrix interferences caused by turbidity because the PDB membrane is not permeable to colloids or other particles larger in diameter than approximately 10 angstroms.

The technology has been evaluated through various studies (Vroblesky and Hyde, 1997; Parsons, 1999; Tunks *et al.,* 2000; Church, 2000; Hare, 2000; McClellan AFB, 2000; Vroblesky *et al.,* 2000; Vroblesky and Peters, 2000; Vroblesky and Petkewich, 2000), and a guidance document for their use has been developed (Vroblesky, 2001). Additionally, the Interstate Technology and Regulatory Council (ITRC) has formed a workgroup to expand on the PDBS guidance document and to address technical and regulatory implementation issues as they arise.

2. SCOPE AND DESCRIPTION OF FIELD ACTIVITIES

A total of 1,199 PDB samples were collected from 332 wells across 14 different DoD installations during this demonstration. The sampled installations are shown on Figure 2-1.

Figure 2.1. Sampled Installations

The standard diffusion sampler used in this demonstration was constructed of a 45-centimeter (cm)-long section of 5.08-cm-diameter, 4-mil-thick, low-density polyethylene (LDPE) tubing with a pore size of approximately 10 angstroms. The sampler is permanently sealed on one end and sealed on the other end with a polyvinyl chloride (PVC) cap. When filled, the sampler holds approximately 350 milliliters (mL) of purified water. The sampler is placed in polyethylene mesh tubing for abrasion protection, attached to a weighted rope, and lowered to a predetermined depth within the screened interval of a well. The rope is weighted to ensure that the sampling devices are positioned at the correct depth and that they do not float upward through the water column. A deployment rate of 1 PDBS per 3 feet of saturated well screen was maintained for each sampled well in order to develop vertical profiles of contamination. An average of 3.6 PDB samplers was deployed in each well.

According to Vroblesky (2001), laboratory and field data suggest that 2 weeks of equilibration probably is adequate for many applications. PDBSs were left in the wells from 13 to 149 days (average 29 days) to allow local groundwater conditions to re-equilibrate following PDBS deployment, to allow adequate time for the diffusion process to occur, and to evaluate long-term deployment of the samplers.

PDB samplers were retrieved as closely in time to the conventional sampling of the same well as feasible. However, the incumbent base environmental sampling contractors typically performed the conventional sampling, and therefore lag times between the two sampling events occurred and ranged from 0 to 49 days with a median of 3.2 days.

Upon retrieval, the end cap on the samplers was removed, and water samples were transferred into pre-acidified 40-mL volatile organics analysis (VOA) vials. The samples were preserved on ice and submitted to a laboratory for analysis of VOCs. The laboratories and analysis methods used for PDB samples were identical to those used for the corresponding conventional samples to minimize analytical variability.

Conventional purging and sampling techniques varied by installation and included 3-casing-volume purge and sample using a bailer, submersible pump, or peristaltic pump; and low-flow (i.e., micropurge) techniques using either peristaltic or submersible pumps.

3. RESULTS AND ANALYSIS

Figure 3-1 is a plot of the maximum PDBS result (e.g., if multiple PDB samplers were deployed in a well) versus the corresponding conventional sample result for the same well for all detected VOCs. Instances where either result (e.g., PDB or conventional sample) was qualified as non-detect were assigned a value equal to the analytical method detection limit (MDL).

Figure 3-1 is a presentation format typically used when comparing two paired data sets that are expected to be similar (e.g., primary and duplicate samples), since it is a visual depiction of the correlation between the two data sets. However, the variance between the results of PDB and conventional samples would be expected to be greater than the variance between a primary and field duplicate sample. This is due in part to the following considerations: PDB and conventional sampling were typically performed by different contractors. Although the same standard operating procedures for tasks common to both sampling approaches (e.g., sample handling and preservation) were followed by both sampling teams, some variability is to be expected simply because different individuals were involved. Conversely, primary and field duplicate samples are collected by the same sampling personnel.

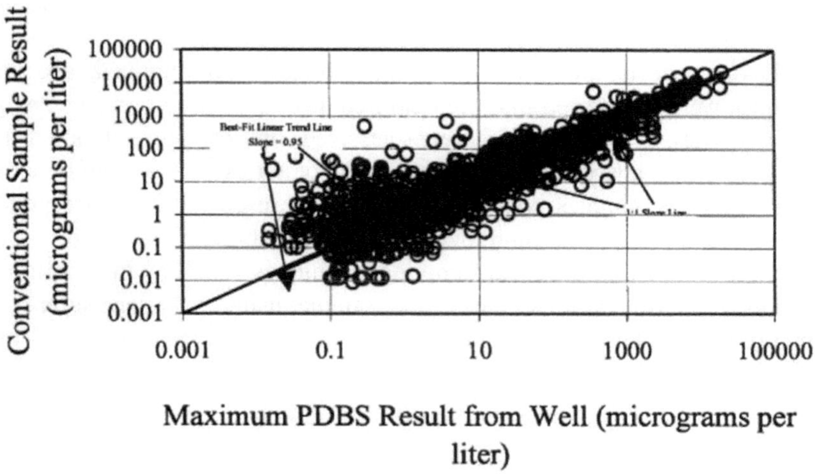

Figure 3.1. Scatter Plot Results for all VOCs

Logistically, it was not always possible to perform PDB sampling immediately prior to conventional sampling. In some instances, a time lag of several weeks occurred between PDB and conventional sampling. Although it is unlikely that groundwater chemistry will change significantly over the course of a few hours, some variability can occur over the course of several days or weeks. In contrast, primary and field duplicate samples are collected at the same time.

PDB and conventional samples were typically not included in the same laboratory sample delivery group (SDG). Therefore, even though the same laboratory was used, the PDB samples may have been analyzed by different instrument operators, on different days, and potentially using different equipment than the conventional samples. Although laboratory QA/QC practices attempt to achieve temporal consistency for analyses, some variability is to be expected. Conversely, primary and field duplicate samples typically are analyzed as part of the same SDG.

The PDB and conventional sampling methods are inherently different. For example, different equipment is used for the different methods (e.g., dedicated [PDBS] versus non-dedicated [conventional] equipment), and different sample collection procedures are employed (i.e., passive sample collection [PDBS] versus active sample collection [pumping/conventional]). These differences are likely to introduce variability in the sample results, whereas primary and field duplicate samples are collected using the same sampling equipment and procedures.

In light of these considerations, a more detailed comparison of the data sets and assessment of correlation was warranted. The approach used is described below.

If the maximum PDBS analytical result obtained for a well was greater than or equal to the conventional sampling result for that well, the PDBS method was inferred to be appropriate for use in that well (Vroblesky, 2001). Nonetheless, comparison to the conventional sampling results was performed in all instances. Specifically, analytical results for all samples collected using the diffusion samplers were compared to results from the conventional sampling using the following relative-percent-difference (RPD) equation:

RPD= 100*[abs(D-C)]/[(D+C)/2]

Where:

abs = absolute value,
D = diffusion sampler result, and
C = conventional sample result.

Based on standard conventions for comparing primary and field duplicate samples (US Environmental Protection Agency [USEPA], 1996, 1999, and 2000; Radian International, 1997 and 1998; EarthTech, 1998; McClellan AFB, 2000; Goad, 2001), and the inherent differences that were expected between the two data sets due to the differences in sample collection procedures, an RPD of 50 was selected as the criterion for this PDBS demonstration. This RPD criterion, in conjunction with five other criteria developed specifically for the PDB/conventional sample comparison, was used to determine the suitability of PDBS for monitoring a specific compound in a specific well. All six correlation criteria are described below.

PDBS ≥ Conventional Criterion: If at least one PDBS result for a given analyte in a given well was equal to or greater than the conventional sampling result, PDB sampling was deemed appropriate for monitoring that analyte in that well.

RPD Criterion: If either the PDB or the conventional sample result was greater than three times the laboratory reporting limit (RL), and the PDBS result was less than the conventional result, then an RPD of 50 was used as the acceptance criterion.

RL Criterion: If both the PDB and conventional sample results were less than or equal to three times the laboratory RL, a value of ± the RL was used as the range of acceptance between the two values. If the RLs for the conventional and PDB samples were different, the lowest RL was used to determine the acceptance range.

Low-Magnitude Concentrations Criterion: In cases where the PDB and conventional results were low-magnitude and differed by less than 5 micrograms per liter (μg/L), an acceptable correlation between sample results was achieved.

Inappropriate Comparisons Criterion: This criterion was developed to exclude compounds from the comparison that may not occur in the groundwater naturally (i.e., they may have been introduced to the sample during sample handling or analysis), or that have been shown experimentally to not diffuse across the PDB membrane completely (Vroblesky, 2001 and 2002). These compounds include: common laboratory contaminants including acetone, 2-butanone, and methylene chloride; common halomethanes generated in water as a result of chlorination of water supplies including chloroform, bromomethane, and bromodichloromethane (these compounds could be present in the PDBS source water and may not be site-related contaminants in the groundwater, or they could also be introduced in the field or at the laboratory as a result of activities that use common tap water); and methyl tert-butyl ether (MTBE), acetone, 4-methyl-2-pentanone (MIBK), and styrene, which have poor diffusiviry across the PDB membrane (Vroblesky, 2001 and 2002). Analytical results for these compounds are not expected to meet the correlation criteria used in this evaluation unless they are present at very low concentrations in the conventional samples and meet the RL criterion described above.

Each VOC concentration from each PDBS in each well evaluated was compared to the conventional sample concentration of the same VOC in the same well using the six correlation criteria. If any one of the correlation criteria was met, then PDBS was considered to be an acceptable alternative sampling method for that analyte in that well.

A correlation ratio was calculated by dividing the number of comparisons in which the correlation criteria were met by the total number of comparisons. For example, if TCE was detected in 227 wells, and at least one of the correlation criteria were met for TCE in 208 of those wells, a correlation ratio of 92 percent was realized.

Figure 3-2 is a graphical presentation of the correlation ratio for each compound that was detected in at least 10 wells. Most of the compounds had correlation ratios of at least 70 percent. Considering all compounds (not just those detected in at least 10 wells) except those that are known to be incompatible with the PDBS method, and compounds for which correlation could not be determined, results for 42 of the 48 VOCs (87.5 percent) detected met the correlation criteria in at least 70 percent of the sampled wells. The six compounds that met the correlation criteria in less than 70 percent of the sampled wells include tert-amyl methyl ether, bromoform, naphthalene, n-propylbenzene, 1,2,4-trimethylbenzene (TMB), and 1,3,5-TMB.

Of the 332 wells included in the demonstration, no VOCs were detected in 12 wells (3.6 percent), correlation could not be determined in 2 wells (0.6 percent), 6 wells (1.8 percent) contained only concentrations of compounds that are not appropriate for comparison, 34 wells (10.2 percent) met the correlation criteria for fewer than 70 percent of the compounds detected, and 278 wells (83.7 percent) met the correlation criteria for at least 70 percent of the compounds detected. Additionally, 239 wells (72 percent of all the wells evaluated) met the correlation criteria for all detected compounds, and could be candidates for immediate PDBS implementation for VOC monitoring. Correlation criteria were met in 1,411 of 1,614 comparison instances (87.4 percent).

Further review of data was performed to identify specific variables that may have led to reduced correlation ratios including review of compound-, well-, and technology-specific variables. A brief description of 14 reasons that were evaluated to explain instances where correlation criteria were not met, in addition to the summary application of these reasons to the data set is provided in Table 3-1.

Of the 332 wells included in the demonstration, no VOCs were detected in 12 wells (3.6 percent), correlation could not be determined in 2 wells (0.6 percent), 6 wells (1.8 percent) contained only concentrations of compounds that are not appropriate for comparison, 34 wells (10.2 percent) met the correlation criteria for fewer than 70 percent of the compounds detected, and 278 wells (83.7 percent) met the correlation criteria for at least 70 percent of the compounds detected. Additionally, 239 wells (72 percent of all the wells evaluated) met the correlation criteria for all detected compounds, and could be candidates for immediate PDBS implementation for VOC monitoring. Correlation criteria were met in 1,411 of 1,614 comparison instances (87.4 percent).

Further review of data was performed to identify specific variables that may have led to reduced correlation ratios including review of compound-, well-, and technology-specific variables. A brief description of 14 reasons that were evaluated to explain instances where correlation criteria were not met, in addition to the summary application of these reasons to the data set is provided in Table 3-1.

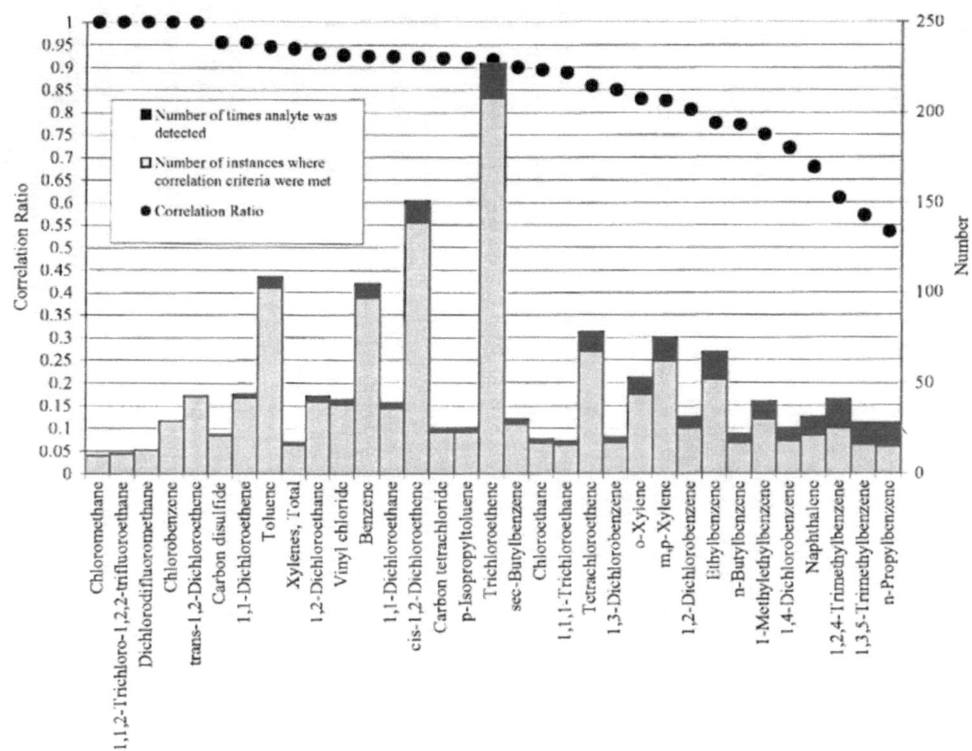

Figure 3.2. Correlation Ratio Overview

Table 3.1. Potential Reasons for Low Correlation between PDB and Conventional Samples

Number	Reason for Low Correlation Between PDB and Conventional Samples	Summary Evaluation of the Impact of this Reason
1	Inappropriate (too short or too long) deployment period.	No apparent effect on correlation
2	Excessive time lag between PDB and conventional sampling events.	No apparent effect on correlation
3	Laboratory- or field- induced contamination that may be present at varying levels in different samples, but are not representative of site-related contamination.	41 of the 294 instances where these compounds were detected did not meet correlation criteria
4	Inherent differences between PDB and conventional sampling methods (i.e., the fact that some conventional samples are drawn from a much larger volume of the aquifer than PDBSs).	Correlation ratio for conventional samples that used a 3-casing-volume purge = 85 percent Correlation ratio for conventional samples that used a micropurge = 93 percent
5	Water table above the top of the well screen. This is a special case of Reason 4, and may be particularly significant at fuel-contaminated sites where fuel concentrations may be highest near the water table. These higher concentrations may be drawn into the well during a conventional purge, but would not be	Correlation ratio when water level was above top of screen = 91 percent Correlation ratio when water level was below top of screen = 84 percent

Number	Reason for Low Correlation Between PDB and Conventional Samples	Summary Evaluation of the Impact of this Reason
	detected by PDBSs deployed in the screen interval.	Correlation ratio for CLDWs = 83 percent Correlation ratio for CMDWs = 91 percent
6	Significant contaminant stratification in which dissolved contaminants are localized in discrete layers that may not be sampled by PDBSs. Also, non-uniform distribution of dissolved contaminants in the aquifer (e.g., among the BTEX compounds) due to variations in solubility, density, sorptive, and degradation properties.	The effects of this reason could not be quantified.
7	Aeration of PDBS during deployment period due to a declining water table or deployment above the water table.	Correlation ratio when upper PDBS was deployed partially or completely above the water surface = 58 percent Correlation ratio when upper PDBS submergence was 0.01 to 4.00 feet = 79 to 92 percent Correlation ratio when upper PDBS submergence was greater than 4 feet = 93 percent 4 wells where upper PDBS was partially or completely exposed to air during deployment

Number	Reason for Low Correlation Between PDB and Conventional Samples	Summary Evaluation of the Impact of this Reason
8	Deployment of the PDBS above or below the screened interval of the well where natural groundwater flow through the well does not occur.	17 wells in which one PDBS was placed either partially or completely above or below the well screened interval 3 of these wells had correlation ratios less than 70 percent
9	Vertical flow in the well resulting from the presence of vertical hydraulic gradients in the screened portion of the aquifer.	The effects of this reason could not be quantified; however, available data suggest that vertical flow did not significantly impact correlation ratios in most instances.
10	Aquifer characteristics such as hydraulic conductivity, groundwater velocity, and lithology that might limit natural flow of contaminated groundwater through the well screen.	The effects of this reason could not be quantified; however, available data suggest that correlation ratios were not significantly lower at sites with low hydraulic conductivities and groundwater velocities.
11	Well-specific conditions that would limit groundwater flow through the well, such as where the filter-pack permeability is lower than that of the aquifer, causing groundwater flow lines to diverge around the well.	13 wells had correlation ratios less than 70 percent and had at least five compounds detected in them, suggesting the possible existence of well-specific conditions (Reasons 10 and/or 11) that are not amenable to the PDBS approach. The relatively low correlation ratio for one of the 13 wells is probably due to PDBS aeration during the deployment period (Reason 7)
12	Compound incompatibility with PDBS. Some	19 instances where one of these compounds did not meet

Number	Reason for Low Correlation Between PDB and Conventional Samples	Summary Evaluation of the Impact of this Reason
	compounds (e.g., MTBE, acetone, 4-methyl-2-pentanone, and styrene) have been shown to have poor diffusivity across the PDB membrane (Vroblesky, 2001 and 2002). Analytical results for these compounds would not be expected to meet the correlation criteria used in this evaluation.	the correlation criteria in a well
13	Physiochemical properties of the compound. This is an expanded case of Reason 12 above that includes compounds not listed in Vroblesky (2001 and 2002), but that have typically met the correlation criteria in fewer than 70 percent of the instances of comparison.	The effects of this reason could not be quantified; however, several physiochemical properties were found to be positively correlated to the correlation ratio, indicating that compound-specific properties do affect compatibility with the PDBS method.
14	Laboratory-induced variability resulting from the PDB and conventional samples being analyzed in different sample delivery groups, and therefore being potentially analyzed on different days, by different operators, on different equipment.	Correlation ratio for PDB/conventional samples analyzed as part of same SDG = 91 percent Correlation ratio for PDB/conventional samples analyzed as part of different SDGs = 87 percent

Notes: PDBS = passive diffusion bag sampler; RPD = relative percent difference; BTEX = benzene, toluene, ethylbenzene, and xylenes; MTBE = methyl tert-butyl ether; CLDW = contaminant less dense than water; CMDW = contaminant more dense than water.

4. COST ANALYSIS

A comparative cost analysis was performed for PDBS and conventional sampling. The four-step cost-analysis procedure consisted of an estimate of PDBS implementation costs, a comparison of long-term PDBS to long-term conventional sampling costs, an analysis of the costs that would be avoided if PDBS were implemented, and a calculation of return on investment (ROI). Table 4-1 summarizes cost analysis information for each of the 14 installations reviewed in this report.

4.1 Estimate of Implementation Costs

The first step of the cost analysis involved estimating the per-well PDBS implementation cost. This step reflected the per-well cost of the PDBS demonstration, assuming it had been performed by the current Base contractor (with normalized labor rates). Therefore, the PDBS implementation costs typically included costs to develop a work plan, procure field equipment, perform field sampling activities, and prepare the PDBS results report.

Additional cost typically included in the implementation costs are associated with development of well-specific recommendations for long-term PDBS deployment, interaction with regulatory agencies, and revision of sampling-related guidance documents such as sampling and analysis plans, records of decision, and quality assurance plans. The magnitude of these future costs typically was not known; however, standard estimates were used for each installation to maintain consistency.

The minimum and maximum implementation costs calculated for this demonstration were $1,597 and $6,878 per well, respectively, with a median of $2,462 per well (Table 4-1).

Table 4.1 Cost Analysis Summary

Installation	Number of Wells Evaluated in PDBS Demonstration	Other Direct Costs for Implementation, per Well	Implementation Labor per Well (hours)[a/]	Total Estimated Implementation Cost per Well[b/]	Number of Wells Sampled Regularly for VOCs Exclusively[c/]	Total Estimated PDBS Implementation Cost	Total Estimated PDBS Implementation Cost for Wells not Included in PDBS Demonstration
Andrews AFB	26	$626	3.5	$2,381	45	$107,145	$45,239
Bolling AFB	10	$370	2.5	$4,256	22	$93,632	$51,072
Buckley AFB	16	$384	3.0	$2,619	45	$117,855	$75,951
Columbus AFB	20	$727	3.0	$2,543	61	$155,123	$104,263
DDJC-Sharpe	25	$649	3.4	$2,308	100	$230,800	$173,100
Dover AFB	20	$520	2.3	$2,267	60	$135,000	$90,680
Edwards AFB	38	$547	2.8	$1,808	75	$135,600	$66,896
George AFB	34	$646	2.1	$1,812	68	$123,216	$61,608
Keesler AFB	17	$504	2.9	$2,809	37	$103,933	$56,180
March ARB	20	$864	2.6	$2,775	38	$94,500	$49,950
Norton AFB	17	$1,046	3.8	$2,900	117	$339,000	$290,000
Shaw AFB	24	$389	1.8	$1,597	300	$479,100	$440,772
Vandenberg AFB	56	$575	2.7	$1,653	72	$119,000	$26,448
Williams AFB	10	$2,298	9.5	$6,878	16	$110,048	$41,268
SUMMARY STATISTICS FOR ALL INSTALLATIONS							
Mean	24	$725	3.3	$2,758	75.4	$167,425	$112,388

Median	20	$600	2.9	$2,462	60.5	$121,108	$64,252
Minimum	10	$370	1.8	$1,597	16	$93,632	$26,448
Maximum	56	$2,298	9.5	$6,878	300	$479,100	$440,772

Table 4.1 **Cost Analysis Summary Cont.**

Installation	Number of Well Sampling Events per Year for VOCs Exclusively[c/d/]	Estimated Per-Well Per-Event Sampling Cost		Cost Difference	Percentage Cost Savings of Switching to PDBS	Return on Investment
		PDBS	Conventional			
Andrews AFB	45	$78	$332	$254	77%	213%
Bolling AFB	32	$68	$186	$118	63%	1%
Buckley AFB	45	$74	$336	$262	78%	200%
Columbus AFB	110	$70	$343	$273	80%	387%
DDJC-Sharpe	160	$60	$227	$167	74%	232%
Dover AFB	120	$68	$323	$255	79%	459%
Edwards AFB	150	$72	$218	$146	67%	324%
George AFB	136	$95	$414	$319	77%	704%
Keesler AFB	37	$106	$474	$367	78%	260%
March ARB	152	$62	$208	$146	70%	423%
Norton AFB	381	$71	$297	$226	76%	500%
Shaw AFB	620	$67	$304	$238	78%	613%

Vandenberg AFB	157	$75	$296	$221	75%	583%
Williams AFB	22.5	$288[e/]	$836	$548	66%	319%
SUMMARY STATISTICS FOR ALL INSTALLATIONS						
Mean	154.8	$90	$342	$253	74%	373%
Median	128	$72	$314	$246	76%	356%
Minimum	22.5	$60	$186	$118	63%	1%
Maximum	620	$288[e/]	$836	$548	80%	704%

Notes:

a/ Using a 2-person field team.

b/ Implementation cost includes the following: work plan development, vertical profiling (1 PDBS per 3 fe et of saturated screen) of each well, field equipment and laboratory analyses, field sampling labor (assumed labor rate of $60 per hour), data analysis/evaluation and reporting, recommendation development, regulatory interaction, and revision of existing s ampling program documents (e.g., SAP, QAPP).

c/ Number of wells used in the installation-specific cost analyses (Attachment 1) and assumed to be sampled for VOCs exclusively at a regular frequency for the duration of the long-term monitoring program.

d/ A well-sampling event is equal to one sample collected from one well.

e/ For Williams AFB, it was assumed that 2 PDB samples would be required per well due to the relatively long well screens at this installation. Therefore, $150 of additional analytical costs are included in this number.

4.2 PDB and Conventional Sampling Cost Comparison

The next step was to estimate and compare long-term cost differences between PDB and conventional sampling methods. A capital cost depreciation period of 20 years (i.e., the assumed duration of the LTM programs) was typically used. Because some dedicated equipment (e.g., submersible pumps, generators, meters) also have limited life expectancies, replacement rates were considered in the analysis where appropriate. Ultimately, a cost per well per sampling event for each sampling method was calculated.

4.2.1 Long-Term PDB Sampling Costs

Two of the primary assumptions made in calculating the long-term costs for PDB sampling were that: only one PDBS per well would be used during each event except at Williams AFB, where two PDBSs per well were assumed due to the particularly long well screens (35 to 80 feet long) at that installation. This is in contrast to the use of multiple PDBSs per well for vertical profiling, which is usually recommended for the initial implementation evaluation only.

Each time a PDBS is retrieved from a well, it would be replaced with a new PDBS so that only one mobilization would be required per sampling event. A PDBS would therefore remain in the well for the entire period between sampling events.

The minimum and maximum long-term PDB sampling costs per well per event calculated for the 14 demonstration sites were approximately \$60 and \$288 (\$288 is for Williams AFB and includes \$150 additional analytical costs due to the use of two PDBSs per well at that installation), respectively, with a median cost of \$72 (Table 4-1).

4.2.2 Long-Term Conventional Sampling Costs

Most of the cost information for conventional sampling was provided by the Base contractors, and any information gaps were filled using professional judgment. The minimum and maximum long-term conventional sampling costs per well per event calculated during this demonstration were \$186 and \$836, respectively, with a median of \$314 (Table 4-1).

4.3 Sampling Cost Avoidance

The cost difference between PDBS and conventional sampling per well per event was calculated based on the per-well, per-event cost estimates for long-term PDB and conventional sampling. This step provided an estimated annual and long-term cost savings that could be realized if conventional sampling was replaced with PDB sampling. Where possible, long-term cost savings were based on only those wells that are being sampled exclusively for VOCs, and in some cases, a frequency of LTM for an extended (i.e., 20-year) period was assumed that was different from the current frequency. For example, if the current LTM program at an installation incorporated a semiannual monitoring frequency, and continuation of this frequency for an extended period was thought to be unrealistic, the average monitoring frequency over the upcoming 20-year period may have been assumed to be less (e.g., annual). The minimum and maximum sampling cost avoidance per well per event calculated for this demonstration were $118 and $548, respectively, with a median of $246 (Table 4-1).

4.4 Return on investment

This step compared the cost to implement a PDBS demonstration (i.e., the up-front investment required to implement long-term PDB sampling) with the estimated cost avoidance over the duration of the LTM program. The ROI is calculated by dividing the potential cost savings by the implementation cost. The ROI values calculated for this demonstration project used an assumed number of wells, frequency of sampling events (i.e., quarterly, semi-annual, or annual), and LTM program duration (typically 20 years), and did not include costs for additional field evaluations of PDB sampling should they be deemed necessary. The minimum and maximum ROIs calculated for this demonstration were 1 and 704 percent, respectively, with a median of 366 percent (Table 4-1).

5. CONCLUSIONS AND RECOMMENDATIONS

The relatively high degree of correlation summarized in Section 3 indicates that the PDBS method is reasonably robust and capable of accurately monitoring concentrations of VOCs dissolved in groundwater in most instances.

Table 3-1 is a summary of potential reasons for instances where correlation criteria were not met. Available information suggests that the most likely causes of non-correlation include:

- The presence of field- or laboratory-introduced contamination that was not indicative of actual site-related contamination.
- Inherent differences in the passive (PDBS) and active (pumping/conventional) sampling approaches.
- The submergence depth of the uppermost PDBS (i.e., PDBSs installed near the water surface in the well tended to have low-biased VOC concentrations to a greater extent than deeper PDBSs).
- Laboratory-induced variability (i.e., analysis of PDBS and conventional samples as part of separate SDGs).
- Compound-specific physicochemical properties that are, in some cases, less conducive to the PDBS method than in other cases.

Correlation ratios for compounds that are the most typical COCs (e.g., BTEX compounds, chlorinated ethenes and ethanes) were generally between 80 and 100 percent. However, even where correlation criteria were not met for a given compound or well, PDBS may still be a viable alternative to conventional sampling depending on the degree to which concentrations of that particular compound must be quantified to achieve LTM objectives. In these instances, it may be desirable to perform additional evaluations to determine whether the instances of reduced correlation were a one-time occurrence, whether correlation of PDB to conventional samples is even reasonable to expect or appropriate in that instance, or whether the wells or compounds are poorly suited to the PDBS method. Additional evaluations could include one or more of the following activities:

- Comparison of PDBS data to historical analytical results to determine whether the PDBS results are anomalously low;
- Review of available stratigraphic and hydrogeologic data to assess whether these wells are screened in low-permeability zones that could restrict groundwater flow through the well;
- Performance of a second PDB sampling event in these wells and comparison of the results to conventional sampling results;
- Use of a borehole flow meter to assess the degree to which vertical flow in the well may be impacting PDBS results; or
- Performing vertical profiling of contaminant concentrations in selected wells using an alternative method (e.g., micropurge or direct-sampling ion-trap mass spectrometry) and comparing the results to PDBS results.

Excluding the costs of additional field testing of PDBSs to clarify outlier (i.e., low-correlation) situations, and disregarding the potential need to analyze samples from a given well for constituents other than VOCs (e.g., SVOCs, pesticides, and/or metals), the PDBS method can provide significant long-term cost savings compared to conventional sampling methods. The median cost to evaluate the implementability of PDBS per well during this demonstration was approximately \$2,500. Furthermore, the median estimated cost difference between LTM using the PDBS method as opposed to the conventional method was approximately \$250 per well per sampling event. Therefore, even if PDBS is implemented on a limited basis, significant long-term cost savings could be realized. The ROI will depend on the number of wells, the frequency of sampling, and the duration of the LTM program. However, using the assumptions detailed in Section 4, the median ROI calculated for the installations included in this demonstration was 356 percent.

REFERENCES

Church, P.E., 2000. Evaluation of a Diffusion Sampling Method for Determining Concentrations of Volatile Organic Compounds in Ground Water, Hanscom Air Force Base, Bedford, Massachusetts: U.S. Geological Survey Water -Resources Investigations Report 00-4242, 20 p.

Earth Tech, 1998. Draft Quality Assurance Project Plan (QAPP) For Operable Units (OUs) 5, 6, 8, And 10 At Edwards Air Force Base, California. Version 2.0. November.

Goad, Linda, 2001. Telephone conversation between John Tunks (Parsons) and Linda Goad (Montgomery Watson) regarding QAPP requirements for March ARB. December.

Hare, P.W., 2000. Passive Diffusion Bag Samplers for Monitoring Chlorinated Solvents in Ground Water: The Second International Conference on Remediation of Chlorinated and Recalcitrant Compounds, Battelle, May 22 -25, 2000, Monterey, California, v.2, no 1, p. 377-386.

McClellan AFB, 2000. Final Passive Diffusion Membrane Samplers Technology Application Analysis Report. National Environmental Technology Test Sites (NETTS). August.

Parsons Engineering Science, Inc. (Par sons), 1999. Final Technical Report for the Evaluation of Groundwater Diffusion Samplers. December.

Radian International (Radian), 1997. Basewide Remedial Investigation /Feasibility Study (RI/FS) Quality Assurance Project Plan (QAPP), Revision 3. Report prepared for McClellan AFB, California. Final, April.

Radian, 1998. Defense Distribution Depot San Joaquin (DDJC) – Sharpe/Tracy Sites, California. Comprehensive Field Work Plan - Volume 2 of 2, Quality Assurance Project Plan. Version 2.0. Report prepared for U.S. Army Corps of Engineers, Huntsville, Alabama. Final, October.

Tunks, J.P., Guest, P.R., Faile, M., and Santillan, J., 2000. Diffusion Sampler Evaluation of Chlorinated VOCs in Groundwater. In: Proceedings of the Second International Conference on Remediation of Chlorinated and Recalcitrant Compounds. May.

USEPA, 1996. Laboratory Data Validation Functional Guidelines for Evaluating Organics Analysis. EPA Order No. EPA540R94082.

USEPA, 1999. SOP No. HW-24, Standard Operating Procedure for the Validation of Organic Data Acquired Using SW-846 Method 8260B (Rev 2, Dec 1996). Region II. Revision 1, June.

USEPA, 2000. Method 526. Determination of Selected Semivolatile Organic Compounds in Drinking Water by Solid Phase Extraction and Capillary Column Gas

Chromotography/Mass Spectrometry (GC/MS). Revision 1.0. National Exposure Research Laboratory, Office of Research and Development. C incinnati, Ohio. June.

Vroblesky, D. A. 2002. Passive Diffusion Bag Sampler Protocol, Training, and Regulatory Considerations. Presented at the AFCEE Cleanup Technology Transfer Workshop. San Antonio, Texas. March 4-7.

Vroblesky, D. A. 2001. User's Guide for Polyethylene -Based Passive Diffusion Bag Samplers to Obtain Volatile Organic Compound Concentrations in Wells. US Geological Survey Water-Resources Investigations Report 01 -4060. Columbia, South Carolina.

Vroblesky, D.A., and Hyde, W.T., 1997. Diffusion samplers as an Inexpensive Approach to Monitoring VOCs in Ground Water. Ground Water Monitoring and Remediation , v. 17, no. 3, p. 177-184.

Vroblesky, D.A., Borchers, J.W., Campbell, T.R., and Kinsey, Willey, 2000. Investigation of Polyethylene Passive Diffusion Samplers for Sampling Volatile Organic Compounds in Ground Water at Davis Global Communication, Sacramento, California. August 1998 to February 1999: U.S. Geological Survey Open -File Report 00-307, 13 p.

Vroblesky, D.A., and Peters, B.C., 2000. Diffusion Sampler Testing at Naval Air Station North Island, San Diego County, California. November 1999 to January 2000: U.S. Geological Survey Water-Resources Investigations Report 00 -4182, 34 p.

Vroblesky, D.A., and Petkewich, M.D., 2000. Field Testing of Passive Diffusion Bag Samplers for Volatile Organic Compound Concentrations in Ground Water, Naval Industrial Reserve Ordnance Plant, Fridley, Minnesota. November 1999 and May 2000: U.S. Geological Survey Water-Resources Investigations Report 00-4246, 10 p.

This page intentionally left blank

Index

MIX
Papier aus verantwortungsvollen Quellen
Paper from responsible sources
FSC® C105338
www.fsc.org

If you have any concerns about our products,
you can contact us on
ProductSafety@springernature.com

In case Publisher is established outside the EU,
the EU authorized representative is:
Springer Nature Customer Service Center GmbH
Europaplatz 3, 69115 Heidelberg, Germany

Printed by Libri Plureos GmbH
in Hamburg, Germany